R Professor E. A. R. son

ADVANCES IN
MOLECULAR STRUCTURE RESEARCH

Volume 4 • 1998

ADVANCES IN
MOLECULAR STRUCTURE RESEARCH

Editors: MAGDOLNA HARGITTAI
Structural Chemistry Research Group
Hungarian Academy of Sciences
Eötvös University
Budapest, Hungary

ISTVÁN HARGITTAI
Budapest Technical University and
Hungarian Academy of Sciences
Budapest, Hungary

VOLUME 4 • 1998

 JAI PRESS INC.

Stamford, Connecticut   *London, England*

CONTENTS

FORMATION OF (*E,E*)- AND (*Z,Z*)-MUCONIC ACID
IN METABOLISM OF BENZENE: POSSIBLE ROLES OF
PUTATIVE 2,3-EPOXYOXEPINS AND PROBES FOR
THEIR DETECTION

SOME RELATIONSHIPS BETWEEN MOLECULAR
STRUCTURE AND THERMOCHEMISTRY

LIST OF CONTRIBUTORS

Ronald J. Gillespie

Department of Chemistry
McMaster University
Hamilton, Ontario, Canada

Jenny P. Glusker

The Institute for Cancer Research
Fox Chase Cancer Center
Philadelphia, Pennsylvania

Arthur Greenberg

Department of Chemistry
University of North Carolina at Charlotte
Charlotte, North Carolina

Amy Kaufman Katz

The Institute for Cancer Research
Fox Chase Cancer Center
Philadelphia, Pennsylvania

Erhard Kemnitz

Institute of Chemistry
Humboldt University
Berlin, Germany

Qi Li

Department of Chemistry
Beijing Normal University
Beijing, P. R. China

Joel F. Liebman

Department of Chemistry and Biochemistry
University of Maryland, Baltimore County
Baltimore, Maryland

Thomas C. W. Mak

Department of Chemistry
The Chinese University of Hong Kong
Shatin, New Territories, Hong Kong

Paul G. Mezey   Mathematical Research Unit
                                 Department of Chemistry
                                 University of Saskatchewan
                                 Saskatoon, Canada

Edward A. Robinson   Department of Chemistry
                                 University of Toronto
                                 Erindale Campus
                                 Mississauga, Ontario, Canada

Camille Sandorfy   Département de Chimie
                                 Université de Montréal
                                 Montréal, Québec, Canada

Suzanne W. Slayden   Department of Chemistry
                                 George Mason University
                                 Fairfax, Virginia

Thomas Steiner   Institut für Kristallographie
                                 Freie Universität
                                 Berlin, Germany

Sergei I. Troyanov   Department of Chemistry
                                 Moscow State University
                                 Moscow, Russia

PREFACE

The review of the frontiers of structural chemistry from time to time helps the structural chemist to delineate the main thrusts of advances in this area of research. What is even more important though, these efforts assist the rest of the chemists to learn about new possibilities in structural research. It is the purpose of the present series to report the progress in structural studies, both methodological and interpretational. We are aiming at making it a "user-oriented" series. Structural chemists of excellence evaluate critically a field or direction including their own achievements, and chart expected developments.

The present volume is the fourth in this series. We always appreciate hearing from those, producing structural information and perfecting already existing techniques, or creating new ones, and from the users of structural information. This helps us gauge the reception of this series and shape future volumes.

We are very pleased by offering the present volume for the quality and breadth of its contributions. As a whole, this volume reflects two increasingly discernible trends in modern structural chemistry. One is that parallel to the ever increasing specialization of various techniques there is a strong interaction between the various techniques, crossing boundaries of the various experiments, those between the experiments and computations, experiments and theory, and organic and inorganic chemistry. The other is the ever increasing penetration of the most modern aspects of structural chemistry in the rest of chemistry, making the delineation of structural chemistry from the rest of chemistry increasingly fuzzy which is the most welcome development from a structural chemist's point of view.

We would like to make, again, a special acknowledgment with respect to Volume 4. Much of the editorial work has been done at the University of North Carolina at Wilmington where we held visiting positions in the spring of 1998 (MH as Visiting Scientist and IH as Distinguished Visiting Professor). We thank the Department of Chemistry of UNCW and our colleagues for excellent working conditions and for helpful interactions that were instrumental in the completion of this volume. We also thank the Hungarian Academy of Sciences, the Loránd Eötvös University, Budapest, and the Budapest Technical University as well as the Hungarian National Science Foundation for continuing support.

Magdolna and István Hargittai
Editors

MOLECULAR GEOMETRY OF "IONIC" MOLECULES:
A LIGAND CLOSE-PACKING MODEL

Ronald J. Gillespie and Edward A. Robinson

Advances in Molecular Structure Research
Volume 4, pages 1–41
Copyright © 1998 by JAI Press Inc.
All rights of reproduction in any form reserved.
ISBN: 0-7623-0348-4

ABSTRACT

A study of the experimentally determined ligand–ligand intramolecular distances in a large number of fluorides, chlorides, oxides, hydroxides, and alkoxides of beryllium, boron, carbon, phosphorus, and sulfur shows that F---F, Cl---Cl, O---O, OH---OH and OX---OX distances are almost constant for a given central atom, independent of the coordination number, bond angles, and bond lengths and the presence of other ligands. From these interligand distances a constant ligand intramolecular contact radius (ligand radius) can be assigned to each ligand that depends only on the central atom. Interligand distances in species with mixed ligands, such as chlorofluorides and oxofluorides, can be accurately predicted with these ligand radii. Calculation of the atomic charges in these molecule shows that they are predominately ionic and are better described by the ionic model than by the covalent model. The ligand charges, although large, are smaller than the full ionic charges and decrease with increasing electronegativity of the central atom and it is the charge on the ligand that determines its radius. On the basis of these results it is proposed that in these molecules anion-like ligands are close-packed around a cation-like central atom and that the observed bond angles and bond distances are a consequence of this close packing. We call this the ligand close-packing (LCP) model. It is shown that this model provides simpler explanations for the bond lengths and bond angles in molecules such as BF_3 and OCF_3^- and for the nontetrahedral bond angles in molecules such as $B(OH)_4^-$ than the covalent model. The relationship between the LCP model and the VSEPR model is discussed and it is shown that they are essentially equivalent if the effect of the ligands on any nonbonding electrons is taken into account. On this basis it is possible to understand why "lone pairs" are sometimes stereochemically inactive or only partially active.

I. INTRODUCTION

The ionic and covalent models of bonding were first proposed in 1916 by Kossel [1] and by Lewis [2], respectively. Since that time the covalent model has been used for essentially all substances that have been shown to consist of individual molecules, as well as for solids that have three-dimensional infinite structures and are electrical nonconductors, such as diamond. In contrast, the ionic model has been used only to describe the bonding in solid substances that are conducting in the molten state or in solution in solvents such as water, and therefore appear to be composed of positive and negative ions. The bonding in solids may range from essentially purely ionic, as in NaCl, through bonding of an intermediate nature (usually referred to as polar covalent) as in SiO_2, to pure covalent, as in diamond. For molecules it is usually considered that the bonding is predominantly covalent but may have some ionic character as, for example, in molecules such as CF_4, CCl_4, PF_3, PCl_3, and SO_2. But the possibility that the bonding in some molecules might be predominantly ionic and best described by the ionic model rather than by the covalent model seems not to have been seriously considered except for the high-temperature vapor phase of ionic crystals such as NaCl(s) where Na^+Cl^- and larger molecules which are fragments of the NaCl(s) lattice are found. Such species are often referred to as ion pairs or ion clusters rather than molecules because of the widespread belief that the bonding in all molecules is predominantly covalent. In this review we discuss the evidence that the bonding in many molecules normally considered to have polar bonds is in fact best described by the ionic model rather than by the covalent model. The structures of many ionic crystals are largely determined by the close packing of large anions around small cations. The purpose of this review is to show that there is a large number of predominantly ionic molecules whose structures are similarly determined by the packing of anion-like ligands around a cation-like central atom, that is by interligand repulsions.

The recognition of the importance of intramolecular interactions between adjacent nonbonded atoms in strongly influencing the structures of many molecules is not new. Bartell [3, 4] proposed in 1960 that for many organic molecules the decrease in the length of C–C bonds with a decrease in the number of attached atoms from 6 down to 2 could be explained as due simply to the decreasing number of nonbonding interactions. He also gave a set of nonbonding radii for H, C, N, O, F, and Cl that rather accurately reproduced the bond angles known at the time in all molecules of the types XX′C=CX″X‴ and XX′C=O [3]. A more extensive set of these nonbonding radii, often referred to as 1,3 radii, was later given by Glidewell [5]. In 1963, Bartell [6] summarized the factors most commonly acknowledged at that time to influence bond lengths and bond angles in molecules as (a) bond order, conjugation, and hyperconjugation, (b) hybridization, (c) partial ionic character and, in a limited number of cases, (d) nonbonded interactions. However, on the basis of his studies Bartell concluded that nonbonded repulsions between ligands are much more important than had usually been assumed and so very probably play an important role in determining the bond lengths and bond angles in many more molecules than has been commonly supposed. Moreover, he pointed out that nonbonding repulsions could account in a simple way for many observed bond lengths and angles that were usually explained at the time in terms of one or more of the factors (a) to (c) above. However, this conclusion has not been widely accepted and bond lengths and bond angles are still commonly interpreted without recourse to the possible importance of nonbonding effects.

That the importance of intramolecular repulsions is still not generally recognized is because intramolecular distances are usually compared with the sum of the Bartell

1,3 radii. These radii are given in Table 1 for the ligands with which we are concerned.

Because observed intramolecular distances in most molecules are larger than the sum of the 1, 3 radii it has usually been concluded that nonbonding interactions are unimportant. However, Bartell deduced his radii for ligands attached to carbon and these radii are not valid for ligands attached to other atoms. We will show that for ligands attached to Be, B, P, and S the Bartell values are too small and that nonbonding repulsions are of major importance in determining the geometry of the molecular fluorides, chlorides, oxides, and hydroxides of these elements, as well as for carbon. The values of the radii of ligands for which evidence is presented in this paper are given in Table 1. Calculated atomic charges show that these molecules are predominantly ionic in nature and we show that they have structures in which the bond lengths and bond angles are largely determined by the close packing of anion-like ligands around a cation-like central atom. The experimental evidence for this model, which we call the ligand close-packing (LCP) model comes from the observation that in many types of AX_n molecules ligand–ligand distances are essentially constant despite considerable variations in bond lengths and bond angles. Hargittai [7] has, for example, previously commented on the strikingly constant O---O distance of 248–249 pm in a large number of sulfones and the importance of taking this into account in the interpretation of molecular geometries.

Table 1. Bartell Ligand Radii, Proposed Ligand Radii for N, O, F, and Cl Bonded to Be, B, C, Si, P, and S, and Anion Radii for N^{3-}, O^{2-}, F^-, and Cl^{-a}

| Ligand X | Bartell 1,3 Radius[b] | Central Atom A | | | | | | Anion Radius | |
		Be	B	C	Si	P	S	Shannon[c]	Pauling[d]
N	119	144	124	119	144	135[e]		150	171
O	113	134	120	112	132	127	124	126	140
F	108	128	113	108	127	118	114	119	136
Cl	144	168[c]	151	144	164	156	154	167	181

Notes: [a] All radii in pm.

[b] Ref. 4.

[c] Effective ionic radii for two-coordinated ions from ref. 11.

[d] Pauling, L. *The Nature of the Chemical Bond*; Cornell University Press: Ithaca, NY, 1960, p. 514.

[e] Tentative radii of *bridging* NX ligands from the results of a very preliminary analysis.

II. ELECTRONEGATIVITY AND IONIC CHARACTER

The electronegativity difference between two atoms is often taken as a rough indication of the ionic character of the bond formed between them. For example, on this basis, bonds such as B–F, Si–F, and P–F, with electronegativity differences between the bonded atoms of 2.1, 2.4, and 2.0, respectively, might reasonably be expected to be as ionic as the bond in NaCl, for which the electronegativity difference is 1.8 [8]. Yet, the bonding in the BF_3, SiF_4, and PF_3 molecules is usually described by writing structures with bond lines, as in I, II, and III (Scheme 1) where the bond lines are usually taken to denote bonds of predominantly covalent character. The polar nature of the bonds is often indicated by adding $\delta+$ and $\delta-$ signs to the atoms as in IV, V, and VI (Scheme 2) but these signs are usually interpreted to mean relatively small charges, implying that the bonds are predominantly covalent and have the directional properties of covalent bonds.

Considering, for example, the period 2 molecular fluorides LiF, BeF_2, BF_3, CF_4, NF_3, and OF_2 for which the electronegativity differences are 3.1, 2.6, 2.1, 1.6, 1.0, and 0.6 [8], respectively, we might reasonably anticipate a gradual change from an almost fully ionic bond in LiF to predominantly covalent bonds in NF_3 and OF_2. However, LiF and BeF_2 are high-melting solids which are considered to be predominantly ionic, while BF_3, which is a gas at room temperature and condenses to a solid that is nonconducting in the molten state, is usually assumed to be predominantly covalent despite the large electronegativity difference (2.1) between boron and fluorine. A similar large change in the nature of the bonding is also generally assumed to occur between the high-melting ionic solid AlF_3 (electronegativity difference 2.8) and supposedly covalent, gaseous, SiF_4 (electronegativity difference 2.4). But such a sudden dramatic change in the nature of the bonding between BeF_2 and BF_3, and between AlF_3 and SiF_4, is surely unexpected and surprising. It seems more reasonable to suppose that the bonding in BF_3 and in SiF_4 more closely resembles that in BeF_2 and AlF_3 and is not nearly as covalent as has generally been supposed.

Although a characteristic property of covalent bonding is its directional nature, it does not follow that a molecule that has a geometry consistent with that predicted for covalent bonds, such as planar triangular for BF_3, must necessarily have covalent bonds. Close packing of anions around a central cation which has an empty valence

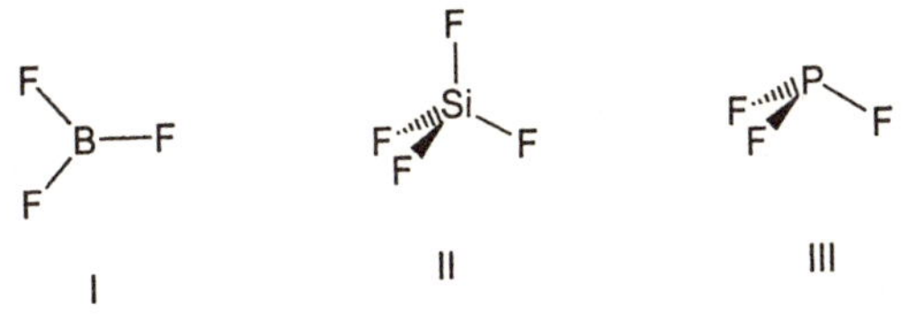

Scheme 1.

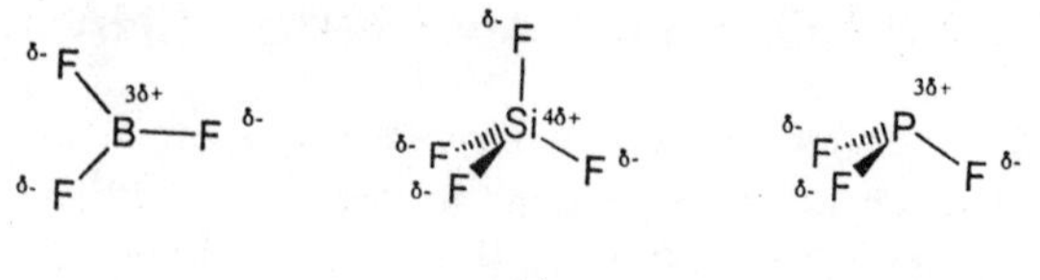

IV   V   VI

Scheme 2.

shell, such as Be^{2+}, B^{3+}, and C^{4+}, gives the same molecular geometries as predicted by the VSEPR model for covalent structures: e.g. BeF_2, linear; BF_3, planar triangular; CF_4, tetrahedral. So no conclusions about the specific nature of the bonding in these molecules can be drawn directly from the observed geometry. Even the structure of a molecule such as PCl_3 in which the phosphorus atom has a lone pair in its valence shell can also be understood in terms of the ionic model, as we will discuss later (Section XI.A).

Any conclusions we might come to from electronegativity values about the ionic nature of bonds are however qualitative, as is the concept of electronegativity itself. To obtain meaningful quantitative information about bond polarity we need to be able to assign charges to the atoms in molecules, and until recently no reliable experimental or theoretical method has been available for doing so. Atomic charges can now be obtained experimentally from the analysis of electron density distributions obtained from X-ray diffraction studies on solids or more readily, as we describe in the next section, from ab initio calculations.

III. THE ANALYSIS OF ELECTRON DENSITY DISTRIBUTIONS AND ATOMIC CHARGES

It is now possible to use ab initio methods to calculate the geometry and the total wavefunction of a molecule to a high degree of accuracy, and hence to calculate its total electron density distribution. Atomic charges can then be obtained by the analysis of either the total wavefunction or the electron density distribution of a molecule. Analysis of the total wavefunction has the disadvantage that the charges obtained depend on the choice of orbitals made in the analysis. However, analysis of the total electron density using the Atoms in Molecules (AIM) theory [9] leads to a set of unique and unambiguous charges. In this method the molecule is divided into its component atoms by means of interatomic surfaces, called zero-flux surfaces. The charge on each atom is then obtained by integrating the electron density over the space surrounding each atomic nucleus as defined by these interatomic (zero-flux) surfaces.

To illustrate the determination of atomic charges using the AIM method we show the electron density distribution in the plane of the BF_3 molecule in Figure 1 in the

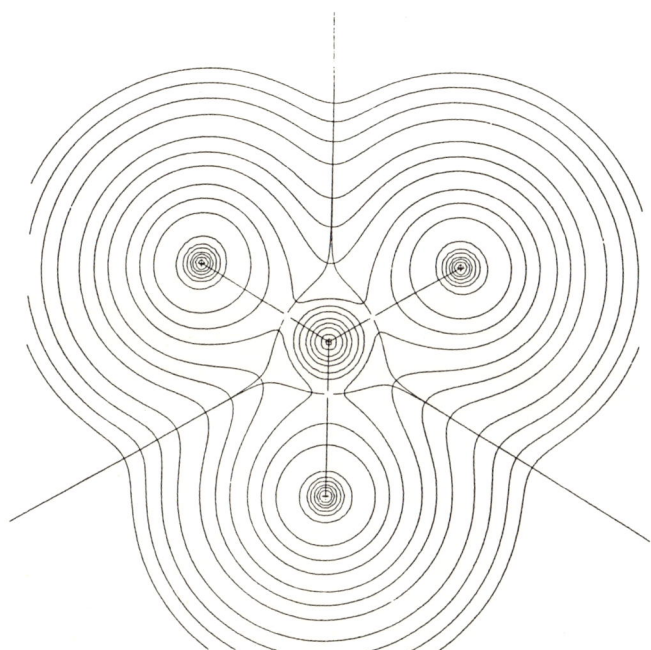

Figure 1. Contour plot of the electron density distribution in the BF_3 molecule. The lines connecting the nuclei are the bond paths along which the electron density is greater than along any other line connecting the two nuclei. The curved lines between the atoms are the lines along which the interatomic (zero-flux) surfaces cut the molecular plane.

form of a contour map [*10*]. In this figure we also see the lines along which the interatomic surfaces cut the molecular plane, and the lines along which the electron density is a maximum with respect to any other line between the atoms (called bond paths). For BF_3, the total density around each atom is very nearly spherical, indicating that this molecule is strongly ionic.

The predominantly ionic nature of the bonding in BF_3 is confirmed by the large atomic charges on boron and fluorine of +2.43 and −0.81, respectively, given in Table 2, which also gives the calculated charges for the other period 2 and 3 molecular fluorides and their fluoro anions [*10*]. The charge on fluorine decreases steadily across each period and is more negative than −0.8 for all the fluorides up to BF_3 in period 2 and up to SiF_4 in period 3, so that the bonds in all these molecules might be regarded as more than 80% ionic. And even for CF_4 and for PF_3 and SF_2 it is larger than −0.5. The positive charge on the central atom increases in the period 2 fluorides to values of 2.43 in BF_3 and 2.45 in CF_4 before decreasing rapidly to 0.83 in NF_3 and 0.27 in OF_2, and in period 3 this charge increases to 3.26 in SiF_4

Table 2. Atomic Charges for the Period 2 and 3 Molecular Flourides from Ab Initio DFT Calculations[a]

	$-q(F)$	$q(A)$		$-q(F)$	$q(A)$
LiF	0.922	0.92	BeF_3^-	0.914	1.75
BeF_2	0.876	1.75	BeF_4^{2-}	0.939	1.76
BF_3	0.808	2.43	BF_4^-	0.856	2.43
CF_4	0.612	2.45	CF_3^+	0.527	2.59
NF_3	0.277	0.83	NF_4^+	0.078	1.32
OF_2	0.133	0.27	AlF_4^-	0.888	2.56
F_2	0	0	SiF_3^+	0.734	3.21
NaF	0.906	0.91	PF_4^+	0.693	3.78
MgF_2	0.878	1.76	AlF_6^{3-}	0.930	2.58
AlF_3	0.845	2.54	SiF_6^{2-}	0.875	3.20
SiF_4	0.813	3.26	PF_6^-	0.782	3.70
PF_3	0.758	2.28			
SF_2	0.579	1.16			
ClF	0.379	0.38			

Note: [a]Complete results of calculations are given in ref. 10.

and then decreases less rapidly to 2.28 in PF_3 and to 1.16 in SF_2 before falling to 0.38 in ClF. It would seem therefore that most of these molecules are better described by an ionic model than by the covalent model, so that BF_3 and SiF_4, for example, are more accurately described by structures such as VII and VIII (Scheme 3) than by structures with bond lines indicating covalent bonds. From the large charges on the central atom in these molecules we expect that the ionic bonds in these molecules will be very strong and indeed the BF bond in BF_3 and the SiF bond in SiF_4 with estimated bond energies of 613 and 565 kJ mol^{-1}, respectively, are among the strongest known bonds [*11*].

VII   VIII

Scheme 3.

IV. THE LIGAND CLOSE-PACKING (LCP) MODEL FOR PREDOMINANTLY IONIC MOLECULES

Because the structures of ionic crystals are very dependent on how the anions and cations can be packed most closely together, it seems reasonable to suppose that the structures of predominantly ionic molecules would be similarly dependent on the packing of their anion-like ligands. Accordingly, in an AF_n molecule ($n \geq 3$) we may imagine that the anion-like fluorine ligands are attracted to the central cation-like A atom until they "touch" each other; in other words, until the repulsive forces between the ligands become sufficiently strong, thus adopting a close-packed arrangement. So AF_3 molecules would be expected to be trigonal planar, AF_4 tetrahedral, AF_5 trigonal bipyramidal, and AF_6 octahedral, as is indeed observed. This simple model predicts that the bond lengths in AX_n molecules depend only on the coordination number and that the ratios of the AX bond lengths in AX_4 and AX_3 molecules should be 1.06 and for AX_6 and AX_4 molecules it should be 1.15.

V. FLUORIDES OF BERYLLIUM, BORON, AND CARBON

Experimental geometrical parameters for a selection of 3- and 4-coordinated fluorides of these elements are given in Table 3 [*10*]. Calculated values are given for molecules for which no experimental data are available. A remarkable feature of this data is that the F---F distance is the same in both 3- and 4-coordinated molecules and that, although there is a great variation in the AF bond lengths and the FAF bond angles, the F---F nonbonding interligand distances remain essentially constant for a given central atom A. For example, the experimental BF bond lengths in BF_3, BF_4^-, and $MeBF_3^-$ are 130.7, 138.6, and 142.4 pm, respectively (Table 3), yet the F---F distances remain constant at 226 pm. This is also the F---F distance in many other XBF_2 and XBF_3 molecules and is consistent with a fluorine ligand radius of 113 pm when it is bonded to boron. A particularly striking example is provided by the BF_4^- ion in the tungsten complex $(Me_3P)W(CO_3)(NO)(BF_4)$ [*12*] where it is behaving as a monodentate ligand and is highly distorted, with BF bond lengths ranging from 131 to 150 pm and FBF angles ranging from 104° to 114°, yet all the F---F distances have values close to the average value of 226 pm found for all the other molecules. Similarly, the F_3CO^- ion has unusually long CF bonds (139.2 pm) relative to those in CF_4 (131.9 pm), yet the intramolecular F---F distances are essentially the same as in CF_4 and other fluorocarbons. This structure is discussed in detail later (Section VII.F).

The average values of the nearly constant intramolecular F---F distances for a given central atom A give interligand radii for fluorine of 128(2) pm in BeF_n molecules, 113(1) pm in BF_n molecules, and 108(1) pm in CF_n molecules (Table 1). The value of the ligand radius for fluorine bonded to carbon agrees with the Bartell radius because he obtained this from molecules in which the central atom is carbon. The radii of the anion-like fluorine ligand are not equal to the radius of

Table 3. Bond Lengths, Bond Angles, and F---F Distances in Some Molecules Containing BeF_n, BF_n, or CF_n Groups[a]

	A-F(pm)	<FAF (°)		F---F (pm)
BeF_3^-	149(1)	120(3)		258(6)
BeF_4^{2-}	155.4(6)	109.5		253.8(10)
			Average:	256(2)
F_2B-F	130.7	120.0		226.4
F_2B-OH	132.3(10)	118.0(10)		226.8(29)
F_2B-NH_2	132.5(12)	117.9(7)		227.0(29)
F_2B-CH_3	131.5(5)	116.8(5)		224.0(15)
F_2B-Cl	131.5(5)	118.1(5)		225.6(14)
F_2B-H	131.1(5)	118.3(10)		225.1(20)
F_2B-BF_2	131.7(2)	117.2(2)		224.8(6)
F_2B-SiF_3	131.2(5)	118.5(5)		225.5(15)
F_3B-F^-	138.2(3)	109.5(3)		225.7(8)
$B_5H_8BF_2$	132.2(3)	115.4(6)		223.5(3)
F_3B-OH_2	138.8	111.5		229.5
F_3B-NH_3	136.7	114.4		229.8
F_3B-NMe_3	137.2(2)	112.9(3)		228.6(8)
$F_3B-CH_3^-$	142.4(5)	105.4(4)		226.6(10)
$F_3B-CF_3^-$	139.1(5)	109.9(5)		227.7(16)
F_3B-PH_3	137.2(2)	112.1(4)		227.6(9)
			Average:	226(2)
CF_3^+ [b]	124.4	120		215.5
CF_4	131.9	109.5		215.4
CF_3^- [b]	141.7	99.5		216.3
F_3C-CF_3	132.6(2)	109.8(1)		217.0(4)
$F_3C-BF_3^-$	134.3(5)	104.9(3)		213.0(12)
F_3C-OF	131.9(3)	109.4(10)		215.3(18)
F_3CO^-	139.2	101.3		215.3
F_3C-Cl	132.5(2)	108.6(2)		215.2(6)
F_3C-Br	132.6(2)	108.8(4)		215.6(9)
$(F_3C)_3CH$	133.6(2)	108.0(2)		216.2(6)
$F_3C-CH(O)$	133.2(7)	108.7(1)		216.5(12)
$F_2C=CF_2$	131.9(1)	112.4(4)		219.2(7)
$F_2C=CCl_2$	131.5(15)	112.1(25)		218.2(31)
$F_2C=CH_2$	132.4(1)	109.4(6)		216.1(9)
$F_2C=CHF$	133.6(1)	109.2(6)		217.8(8)
			Average:	**216(1)**

Notes: [a]For more extensive tables and references for structures see ref. 10.

[b]Calculated structure.

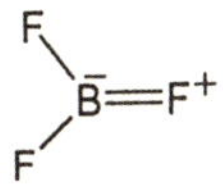

$$\underset{F}{\overset{F}{\diagdown}}\bar{B}\!\!=\!\!F^{+}$$

IX

Scheme 4.

the fluoride ion (129 pm) but are smaller and decrease as the charge on fluorine decreases with increasing electronegativity of A. That the F---F interligand distances for a given central atom are not exactly constant from molecule to molecule reflects small differences in the ligand radius due to small differences in the charge on the ligand, as well as errors in the experimental data. However, these differences in the F---F contact distance are considerably smaller than the differences due to a change in the central atom. The bonds in molecules in which the A atom is 3-coordinated are consistently shorter than the bonds in 4-coordinated molecules and the bond length ratios BeF_4^{2-}/BeF_3, BeF_4^-/BF_3, and CF_4/CF_3^+ have the values 1.04, 1.06, and 1.07 respectively in reasonable agreement with the theoretical value of 1.06 for the close-packed ions model. The LCP model provides a simple explanation for the decrease in A–F bond lengths with decreasing coordination number that renders the frequently quoted π back-bonding explanation unnecessary, although it does not directly disprove it. According to this model the apparently short BF bond in BF_3 (130.7 pm) compared to that in the BF_4^- ion (138.6 pm) is attributed to a significant overlap of fluorine $2p$ orbitals with the "empty" boron $2p$ orbital, which allows electron density to move from the singly bonded fluorine ligands into bonding π orbitals, thus increasing the bond order and correspondingly decreasing the bond length. This is also called back-bonding and can be described in terms of resonance structures such as IX (Scheme 4), which imply partial donation of lone pair electrons from highly electronegative fluorine to the much less electronegative boron, giving partial double bond character to the BF bonds, which are therefore shorter than the supposed single bonds in the BF_4^- ion.

VI. CHLORIDES AND CHLOROFLUORIDES OF BORON AND CARBON

The structural data given in Table 4 for a variety of molecules containing BCl_n and CCl_n groups in both 3- and 4-coordinated molecules shows that the intramolecular Cl---Cl distances are essentially constant giving values of 151 and 144 pm for the ligand radius of Cl on B and C, respectively. The BCl_4^-/BCl_3 and CCl_4/CCl_3^+ bond length ratios have the values of 1.05 and 1.06, respectively, which compare well with the ideal value of 1.06 for close-packed 4 and 3 coordination.

Table 4. Bond Lengths, Bond Angles, and Cl---Cl Distances in Some Molecules Containing BCl_n or CCl_n Groups

	A–Cl (pm)	<ClACl (°)	Cl...Cl (pm)	Ref.
BCl_3	174.2(2)	120.00	301.7(3)	a
$Cl_2B–BCl_2$	175.0(5)	118.65(33)	301.1(4)	b
$Cl_2BB_5H_8$	172.0(15)	127.7(28)	300.4(17)	c
$H_3N–BCl_3$	183.8(7)	111.2(2)	303.3(15)	d
$C_5H_5N–BCl_3$	183.7(4)	110.1(3)	301.1(12)	e
$Me_3N–BCl_3$	183.1(5)	109.3(5)	298.7(17)	f
$Me_3P–BCl_3$	185.5(7)	110.9(2)	305.6(15)	g
BCl_4^-	183.3(13)	109.47	299.3(21)	h
$(PhO)_3PO–BCl_3$	182.4(6)	110.9(1)	300.5(11)	i
$[Cl_2B(NPPh_3)]_2$	188.4(4)	105.4(2)	299.7(10)	j
$[Cl_2B(NPEt_3)_2BCl]^+$	185.8(9)	108.6(1)	301.8(16)	j
		Average:	**301(2)**	
CCl_4	177.1(2)	109.5	289.3(3)	k
$FCCl_3$	176.0	109.7	287.8	l
F_2CCl_2	174.4	112.5(5)	290.0(19)	m
Me_2CCl_2	179.9	108.3(3)	291.7(11)	n
$Cl_3C–CCl_3$	176.9(3)	108.9(3)	287.9(4)	o
CCl_3^{+*}	166.3	120	288	p
$Cl_2C=CCl_2$	171.9(3)	115.6(3)	290.9(10)	q
$Cl_2C=CH_2$	171.8(7)	112.4(10)	285.5(29)	r
$Cl_2C=CF_2$	170.6(8)	119.0(9)	294(3)	s
Cl_2CO	173.8(2)	111.8(2)	287.8(7)	t
Cl_2CS	172.9(3)	111.2(3)	285.3(9)	u
$Cl_3C–NO_2$	172.6(5)	111.8(7)	285.9(10)	v
		Average:	**288(2)**	

Notes: [*]Calculated structure.

[a] Konaka, S.; Murata, Y.; Kuchitsu, K.; Morino, Y. *Bull. Chem. Soc. Jpn.* **1966**, *39*, 1134.

[b] Ryan, R. R.; Hedberg, K. *J. Chem. Phys.* **1969**, *50*, 4986.

[c] Rankin, D. W. H.; Robertson, H. E.; Alberts, I. L.; Downs, A. J.; Greene, T. M.; Hofmann, M.; Schleyer, P. R. *J. Chem. Soc., Dalton Trans.* **1995**, 2193.

[d] Avent, A. G.; Hitchcock, P. B.; Lappert, M. F.; Liu, D.-S.; Mignani, G.; Richard, C.; Roche, E. *J. Chem. Soc., Chem. Commun.* **1995**, 855.

[e] Töpel, K.-H.; Hensen, K.; Trömel, M. *Acta Crystallogr.* **1981**, *B25*, 2338.

[f] Hess, H. *Acta Crystallogr.* **1969**, *B25*, 2338.

[g] Black, D. L.; Taylor, R. C. *Acta Crystallogr.* **1975**, *B31*, 1116.

Table 4. Continued

[h]Allen, F. H.; Kennard, O.; Watson, D. G.; Brammer, L.; Orpen, A. G.; Taylor R. *J. Chem. Soc., Perkin Trans. II.* **1987**, S1.

[i]Levin, M. L.; Fieldhouse, J. W.; Allcock, H. R. *Acta Crystallogr.* **1982**, *B38*, 2284.

[j]Moehlen, M.; Harms, K.; Dehnicke, K.; Magoll, J.; Goesmann, H.; Fenske, D. *Z. Anorg. Allg. Chem.* **1996**, *622*, 1692.

[k]Haase, J.; Zeil, W. *Z. Phys. Chem. (Frankfurt)* **1965**, *45*, 202.

[l]Long, M. W.; Williams, Q.; Weatherly, T. *J. Chem. Phys.* **1960**, *33*, 508.

[m]Takeo, H.; Matsumura, C. *Bull. Chem. Soc. Jpn.* **1977**, *50*, 636.

[n]Hirota, M.; Iijima, T.; Kimura, M. *Bull. Chem. Soc. Jpn.* **1978**, *51*, 1594.

[o]Almenningen, A.; Andersen, B.; Trætteberg, M. *Acta Chem. Scand.* **1964**, *18*, 603.

[p]Olah, G. A.; Rasul, G.; Yudin, A. K.; Burrichter, A.; Surya Prakash, G. K.; Chistyako, A. L.; Stankevich, I. V.; Akhrem, I. S.; Gambaryan, N. P.; Vol'pin, M. E. *J. Am. Chem. Soc.* **1996**, *118*, 1446.

[q]Strand, T. *Acta Chem. Scand.* **1967**, *21*, 2111.

[r]Davis, M. I.; Hanson, H. P. *J. Phys. Chem.* **1965**, *69*, 4091.

[s]Lowrey, A. H.; D'Antonio, P.; George, C. *J. Chem. Phys.* **1976**, *64*, 2884.

[t]Nakata, M.; Fukuyama, T.; Kuchitsu, K.; Takeo, H.; Matsumura, C. *J. Mol. Spectrosc.* **1980**, *83*, 118; Nakata, M.; Fukuyama, T.; Wilkins, C. J.; Kuchitsu, K. *J. Mol. Struct.* **1981**, *71*, 195.

[u]Carpenter, J. H.; Rimmer, D. F.; Smith, G.; Whiffen, D. H. *J. Chem. Soc., Faraday Trans. II.* **1975**, *71*, 1752.

[v]Knudsen, R. E.; George, C. F.; Karle, J. *J. Chem. Phys.* **1966**, *44*, 2334.

We expect therefore that chlorofluoro molecules of boron and carbon will also be close packed with F---Cl distances of $113 + 151 = 264$ pm for boron molecules and of $113 + 144 = 257$ pm for carbon molecules. In F_2BCl, for example, the observed bond lengths are BF 131.5(6) pm and BCl 172.8(9) pm, with <FBF 118.1(5)° [*13*], giving experimental F---F and F---Cl distances of 226(2) pm and 266(2) pm, respectively. In CF_3Cl [*14*], CF_2Cl_2 [*15*], and $CFCl_3$ [*16*] the F---Cl distances are 254, 253, and 253 pm, respectively, in reasonable agreement with the predicted value of 257 pm.

VII. OXIDES, HYDROXIDES, AND ALKOXIDES OF Be, B, AND C

A. Oxoboron Molecules

We have recently surveyed the experimental data for some 50 molecules containing BO_3 and BO_4 groups and have shown that the O---O intramolecular contact distances have an almost constant value close to 238 pm though the BO bond lengths cover a wide range from 132 to 150 pm and individual OBO bond angles often deviate considerably from the ideal angles of 120.0° and 109.5° expected for regular trigonal planar BO_3 and tetrahedral BO_4 groups [*17*]. A selection of this data is given in Table 5. The BO bond lengths in the 4-coordinated molecules are consis-

RONALD J. GILLESPIE and EDWARD A. ROBINSON

Table 5. Examples of Average Bond Lengths, Bond Angles, and O---O Distances in Some Molecules Containing BO_n Groups[a]

	B–O (pm)	<OBO (°)	O---O (pm)	
BO₃ Groups				
Li_3BO_3	137.7(6)	120.0(5)	238(2)	O_t--O_t [b]
$BaNaBO_3$	138.2(12)	120.0(3)	239(2)	O_t--O_t
Ag_3BO_3	137.8(5)	120.0	238.7(7)	O_t--O_t
$FeBO_3$	137.9(2)	120	238.8(4)	O_t--O_t
$K_3B_3O_6$	1 133.0	1–2 121.3(4)	238.1(5)	O_t--O_b
	2 139.8	2–2 117.3(8)	238.9(9)	O_b--O_b
$[HO\text{-}BO]_3$	1 135.5(8)	1–2 120.0(18)	236(2)	O_b--OH
	2 137.3(7)	2–2 120.1(5)	238(2)	O_b--O_b
$B(OH)_3$	136.1(4)	120	235.7(7)	HO--OH
$B(OCH_3)_3$	136.8(2)	120	236.9(4)	XO--OX
$B(OTeF_5)_3$	135.8(6)	120	235.2(11)	XO--OX
BO₄ Groups				
$Ca[B(OH)_4]_2$	147.8(4)	106.1(5)	236.2(14)	HO--OH
		111.0(4)	243.6(12)	HO--OH
$Na_2[B(OH)_4]Cl$	148.1(2)	105.1(1)	235.1(5)	HO--OH
		111.7(1)	245.1(5)	HO--OH
$B(OMe)_4^-$	145.8(11)	101.7(3))	226.1(22)	XO--OX
		113.5(6)	243.9(26)	XO--OX
$B(OTeF_5)_4^-$	147(1)	109.5	240(1)	XO--OX
BO₂ Groups				
$(MeO)_2B\text{–}B(OMe)_2$	136.9(3)	119.9(4)	237.0(10)	XO--OX
$(MeO)_2B\text{–}Me$	137.5(4)	120.9(5)	239.2(13)	XO--OX
$[HBO]_3$	137.6(2)	120.0(6)	238.3(11)	O_b--O_b
$[EtBO]_3$	138.4(4)	118.4(1)	237.7(9)	O_b--O_b
$[PhBO]_3$	138.6(1)	118.0(4)	237.6(7)	O_b--O_b
$BrC_6H_4B(OH)_2$	136	122	238	HO--OH

Notes: [a]For more extensive tables, including complex borates, and structural references see ref 17.

[b]O_t terminal oxygen; O_b binding oxygen.

tently longer than those in the 3-coordinated molecules and they have average values of 148 and 138 pm, respectively, giving a d_4/d_3 ratio of 1.07 close to the predicted value for the LCP model of 1.06. The oxygen ligand radius of 119 pm obtained from the O---O contact distance is smaller than the 132 and 135 pm crystal radii of the OH$^-$ and O^{2-} ions [*18*], but consistent with the somewhat smaller calculated charges of the OH and O ligands discussed below. It is not usually possible to clearly differentiate between the ligand radii for the OH and O ligands, which in any case would be expected to differ by only about 3 pm or less. None of the B(OX)$_4$ molecules are exactly tetrahedral but have two different bond angles and two different O---O contact distances, one smaller and the other larger than the average O---O distance. The reason for this is discussed in Section VII.

Calculated charges, bond lengths, and bond angles for BO$^+$, BO, BO$_2^-$, B(OH)$_2^+$, BO$_3^{3-}$, B(OH)$_3$, (HO)$_2$BOH$_2^+$, B(OH)$_4^-$, B$_3$O$_6^{3-}$, and (HOBO)$_3$ are given in Table 6 [*17*]. The charges on the O and OH ligands, which range from -1.72 to -1.53 and from -0.81 to -0.72, respectively, are much closer to the ionic limits of -2 and -1 than to the zero charges expected for fully covalent bonds. The oxygen ligand radius of calculated bond lengths and bond angles are in good agreement with the experimental values where these are available.

B. 1 and 2 Coordinated Oxoboron Molecules

Further support for the close-packed ligand model is provided by the observation that when the constraint of close packing is removed in 1- and 2-coordinated molecules, the BO bonds are found to be still shorter than in 3- and 4-coordinated molecules, which is also the case for molecules with both BO and BF bonds (Table 6). Although the atomic charges in these molecules are somewhat smaller than for a *fully* ionic model, this model is nevertheless useful for understanding the observed variations in the bond lengths simply in terms of electrostatic repulsions and attractions between the component ions. Figure 2 shows the calculated bond lengths and charges for FBO, BO$_2^-$, HOBO, B(OH)$_2^+$, BO, and BO$^+$ as well as the charges based on the fully ionic model [*17*]. In terms of this model, the BO bond in BO$_2^-$ is longer than in FBO or in HOBO because the O^{2-} ligand is repelled more strongly by another O^{2-} ligand than by either a F$^-$ or a OH$^-$ ligand, while it is still shorter in BO$^+$ because there is no second ligand; the bond in the BO radical is longer than in BO$^+$ because its boron atom has only a $+2$ charge rather than the $+3$ charge in BO$^+$, and the B–OH bond in HOBO is longer than in (HOBOH)$^+$ because the O^{2-} ligand in the former repels the OH$^-$ ligand more strongly than does the second OH$^-$ in the latter. Finally, even the slightly longer BO bond length in FBO compared to that in HOBO (120.6 versus 119.5 pm) can be accounted for in terms of the *actual* charges because the charge on the F ligand in FBO is slightly greater than that on the OH ligand in HOBO.

C. The Cyclic B$_3$O$_6^{3-}$ and B$_3$O$_3$(OH)$_3$ Molecules

We can also use the ionic model to account for the differences in the bond lengths in the cyclic B$_3$O$_6^{3-}$ anion, XII, and the cyclic acid B$_3$O$_3$(OH)$_3$, XIII (Table 5). As shown in Scheme 5, the terminal B–O$_t$ bonds (132.8 pm) in XII are shorter than the terminal B–OH bonds in XIII (135.3 pm) because the charge on the terminal O

Table 6. Results of Ab Initio Calculations for Some Boron Oxo and Hydroxo Molecules[a]

	B–O (pm)[b]		<OBO (°)	−q(O)	q(H)	−q(OH)	q(B)
BO⁺	118.8			1.043			2.043
BO	120.3 (120.5)			1.553			1.553
BO₂⁻	126.4 (125.3)		180	1.576			2.15
OBOH	121.2	–O	180 180	1.436			2.179
	132.4	–OH		1.333	0.59	0.741	
B(OH)₂⁺	125.5		180	1.343	0.69	0.652	2.305
	124.5						
OBF	120.6	–O	180	1.449			2.262
	128.4	–F					
BO₃³⁻	141.8 (137.7(4))		120	1.724			2.192
OBF₂⁻	1 127.0	–O		1.593			2.325
	2 140.5	–F	2–2 106.4				
B(OH)₃	136.9 (136.1(4))		120	1.316	0.56	0.761	2.282
B(OH)₃	135.8						
(HO)₂BOH₂⁺	132.2	–OH	141.2	1.331	0.618	0.713	2.296
	149.9	–OH₂	109.4	1.200	0.665		
(HO)₂BOH₂⁺	131.2	–OH	132				
	150.2	–OH₂	114				
[OBO⁻]₃	1 132.8	–Oₜ	122.9 b-t	1.644			2.239
	2 143.2	–Oᵦ–	114.2 b-b	1.592			
[OBOH]₃	1 135.3 (135.5(8))	–OₜH	119.8 b-b	1.301	0.56	0.742	2.282
	2 138.2 (137.3(7))	–Oᵦ–	120.1 b-t	1.546			
	3 137.9	–Oᵦ–	1–2 121.1 b-t	1.546			
B(OH)₄⁻	148.7 (147.7(6))		106.2 111.1	1.3	0.48	0.819	2.275

Notes: [a]For references to experimental bond lengths, see ref. 17.

       [b]Experimental bond lengths in parentheses ( ).

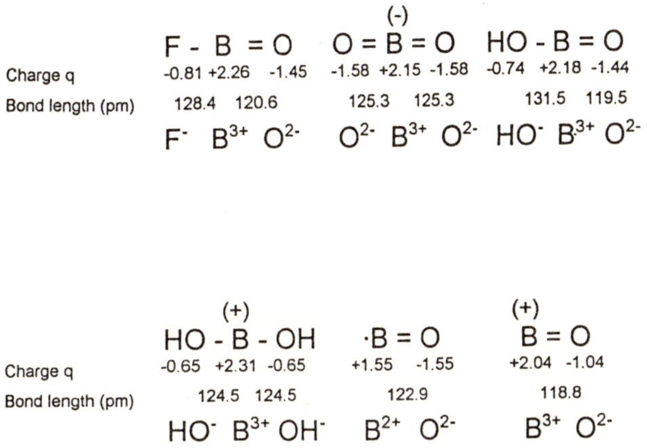
(-)
F - B = O
O = B = O
HO - B = O
Charge q
-0.81 +2.26 -1.45
-1.58 +2.15 -1.58
-0.74 +2.18 -1.44
Bond length (pm)
128.4 120.6
125.3 125.3
131.5 119.5
F⁻ B³⁺ O²⁻
O²⁻ B³⁺ O²⁻
HO⁻ B³⁺ O²⁻
(+)
HO - B - OH
·B = O
(+)
B = O
Charge q
-0.65 +2.31 -0.65
+1.55 -1.55
+2.04 -1.04
Bond length (pm)
124.5 124.5
122.9
118.8
HO⁻ B³⁺ OH⁻
B²⁺ O²⁻
B³⁺ O²⁻

Figure 2. Covalent and ionic models for some 1 and 2 coordinated molecules of boron.

ligand in XII is larger than the charge on the terminal OH ligand in XIII, and the bridging BO bonds are longer in XII than in XIII for the same reason; the bridging O in XII is more strongly repelled by its terminal O ligand than is the bridging O in XIII by its terminal OH ligand.

D. Oxoberyllium Molecules

Although the available experimental data summarized in Table 7 is not extensive, tetrahedral BeO_4 groups have bond lengths ranging from 162 to 165 pm (average 163.2 pm) except for the calculated structure of $Be(OH)_4^{2-}$ and the two examples of BeO_3 groups both have a bond length of 154.3 pm, giving an average bond length ratio of 1.06, as expected for ligand close-packing. The average O---O contact

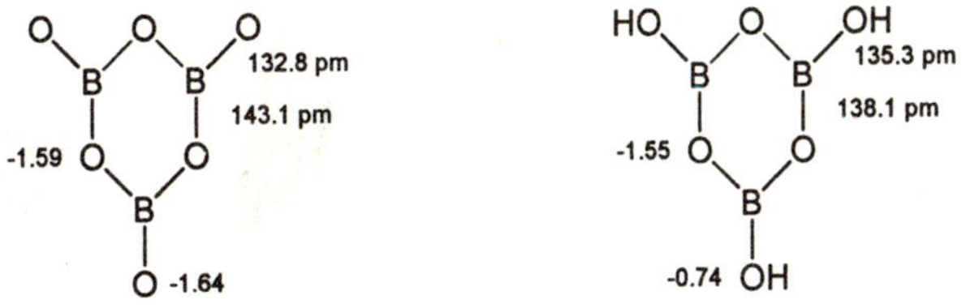
132.8 pm
143.1 pm
-1.59
-1.64
135.3 pm
138.1 pm
-1.55
-0.74

XII

XIII

Scheme 5.

Table 7. Average Bond Lengths, Bond Angles, and O---O Nonbonding Distances in Compounds with BeO_n Groups[a]

	Be–O (pm)	<OBeO (°)	O---O (pm)
BeO_3 Groups:			
Y_2BeO_4	154.3(11)	120	267(2)
$SrBe_3O_4$	154.3(2)	120.0	266
$Be(OH)_3^-$	154.3	120.0	267
BeO_4 Groups:			
BeO(s)	164	109.5	268
$Li_{14}Be_5B(BO_3)_9$	162(2)	109.5	265(3)
$SrBe_3O_4$	164(22)	109.5	268(3)
$LiBePO_4 \cdot H_2O$	163(2)	109.5	266(3)
$Be_2AsO_4(OH)$	162(2)	109.5	264(5)
	163(2)	109.5	266(3)
$Be_2BO_3(OH)$	163(1)	109.5	266(2)
$Be(OH_2)_4 \cdot SO_4$	161.8(4)	109.5	264(1)
$\gamma\text{-}Li_2BeSiO_4$	164.7	109.5	269
$Be(OH)_4^{2-\,b}$	168.8	109.5	273
		Mean:	**265(3)**

Notes: [a]For references to structures see ref. 17.

       [b]Calculated structure.

Table 8. Results of Ab Initio Calculations for Some Beryllium Oxo and Hydroxo Molecules[a]

	Be–O (pm)[b]	<OBeO (°)	–q(O)	q(H)	–q(OH)	q(Be)
BeO_2^{2-}	145.9	180	1.80			1.64
$Be(OH)_2$	142.3	180	1.42	0.57	0.85	1.70
$Be(OH)_3^-$	154.6 (154.3)	120	1.37	0.47	0.90	1.69
$Be(OH)_4^{2-}$	168.8	107.8	1.34	0.42	0.93	1.70
		110.3				
$(HBe)_2O$	139.6	180	1.79			1.74

Notes: [a]For references to experimental bond lengths, see ref. 17.

       [b]Experimental bond lengths in parentheses ( ).

distance in all of the molecules is 265 pm, giving an oxygen ligand radius of 133 pm essentially equal to the radii of 132 pm for OH^- and 135 pm for O^{2-} [18]. This radius is consistent with the large ligand charges given in Table 8 for BeO, BeO_2^{2-}, $Be(OH)_2$, $Be(OH)_3^-$, $Be(OH)_4^{2-}$, and BeO_3^{4-} which are in the range -0.93 to -0.85 for OH ligands and -1.79 to -1.69 for O ligands. As for oxoboron molecules, the two coordinated molecules in which the constraint of close packing is removed have still shorter bonds. For example, in $Be(OH)_2$ the bond length is 142.3 pm and in BeO_2^{2-} it is 145.9 pm. These bonds are longer than in the corresponding boron molecules, $B(OH)_2^+$ (124.5 pm) and BO_2^- (123.5 pm), because in the fully ionic model the charge on Be (+2) is smaller than that on B (+3).

E. Oxocarbon Molecules and the Distortion of A(OX)₄ Molecules from Tetrahedral Symmetry

Experimental data for both CO_4 and CO_3 groups are given in Table 9 [17]. The most striking and at first sight somewhat surprising feature of this data is that there are two distinct O---O contact distances in all the $C(OX)_4$ molecules. Thus none of these molecules are truly tetrahedral but have either two angles smaller and four larger, or four smaller and two larger, than the tetrahedral angle and have either S_4 or D_2 symmetry. In Section VII.B we noted that $B(OX)_4$ molecules are similarly not exactly tetrahedral. Table 10 shows that this is a common property of $A(OX)_4$ and also some $A(NX_2)_4$ molecules. We have previously attributed these deviations from tetrahedral symmetry to the noncylindrical electron density distribution of the OX ligand and NX_2 ligands [17]. Whereas a ligand such as O or F with a cylindrical electron density distribution has a single ligand radius independent of the direction of the contacts that it makes, unsymmetrical ligands such as OH and NH_2 have a different ligand radius in different directions giving rise to different O---O and N---N contact distances depending on the direction of the contact. The summary of the calculated data [17] for $Be(OH)_4^{2-}$, $B(OH)_4^-$, and $C(OH)_4$ given in Table 11 shows that the difference in the two contact distances increases and thus the deviation from tetrahedral symmetry increases from Be, to B to C with increasing covalency of the AO bond.

In planar $A(OH)_3$ molecules all three OH groups have the same relative orientation and hence the same O---O contact distance. In contrast, in a tetrahedral molecule, because the electron density distribution of the OX and similar ligands does not have cylindrical symmetry, they cannot be oriented so that they all make the same contacts with their neighbors, but they must give at least two different contact distances.

Previously, the distortion of $A(OX)_4$ molecules from T_d symmetry has been attributed to interactions between the OX dipoles and to the anomeric effect (back-bonding from O to A involving negative hyperconjugation) [19]. However, in order for back-bonding to account for the observed angles it would have to produce different amounts of double-bond character in at least two of the bonds in

Table 9. Bond Lengths, Bond Angles, O---O Distances in Some Molecules Containing CO_n Groups[a]

	Symmetry	C–O (pm)	<OCO (°)	O---O (pm)	<OCO (°)	O---O (pm)
CO$_4$ Groups:						
C(OMe)$_4$	S_4	139.6(1)	106.9(5) × 4	224(1)	114.6(5) × 2	235(1)
C(OPh)$_4$	D_2	139.4(2)	101.2(8) × 2	216(1)	113.8(2) × 4	234(1)
C(OC$_6$H$_4$Me$_2$-3,5)$_4$	D_2	139.6(15)	101.3(12)	216(4)	114.3(14)	235(4)
C(OH)$_4$ [b]	S_4	138.8	107.2 × 4	223	114.2 × 2	233
HC(OMe)$_3$		138.2(6)	109.2(6) × 2	225(2)	115.0(10)	233(2)
H$_3$C–C(OMe)$_3$		139.8(6)	106.7(9)	224(2)	110.8(9)	230(2)
					108.5(9)	227(2)
HC(OH)$_3$ [b]	C_3	140.8	108	229		
H$_2$C(OMe)$_2$	C_2	138.2(4)	114.3(7)			232(2)
Me$_2$C(OMe)$_2$	C_2	142.3(6)	117.4(22)			243(4)
H$_2$C(OH)$_2$ [b]	C_s	142	114.4			239
CO$_3$ Groups:						
(HO)$_2$CO[b]	1	131.5	1–1	109.2	214[d]	
	2	118.8	1–2	125.4	223	
(H$_3$CO)$_2$CO	1	134.3(10)	1–1	107.0(1)	216(2)[d]	
	2	120.3(9)	1–2	126.5(1)	227(2)	
Ca^{2+} CO$_3^{2-}$ [c]		128.2(2)		120.0	222	
		129.4(4)		120.0	224(1)	
Na$^+$ HO–CO$_2^-$	1	134.6	1–2	125.0	224	
	2	126.4	1–3	118.8	225	
	3	126.3	2–3	116.3	222	
NH$_4^+$ HCO$_2^-$		124.2(7)			222(2)	
Li$^+$MeCO$_2^-$		124.5(5)		125.7(3)	222(1)	

Notes: [a]For more extensive tables and references to structures see ref. 17.

[b]Calculated structure.

[c]Two different determinations.

[d]The HO---OH contact distance.

Table 10. Symmetries and Average Bond Angles in Some $A(XY)_4$ Molecules with Distorted AX_4 Tetrahedral Structures[a]

	Symmetry	<XAX (°)	<XAX (°)
$Be(OH)_4^{2-}$ [b]	D_2	107.8 × 2	110.3 × 4
$B(OH)_4^-$ [b]	D_2	106.2 × 2	111.1 × 4
$Na_2B(OH)_4Cl$	D_2	105.1 × 2	111.7 × 4
$B(OMe)_4^-$	D_2	101.7 × 2	113.5 × 4
$KB(OSO_2Cl)_4$	S_4	107.4 × 4	113.8 × 2
$C(OH)_4$ [b]	D_2	103.6 × 2	112.5 × 4
	S_4	107.2 × 4	114.2 × 2
$C(OMe)_4$	S_4	106.9 × 4	114.6 × 2
$C(OC_6H_5)_4$	D_2	101.2 × 2	113.8 × 4
$C(OC_6H_3Me_2\text{-}3,5)_4$	D_2	100.9 × 2	114.0 × 4
$C(SC_6H_5)_4$	S_4	106.3 × 4	116.0 × 2
$C(CH_2OH)_4$	S_4	106.7 × 2	110.9 × 4
$C(CH_2Cl)_4$	S_4	106.1 × 2	112.9 × 2
	D_2	108.3 × 4	111.9 × 2
$Si(OH)_4$ [b]	D_2	104.8 × 2	111.8 × 4
	S_4	107.1 × 4	114.2 × 2
$Ti(NH_2)_4$ [b]	S_4	107.2 × 4	114.2 × 2
$Ti(NMe_2)_4$	S_4	107.2 × 4	114.2 × 2
$V(NMe_2)_4$	D_2	100.6 × 2	114.1 × 4
$V(O^tBu)_4$	S_4	106.7 × 4	115.1 × 2

Notes: [a] For references to structures, see ref. 17.

[b] Calculated structure.

order to affect the angle between these bonds, and this would produce at least two different bond lengths. But in all these molecules all four AO bonds have the same length. That the noncylindrical symmetry of the ligand atom causes the deviations of the OAO bond angles from tetrahedral is also supported by the similar quite large deviations in the bond angles from regular tetrahedral observed in tetrachloromethyl methane, $C(CH_2Cl)_4$ [20], and in pentaerythritol, $C(CH_2OH)_4$ [21]. In these two molecules all four CC bonds have normal CC single-bond lengths of 154.8 and 153.9 pm, respectively, and the ligand CH_2 groups in these molecules have no lone pairs to take part in back-bonding, but they must have a nonspherical symmetry. It is also interesting to note that the same type of distortion has been observed in some transition metal molecules (Table 10).

Table 11. Ab Initio Structural Data for $A(OH)_3$ and $A(OH)_4$ Molecules

	$Be(OH)_3^-$	$B(OH)_4^{2-}$	$B(OH)_3$	$B(OH)_4^-$	$C(OH)_3^+$	$C(OH)_4$
A–O (pm)	154.6	168.8	136.9	148.7	128.1	139.3
$-q_{OH}$ (au)	0.90	0.93	0.76	0.82	0.43	0.50
<OAO (°)	120.0×3	107.8×2	120.0×3	106.2×2	120.0×3	103.6×2
<OAO (°)	—	110.3×4	—	111.1×4	—	112.5×4
O---O (pm)	268×3	273×2	237×3	238×2	222×3	219×2
O---O (pm)	—	277×4	—	245×4	—	232×4
Δ(O---O) (pm)*	0	4	0	7	0	13

Note: *Difference in the two contact distances in the $A(OH)_4$ molecules.

Because there are two significantly different O---O contact distances in the $C(OX)_4$ molecules the experimental O---O contact distances in the CO_3 and CO_4 groups vary over a wider range than in the corresponding boron and beryllium molecules (Table 9). They have an overall average value of 227 pm, while the average value of the shorter contacts in the $C(OX)_4$ molecules is 222 pm and that of the longer contacts is 234 pm. The CO bond distances in the CO_3 groups (average value 126 pm) are shorter than those in the CO_4 groups (average value 140 pm), but the average value of the O---O contacts in the $C(OX)_3$ molecules is 222 pm which is the same as the lower value of 222 pm in the $C(OX)_4$ molecules, consistent with the LCP model.

From the contact distances we conclude that the ligand radius of the oxygen atom in COX varies from 111 to 117 pm with an average value of 114 pm depending on the direction of the contact that it is making. Considering the experimental errors in the bond distances which are relatively large in some cases, the small variations in the charge on the oxygen from one molecule to another which are greater in these more covalent molecules, and the expected small difference in the radii of the oxygen in an OX group and that of a terminal O, the extent of the scatter of the observed values around the average values is very reasonable. This small radius of 114 pm for oxygen bonded to carbon compared to the radius of oxygen when bonded to B and Be is consistent with the relatively small calculated charges of -1.0 to -1.4 on the O ligands and 0.41 to 0.6 on the OH ligands (Table 12). These charges indicate that the bonds in these molecules can be regarded as only about 50% ionic.

F. Oxofluoroboron and Oxofluorocarbon Molecules

The ligand radii of O and F bonded to boron of 119(1) and 113(1) pm, respectively, predict intermolecular O---F distances in oxofluoroboron molecules of 232

Table 12. Results of Ab Initio Calculations for Some Carbon Oxo and Hydroxo Molecules[a]

	C–O (pm)[b]		<OCO (°)	–q(O)	q(H)	–q(OH)	q(C)
CO	111.4 (112.8)			1.346			1.346
	110.3			1.357			
CO_2	114.3 (116.0)		180	1.298			2.595
	116.0			1.076			2.151
CO_3^{2-}	130.8 (129.4)		120	1.337			2.013
$HOCO_2^-$	1 123.3 (126.4)	–O	1–2 132.8	1.239			2.053
	2 125.1 (126.3)	–O	1–3 113.9	1.258			
	3 145.4 (134.6)	–OH	2–3 113.3	1.046	0.5	0.55	
$(HO)_2CO^b$	1 120.4 (120.3(9))	–O	2–2 108.6	1.047			2.129
	2 133.9 (134.3(10))	–OH	1–2 125.7	1.166	0.57	0.598	
$(HO)_2CO$	1 118.8	–O	2–2 109.2				
	2 131.5	–OH	1–2 125.4				
$C(OH)_4$ [b]	139.3 (139.6(1))		103.6	1.04	0.54	0.496	1.985
			112.5				
CO_4^{4-}	145.2		109.5	1.405			1.617
$C(OH)_3^+$	128.1		120	1.05	0.64	0.41	2.228
H_2CO	118.3 (120.9(1))			1.240			1.245
	117.8			1.271			1.292
Cl_2CO	117.2 (117.6(3))		124.1	1.052			1.248
F_2CO	117.1 (117.2(1))	–O		1.088			2.297
	132.0 (131.6(1))	–F	107.7				
F_3CO-	122.7 (121.4)	–O		1.26			2.16
	139.2 (139.4)	–F	100.6				

Notes: [a]Experimental bond lengths in parentheses ( ); for references see ref. 17.

[b]These experimental bond lengths are for $(MeO)_2CO$ and $C(OMe)_4$, respectively.

pm. The experimental data in Table 13 show that this is the case despite considerable variations in BO and BF bond lengths between molecules. For example, F_2BOH and its anion, F_2BO^-, both have O---F distances of 234 pm, close to the expected value of 232 pm, but remarkably different structures. Compared to F_2BOH, its anion, F_2BO^-, has a very short BO bond length (120.7 versus 134.4 pm) and a very long BF bond length (140.5 versus 132.3 pm). However, these differences are compatible with the ionic model, because an O ligand has a much larger charge than an OH ligand, so that the oxygen atom in the neutral molecule is more strongly

Table 13. 1,3-Nonbonding O---F Distances in Some Oxofluoroboron Compounds[a]

	Bond Lengths (pm)	<FBO (°)	O---F (pm)
F_3B-OH_2	BF(1) 138.2	1–4 105.9	233
	BF(3) 138.3	3–4 106.5	234
	BO(4) 153.2		
$F_3B-O(H)Me$	BF(1) 139.9	1–4 105.7	233
	BF(3) 135.5	3–4 106.0	230
	BO(4) 152.4		
$F_3B-OPPh_3$	BF(1) 135.7(5)	1–4 105.7(3)	229(1)
	BF(2) 135.3(6)	2–4 108.1(4)	233(2)
	BF(3) 133.4(6)	3–4 109.2(4)	233(2)
	BO(4) 151.6(6)		
$F_3B-OAsPh_3$	BF(1) 135.4(5)	1–4 106.4(3)	228(1)
	BF(2) 136.2(5)	2–4 109.0(3)	232(1)
	BF(3) 135.2(5)	3–4 109.0(3)	231(1)
	BF(4) 148.6(5)		
F_2B-OH	BF 132.3	122.8	234
	BO 134.4		
F_2B-O^{-} [b]	BF 140.5	126.8	234
	BO 120.7		
		Average:	**232(2)**

Notes: [a]For structural references, see ref. 17.

[b]Calculated structure.

attracted to boron than is OH in the anion giving a shorter BO bond. In order for the oxygen to move closer it must push the F ligands away giving very long BF bonds but nevertheless maintaining the characteristic O---F distances.

From the ligand radii of F and O to carbon of 108(1) and 114(1) pm, respectively, we expect the O--F intermolecular nonbonding distance in oxofluorocarbon molecules to have a value of 222 pm. The experimental data in Table 14 gives an average value of 222(2) pm, in agreement with the predicted value.

The structure of the F_3CO^{-} ion is of particular interest because the CO bond length (121.4 pm) is very similar to that in formaldehyde (120.9 pm), which is usually taken as that appropriate for a CO double bond, apparently making carbon pentavalent in this ion (XIV). However, the CF bonds (139.4 pm) are considerably longer than in COF_2 (131.7 pm) or CF_4 (131.9 pm). This geometry is usually rationalized by writing resonance structures such as XV and XVI in which the octet

Table 14. 1,3-Nonbonding O---O, F---F, and O---F Contact Distances in Some Oxofluorocarbon Molecules[a]

	Bond Length (pm)	Bond Angle (°)	$[X\text{---}X']_{obs}$ (pm)	$[X\text{---}X']_{pred}$ (pm)[a]	r_F (pm)	r_o (pm)
CF$_3$OCF$_3$	CF 132.7	FCF 108.7	215.6 F---F	216	107.8	
	CO 136.9	OCF 110.2	221.1 O---F	223		113.3
CF$_3$CO$_2^-$	CF 131.4	FCF 107.2	216.7 F---F	216	108.4	
	CO 126.9	OCO 128.2	228.3 O---O	230		114.2
CF$_3$O$^-$	CF 139.2	FCF 101.3	215.3 F---F	216	107.7	
	CO 122.7	OCF 116.2	222.5 O---F	223		114.8
CF$_3$OF	CF 131.9	FCF 109.5	215.3 F---F	216	107.7	
	CO 139.5	OCF 109.6	221.9 O---F	223		114.2
COF$_2$	CF 131.7	FCF 107.6	212.6 F---F	216	106.3	115.6
	CO 117.0	OCF 126.2	221.9 O---F	223		
MeC(O)F	CF 134.8	OCF 121.4	220.7 O---F	223		
	CO 118.1					
FC(O)OF	CF 132.4	OCF 126.5	222.8 O---F	223		
trans	CO 117.0					
FC(O)OF	CF 132.0	OCF 126.4	222.5 O---F	223		
cis	CO 117.2					
FC(O)NO$_3$	CF 132.0	OCF 128.8	224.2 O---F	223		
	CO 116.5					
F(O)C–C(O)F	CF 132.9	OCF 124.2	221.8 O---F	223		
	CO 118.0					
				Mean:	**108(1)**	**115(1)**

Note: [a]For structural references, see ref. 17.

rule is obeyed (Scheme 6). Although these structures are consistent with the observed bond lengths they do not explain for example why XV is apparently more important than XVI.

A more satisfactory explanation can be given in terms of the ionic model, XVII. According to this model, the O ligand forms the stronger shorter bond because of its higher charge, pushing the F ligands away and giving longer CF bonds and small FCF angles, just as in the F$_2$BO$^-$ ion discussed above. The actual charges (O −1.26 and F −0.63) are considerably less than the fully ionic charges but the charge on O is nevertheless almost exactly twice the charge on F. Moreover, the O--O and O--F

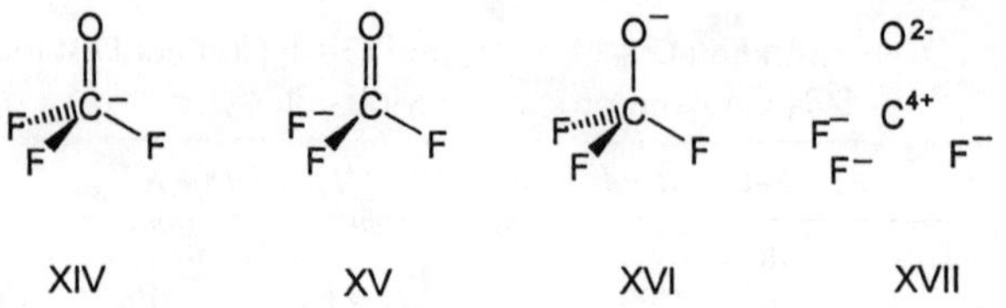

XIV   XV   XVI   XVII

Scheme 6.

distances in F_3CO^- have values of 215.3 and 222.5 pm, respectively, in excellent agreement with the predicted values of 216 and 223 pm.

VIII. OXOSULFUR MOLECULES

Hargittai has previously drawn attention to the remarkably constant interligand distances found in sulfones [7]. For example, he noted that the O---O distance in the SO_2 groups in more than 30 organic sulfones in the crystal phase is 247(2) pm, and 248–249 pm in a variety of X_2SO_2 and XSO_2Y molecules. Table 15 gives experimental data for the SO_2 and SO_3 groups in some $XYSO_2$ and XSO_3 groups, and for the SO_2 and SO_3 molecules as well as some calculated values of the charge on oxygen for a few of the molecules [22]. All the neutral X_2SO_2 and $XYSO_2$ molecules in the table have O---O distances in the range of 247 to 250 pm. In the anions, slightly smaller values of 240 to 242 pm are observed. The values for anions are all for the crystalline state and are dependent to some extent on the nature of the accompanying cation and on hydrogen bonding when this is present. The overall average O--O distance is 248(2) pm. This nearly constant interligand distance is consistent with the close packing of oxygen ligands with an almost constant radius of 124(1) pm. The calculated charges on oxygen are all in the range of -1.1 to -1.5, indicating substantial ionicity of the SO bonds in these molecules.

Further support for the LCP model is provided by the O---F and O---Cl contact distances in oxofluorosulfur and oxochlorosulfur molecules. Hargittai [23] noticed that the O---Cl intramolecular distances in some 11 molecules containing the SO_2Cl group are all in the range of 276 to 282 pm, and the O--F distances in five molecules containing the SO_2F group are all close to 242 pm. Table 16 gives structural data for some oxofluoro and oxochlorosulfur molecules. The O---F distances are in the range of 236 to 245 pm with an average value of 239 pm, which agrees well with the predicted value of 238 pm obtained from the sum of the radii of oxygen (124 pm, Table 16) and the radius of fluorine (114 pm obtained from the data for SO_2F_2 and SOF_2). The O---Cl distances are in the range of 278 to 284 pm, in good agreement with the value of 279 pm obtained from the radius of oxygen (124 pm, Table 16) and the radius chlorine of 155 pm [obtained from the Cl---Cl distances in SO_2Cl_2 and $SOCl_2$ of 308.9(11) and 309.2(16) pm, respectively].

Table 15. Average Bond Lengths, Bond Angles, and O---O Distances for SO_2 and SO_3 Groups in Some Oxosulfur Molecules

	S–O (pm)	<OSO (°)	O---O (pm)	−q(O)	Ref.
F_2SO_2	140.5(3)	124.0(2)	248.1(8)	−1.18	a
	139.7(2)	122.6(12)	245(2)		b
Cl_2SO_2	141.8(3)	123.5(2)	249.8(8)	−1.15	c
$ClSO_2F$	140.9(2)	122.7(1)	247.3(4)		d
FSO_2OH	141.6(4)	120.9(1)	246.4(3)		e
CF_3SO_2OH	142.1(4)	121.5(2)	247.8(9)		e
$(HO)_2SO_2$	142.2(10)	123.3(10)	250(3)	−1.47	f
$MeOSO_2F$	141.0(2)	124.4(7)	249.5(11)		g
$MeOSO_2Cl$	141.9(3)	122.2(15)	249(2)		h
$(MeO)_2SO_2$	141.9(4)	122.3(8)	249(2)		i
CF_3SO_2Cl	141.5(7)	122.4(10)	248(2)		j
$MeSO_2F$	141.0(3)	123.1(15)	248(1)		k
$MeSO_2Cl$	142.4(3)	120.8(8)	248(2)		l
CCl_3SO_2Cl	142.0(3)	121.5(9)	248(2)		m
Me_2SO_2	143.5(3)	119.7(11)	248(2)		n
FSO_3^- Cs^+	144.4(9)	113.4(2)	241.0(8)		o
FSO_3^- Li^+	143.4	114.8(14)	241.6(11)		p
$HOSO_3^-$ H_3O^+	144.8(9)	112.5(3)	240.3(10)		q
$HOSO_3^-$ Cs^+	144.5(16)	113(2)	241(2)		r
SO_4^{2-}	147.2(13)	109.5	240(2)		s
SO_3	141.98(2)	120.00	245.9(1)	−1.11	t
SO_2	143.08(1)	119.33(1)	247.0	−1.16	u
$(Ph_3P)_3Ni.SO_2$	144.8(7)	113.4(4)	242(2)		v
$Fe_3(CO)_7(CCO)(SO_2)_2$	147.1(9)	111.8(5)	244(2)		w
	146.8(9)	113.1(5)	245(2)		
SO_3^{2-} $(NH_4^+)_2$	152.4(6)	104.8	242(1)		x
$^-O_2S-SO_2^-$ $(Na^+)_2$	150.6(10)	108.2	244(2)		y
$^-O_3S-SO_2^-$	149.5 (SO_3)	111.2	247		z
	145.4 (SO_2)	112.7	242		

Notes: [a]Lide, D. R.; Mann, D. E.; Fristrom, R. M. *J. Chem. Phys.* **1957**, *26*, 734.

[b]Hagen, K.; Cross, V. R.; Hedberg, K. *J. Mol. Structure*, **1978**, *44*, 187.

[c]Hargittai, I.; Hargittai, M. *J. Mol. Structure*, **1981**, *73*, 253.

[d]Mootz, D.; Merschenz-Quack, A. *Acta Crystallogr.* **1988**, *C44*, 924.

[e]Bartmann, K.; Mootz, D. *Acta Crystallogr.* **1990**, *C46*, 320.

[f]Kuczkowski, R. L.; Suenram, R. D.; Lovas, F. J. *J. Am. Chem. Soc.* **1981**, *103*, 2561.

Table 15. Continued

---

[g]Hargittai, I.; Seip, R.; Nair, K. P. R.; Britt, Ch. O.; Boggs, J. E.; Cyvin, B. N. *J. Mol. Structure.* **1977**, *39*, 1.

[h]Hargittai, I.; Schutz, Gy.; Kolonits, M. *J. Chem. Soc., Dalton Trans.* **1977**, 1299.

[i]Brunvoll, J.; Exner, O.; Hargittai, I. *J. Mol. Structure*, **1981**, *73*, 99.

[j]Brunvoll, J.; Hargittai, I.; Kolonits, M. *Z. Naturforsch.* **1978**, *33A*, 1239.

[k]Hargittai, I.; Hargittai, M. *J. Mol. Structure*, **1973**, *15*, 399.

[l]Hargittai, I.; Hargittai, M. *J. Chem. Phys.* **1973**, *59*, 2513.

[m]Brunvoll, J.; Hargittai, I.; Seip, R. *Z. Naturforsch.* **1978**, *33A*, 222.

[n]Hargittai, M.; Hargittai, I. *J. Mol. Structure*, **1974**, *20*, 283.

[o]Zhang, D.; Rettig, S. J.; Trotter, J.; Aubke, F. *Inorg. Chem.* **1996**, *35*, 6113.

[p]Zák, Z.; Kosicka, M. *Acta Crystallogr.* **1978**, *B34*, 38.

[q]Taesler, J.; Olovsson, I. *Acta Crystallogr.* **1968**, *B24*, 299.

[r]Itoh, K.; Ukeda, T.; Ozaki, T.; Nakamura, E. *Acta Crystallogr.* **1990**, *C46*, 358.

[s]Allen, F. H.; Kennard, O.; Watson, D. G.; Brammer, L.; Orpen, A. G.; Taylor, R. *J. Chem. Soc. Perkin Trans. II.* **1987**, S1.

[t]Kaldov, A.; Maki, A. G. *J. Mol. Structure*, **1970**, *36*, 448.

[u]Saito, S. *J. Mol. Spectrosc.* **1969**, *30*, 1.

[v]Moody, D.; Ryan, R. R. *Inorg. Chem.* **1979**, *18*, 223.

[w]Bogdan, P. L.; Sabat, M.; Sunshine, S. A.; Woodcock, C.; Shriver, D. F. *Inorg. Chem.* **1988**, *27*, 1904.

[x]Batelle, L. F.; Trueblood, K. N. *Acta Chem. Scand.* **1965**, *19*, 531.

[y]Dunitz, J. D. *Acta Crystallogr.* **1956**, *9*, 579.

[z]Baggio, S. *Acta Crystallogr.* **1971**, *B27*, 517.

Although Hargittai recognized the constancy of the O---O, O---F, and O---Cl intramolecular distances in $XYSO_2$ and $XYSO$ molecules, he was of the opinion [23] that nonbonding interactions between ligands, although important, cannot be the only factor determining the structures on the grounds that these distances are larger than predicted by the Bartell radii. However, as we can see from Table 1, the radii for ligands bonded to sulfur are consistently larger than those for the same ligands bonded to carbon.

IX. OXOPHOSPHORUS MOLECULES

Because phosphorus is less electronegative than sulfur we expect oxophosphorus molecules to be more ionic than oxosulfur molecules and therefore their geometry should also be consistent with the LCP model. Table 17 gives experimental data for some $XYPO_2$, XPO_2, and XPO_3 molecules. The O---O distances are all in the range of 251 to 256 pm with an average value of 254 pm giving a ligand radius for oxygen of 127 pm. Table 18 gives experimental data for some P_4O_x and MP_3O_9 cage molecules. The nearly constant O---O distances in these molecules also give an

Table 16. Oxygen---Fluorine and Oxygen---Chlorine Distances in Some Oxofluoro and Oxochlorosulfur Molecules

	$(O\text{---}F)_{exp}$ (pm)	Ref.		$(O\text{---}Cl)_{exp}$ (pm)	Ref.
SO_2F_2	239(1)	b	SO_2Cl_2	280(1)	b
FSO_2Cl^a	236	b	FSO_2Cl	278	b
F_2SO	240	c	Cl_2SO	284(2)	d
$MeSO_2F$	238(1)	b	$MeSO_2Cl$	281(2)	b
$S_2O_5F_2$	245(4)	e	CF_3SO_2Cl	280(2)	b
$MeOSO_2F$	237(1)	b	CCl_3SO_2Cl	283(2)	b
$HOSO_2F$	237	b	$MeOSO_2Cl$	278(2)	b
$HNSOF_2$	239	f	$CH_2{=}CHSO_2Cl$	279(2)	g
$ClNSOF_2$	239	h	$C_6H_5SO_2Cl$	278(5)	i
CF_3SOF	238(3)	j			
Predicted:	**238(2)**		Predicted:	**279(2)**	

Notes: [a]For SO_2ClF the experimental Cl---F distance is 267.6(3) pm compared to 269 pm from the radii for F and Cl of 114 and 155 pm.

[b]See Table 15.

[c]Hargittai, I.; Mijlhoff, F. C. *J. Mol. Structure*, **1973**, *16*, 69.

[d]Hargittai, I. *Acta. Chim. Acad. Sci. Hung.* **1969**, *60*, 231.

[e]Hencher, D. L.; Bauer, S. H. *Can. J Chem.* **1973**, *51*, 2047.

[f]Cassoux, P.; Kuczkowski, R. L.; Cresswell, R. A. *Inorg. Chem.* **1977**, *16*, 2959.

[g]Oberhammer, H.; Glemser, O.; Klüver, H. **1974**, *29a*, 901.

[h]Brunvoll, J.; Hargittai, I. *Acta Chim. (Budapest)* **1977**, *94*, 333.

[i]Minkwitz, R.; Molsbeck, W.; Oberhammer, H.; Weiss, I. *Inorg. Chem.* **1992**, *31*, 2104.

[j]Brunvoll, J.; Hargittai, I. *J. Mol. Structure*, **1976**, *30*, 361.

average oxygen ligand radius of 127 pm with an apparent slight difference between bridging and terminal oxygen atoms. The validity of the LCP model for these oxophosphorus molecules is also supported by the O---F, O---N, and O---Cl distances given in Table 19. There is a little more scatter in these values about the average than for most of the ligand distances reported in this review. This is, however, not unexpected since we do not expect the radii for O and N to be completely constant, for example, when these atoms are in a bridging position as in many of these molecules, but to vary a little with the contact direction. Moreover, there may be appreciable errors in some of the X-ray structure determinations which are not very recent. However, the average values agree well with the values predicted from the sum of the ligand radii given in Table 1.

Table 17. Bond Lengths, Bond Angles and O--O Distances in Some PO_n Groups

	P–O (pm)	<OPO (°)	O···O (pm)	Ref.
$(NH_4)_3PO_4 \cdot 3H_2O$	153.6(4)	109.5(4)	251(1)	a
$(NH_4)_2[HOPO_3]$	152.4(4)	111.7(10)	252(1)	b
$Na_2[HOPO_3]$	153	111	254	c
$Na_2[HOPO_3] \cdot 7H_2O$	152.8(5)	113.0(8)	255(2)	d
$(NH_4)_2[MeOPO_3] \cdot 2H_2O$	150.7(12)	113.1(14)	251(1)	e
$K[H_3N–PO_3]$	150.9(3)	115.1(4)	255(1)	f
$Ca(MePO_3) \cdot H_2O$	153.7(9)	110.5(18)	252(2)	g
$Na_3[O_2C–PO_3] \cdot 6H_2O$	152.8(4)	112.4(6)	254(1)	h
$Na_4[OC(PO_3)_2]$	152.4(6)	113.5(6)	255	i
$(NH_4)_2[FPO_3] \cdot H_2O$	150.5(3)	114.0(7)	253(1)	j
$CaFPO_3 \cdot 2H_2O$	150.8(3)	113.9(29)	253(4)	k
$[C_6H_5NH_3]_2[FPO_2(OH)]_2$	148.5(4)	117.2(5)	254(1)	l
$(N_2H_6)[(HO)_2PO_2]_2$	151.0(6)	115.5(2)	255	m
$[C_9H_{14}N_3O][(HO)_2PO_2]$	150.3(6)	115.1(3)	254(1)	n
$[HO(O_2)P–PO_2(OH)]$	150.7(4)	114.9(1)	254	o
$K[F_2PO_2]$	147.0(5)	122.4(4)	255	p
$Rb[F_2PO_2]$	146.1	122	256	q
$(NH_4)_2[O_2P(S)NH_2]$	152.1(11)	114.5(3)	256(1)	r
$Na_4P_2O_7 \cdot 10H_2O$	151.3(9)	113.0(14)	252(2)	s
$Na_2[H_2P_2O_7] \cdot 6H_2O$	149.4(3)	117.7(1)	253(1)	t
$Na_2K[P_3O_9]$	148.2(2)	120.5(4)	257(1)	u
$Ag_3P_3O_9 \cdot H_2O$	148.9(4)	119.7(4)	257(1)	v
$Cu_2Li_2[P_6O_{18}]$	148.5	118.3(9)	255(1)	w
		Average:	**254(1)**	

Notes: [a]Mootz, D.; Wunderlich, H. *Acta Crystallogr.* **1970**, *B26*, 1826.

[b]Khan, A. A. ; Roux, J. P.; James, W. J. *Acta Crystallograph.* **1972**, *B28*, 2085.

[c]Garbassi, F.; Giarda, L.; Fagherazzi, G. *Acta Crystallogr.* **1972**, *B28*, 1665.

[d]Baur, W. H.; Khan, A. A. *Acta Crystallogr.* **1970**, *B26*, 1584.

[e]Baldus, M.; Meier, B. H.; Ernst, R. R.; Kentgens, A. P. M.; Meyer zu Altenschildesche, H.; Nesper, R. *J. Am. Chem. Soc.* **1995**, *117*, 5141.

[f]Cameron, T. S.; Chan, C.; Chute, W. J. *Acta Crystallogr.* **1980**, *B36*, 2391.

[g]Cao, G.; Lynch, V. M.; Swinnea, J. S.; Mallouk, T. E. *Inorg. Chem.* **1990**, *29*, 2112.

[h]Naqui, R. R.; Wheatley, P. J.; Foresti-Serantoni, E. *J. Chem. Soc. (A).* **1971**, 2751.

[i]Uchtman, V. A.; Jandacek, R. J. *Acta Crystallogr.* **1976**, *B32*, 488.

[j]Berndt, A. F.; Sylvester, J. M. *Acta Crystallogr.* **1972**, *B38*, 2191.

[k]Perloff, A. *Acta Crystallogr.* **1972**, *B28*, 2183.

[l]Idrissi, A. K.; Rafiq, M.; Gougeon, P.; Guerin, R. *Acta Crystallogr.* **1995**, *C51*, 1359.

Table 17. Continued

[m]Liminga, R. *Acta Chem. Scand.* **1965**, *19*, 1629.

[n]Chieh, P. C.; Palenik, G. J. *J. Chem. Soc. (A).* **1971**, 576.

[o]Mootz, D.; Altenburg, H. *Acta Crystallogr.* **1971**, *B27*, 1520.

[p]Harrison, R. W.; Thompson, R. C.; Trotter, J. *J. Chem. Soc. (A).* **1966**, 1775.

[q]Granier, W.; Durand, J.; Cot, L.; Galigné, J. L. *Acta Crystallogr.* **1975**, *B31*, 2506.

[r]Mootz, D.; Goldmann, J. *Acta Crystallogr.* **1969**, *B25*, 1256.

[s]McDonald, W. S.; Cruickshank, D. W. J. *Acta Crystallogr.* **1967**, *22*, 43.

[t]Collin, R. L.; Willis, M. *Acta Crystallogr.* **1971**, *B27*, 291.

[u]Tordjman, I.; Durif, A.; Cavero-Ghersi, C. *Acta Crystallogr.* **1974**, *B30*, 2701.

[v]Bagieu-Beucher, M.; Durif, A.; Guitel, J. C. *Acta Crystallogr.* **1975**, *B31*, 2264.

Table 18. Summary of O...O Distances and Ligand Radii (O_t, Terminal, and O_b, Bridging) in Some P_4O_x and MP_3O_x Cages

	O---O (pm)		
	t-b	*b-b*	*Ref.*
$P_4O_{10}(g)$	256(5)	251(5)	a
P_4O_9	253(3)	246(2)	b
P_4O_8	254(4)	249(4)	c
P_4O_6	—	251(1)	c
$P_4O_6(g)$	—	253(1)	d
$P_4O_6S_4$	—	249(5)	a
$[(Me_3CN)_3Ru(P_3O_9)]^-$	252(1)	248(1)	e
	255(1)	253(1)	
$[(C_8H_{12})(NH_2NCMe_2)Ru(P_3O_9)]^-$	253	249	e
	255	252	
$[(CO)_4Ru_2(P_3O_9)_2]^{4-}$	251(1)	248(1)	f
	256(1)	252(1)	
$(CO)_5Co-P(OCH_2)_3CMe$	—	251	g
Averages	**254(1)**	**250(2)**	
Ligand	O_t	O_b	
Radius	**129(2)**	**125(1)**	

Notes: [a]Hampson, G. C.; Stösick, A. J. *J. Am. Chem. Soc.* **1938**, *60*, 1814.

[b]Beagley, B.; Cruickshank, D. W. J.; Hewitt, T. G.; Haaland, A. *Trans. Farad. Soc.* **1967**, *63*, 836.

[c]Beagley, B.; Cruickshank, D. W. J.; Hewitt, T. G.; Jost, K. H. *Trans. Farad. Soc.* **1969**, *65*, 1219.

[d]Maxwell, L. R.; Hendricks, S. B.; Deming, L. S. *J. Chem. Phys.* **1937**, *5*, 626.

[e]Day, V. W.; Eberspacher, T. A.; Klemperer, W. G.; Planalp, R. P.; Schiller, P. W.; Yagasaki, A.; Zhong, B. *Inorg. Chem.* **1993**, *92*, 1629.

[f]Klemperer, W. G.; Zhong, B. *Inorg. Chem.* **1993**, *32*, 5821.

[g]Caffery, M. L.; Brown, T. L. *Inorg. Chem.* **1991**, *30*, 2907.

X. PHOSPHAZENES

The nature of the bonding in phosphazenes has been the subject of much discussion which does not seem to have led to any definite and widely accepted conclusions [24]. Therefore, it was of interest to see if the LCP model could be applied to these molecules also. It can be seen from Table 19 that the N---N distances in a variety of phosphazenes are nearly constant with an average value of 270 pm giving a nitrogen ligand radius of 135 pm. This is consistent with the other radii for nitrogen given in Table 1. Again further confirmation of the validity of the LCP model for these molecules is provided by the N---F, N---O, and N---Cl distances given in Table 20, which are close to the values predicted by the sum of the ligand radii given in Table 1. Clearly a more detailed study of this class of molecules is needed, including calculation of the atomic charges.

XI. THE LIGAND CLOSE-PACKING (LCP) MODEL AND THE VSEPR MODEL

The VSEPR model is based on the assumption that interactions between non-bonding and bonding electron pairs in the valence shell of the central atom A in an AX_nE_m molecule are the dominant factor in determining geometry. In contrast the LCP model assumes that interaction between the ligands is the major factor in determining geometry. For AX_n molecules in which the central atom does not have any lone pair electrons in its valence shell, the two models lead to exactly the same geometries. It might appear, however, that the LCP model would predict that molecules such as PF_3 and PCl_3 would have a planar triangular geometry resulting from the packing of the ligands around a spherical P^{3+} central ion rather than the observed pyramidal geometry. However, in such molecules the P^{3+} ion is not spherical because of the interaction of the negatively charged ligands with the two nonbonding electrons. This interaction is a consequence of electrostatic repulsions and the operation of the Pauli exclusion principle causing the two nonbonding electrons to form a partially localized lone pair. We can think of this pair of electrons as being forced out of the P^{3+} ion to form a pseudo ligand E^{2-} which together with the halogen ligands surround the resulting P^{5+} spherical core in a tetrahedral arrangement (Figure 3). This description gives the same predicted geometries as the VSEPR model which relates the geometry to the tetrahedral arrangement of three bonding pairs and a lone pair in the valence shell of the phosphorus atom.

A.   Sterically Inactive Lone Pairs

The $SeCl_6^{2-}$ ion and several related ions have a regular octahedral geometry which is not in accord with the prediction of the VSEPR model for an AX_6E molecule [25]. Therefore these molecules are often cited as exceptions to the model and are said to have a stereochemically inert lone pair. The LCP model, however, allows us

Table 19. Average PN Bond Lengths, NPN Bond Angles, and N---N (*trans*-Ring) Distances in Some Cyclic Phosphazenes and the $P_4N_{10}^{10-}$ Ion

	P–N (pm)	<NPN (°)	N---N (pm)	Ref.
$N_3P_3F_6(g)$	158.6(5)	120(1)	275(1)	a
$N_3P_3F_6(s)$	156.0(12)	119.5(9)	271(3)	b
$N_3P_3Cl_6(g)$	158.3(5)	118.8(9)	273(2)	c
$N_3P_3Cl_6(s)$	157.5(4)	118.4(3)	271(1)	d
$N_3P_3Br_6$	157.6(8)	118.2(8)	270(2)	e
$N_3P_3(OPh)_6$	157.5(5)	117.3(5)	269(1)	f
$N_3P_3(OC_6H_4Me\text{-}4)_6$	158.4(3)	117.2(3)	269(1)	g
$N_3P_3(OC_6H_4OH\text{-}4)_6$	157.9(10)	117.1(6)	269(1)	g
$N_3P_3(O_2C_6H_3Cl_2)_3$	157.4(10)	117.7(12)	269(2)	h
$N_3P_3(NMe_2)_6$	158.8(3)	116.7(5)	270(1)	i
$N_3P_3(NCS)_6$	154	120	267	j
$N_3P_3[(Cl)NMe_2]_3$	157.3(7)	119.0(3)	271(1)	k
$N_3P_3F_4(2,3\text{-}O_2C_{10}H_6)$	157.3(2)	119.8(2) × 2	271(1)	l
	156.4(4)	116.0(1) × 1	267(1)	
$N_3P_3F_4[O(CH_2)_2S]$	156.4(6)	120.5(5) × 2	272(1)	l
	159.7(2)	114.6(1) × 1	269(1)	
$N_3P_3F_4[O(CH_2)_3S]$	155.6(5)	120.2(3) × 2	270(1)	l
	159.4(3)	115.9(2) × 1	270(1)	
$N_3P_3F_4[S(CH_2)_2S]$	155.6(8)	120.5(4) × 2	270(1)	l
	160.2(4)	113.6(1) × 1	268(1)	
$N_4P_4F_8$	150.7(16)	122.3(10)	265(4)	m
$N_4P_4Cl_8$	157.0(9)	121.2(5)	273(1)	n
$N_4P_4Br_8$	157.5(6)	120.1(4)	273(1)	o
$N_4P_4(OMe)_8$	157.0(18)	121.0(6)	274(3)	p
$N_4P_4(NMe_2)_8$	157.8(10)	120.1(5)	274(2)	q
$(HN)_3P_3O_6^{3-}$	165.4(8)	104.5(3)	266(1)	r
$(HN)_4P_4O_8^{4-}$	166.1(10)	107.4(5)	268(3)	s
		Average:	**270(2)**	
$Li_{10}P_4N_{10}$	167.6(3)	105.9(1)	268(1)	t
	158.1(3)	112.7(1)	271(1)*	

Notes: *This is the contact distance between bridging and terminal N atoms, corresponding to N ligand radii of 134 pm and 137 pm for bridging and terminal nitrogens, respectively.

[a]Davis, M. I.; Paul, J. W. Jr. *J. Mol. Structure*, **1971**, *9*, 478.

[b]Dougill, M. W. *J. Chem. Soc. (A).* **1963**, 3211.

[c]Davis, M. I.; Paul, J. W. Jr. *J. Mol. Structure.* **1972**, *12*, 249.

[d]Bullen, G. J. *J. Chem. Soc. (A).* **1971**, 1450.

Table 19. Continued

^{e}Zoer, H.; Wagner, A. J. *Acta Crystallogr.* **1970**, B26, 1812.

^{f}Marsh, W. C.; Trotter, J. *J. Chem. Soc. (A).* **1971**, 169.

^{g}Allcock, H. R.; Ali-Shali, S.; Ngo, D. C.; Visscher, K. B.; Parvez, M. *J. Chem. Soc., Dalton Trans.* **1996**, 3549.

^{h}Allcock, H. R.; Ngo, D. C.; Parvez, M. J.; Visscher, K. B. *Chem. Soc., Dalton Trans.* **1992**, 1687.

^{i}Retting, S. J.; Trotter, J. *Can. J. Chem.* **1973**, *51*, 1295.

^{j}Faught, J. B. *Can. J. Chem.* **1972**, *50*, 315.

^{k}Ahmed, F. R.; Gage, E. J. *Acta Crystallogr.* **1975**, *B31*, 1028.

^{l}Vij, A.; Geib, S. J.; Kirchmeier, R. L.; Shreeve, J. M. *Inorg. Chem.* **1996**, 2915.

^{m}McGeachin, H. McD.; Tromans, F. R. *J. Chem. Soc.* **1961**, 4777.

^{n}Hazekamp, R.; Migchelsen, T.; Vos, A. *Acta Crystallogr.* **1962**, *15*, 539.

^{o}Zoer, H.; Wagner, A. J. *Acta Crystallogr.* **1972**, B28, 252.

^{p}Ansell, G. B.; Bullen, G. J. *J. Chem. Soc. (A).* **1971**, 2498.

^{q}Bullen, G. J. *J. Chem. Soc.* **1962**, 3193.

^{r}Olthof, R.; Migchelsen, T.; Vos. A. *Acta Crystallogr.* **1965**, *19*, 596.

^{s}Migchelsen, T.; Olthof, R.; Vos. A. *Acta Crystallogr.* **1965**, *19*, 603.

^{t}Schnick, W.; Berger, V. *Angew. Chem. Int. Ed. Engl.* **1991**, *30*, 830.

to better understand the geometry of this and related molecules. The relative sizes of the central atom and the ligands allows a maximum coordination number of only four for the period 2 elements while for the period 3 and 4 elements it is six and it can be even higher for the elements in the later periods. Given that the maximum coordination number for a period 3 or 4 element is six and that a lone pair behaves like a pseudo ligand and occupies one of the available coordination sites, then if there is one lone pair there can be a maximum of five ligands. These five ligands will have a square pyramidal geometry with the lone pair occupying the sixth completing an octahedral arrangement as in SeF_5^-, BrF_5, and XeF_5^+. However, if there are six ligands there is no coordination site available for a lone pair so the nonbonding electrons remain on the central atom which, in the case of the $SeCl_6^{2-}$ ion, can be regarded as an Se^{4+} ion rather than an Se^{6+} ion as in SeF_5^-, and the six ligands adopt an octahedral geometry around the Se^{4+} ion. In such cases the two nonbonding electrons remain on the central spherical ion and do not form a localized lone pair in the valence shell. However, these electrons are not without an influence on the geometry; the Se–Cl bonds are longer than normal [25] because the central Se^{4+} core is larger than the Se^{6+} core. That the chlorine ligands are close-packed is shown by the Cl---Cl intramolecular distance of 340 pm giving a ligand radius of 170 pm. This Cl ligand radius is smaller than the ionic radius of 181 pm but larger than the radius of Cl on B and C which are 151 and 144 pm, respectively, consistent with the expected considerably higher charge on the Cl

Table 20. Ligand–Ligand Distances for Some Oxofluoro and Oxochloro Phosphorus Molecules and Phosphazenes

	O---F (pm)	Ref.		N---F (pm)	Ref.
F_3PO	253	c	F_2PNCO	248(2)	o
$F_3PO-Sb_2F_{10}$	244(2)	d	$C(NPF_2)_2$	255(2)	p
KPO_2F_2	245(2)	a	H_2NPF_2	250(2)	q
$RbPO_2F_2$	246	e	$H_3SiN(H)PF_2$	249(3)	r
$NH_4PO_2F_2$	241(2)	f	$N_3P_3F_6(g)$	256(1)	b
$NaK_3(PO_3F)_2$	245	g	$N_3P_3F_6$	251(2)	b
$(NH_4^+)_2PO_3F·H_2O$	245(4)	a	$N_4P_4F_8$	244(4)	b
$CaPO_3F^{2-}·H_2O$	245(1)	h	$[\{(Me_2N)_2PN\}(F_2PN)]_2$	252(3) × 2	s
				242(3) × 2	
$(PhNH_3^+)PO_3F^-$	243(1)	i			
$MeOPF_2$	245(3)	j			
Predicted:	**245(2)**		Predicted:	**253(2)**	

	O---N (pm)	Ref.		O---Cl (pm)	Ref.
$K^+ {}^+H_3NPO_3^-$	260(1)	k	Cl_3PO	293	c
$(Me_2N)_3PO$	256(3)	l	$Cl_3PN-P(O)Cl_2$	288(1)[1]	t
$N_3P_3(OPh)_3$	258(2)	m	$[Cl_2PO]_2CH_2$	288(1)	u
$N_4P_4(OMe)_4$	258(2) × 4	n	$Ti_2Cl_6(OPCl_3)_2(O_2PCl_2)$	286(3)[2]	v
	249(2) × 4			277(3)[2]	
				282(2)[3]	
$Na_3[(HN)_3P_3O_6]$	258(1)	b			
$K_4[(HN)_4P_4O_8]$	261(1) × 4	b	$Sn_2Cl_6(OPCl_3)_2(O_2PCl_2)_2$	283(3)[2]	w
	252(1) × 4			283(3)[3]	
Predicted:	**253(2)**		Predicted:	**283(2)**	

	N---Cl (pm)	Ref.		N---Cl (pm)	Ref.
$Cl_3PN-P(O)Cl_2$	291(1)[4]	t	$P_4N_4[(NMe_2)Cl]_4$	293(1) × 4	z
	293(2)[5]			291(3) × 4	
$Cl_3PN^+PCl_3$	291(1)	x	$P_4N_4[(Ph)Cl]_4$	293(1)	aa
$P_2N_2Cl_4(NSO_2Cl)$	291(3)	y	$\{N[S(O)F]\}_2NPCl_2$	290(1)	ab
$P_3N_3Cl_6$	292	b	$P_4N_4Cl_8$	289(1)	b
$P_3N_3[(NMe_2)Cl]_3$	292(1) × 4	b	$P_4N_4Cl_8$	292(2) × 4	b
	291(1) × 4			282(2) × 4	
			Predicted:	**291(2)**	

Table 20. Continued

Notes: [1]For $P(O)Cl_2$.

[2]For the two monodentate $OPCl_3$ groups.

[3]For the bidentate bridging O_2PCl_2 group(s).

[4]For the $NP(O)Cl_2$ group.

[5]For the $NPCl_3$ group.

[a]See Table 17.

[b]See Table 19.

[c]Moritani, T.; Kuchitsu, K.; Morino, Y. *Inorg. Chem.* **1971**, *10*, 344.

[d]Edwards, A. J.; Khallow, K. I. *J. Chem. Soc., Dalton Trans.* **1984**, 2541.

[e]Granier, W.; Durand, J.; Cot, L.; Galigné, J. L. *Acta Crystallogr.* **1975**, *B31*, 2506.

[f]Harrison, J. W.; Trotter, J. *J. Chem. Soc. (A).* **1969**, 1783.

[g]Durand, J.; Granier, W.; Cot, L.; Galigné, J. L. *Acta Crystallogr.* **1975**, *B31*, 1533.

[h]Berdt, A. F.; Sylvester, J. M. *Acta Crystallogr.* **1972**, *B28*, 2191.

[i]Idrissi, A. K.; Rafiq, M.; Gougeon, P.; Guerin, R. *Acta Crystallogr.* **1995**, *C51*, 1359.

[j]Zeil, W.; Kratz, H.; Haase, J.; Oberhammer, H. *Z. Naturforsch.* **1973**, *28a*, 1717.

[k]Cameron, T. S.; Chan. C.; Chute, W. J. *Acta Crystallogr.* **1980**, *B36*, 2391.

[l]Belderrain, T. R.; Espinós, J. P.; Fernández, A.; González-Elipe, A. G.; Leinen, D.; Monge, A.; Paneque, M.; Ruiz, C.; Carmona, E. *J. Chem. Soc., Dalton Trans.* **1995**, 1529.

[m]Marsh, W. C.; Trotter, J. *J. Chem. Soc. (A).* **1971**, 169.

[n]Ansell, G. B.; Bullen, G. J. *J. Chem. Soc. (A).* **1971**, 2498.

[o]Rankin, D. W. H.; Cyvin, S. J. *J. Chem. Soc., Dalton Trans.* **1972**, 1277.

[p]Rankin, D. W. H. *J. Chem. Soc., Dalton Trans.* **1972**, 869.

[q]Holywell, G. C.; Rankin, D. W. H.; Beagley, B.; Freeman, J. M. *J. Chem. Soc. (A).* **1971**, 785.

[r]Arnold, D. E. J.; Ebsworth, E. A.; Jessep, H.; Rankin, D. W. H. *J. Chem. Soc., Dalton Trans.* **1972**, 1681.

[s]Marsh, W. C.; Trotter, J. *J. Chem. Soc. (A).* **1971**, 569.

[t]Beluj, F. *Acta Crystallogr.* **1993**, *B49*, 254.

[u]Sheldrick, W. S. *J. Chem. Soc., Dalton Trans.* **1975**, 943.

[v]Moras, D.; Mtschler, A., Weiss, R. *Acta Crystallogr.* **1969**, *B25*, 1720.

[w]Clair, P. P. K.; Willey, G. R.; Drew, M. G. B. *J. Chem. Soc., Dalton Trans.* **1989**, 57.

[x]Faggiani, R.; Gillespie, R. J.; Sawyer, J. F.; Tyrer, J. D. *Acta Crystallogr.* **1980**, *B36*, 1014.

[y]van Bolhuis, F.; van de Grampel, J. C. *Acta Crystallogr.* **1976**, *B32*, 1192.

[z]Bullen, G. J.; Tucker, P. A. *J. Chem. Soc., Dalton Trans.* **1972**, 2437.

[aa]Bullen, G. J.; Tucker, P. A. *J. Chem. Soc., Dalton Trans.* **1972**, 1651.

[ab]Tucker, P. A.; van de Grampel, J. C. *Acta Crystallogr.* **1974**, *B30*, 2795.

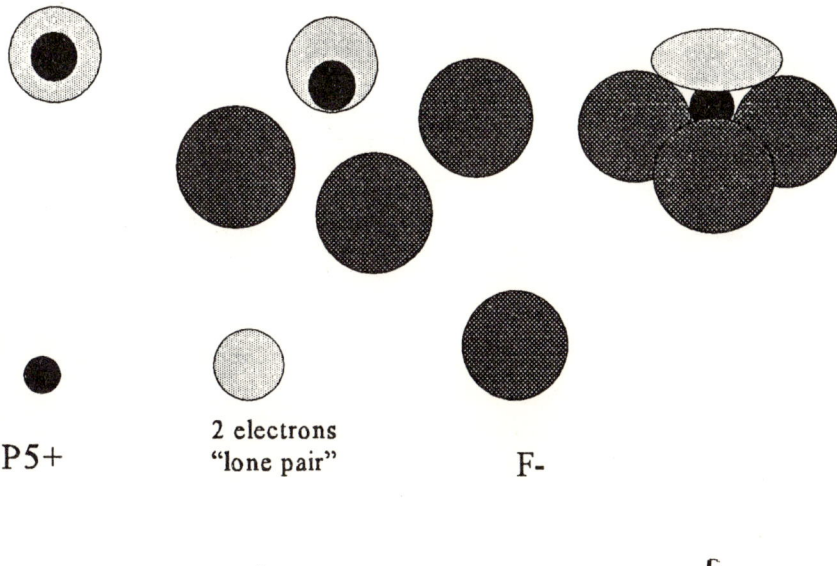

P5+   2 electrons "lone pair"   F-

a   b   c

Figure 3. Diagrammatic representation of the formation of the lone pair in the PCl_3 molecule. (a) An isolated P^{3+} ion consisting of a P^{5+} core surrounded by two valence shell electrons in a spherical shell. (b) Three approaching Cl^- ions distort the distribution of the two valence shell electrons pushing them to one side of the P^{5+} core. (c) When the Cl^- ions reach their equilibrium positions the two valence shell electrons are localized into a lone pair which acts as a pseudo ligand giving the PCl_3 molecule its pyramidal shape.

ligands in $SeCl_6^{2-}$. The corresponding fluoride SeF_6^{2-}, however, has a distorted octahedral C_{3v} geometry which can be described as a monocapped octahedron with the lone pair in the capping position [26]. In this case the lone pair appears to be having only a weak effect on the geometry because a fully stereochemically active lone pair would be expected to give an AX_6E pentagonal pyramidal geometry for the molecule analogous to that of the geometry of the F ligands in the pentagonal bipyramidal molecule IOF_6 [27]. This small distortion from an octahedral geometry is presumably possible because fluorine is a smaller ligand than chlorine, thus there is space in the valence shell for some nonbonding density but not enough to allow the two nonbonding electrons to become a true localized lone pair—in other words an E^{2-} pseudo ligand. So the two nonbonding electrons remain largely within the central Se atom which is therefore closer to an Se^{4+} ion than an Se^{6+} ion and has a shape that is only slightly distorted from spherical. Accordingly the geometry of the six surrounding ligands is only slightly distorted from octahedral (Figure 4). The SeF_6^{2-} ion is a borderline example between the majority of molecules, in which lone pairs are fully active, and the less common cases such as $SeCl_6^{2-}$ with observed

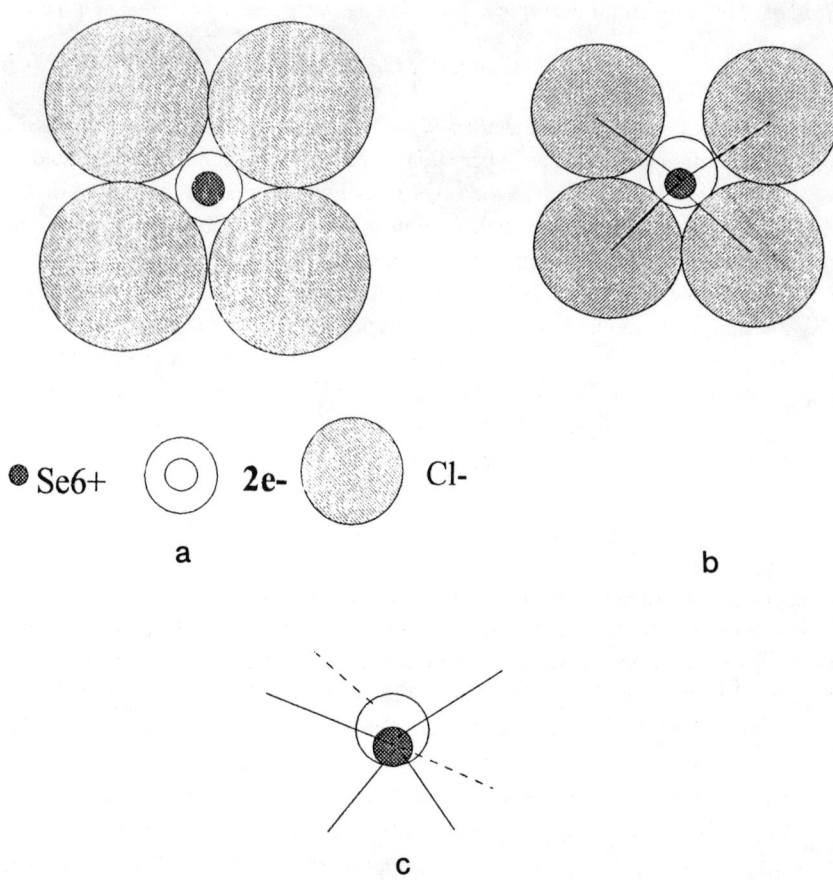

Figure 4. Stereochemically inactive (inert) and partially active "lone-pairs." (a) Diagrammatic representation of the $SeCl_6^{2-}$ ion in a plane perpendicular to a fourfold axis. The Cl^- ions have a close-packed arrangement around an Se^{4+} ion consisting of an Se^{6+} surrounded by a spherical shell of two valence electrons. (b) Diagrammatic representation of the SeF_6^{2-} ion in which the distortion of the Se^{4+} ion is allowed to occur only in the plane of four F ligands showing how two long bonds with a bond angle >90° and two short bonds with a bond angle <90° are formed. (c) Diagrammatic representation of the SeF_6^{2-} ion showing the distortion of the molecule along the C_3 axis giving the molecule the observed C_3 symmetry with three long bonds with bond angles >90° and three short bonds with bond angles <90°.

coordination numbers of six and higher, in which the lone pair is stereochemically inactive. Another similar example is provided by the isoelectronic XeF_6 molecule.

B. Hypervalent Molecules

Five and six coordinated molecules of the main group nonmetals, such as SF_6 and PF_5, are often described as hypervalent because, if the bonding is regarded as covalent, they are exceptions to the octet rule. For this reason the nature of the bonding in these molecules, which is almost always described in terms of the covalent model, has been the subject of endless discussion for many years. However, these molecules invariably have very electronegative ligands such as fluorine for which the ionic model is particularly appropriate. The bonding in these molecules can be simply described as predominately ionic and their geometry is in accordance with the LCP model and also with the VSEPR model. Moreover, there is no need to consider these molecules as exceptions to the octet rule as the central cation and the ligand anions all obey the octet rule.

XII. SUMMARY AND CONCLUSIONS

We have shown in this review that, at least for the fluorides, chlorides, oxides, and hydroxides of Be, B, C, P, and S, and probably for all but the most electronegative elements, ligand–ligand distances are almost constant from one molecule to another, independent of the presence of other ligands and lone pairs, and of the coordination number. These constant interligand distances strongly suggest that the ligands have a constant interligand radius and that they are close-packed around the central atom. We have also shown that the charges on these ligands are larger than has commonly been supposed and we suggest that the model of ions close packed around a central atom is a better first approximation than the covalent model for these molecules. However, the ligand charges are not as large as the full ionic charges, thus the ligands are not as large as the corresponding ions. The charge on a ligand depends primarily on the electronegativity of the atom to which it is attached and varies only slightly from molecule to molecule. Thus the ligand radius varies with the nature of the central atom, but is reasonably constant for a given central atom so that a useful fixed ligand radius can be assigned to a ligand attached to a particular central atom. The close packing of the ligands around a given central atom then determines the bond angles and bond lengths in a molecule. It should be stressed that the conclusion that ligands attached to a given central atom have an almost constant ligand radius is based only on observed experimental data. That this radius is only slightly less than the ionic radius for the least electronegative central atoms and decreases with increasing electronegativity of the central atom suggests that the ligand radius depends on the ligand charge which was confirmed by the calculated ligand charges.

Although Bartell, Hargittai, and others have claimed much earlier that non-bonding interactions are an important factor in determining the geometry of certain classes of molecules, their claim has not been widely accepted. One reason is that the interligand radii suggested by Bartell are too small for most molecules because they were determined for ligands attached to carbon only and so are not appropriate for other atoms for which the ligand radii are larger in most other cases. Thus ligand–ligand intramolecular distances calculated from the Bartell radii are generally smaller than the observed distances, giving the impression that nonbonding interactions are not important in most molecules.

The LCP model was initially based on the ionic model, but it appears to be valid for molecules that are at least up to 50% covalent, and probably even more covalent. Indeed Bartell [6] showed that the bond angles in isobutene and related molecules can be explained in terms of a constant ligand radius. Thus it would appear that the LCP model is independent of the nature of the central atom–ligand bonding. The VSEPR model was based on the covalent model, but it is well known that it applies particularly well to fluorides and other halides which we have shown are predominately ionic. Moreover, both models lead to correct predictions of geometry and are equivalent if we replace the bond–bond repulsions of the VSEPR model by the ligand–ligand repulsions of the LCP model. The interactions that determine geometry are not only those just between the bond pairs and lone pairs, as assumed in the VSEPR model, but between all the electrons on one ligand and those on another—in other words ligand–ligand interactions and the interaction of the ligands with any lone pairs. An advantage of the LCP model is that it enables us to understand that lone pairs are not always formed in molecules which have nonbonding electrons on the central atom and hence to understand molecules that are said to have stereochemically inactive or weakly active "lone pairs" because in these molecules nonbonding electrons are not localized into pairs or are only weakly localized. It seems likely that ligand–ligand interactions are important in many predominately covalent molecules. Another advantage is that by replacing bond–bond repulsions by repulsions between ligands whose size can be approximately determined, it puts the VSEPR model on a somewhat more quantitative basis.

We conclude that ligand–ligand interactions must be considered seriously in all discussions of molecular geometry and that they are particularly important in predominantly ionic molecules. That they have been largely ignored in the past has given rise to incorrect, or at least unnecessarily complicated, bonding descriptions and explanations of geometry as in the cases of the BF_3 and OCF_3^- molecules.

Monatomic ligands such as F, Cl, and O have a cylindrically symmetric electron density distribution and so they have a ligand radius that is independent of the direction of the contact that the ligand makes with adjacent ligands. However, the electron density distribution around oxygen in an OH or OX ligand is not cylindrically symmetric and so their radii vary with the direction in which it makes contact with another ligand. This slight variation with direction of the radius of these ligands provides an explanation of the nontetrahedral bond angles in $A(OH)_4$ and $A(OX)_4$

molecules and probably other related molecules, with NX_2 and CX_3 ligands for example.

It is remarkable that the LCP model based on "hard" contacts works so well. Although it could possibly be improved by using "soft" contacts based on some appropriate interatomic potential, this would unnecessarily complicate what appears to be a simple but useful model.

REFERENCES

1. Kossel, W. *Ann. Phys. (Leipzig)* **1916**, *14*, 677.
2. Lewis, G. N. *J. Amer. Chem. Soc.* **1916**, *38*, 762.
3. Bartell, L. S. *J. Chem. Phys.* **1960**, *32*, 827.
4. Bartell, L. S. *J. Chem. Educ.* **1968**, *45*, 754.
5. Glidewell, C. *Inorg. Chim. Acta Rev.* **1973**, *7*, 69.
6. Bartell, L. S. *Tetrahedron* **1962**, *17*, 177.
7. Hargittai, I. *The Structure of Volatile Sulfur Compounds*; Reidel: Dordrecht, Boston, Lancaster, 1985, p. 217.
8. These values are based on the Allred-Rochow electronegativity scale but almost any other scale such as that due to Pauling gives essentially the same differences. See, for example, Huheey, J. E.; Keiter, E. A.; Keiter, R. L. *Inorganic Chemistry*, 4th ed.; Harper Collins: New York, 1993, p. 211.
9. Bader, R. F. W. *Atoms in Molecules: A Quantum Theory*; Clarendon Press: Oxford, 1990.
10. Robinson, E. A.; Johnson, S. A.; Tang, T. -H.; Gillespie, R. J. *Inorg. Chem.* **1997**, *38*, 3022.
11. Huheey, J. E.; Keiter, E. A.; Keiter, R. L. *Inorganic Chemistry*, 4th ed.; Harper Collins: New York, 1993, p. A30–A31.
12. Honeychuck, R. V.; Hersch, H. *Inorg. Chem.* **1989**, *28*, 2869.
13. Kroto, H. W.; Maier, M. *J. Mol. Spectrosc.* **1977**, *65*, 280.
14. Bartell, L. S.; Brockway, L. O. *J. Chem. Phys.* **1955**, *23*, 1960.
15. Takeo, H.; Matsumura, C. *Bull. Chem. Soc. Jpn.* **1977**, *50*, 676.
16. Long, M. W.; Williams, P.; Weatherly, T. *J. Chem. Phys.* **1960**, *33*, 508.
17. Gillespie, R. J.; Bytheway, I.; Robinson, E. A. *Inorg. Chem.* **1998**, *39*, xxxx.
18. Shannon, R. D. *Acta Crystallogr.* **1976**, *A32*, 751.
19. Reed, A. E.; Schade, C.; Schleyer, P. v. R. J.; Kamath, P. V.; Chandrasekhar, *J. Chem. Soc., Chem. Commun.* **1988**, 67 and references therein.
20. Stølevik, R. *Acta Chem. Scand.* **1974**, *28A*, 327.
21. Shiono, R.; Cruichshank, D. W. J.; Cox, E. G. *Acta Crystallogr.* **1958**, *11*, 389.
22. Gillespie, R. J.; Robinson, E. A.; Tang, T-H.; Heard, G. Unpublished.
23. Reference 7, pp. 216–218.
24. See, for example, Greenwood, N. N.; Earnshaw, A. *Chemistry of the Elements*; Pergamon: Oxford, 1984, pp. 627–630, and reference 11 pp. 769–775.
25. Gillespie, R. J.; Robinson, *Angew. Chem. Int. Ed. Engl.* **1996**, *35*, 495.
26. Mhajoub, A. R.; Zhang, X.; Seppelt, K. *Eur. J. Chem.* **1995**, *4*, 261.
27. Christe, K. O.; Saunders, J. C. P.; Schrobilgen, G. J.; Wilson, W. W. *J. Chem. Soc., Chem. Commun.* **1991**, 837.

THE TERMINAL ALKYNES:
A VERSATILE MODEL FOR WEAK DIRECTIONAL
INTERACTIONS IN CRYSTALS

Thomas Steiner

Advances in Molecular Structure Research
Volume 4, pages 43–77
Copyright © 1998 by JAI Press Inc.
All rights of reproduction in any form reserved.
ISBN: 0-7623-0348-4

ABSTRACT

Terminal alkynes, $C \equiv C-H$, are the currently best studied type of C–H hydrogen bond donors, and serve as a model to explore weak hydrogen bond effects in general. This is because of their strong donor and acceptor potentials, and their good suitability for infrared spectroscopic experiments. An overview is given on the hydrogen bond functions of the terminal alkyne residue. It is shown that in crystal structures, terminal alkynes form hydrogen bonds with strong acceptors like O=P as well as with weak acceptors like phenyl rings and $C \equiv C$ and $C = C$ moieties. Geometric data is given for these types of interactions. Terminal alkynes can, furthermore, accept π-type hydrogen bonds from strong and also from weak donors, including such from other alkyne groups. The ability to operate as donor and acceptor simultaneously enables terminal alkynes to participate in cooperative hydrogen bond networks with the same functions as hydroxyl groups. Insights into weak hydrogen bond effects obtained from terminal alkynes serve for a better understanding of the hydrogen bond phenomenon as such.

I. INTRODUCTION

A. General

It is long known from infrared spectroscopic experiments that polar C–H groups may act as weak hydrogen bond donors [1–4] and π-bonded moieties as weak hydrogen bond acceptors [3, 4]. Structural studies of these phenomena have been performed already in the 1950s and 1960s, but remained without substantial impact. Only in recent years, weak hydrogen bonding effects became a major subject of structural investigation, in particular the ubiquitous C–H···O hydrogen bonds [5–10]. Hydrogen bonds with π-acceptors (often called "π-hydrogen bonds") occur in various biological systems, in particular with phenyl acceptors, and have been extensively described for these [11–17]. In organic [18–21] and organometallic [22] crystals, π-hydrogen bonds have also been structurally characterized. More recently, hydrogen bonds involving metal centers [23, 24] and even the so-called di-hydrogen bonds $X-H^{\delta+}···H^{\delta-}-Y$ have been described [25]. It has now become clear that weak hydrogen bonds frequently play crucial roles in determining crystal structures (not always, though), and sometimes even of molecular conformation. The practice to neglect weak hydrogen bonds if they occur together with stronger O/N–H···O hydrogen bonds or ionic interactions, as was formerly commonly done, is not justified and in many instances misleading.

Even though substantial progress has been made in the understanding of weak hydrogen bonds, the level of insight has not yet reached that of the conventional O–H···O and N–H···O hydrogen bonds. One way to gain further and deeper insight into complex phenomena, are systematic studies of a suitable model system. For investigations of C–H···A and X–H···π-hydrogen bonds, several model systems can be used (and are used by different groups of authors). Currently, the best studied C–H donor group are the terminal alkynes, C≡C–H. For these, a larger variety of hydrogen bond phenomena has been explored and quantitatively characterized than for any other C–H type. In the present chapter, experimental findings on hydrogen bonding terminal alkynes (often called terminal acetylenes) are overviewed.

B. Why Study Terminal Alkynes?

There are several reasons why the terminal alkynes are a particular useful model system to study hydrogen bond effects. These are, in short: (a) their high C–H acidity, (b) their ability to donate and accept hydrogen bonds simultaneously, (c) their easy structural characterization from X-ray diffraction data even of only moderate quality, and (d) their exceptional suitability for infrared spectroscopic experiments. These points shall be discussed in more detail:

(a)   The hydrogen bond donor strength of C–H groups depends on the carbon hybridization as $C(sp)$–H > $C(sp^2)$–H > $C(sp^3)$–H and is enhanced by electron-withdrawing groups attached to C [1]. Structurally, this reflects in shortening of the average hydrogen bond distances, which are correlated with the conventional C–H acidities [26], and sharpening of the hydrogen bond directionalities [27]. Figure 1 shows the directionality characteristics of C–H···O=C interactions formed by hydroxyl, alkynyl, vinyl, and methyl groups, as compared to pure van der Waals contacts: it is obvious that the preference for linearity of C≡C–H···O=C hydrogen bonds comes close to that for hydroxyl groups, indicating a considerable hydrogen bond strength. Although less acidic C–H groups can also donate hydrogen bonds [6–9, 27] (note that in Figure 1, the histogram for methyl groups is not yet flat as that for van der Waals contacts), it is clear that the more acidic C–H groups will be particularly interesting systems. Unfortunately, molecules carrying very acidic C–H groups are typically problematic to handle (such as HCN and trinitromethane [28]). This points at the terminal alkynes (and also at the haloforms), which are appreciably strong hydrogen bond donors, and at the same time "well behaved" so that they can be explored with reasonable experimental effort.

(b)   Terminal alkynes can not only donate, but they can also *accept* hydrogen bonds directed at the C≡C moiety. The ability to act as donor and acceptor simultaneously distinguishes the terminal alkynes from most other weak hydrogen bonding functional groups, and leads to a unique richness of observed hydrogen bond effects.

(c)   In the linear C≡C–H group, the hydrogen atom position is unambiguously defined by the positions of the two C-atoms, and can therefore be reliably calculated even if it has not been located experimentally. This allows to characterize C≡C–H···X hydrogen bonds even from structural data of moderate quality. A good example is the first crystallographic study of a hydrogen bonding alkyne, cyanoace-

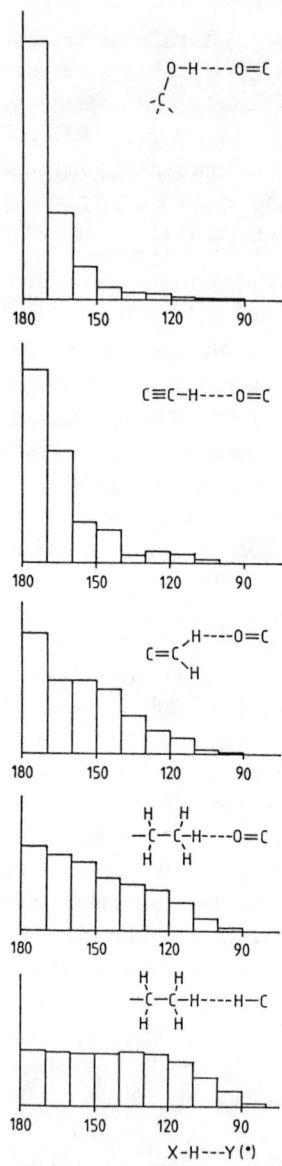

Figure 1. Directionality characteristics of different hydrogen bond types: angular frequencies of X–H···O=C contacts of different donor types, and of C–H···H–C van der Waals contacts. From top to bottom: hydroxyl, ethynyl, vinyl, and ethyl donors, van der Waals contacts. The distributions are "cone-corrected," i.e. weighted by the factor 1/sin θ (Kroon, J.; Kanters, J. A. *Nature*, **1974**, *248*, 667–669). Drawn based on published data [*27*].

tylene [29], where a C–H···N interaction is conclusively discussed based on only non-H atom positions.

(**d**) The C–H stretching vibration of terminal alkynes possesses a combination of properties which makes it an exceptionally useful and robust infrared (IR) spectroscopic probe for hydrogen bond effects. This is easiest illustrated in an example: Figure 2 shows the crystal structure of *N*-propargyloxy-phthalimide [30] together with the IR absorption spectrum of the crystalline compound. In the crystal, the molecules form centrosymmetric dimers linked by C≡C–H···O=C hydrogen bonds (Figure 2, top). In IR absorption spectra, acetylenic C–H stretching vibrations $\nu_{\equiv C-H}$ typically absorb around wavenumbers of 3300 cm^{-1} (with environmental variations), whereas all other C–H groups absorb well separated at lower wavenumbers (2700–3100 cm^{-1}). This makes band assignment easy, often even trivial. Furthermore, $\nu_{\equiv C-H}$ absorptions are normally much intenser than those from the other C–H groups. In Figure 2, bottom, the sharp and intense band peaking at 3253 cm^{-1} is the $\nu_{\equiv C-H}$ absorption band, whereas the small and poorly separated bands around 3000 cm^{-1} originate from the other C–H oscillators. If the compound is dissolved in the apolar solvent CCl$_4$, the acetylenic C–H absorption band peaks at 3312 cm^{-1}, representing an alkyne group which is "free" from hydrogen bonding. The red-shift of 59 cm^{-1} of the crystal absorption compared to that of the CCl$_4$ solution is caused by weakening of the C–H bond in the C–H···O interaction, and is in the typical range observed for this type of hydrogen bond [*31, 32*]. This kind of red-shift is a very sensitive probe for hydrogen bonding (wavenumbers are measured with accuracies <1 cm^{-1}, often even <0.1 cm^{-1}). For terminal acetylenes, $\nu_{\equiv C-H}$ is easy to determine, and by experience it is robust against effects from vibrational coupling. For hydrogen bonds donated by all other C–H groups, one would have to analyze spectral regions around 3000 cm^{-1} like the one in Figure 2, bottom, which is obviously much more difficult, and normally not possible in a routine fashion. In consequence, solid-state IR spectroscopy can be conveniently used as a powerful analytic tool in combination with X-ray crystallography.

The reasons given above make the terminal acetylenes a model system for weak hydrogen bonds that exhibit a multitude of interesting effects, and can be studied conveniently and in great detail by complementary experimental methods. Furthermore, it is well suitable for theoretical studies, such as ab initio molecular orbital calculations on realistic systems. In consequence, the work of different research groups on hydrogen bonding terminal alkynes has accumulated to a volume far exceeding that for any other C–H group. Insights into weak hydrogen bond effects obtained from terminal alkynes are not only interesting in themselves (this would hardly justify the amount of work spent on the subject), but serve for a better understanding of the hydrogen bond phenomenon as such.

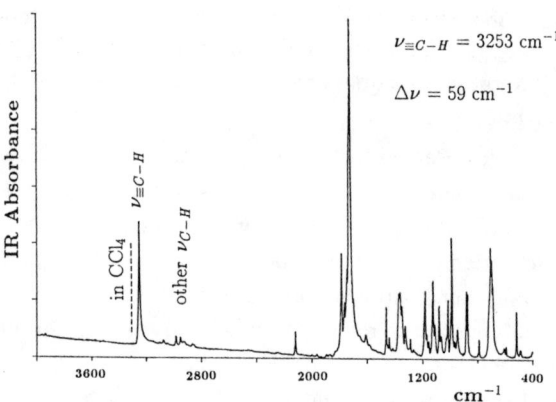

Figure 2. Crystal structure of *N*-propargyloxy-phthalimide (top, drawn based on a published X-ray structure [*30*]) and the infrared absorption spectrum of a polycrystal-line sample in a KBr pellet (bottom, unpublished spectrum recorded in the course of preparing ref. 30); $\nu_{\equiv C-H}$ in a dilute CCl_4 solution, 3312 cm^{-1}, is shown by a dashed line. Distances are given in Å.

II. O-ACCEPTORS

Of the hydrogen bonds from terminal alkynes to various acceptors, best studied are those to oxygen. Not long after Sutor's classical general study on C–H···O hydrogen bonds in crystals [*33*], Ferguson published a series of crystal structures of terminal alkynes and supported the bonding nature of the occurring C–H···O contacts by IR spectroscopy [34–36]. The example of *o*-chlorobenzoacetylene is shown in Figure 3 (IR wavenumber red-shift of $\nu_{\equiv C-H}$ compared to solution in cyclohexane: 69 cm⁻¹). Part of this series is the crystal structure of the natural compound laurencin (isolated from the seaweed *Laurentia glandulifera*), which contains the first re-ported example of a three–centered ("bifurcated") C≡C–H···O hydrogen bond (Figure 4) [36]. This figure gives an impression of how hydrogen bonds have formerly been characterized without knowledge of the H-atom position. This is also an early example that in a crystal structure, not necessarily the strongest accepor must accept the strongest hydrogen bond: in crystalline laurencin, the two relatively weak C–O–C acceptors are involved in hydrogen bonding, whereas the far stronger C=O acceptor is unsatisfied [36].

Following these early studies, an increasing number of crystal structures contain-ing C≡C–H···O interactions has been published and are readily available in the Cambridge Structural Database [37] (CSD). Most have simple topologies similar to the examples shown in Figures 2 to 4, and need not be discussed here further. More interesting are C≡C–H···O hydrogen bonds which are part of finite or "infinite" arrays, such as chains C≡C–H···O–H···O or O–H···C≡C–H···O, and related. Systems of this kind will be discussed in Section V ("hydrogen bond networks"). In the present section, only some general properties of C≡C–H···O contacts will be described.

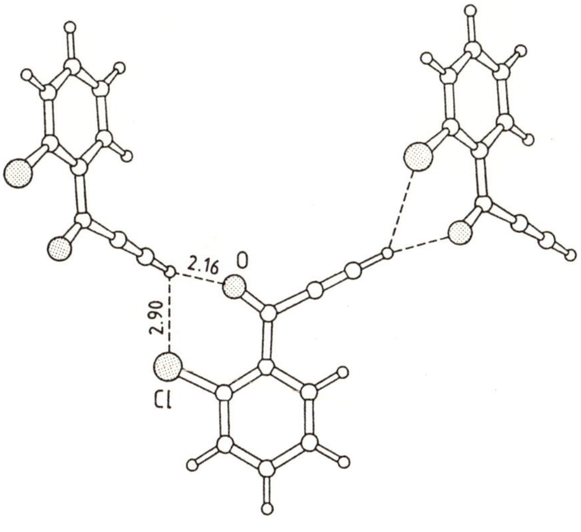

Figure 3. The C≡C–H···O hydrogen bonds in *o*-chlorobenzoacetylene (drawn using coordinates from Ferguson and Islam [35]). The additional C–H···Cl interaction is not mentioned in the original paper and is certainly much weaker than the C–H···O hydrogen bond.

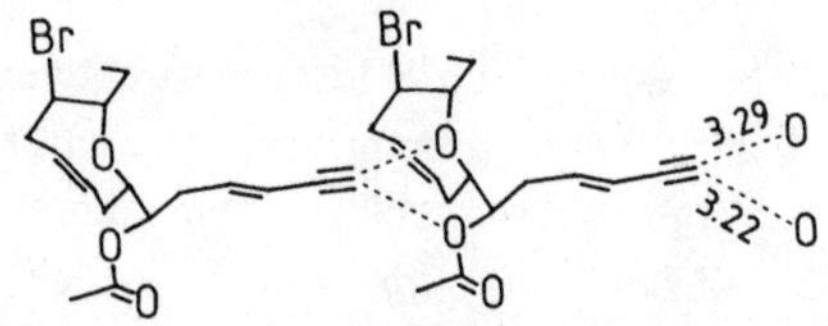

Figure 4. The three-centered ("bifurcated") C≡C–H⋯O hydrogen bond in the natural compound Laurencin (drawn using a figure of the original publication [36] as a template). H-atom positions have not been located.

The directionality characteristics of C–H⋯O hydrogen bonds is clearly shown in Figure 1. C≡C–H⋯O donor directionality is very pronounced, i.e. close to linear contacts are strongly preferred over bent ones [27]. Acceptor directionality, on the other hand, such as the preference for in-plane contacts to carbonyl acceptors, is only weak and detectable not without some effort [38]. Energies of C≡C–H⋯O hydrogen bonds are typically calculated to be in the range of 1 to slightly above 2 kcal/mol [9], with bonds involving the strongest acceptor types (see below) certainly being stronger.

In hydrogen bonds, average donor–acceptor separations generally reduce with increasing donor acidity and acceptor basicity [5–9]. For different types of C–H

Table 1. Mean H⋯O and C⋯O Separations (in Å) for C–H⋯O Hydrogen Bonds from Small Solvent Molecules to Three Common Acceptor Types (with H⋯O < 2.8 Å)

	Acceptor		
	$C{=}O$	$C{-}OH$	$C{-}O{-}C$
Mean H⋯O			
$CHCl_3$	2.22(5)		2.31(4)
CH_2Cl_2	2.27(4)		2.50(4)
$C(sp^3){-}C{\equiv}CH$	2.24(3)	2.41(5)	2.44(6)
$Me{-}C{\equiv}N$	2.44(7)	2.44(7)	2.46(4)
Me_2SO	2.51(2)	2.61(4)	2.63(6)
Me_2CO	2.60(2)	2.69(2)	2.69(3)
Mean C⋯O			
$CHCl_3$	3.16(3)		3.26(4)
CH_2Cl_2	3.21(3)		3.43(4)
$C(sp^3){-}C{\equiv}CH$	3.28(2)	3.37(4)	3.38(4)
$Me{-}C{\equiv}N$	3.33(6)	3.42(9)	3.37(3)
Me_2SO	3.41(2)	3.44(3)	3.59(6)
Me_2CO	3.49(2)	3.55(5)	3.61(4)

Note: [a]Data from ref. 39.

groups, mean H···O and C···O distances to C=O and C–O–C acceptors are listed in Table 1 [*39*], showing that alkynes are much stronger donors than methyl groups (compare Figure 1), but not as strong as chloroform. For methyl groups bonded to different atoms, R–CH_3, variations of the mean C/H···O distances are observed which show that the donor strength of CH_3 increases as N≡C–CH_3 > O=S(CH_3)$_2$ > O=C(CH_3)$_2$. For terminal alkynes R–C≡C–H, analogous acidity variations must be expected for different groups R which can influence the alkyne polarity, such as for P–C≡C–H, Si–C≡C–H, C(sp^2)–C≡C–H, C(sp^3)–C≡C–H, H–C≡C–H, and so on. This should lead to different average hydrogen bond distances with a constant acceptor type (such as carbonyl groups). To quantify this effect, which is certainly only small, relatively large and well-defined structural data samples would be required. Since these data quantities are not available at this time, an acidity effect on hydrogen bond lengths for different terminal alkynes is as yet unproven.

The effect of acceptor basicity on average C≡C–H···O hydrogen bond distances is easier to show [*39*]. In particular, for carbonyl and hydroxyl acceptors, data quantities are available which are sufficient for statistical analysis: Figure 5 shows the frequency distributions of hydrogen bond lengths with alkyne donors and C=O acceptors (top) and C–OH acceptors (bottom) (technical details are given in the figure legend). The distributions are relatively broad, reflecting the typical geometrical variability of weaker hydrogen bond types. The mean value of the distribution with carbonyl acceptors is 2.25(2) Å, and that with hydroxyl acceptors is 2.38(4) Å. The difference of 0.13 Å clearly shows that C=O is a stronger acceptor than C–OH. Similar differences are found for other C–H donor types, as is shown for a different data sample in Table 1 [*39*]. It should be noted that the distance distribution for the weaker acceptor is broader, and has no defined end at long distances, indicating that the interaction as a whole is "softer." For acceptor types other than C=O, C–OH, and C–O–C, data quantities are much smaller, not allowing analysis like that shown in Figure 5. However, it is remarked that the shortest C≡C–H···O hydrogen bond found as yet is formed with the very strong P=O acceptor in triphenylphosphinoxide [*40*] (Figure 6) ("crystal engineering" is not a topic of this article, but it is mentioned that this dimeric structure was designed with the particular aim to produce a short C–H···O hydrogen bond).

Figure 6 shows the shortest C≡C–H···O hydrogen bond known. It is of interest to point also at the longest one: in the organometallic compound shown in Figure 7, the ethynyl group forms a very long contact to a carbonyl ligand, with H···O = 2.92 Å and C···O = 3.71 Å. For this compound, careful IR and Raman spectroscopic experiments under different sampling conditions showed a red-shift of $\nu_{≡C-H}$ compared to apolar solvents of ca. 20 cm^{-1}, which is clearly above possible sampling errors and indicative of weak hydrogen bonding [*41*]. It is of importance to note that the distance appreciably exceeds the sum of van der Waals radii (ca. 2.7 Å for H···O), demonstrating again [*5*] the known long range character of the hydrogen bond. When compared to the "normal" C≡C–H···O hydrogen bond in Figure 2 ($\nu_{≡C-H}$ red-shift = 59 cm^{-1}) and the unusually strong one in Figure 6

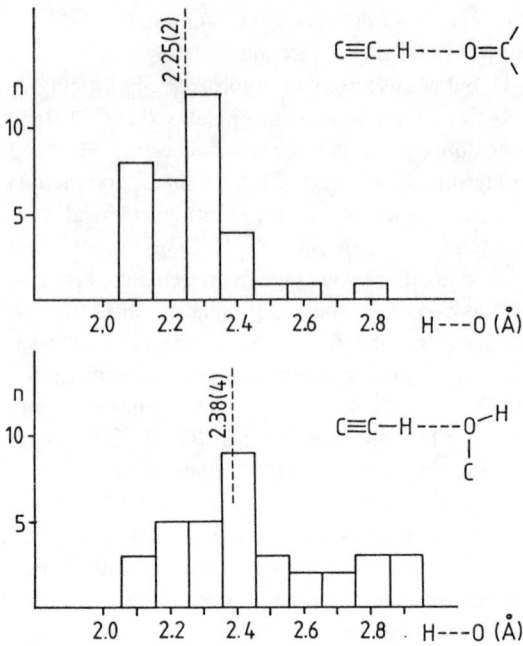

Figure 5. H···O frequency distributions in crystals for hydrogen bonds from C≡C–H donors to (top) C=O acceptors, $n = 33$, and (*bottom*) C–OH acceptors, $n = 31$. Database analysis performed for this article [CSD, June 1997 update with 167797 entries, ordered and error–free crystal structures with $R < 0.10$, normalized H-atom position based on a linear C≡C–H group, and a C–H bond distance of 1.08 Å; neither donor nor acceptor group directly bonded to a metal atom, H···O < 2.95 Å; for three-center hydrogen bonds, only the short component is considered].

($\nu_{\equiv C-H}$ red-shift = 145 cm^{-1}), it is obvious that the long and bent ethynyl–carbonyl contact in Figure 7 represents the "lower limit" of C≡C–H···O hydrogen bonding. The weakness of this interaction is certainly associated with the poor (but non-zero [*10*]) acceptor strength of carbonyl ligands.

It is known that hydrogen bonding reduces the thermal vibrations of the involved groups. For alkynes, this was demonstrated in a statistical study, where H···X distances (X = any hydrogen bond acceptor) were correlated with the displacement parameters of the donor groups. Since absolute values of U cannot be reasonably compared between different crystal structures, the ratio U_{C2}/U_{C1} was analyzed instead (atom labeling: C1≡C2–H) [*42*]. In Figure 8, the reduction of thermal vibrations due to hydrogen bonding can be clearly seen.

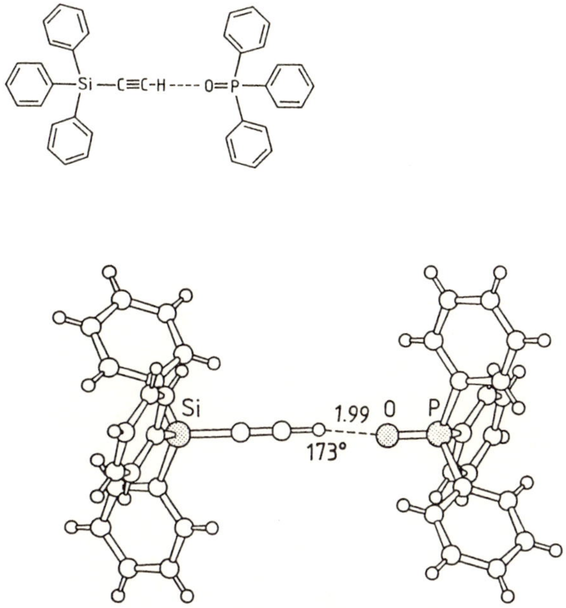

$\Delta\nu_{\equiv C-H} = -145 \text{ cm}^{-1}$

Figure 6. The short C≡C–H···O=P hydrogen bond in the crystalline adduct of triphenylsilylacetylene and triphenylphosphinoxide (drawn using published atomic coordinates [40]). In the crystal structure, there are four symmetry-independent dimers of this kind. Since one is placed on an inversion center, the content of the asymmetric crystal unit is 3.5 formula units ($Z' = 3.5$), which is an extremely unusual value (only one other case is known).

III. OTHER CONVENTIONAL ACCEPTORS

A. N-Acceptors

The first crystal structure where C–H···X hydrogen bonding has been discussed is not with an O-acceptor but with an N-acceptor: solid hydrogen cyanide, published by Dulmage and Lipscomb in 1951 (Figure 9) [43]. HCN is not a terminal alkyne, but the C-atom is *sp*-hybridized, making a chemical relation obvious. Substitution of H by an acetylene group leads to cyanoacetylene, N≡C–C≡C–H, which has an analogous chain-type crystal structure as HCN (Figure 9, bottom) [29]. It is worthwhile to note that HCN molecules also associate in chains in solution; based on the dielectric constant of liquid HCN, this was correctly proposed by Kumler in 1935 [44], which is the earliest literature example of hydrogen bonding C–H groups known to the author.

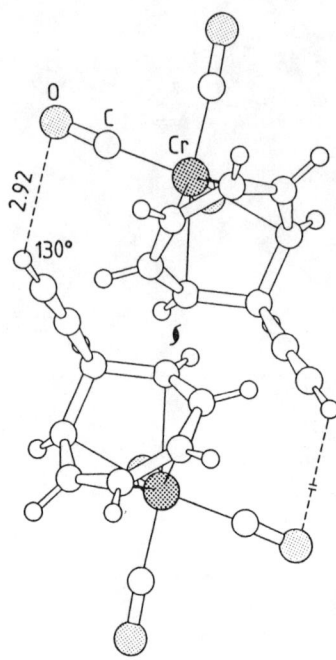

Figure 7. The very long C≡C–H···OC–Cr hydrogen bond in crystalline [Cr(CO)$_3$[η^6-(7-*exo*-(C≡CH)C$_7$H$_7$]. Drawn using published atomic coordinates [*41*].

Although C≡C–H···N hydrogen bonding has been established in crystal structures earlier than C≡C–H···O hydrogen bonding, it is today much less explored. The number of reported crystal structures is too small for comprehensive analysis, so that only some representative examples can be shown below. In Figure 10, C≡C–H···N hydrogen bonds in crystals are schematically shown for four different N-acceptor types (compound code names are as in the Cambridge Structural Database). In DOJHAL [45], molecules are arranged in chains linked by C≡C–H···N≡C hydrogen bonds. In FUCVOO [46], molecular dimers are formed which are linked by a pair of C≡C–H···N hydrogen bonds accepted by tertiary amino N-atoms. In the dialkyne JELTUP [47], one C≡C–H group donates a hydrogen bond to an N(sp^2) atom in a heterocycle, and the other one to a carbonyl group. And finally, in the synthetic sex steroid SORCEH [48], the ethynyl group donates a short hydrogen bond to an oxime N-atom. It must be admitted that the examples in Figure 10 have been selected so as to have close to ideal hydrogen bond geometries. It is not untypical for C≡C–H···N hydrogen bonds to have more elongated hydrogen bond distances and bent angles at the H-atom.

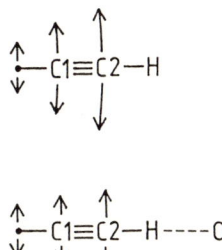

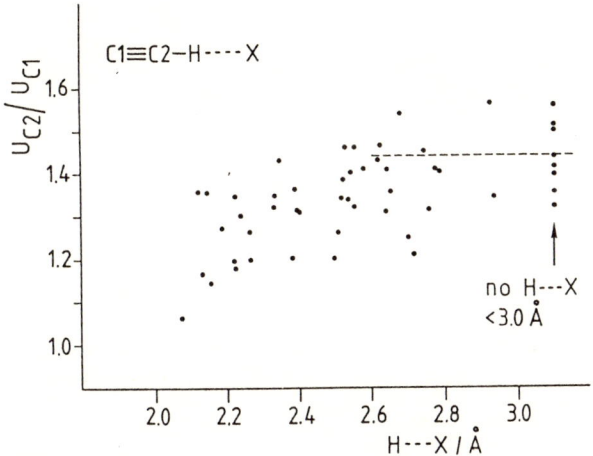

Figure 8. Correlation of the quantity U_{C2}/U_{C1} with the H$\cdots$X distance in C$\equiv$C–H$\cdots$X hydrogen bonds (X = any acceptor), for normalized hydrogen atom positions ($n = 51$). Drawn using published data [42].

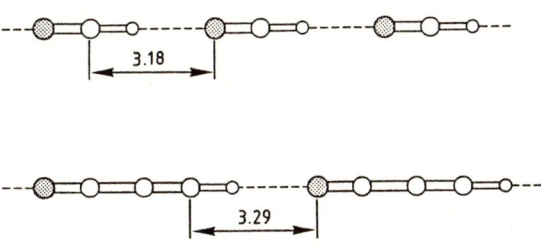

Figure 9. Infinite molecular chains linked by C–H$\cdots$N hydrogen bonds: crystalline hydrogen cyanide (N$\equiv$C–H, *top*, drawn based on published atomic coordinates [43]) and cyanoacetylene (N$\equiv$C–C$\equiv$C–H, *bottom*, drawn using published atomic coordinates [29]).

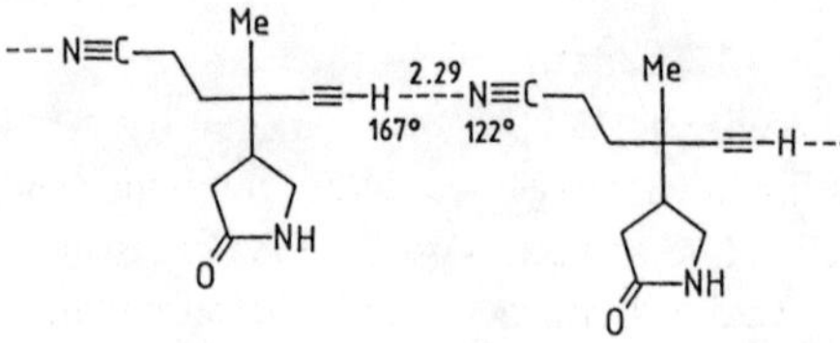

DOJHAL

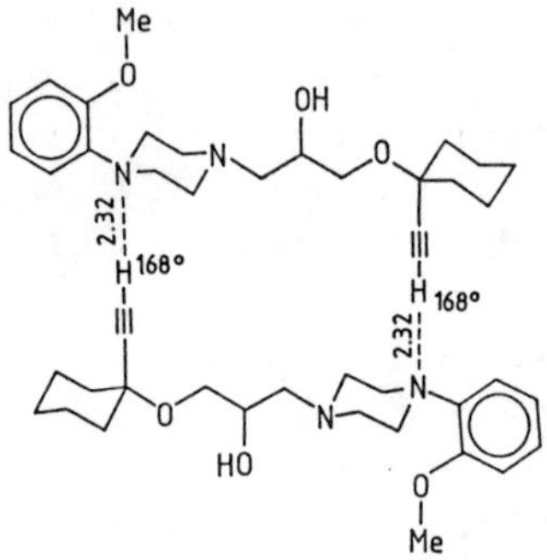

FUCVOO

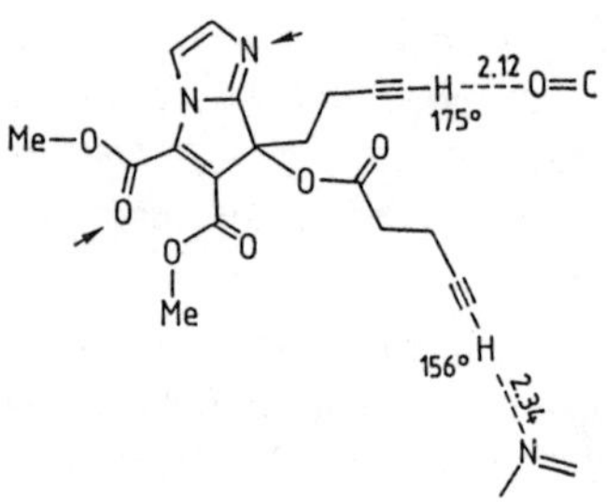

JELTUP

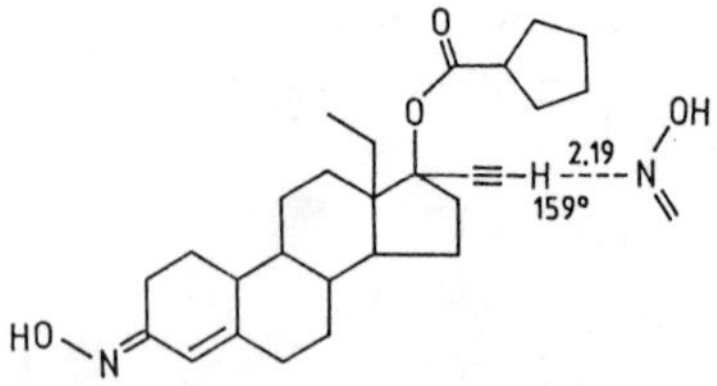

SORCEH

Figure 10. Examples of C≡C–H···N hydrogen bonds for different acceptor types. Geometries are based on normalized H-atom positions. (**a**) With a cyano acceptor: DOJHAL [*45*]. (**b**) With a tertiary amino acceptor: FUCVOO [*46*]. (**c**) With a C–N=C acceptor: JELTUP [*46*]; the dialkyne also forms a C≡C–H···O hydrogen bond, the acceptor atoms are indicated by arrows. (**d**) With an oxime N-acceptor: SORCEH [*48*]. Figure drawn for this article.

B. S-Acceptors

Only a single example with a C≡C–H···S hydrogen bond is found in the CSD: compound VITYUS [49], with a fairly linear hydrogen bond accepted by an S-atom which is bonded to P. The H···S separation is 2.58 Å (Figure 11). In general, sulfur is known to be a much weaker hydrogen bond acceptor than oxygen or nitrogen [50], so that the rareness of C≡C–H···S interactions is not surprising.

C. Halide Anion Acceptors

Halide ions are among the strongest hydrogen bond acceptors, and hydrogen bonds of terminal acetylenes to halides are frequently observed. The distribution of H···Cl⁻ distances in crystals is shown in Figure 12 in comparison with related distributions for other C–H donors. For chloride ions, the mean H···Hal⁻ distance is 2.56(4) Å, and for bromide ions, it is 2.70(6) Å [the mean C···Hal⁻ distances are 3.58(4) and 3.72(5) Å, respectively] [51].

Most of the C≡C–H···Hal⁻ hydrogen bonds occur in halide salts of protonated amines, three representative examples with chloride acceptors are shown in Figure 13 (compounds DHISNX10 [52], HEZVOX [53], and SOXVAC [54]). In these cases, the chloride ions typically accept also stronger N⁺–H···Cl⁻ hydrogen bonds. However, it would be misleading to regard the weaker C≡C–H···Cl⁻ interactions in such configurations only as bystanders: in the crystal structure of propargylammonium chloride, the anions accept three N⁺–H···Cl⁻ and one C≡C–H···Cl⁻ hydrogen bond, which only in their concert make a complete coordination sphere to the chloride ion (Figure 14) [55].

Although C≡C–H···Hal⁻ hydrogen bonds may participate in anion coordination with similar function as those from N⁺–H donors, they are certainly suffering from competition effects: normally, the stronger N⁺–H···Hal⁻ hydrogen bonds will optimize their geometry at the price of rather distorted C≡C–H···Hal⁻ interactions. This is illustrated in Figure 15 for the example of propargylammonium bromide [55] (top) and triphenylpropargylphosphonium bromide [56] (bottom). In the

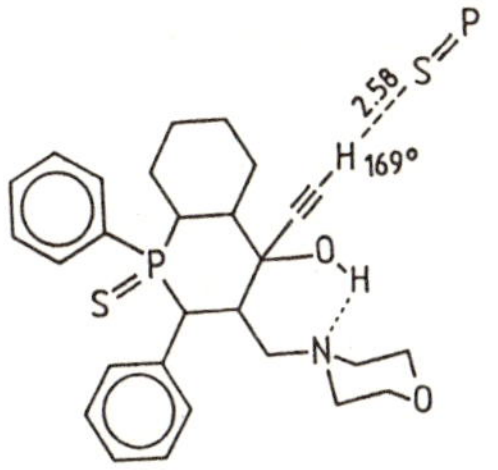

VITYUS

Figure 11. The only example of a C≡C–H···S hydrogen bond: compound VITYUS [49]. The geometry is given for a normalized H-atom position. Figure drawn for this article.

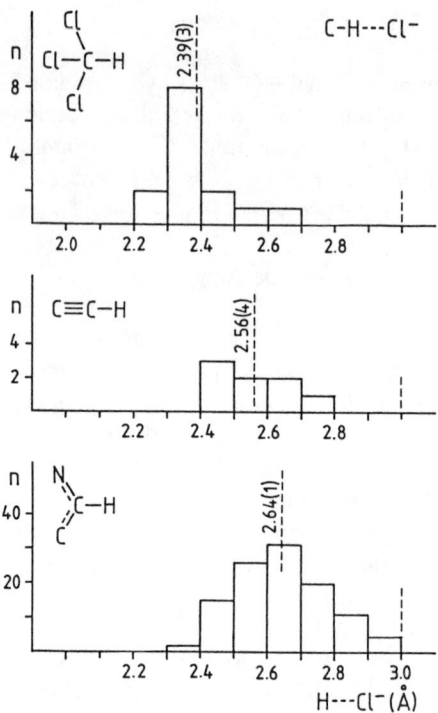

Figure 12. Distributions of H···Cl⁻ separations for hydrogen bonds (H···Cl⁻ < 3.0 Å) from three different types of C–H donors. Drawn using published data [51].

propargylammonium compound, which crystallizes isomorphously as the related chloride salt (Figure 14), the alkyne donor has to compete with three N^+–H donors, and the H···Br⁻ distance is long with 2.75 Å. In the triphenylphosphonium salt, there are no stronger donors present than the alkynyl group, and the C≡C–H···Br⁻ hydrogen bond can be optimized to a much better geometry [55].

D. Halogen Acceptors

Whereas halide ions are strong hydrogen bond acceptors, carbon-bonded halogen atoms are only very weak acceptors. Metal-bonded halogen is of intermediate acceptor strength. Only very few examples are available for C≡C–H···Hal–X hydrogen bonds. In Figure 16, an example of each is shown for a Cl–C and for a Cl–M acceptor (compounds BEVNIZ [57] and NADBID [58], respectively). Note that the C≡C–H···Cl–Pt hydrogen bond is relatively short (the supramolecular array is nicely discussed in the original publication [58]). Analogous hydrogen bonds with the other halogen atoms than Cl should be possible, but have not been observed as yet.

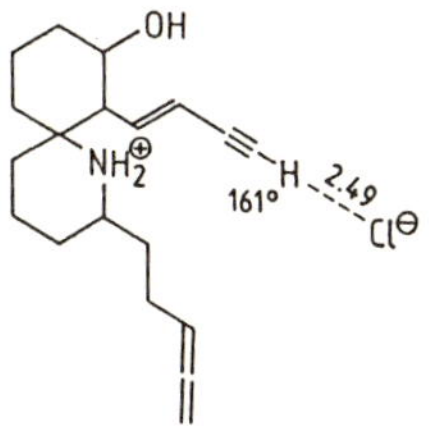

DHISNX10

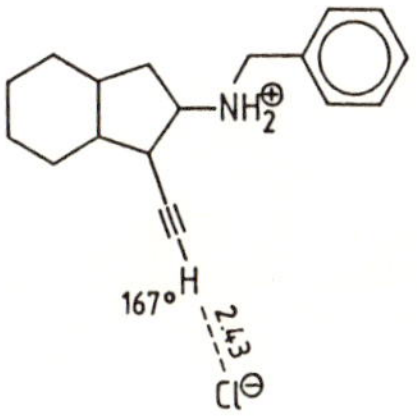

HEZVOX

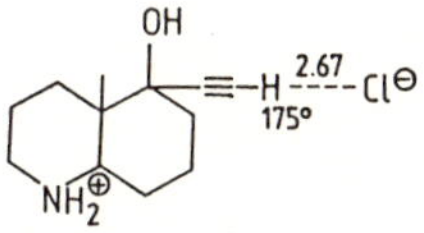

SOXVAC

Figure 13. Three examples of protonated amines forming C≡C–H···Cl⁻ hydrogen bonds (compounds DHISNX10 [52], HEZVOX [53], and SOXVAC [54]). Geometries are given for normalized H-atom positions. Figure drawn for this article.

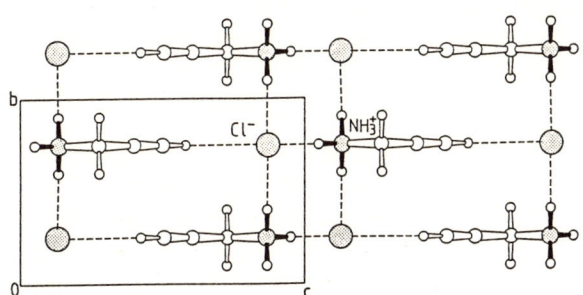

Figure 14. Crystal structure of propargylammonium chloride, exhibiting three N⁺–H···Cl⁻ and one C≡C–H···Cl⁻ hydrogen bonds (drawn using published atomic coordinates [55]).

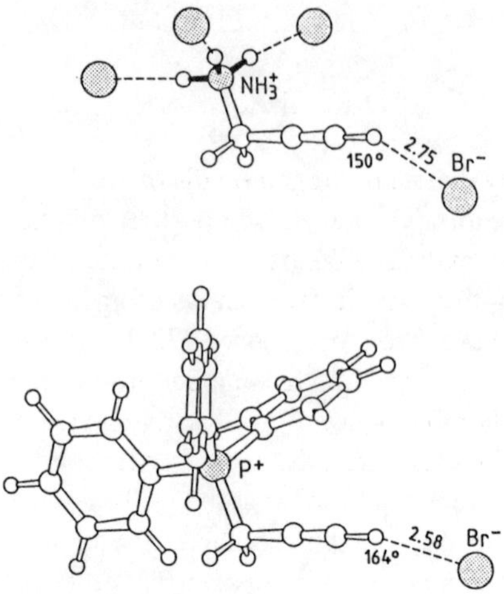

Figure 15. Crystal structure of propargylammonium bromide (*top*, drawn using published atomic coordinates [55]), exhibiting three N^+–H···Br^- and one C≡C–H···Br^- hydrogen bonds, and triphenylpropargylphosphonium bromide (*bottom*, drawn using published atomic coordinates [56]), exhibiting only a C≡C–H···Br^- hydrogen bond.

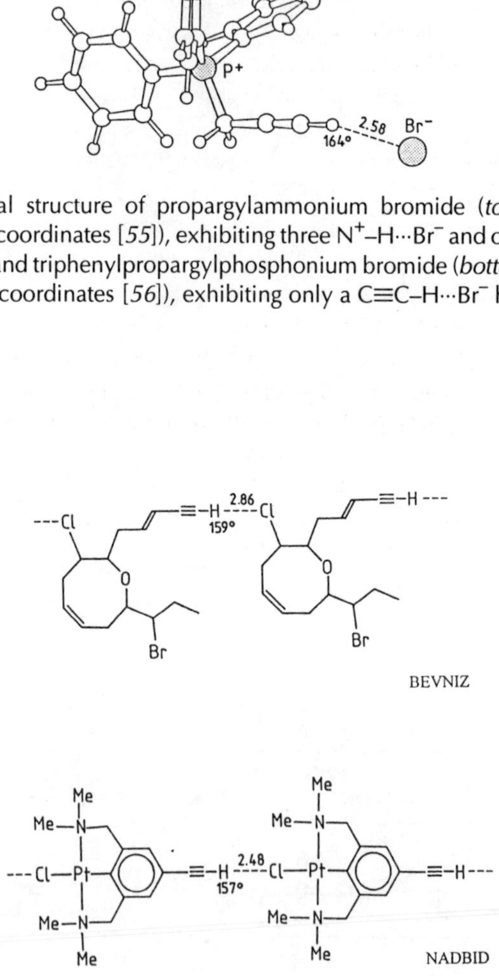

Figure 16. Examples for a C≡C–H···Cl–C hydrogen bond: compound BEVNIZ [57] (*top*) and for a C≡C–H···Cl–M hydrogen bond: compound NADBID [58] (*bottom*). Figure drawn for this article.

IV. π-ACCEPTORS

A. C≡C-Acceptors

Terminal alkynes may donate hydrogen bonds of weak to moderate strengths to O-, N-, and halide-acceptors. This raises the question whether in crystal structures, this donor can also form hydrogen bonds with the weaker π-acceptors like phenyl rings and C≡C and C=C bonds. In fact, chains of apparently directional C≡C–H···C≡C–H contacts have been observed 1972 by Leiserowitz in crystalline but-3-ynoic acid [59], and a two-dimensional pattern of related contacts links the molecules in the orthorhombic phase of solid acetylene (Figure 17) [60]. Whether such contacts deserve the classification as "hydrogen bonds," or if they are better termed "herringbone interactions" similar to the well-known edge-to-face contacts between phenyl rings [61, 62], remained unclear and authors cautiously preferred to use unspecific terms to characterize the contacts. A further possible interpretation could consider these contacts as nothing but a consequence of close-packing requirements, i.e. neither hydrogen bond nor herringbone interactions, but simply van der Waals contacts.

At a closer look, T-shaped alkyne–alkyne contacts are a very common pattern in crystal structures: two examples with chains of such contacts are shown in Figure 18 (top: prop-2-ynylglycine [63]; bottom: pent-4-ynoic acid [64]; in the lower homologues but-3-ynoic acid [59] and propynoic acid [65] analogous patterns are formed). In Figure 19, the recent example of 1,4-diethynylbenzene is shown, which contains antiparallel chains of short alkyne–alkyne interactions [66]. In a database search performed in 1995, 18 examples for T-shaped C≡C–H···C≡C–H contacts were found [64], many of which form interconnected chains, but some are "iso-

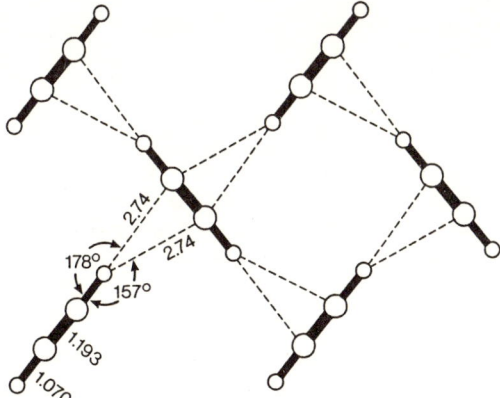

Figure 17. Neutron diffraction study of the orthorhombic phase of acetylene [60]; one layer of molecules is shown. Figure drawn for this article.

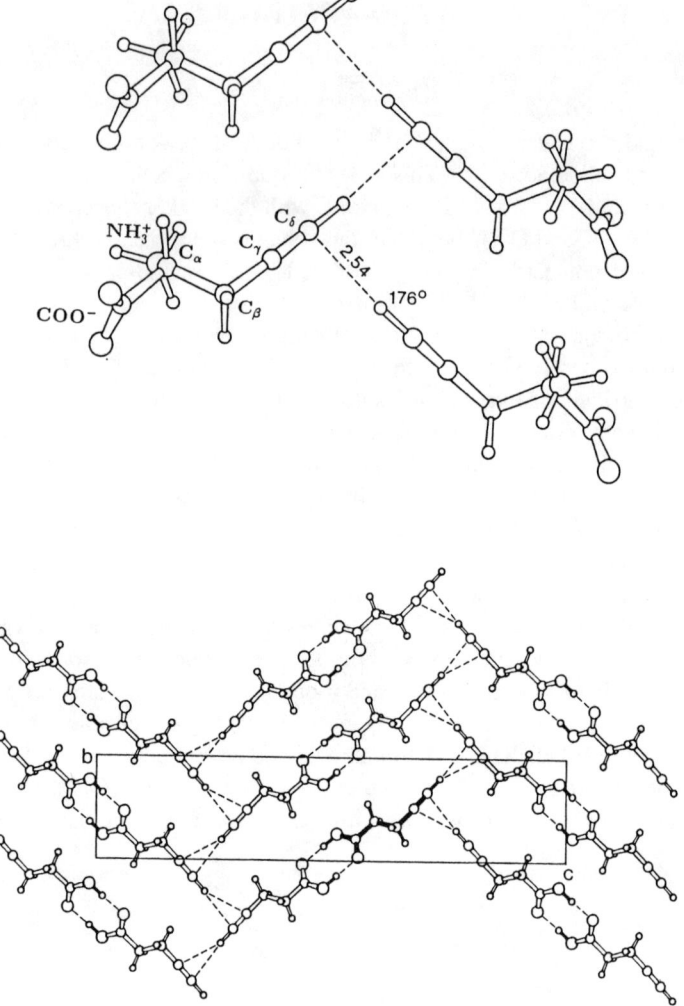

Figure 18. Two examples of interconnected C≡C–H···C≡C–H hydrogen bonds. *Top*: in prop-2-ynylglycine (drawn using published atomic coordinates [53]). *Bottom*: in pent-4-ynoic acid; the contact is directed at the C≡C midpoint with H···C distances of 2.71 and 2.75 Å (drawn using published atomic coordinates [54]).

lated" or part of networks with other hydrogen bond types (see below). The geometry of C≡C–H···C≡C–H contacts is variable: the donor group may point more or less at the midpoint of the C≡C triple bond (as in Figure 18, bottom, and Figure 19), or at an individual C-atom (as in Figure 18, top). In the latter case, each of the C-atoms in the $C_2≡C_1$–H moiety can act as the acceptor. This is a behavior

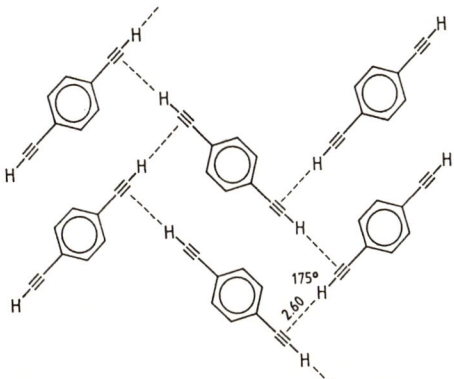

Figure 19. A layer of crystalline 1,4-diethynylbenzene [*66*], schematically drawn for this article.

that is observed for X–H···C≡C hydrogen bonding in general [*18*]. The shortest distances from the H-atom to the C≡C midpoint are about 2.5 Å, but this distance can also be much longer (see below).

To answer the question if T-shaped alkyne–alkyne contacts can be regarded as "hydrogen bonds," IR absorption spectra of a number of relevant crystalline compounds have been recorded. For pent-4-ynoic acid (Figure 18, bottom), the red-shift of $\nu_{\equiv C-H}$ compared to a solution in CCl_4 is 33 cm^{-1}, which is more than half of the red-shift associated with the decent C≡C–H···O=C hydrogen bond shown in Figure 2. This is good evidence that the interaction is associated with appreciable weakening of the covalent C–H bond, and therefore has in fact hydrogen bond character. In subsequent quantum-chemical calculations, bond energies of such contacts were calculated to be in the range 1–2 kcal/mol, similar to the energies of C–H···O interactions [*64*]. The largest bond energies (ca. 2 kcal/mol) are calculated for interconnected arrangements, which experience cooperativity effects.

Since C≡C–H···C≡C–H hydrogen bonds can be connected to form infinite chains, it must be expected that they can also form rings. In fact, the contact pattern in orthorhombic acetylene (Figure 17) can be decomposed into four-membered rings (a "member" here is a functional group, not a single atom). A six-membered ring was first observed in crystalline ethynylferrocene (Figure 20) [*67*]. Of the six ethynyl groups, three are symmetry-independent, and in the IR spectrum, three absorption bands at different wavenumbers are found (Figure 20, bottom). The ring in Figure 20 is also of importance because it shows that the C≡C–H···C≡C–H hydrogen bond has a very long range: even for long H···C separations of 2.9 Å and more, a bonding effect can be inferred from the IR absorption spectrum. A long

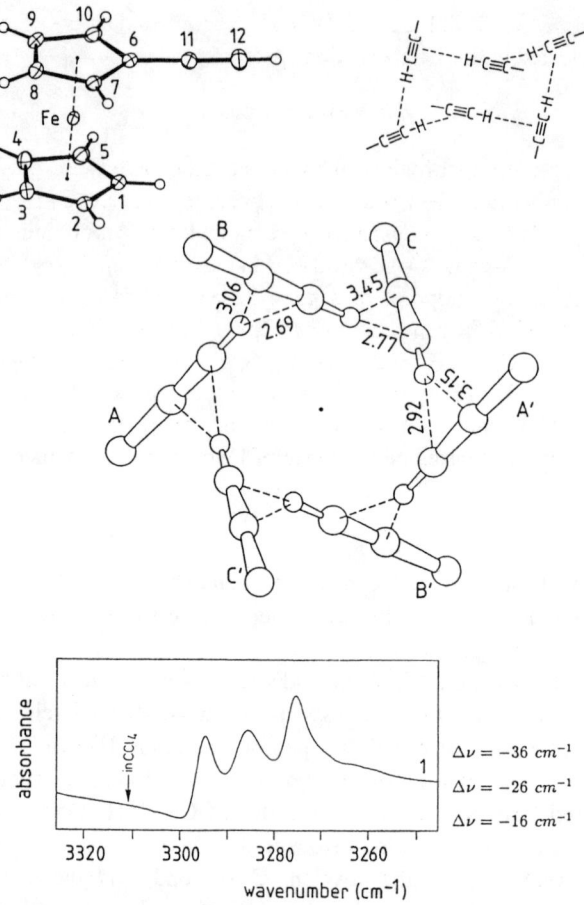

Figure 20. The intermolecular contacts of the ethynyl groups in ethynylferrocene forming a cooperative puckered cycle of six C≡C–H···C≡C–H hydrogen bonds. Note the long H···C distances. In the midpoint of the ring, there is a crystallographic inversion center. The region of ≡C–H stretching frequencies in the IR absorption spectrum is shown at the bottom (drawn using published crystallographic and spectroscopic data [67]).

range is a general property of hydrogen bonds [5], and this property can here be directly shown (compare Figure 7).

Hydrogen bond donors can interact with more than one acceptor simultaneously (Figure 4), and one must expect to find such behavior also for X–H···π hydrogen bonds. For N+–H···Ph hydrogen bonding, bi- and trifurcated arrangements have been reported by Knop and coworkers in tetraphenylborate salts [19]. For C≡C–H···C≡C–H hydrogen bonding, a bifurcated arrangement was found in 3-

phenylpenta-1,4-diyn-3-ol, Figure 21 [67]. Hydrogen bond nature of these contacts was found reflected in the IR absorption spectrum.

B. Ph-Acceptors

Phenyl rings are stronger hydrogen bonds acceptors than $C\equiv C$ groups (because of the larger number of π-electrons), and the occurrence of $C\equiv C-H\cdots Ph$ hydrogen bonding in crystals must in consequence be anticipated. An obvious example has been found in ($\pm$)-3-phenylbut-1-yn-3-ol [64], where an alkynyl group points almost directly at the midpoint of a phenyl ring. This alkyne group accepts a $C\equiv C-H\cdots C\equiv C-H$ hydrogen bond from another alkyne group, so that the hydrogen bond pattern is $C\equiv C-H\cdots C\equiv C-H\cdots Ph$. In a subsequent database analysis, six examples of $C\equiv C-H\cdots Ph$ hydrogen bonding in published crystal structures were found. The H$\cdots$M distances (M = aromatic midpoint) are typically in the range 2.5 to 2.85 Å [64]. In most cases, the distance of H to M is shorter than that to any of

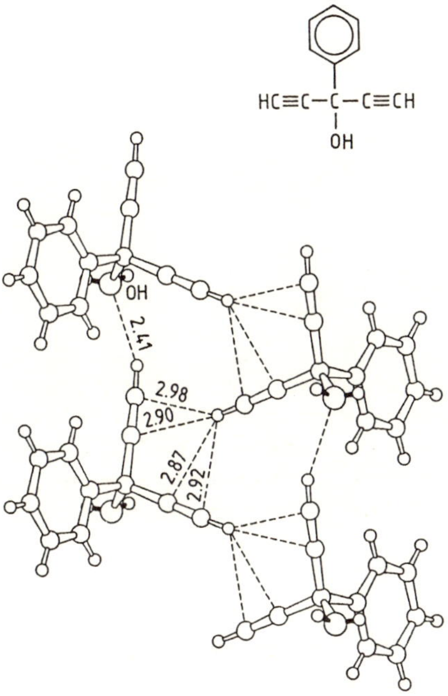

Figure 21. The intermolecular contact pattern of the ethynyl groups in 3-phenyl-1,4-diyn-3-ol, representing a chain of interconnected bifurcated $C\equiv C-H\cdots C\equiv C-H$ hydrogen bonds (drawn using published atomic coordinates [67]). There are also $C\equiv C-H\cdots O$ hydrogen bonds formed. The H-atom of the hydroxyl group is twofold disordered; both alternative positions are shown.

the individual C-atoms, but there are also exceptions where one of the H···C distances is shorter (down to 2.50 Å) than H···M.

Two relevant examples, which have been characterized later in time, are shown in Figures 22 and 23. The molecular dimer in Figure 22 is linked by a remarkable system of cooperative O–H···C≡C–H···Ph hydrogen bonds; both hydrogen bond types are directed at the midpoints of the acceptor groups [68]. The crystal structure shown in Figure 23 contains a highly peculiar system of O–H···Ph···H–C≡C hydrogen bonds, in which the ethynyl group points roughly at the midpoint of the phenyl ring, whereas the hydroxyl group points at an individual C-atom. To make sure that this is a true observation, the original X-ray study [69] was repeated (and verified) with low-temperature neutron diffraction [70]. In quantum chemical calculations, bond energies of 1.3 kcal/mol were obtained for both interactions. Since O–H is the far stronger donor than C≡C–H, this means that the off-centered geometry of the O–H···Ph bond is much less favorable than the centered geometry of the C≡C–H···Ph interaction [69]. This may be surprising because the center of the phenyl ring is the point of lowest electron density, but in centered contacts, an acceptor may interact with the whole π-cloud. These results are in line with

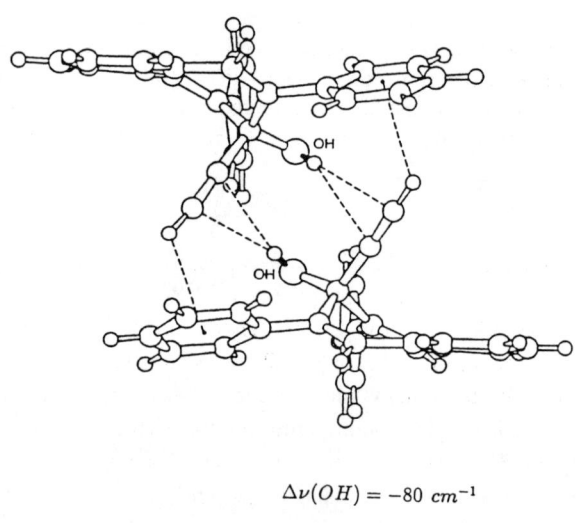

$$\Delta\nu(OH) = -80 \ cm^{-1}$$

$$\Delta\nu(\equiv CH) = -41 \ cm^{-1}$$

Figure 22. Molecular dimer of 7-ethynyl-6,8-diphenyl-7H-benzocyclohepten-7-ol, linked by two cooperative chains of O–H···C≡C–H···Ph hydrogen bonds (drawn using published atomic coordinates [68]). The red-shifts of $\nu_{\equiv C-H}$ and of ν_{O-H} compared to a dilute solution in CCl_4 are given. The distances of the H-atoms to the acceptor midpoints are $H_C···M_{Ph} = 2.58$ Å and $H_O···M_{C\equiv C} = 2.58$ Å.

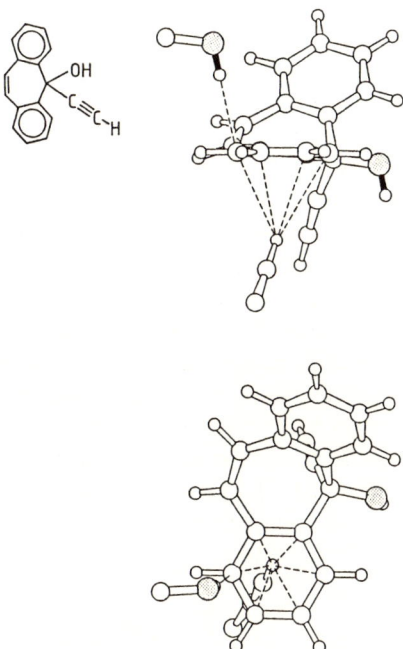

Figure 23. The C≡C–H⋯Ph and O–H⋯Ph hydrogen bonds in 5-ethynyl-5*H*-dibenzo[*a,d*]cyclohepten-5-ol as characterized in a neutron diffraction study at 20 K (drawn using published atomic coordinates [*70*]). The hydroxyl group donates a hydrogen bond which is almost linearly directed at an individual C-atom with H⋯C = 2.34 Å and an angle at H of 174°. The C–H⋯Ph hydrogen bond has a H⋯M separation of 2.59 Å.

theoretical calculations on N–H⋯Ph hydrogen bonds, which have shown that the centered contact geometry is the optimal one [*13*].

A different type of Ph-acceptors is found in tetraphenylborate anions [B(Ph)₄]⁻, which due to their negative charge are π-acceptors *par excellence* [*19*]. Alkyne–Ph hydrogen bonds with tetraphenylborate acceptors have been constructed by crystallization of the salt propargylammonium tetraphenylborate, NH_3^+–CH_2–C≡CH [B(Ph)₄]⁻ (Figure 24) [*71*]. Note that the N⁺–H⋯Ph hydrogen bonds are almost ideally centered with very short distances around 2.1 Å of the H-atoms to the aromatic midpoints.

C. C=C-Acceptors

The C=C bond is only a weak π-acceptor, and in consequence, it was the last for which C≡C–H⋯π hydrogen bonding has been found. This interaction was characterized from a sample of crystalline ethynyl steroids [*72*], one of which is shown

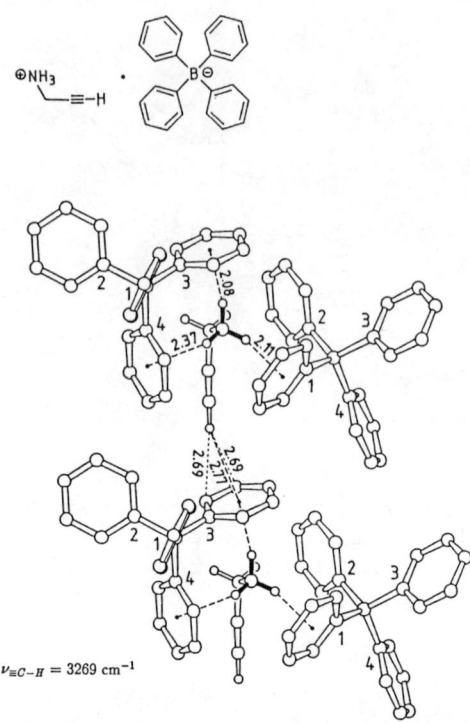

Figure 24. Section of the crystal structure of propargylammonium tetraphenylborate, showing N⁺–H···Ph and C≡C–H···Ph hydrogen bonds (drawn published atomic coordinates [71]).

in Figure 25. In this crystal structure ("lynestrenol" [73]), the hydroxyl group of the steroid points at one side of the C=C bond, and the ethynyl group of a symmetry-related molecule points at the other side. IR absorption spectra clearly show that both types of interactions represent weak hydrogen bonds. In the sample of 10 C≡C–H···C=C hydrogen bonds, the H···M distances are in the range 2.52–2.98 Å, and a correlation of H···M separations with red-shifts of the infrared ≡C–H stretching frequencies is evident [72].

D. Hydroxyl Groups Which Apparently Donate No Hydrogen Bonds

In crystal structure publications, hydroxyl groups are occasionally reported to "form no hydrogen bonds." More frequently, hydroxyl groups are missing in tabulations of hydrogen bonds, without a remark given in the text what they might be doing with their donor potential. A look at the crystal structures discussed above lets assume that in many of these cases the possibility of π-hydrogen bonding has simply not been considered. In the crystal structures shown in Figures 22, 23, and

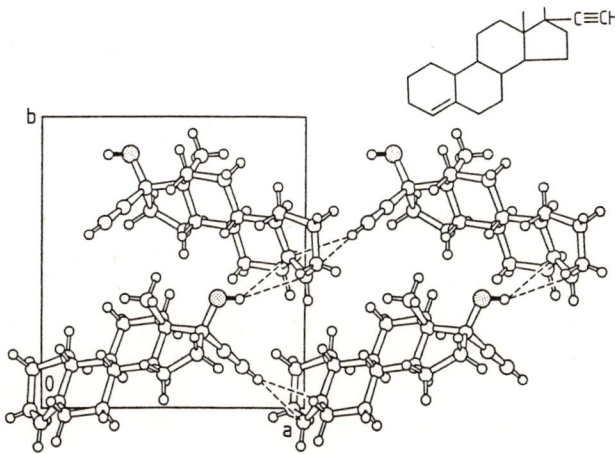

Figure 25. Crystal structure of the ethynyl steroid lynestrenol (drawn using atomic coordinates published by Duax and coworkers [73]), exhibiting O–H···C≡C and C≡C–H···C≡C hydrogen bonds with H···M separations of 2.64 and 2.62 Å, respectively. The red-shift of $\nu_{\equiv C-H}$ compared to a solution in CCl_4 is 39 cm^{-1} [72].

25, there are no O–H···O, but only O–H···π-hydrogen bonds. If only conventional O–H···O hydrogen bonding is considered, in none of the three crystal structures, there would be *any* hydrogen bonding.

A different reason why O–H···X hydrogen bonds can be overlooked is the use of too restrictive distance cutoff criteria, such as the criterion that the H···O/N distance must be shorter than 2.4 Å [17].

V. HYDROGEN BOND NETWORKS

Hydrogen bonds of terminal alkynes can have very different functions in hydrogen bond networks. Very simple ones are "isolated" hydrogen bonds of the types,

$$C\equiv C-H\cdots O=C \qquad (1)$$

$$C\equiv C-H\cdots N\equiv C \qquad (2)$$

$$C\equiv C-H\cdots Ph \qquad (3)$$

$$C\equiv C-H\cdots C=C \qquad (4)$$

and many related, which are all topologically equivalent. These isolated hydrogen bonds occur very frequently in crystal structures (e.g. Figures 2, 3, 6, 7, 9, 10). A further frequent but simple arrangement are hydrogen bonds to double π-acceptors, such as:

$$C\equiv C-H\cdots Ph\cdots H-O \tag{5}$$

$$C\equiv C-H\cdots C=C\cdots H-O \tag{6}$$

Examples are shown in Figures 23, 24, and 25. Quantum chemical calculations for the example in Figure 23 suggested that the two hydrogen bonds do not (or only marginally) influence each other [69]. This might be because the approach is from opposite directions so that the two donors are not in steric conflict with each other. Somewhat different are two hydrogen bonds directed to a carbonyl acceptor.

$$\underset{\underset{\displaystyle C}{\displaystyle \|}}{C\equiv C-H\cdots O}\cdots H-O \tag{7}$$

In this configuration, the two hydrogen bonds are typically directed more or less at the electron lone pair lobes which form an angle of 120°. This leads to relatively short contacts of the two donors and to mutual repulsion. The arrangement should therefore be slightly anticooperative (i.e. the total energy is smaller than the sum for two isolated hydrogen bonds). Topologically related (but chemically different) are the hydrogen bond arrangements around halide ions, which accept n hydrogen bonds with n from 1 to over 5 (e.g. Figures 13, 14).

In the hydrogen bond patterns shown above, all constituents operate only as donor or only as acceptor. This leads to a very limited number of arrangements. Patterns become more complex if they involve groups which act as donor and acceptor simultaneously. The simplest possible "network" of this kind is a short chain,

$$X-H\cdots Y-H\cdots Z \tag{8}$$

in which the central group is any group possessing the required hydrogen bond potentials (such as a hydroxyl group, a secondary amine or a terminal alkyne), and the terminal group is a "chain stopper" like C=O, C–O–C, C=C, and so on. In crystals, these chains can be of the types,

$$O-H\cdots O-H\cdots O=C \tag{9}$$

$$C\equiv C-H\cdots O-H\cdots O=C \tag{10}$$

$$O-H\cdots C\equiv C-H\cdots O=C \tag{11}$$

$$O-H\cdots C\equiv C-H\cdots Ph \tag{12}$$

$$C\equiv C-H\cdots C\equiv C-H\cdots Ph \tag{13}$$

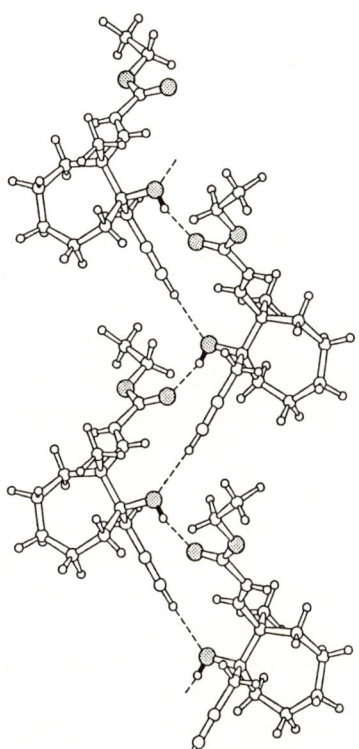

Figure 26. Finite cooperative chains of hydrogen bonds in 1β-hydroxy-1α-propargyl-2β-methyl-2-(2-ethoxycarbonylvinyl)-cycloheptane (drawn using published atomic coordinates [74]).

or any related arrangement. Examples are shown in Figures 22 and 26. All these chains are *cooperative*: due to mutual polarization of the involved groups, the two hydrogen bonds enhance the strengths of each other, and the total hydrogen bond energy is larger than the sum for two isolated hydrogen bonds.

If hydrogen bonding groups can act as donor and acceptor simultaneously, they can easily be linked into cycles or "infinite" chains. The classical example are the chains and cycles of O–H···O–H···O–H hydrogen bonds [5]. From the point of topology, such chains are completely analogous if they are composed of any other groups which can donate and accept hydrogen bonds at the same time, and it is irrelevant if the chain (or ring) consists of only one type of chemical groups or of different types. This means that the chains,

$$\cdots\text{O–H}\cdots\text{O–H}\cdots\text{O–H}\cdots\text{O–H}\cdots \tag{14}$$

$$\cdots\text{C}\equiv\text{C–H}\cdots\text{C}\equiv\text{C–H}\cdots\text{C}\equiv\text{C–H}\cdots\text{C}\equiv\text{C–H}\cdots \tag{15}$$

$$\cdots C\equiv C-H\cdots O-H\cdots C\equiv C-H\cdots O-H\cdots \tag{16}$$

$$\cdots C\equiv C-H\cdots O-H\cdots O-H\cdots C\equiv C-H\cdots \tag{17}$$

are all analogous, and have analogous cooperativity properties. The examples shown above have all been observed in crystal structures: chains and rings composed of only alkyne groups have already been shown in Figures 17 to 21. Cooperative chains with mixed $C\equiv C-H$ and $O-H$ constituents were described by Desiraju and coworkers [18, 75]. Rings with mixed constituents were found in a hydrated dialkyne [76] (Figure 27, and more recently in a urea derivative, Figure 28) [77]. The crystal structure in Figure 27 also provides the first example of a water molecule hydrogen bonding to an alkyne group (bottom of Figure 27).

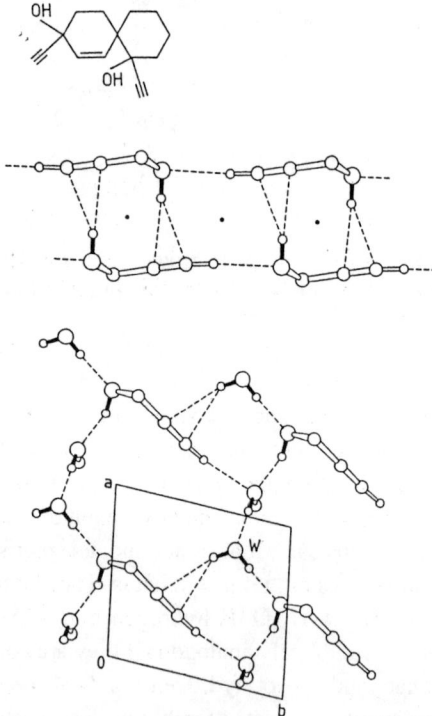

Figure 27. Two sections of the complex hydrogen bond pattern in a hydrated dialkyne (drawn using published atomic coordinates [76]). The molecules carry two ethynyl and two hydroxyl groups. Since there are two dialkyne molecules and an additional water molecules in the asymmetric crystal unit, there are four independent ethynyl and six O–H donors in the crystal lattice.

It is stated above that chains of hydrogen bonds are cooperative. This has been shown repeatedly for O–H···O–H···O–H chains [5]. It is easy to conclude from analogies that all related chains (e.g. those in Figures 14 to 17) should have similar cooperativity properties, but this is difficult to prove experimentally. Unfortunately, only few real systems are suitable for corresponding experimental studies. For the example of a O–H···C≡C–H···O–H chain, however, a suitable model system for a combined structural and IR spectroscopic study was found in the crystalline sex steroid, mestranol [78]. This compound crystallizes with two molecules in the asymmetric crystal unit, A and B. The independent molecules are stacked along the *a*-axis, and C≡C–H···O hydrogen bonds are formed within these stacks (Figure 29; only the relevant groups are shown, whereas the steroid core is omitted). Because the separation between molecules A–A and B–B are constrained to be equal (i.e. one unit cell dimension *a*), the two hydrogen bond distances must be very similar. For molecule A, the ethynyl group accepts an O–H···C≡C hydrogen bond, so that an infinite chain O–H···C≡C–H···O–H···O–H···C≡C–H is formed, for which cooperativity is anticipated. The ethynyl group of molecule B accepts no such interaction, but it donates a C≡C–H···O hydrogen bond which is presumably anticooperative (similar to the pattern 7 shown above). In the IR absorption spectrum, the red-shifts of the ≡C–H stretching frequencies compared to solution in CCl_4 differ by a factor of 3, which is a very dramatic effect (Figure 29, bottom). This is good evidence that the two C≡C–H···O hydrogen bonds are, despite the similar H···O separations, very different in strengths: one is strengthened by cooperativity, and the other one presumably weakened by anticooperativity.

In essence, the above observations show that in hydrogen bond networks, the terminal alkyne group can play the same role as the hydroxyl group, and is in many instances found involved in mixed networks where hydroxyl and alkynyl groups are isofunctional. This is a behavior that distinguishes terminal alkynes from all other C–H donors and π-acceptors.

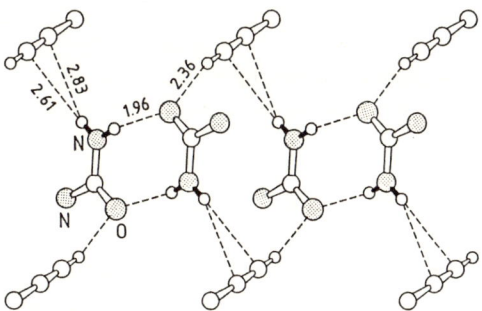

Figure 28. The hydrogen bond network in *N*-(4-methyl-phenyl)-*N*-prop-2-ynyl-urea, which is composed of interwoven N–H···O, N–H···C≡C, and C–H···O hydrogen bonds. Drawn using published atomic coordinates [77].

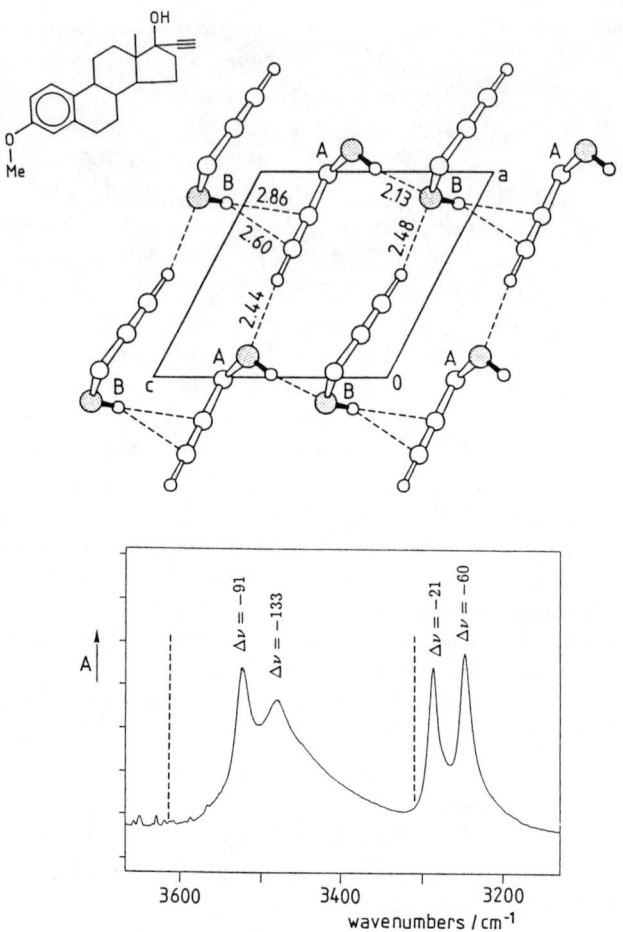

Figure 29. The hydrogen bond network in the crystalline ethynyl steroid mestranol (*top*) and a section of ist IR absorption spectrum; the two left bands represent hydrogen bonding O–H, and the two right bands hydrogen-bonding C≡C–H groups (*bottom*). Drawn based on published crystallographic and spectroscopic data [*78*].

VI. CONCLUSIONS

Terminal alkynes exhibit a variety of hydrogen bond interactions that is unparalleled by other weak hydrogen bonding functional groups: in crystal structures, they are found donating hydrogen bonds to strong acceptors like O=P, O=C and halide ions, and to weak acceptors like OC–M (M = transition metal), Cl–C, Ph, C≡C, and C=C. In addition, they can accept hydrogen bonds from strong and also from weak donors, including hydrogen bonds from other alkyne groups. This ability to act as donor and acceptor simultaneously enables terminal alkynes to participate in cooperative hydrogen bond networks in the same way as hydroxyl groups.

This chapter is not a review on a closed subject, but the matters discussed here are currently under intense investigation by several research groups. The variability of terminal alkynes in hydrogen bond arrangements and their unusual suitability for IR spectroscopic investigations is used to look into hydrogen bond phenomena as such (i.e there is relevance for hydrogen bonding in general, for "The" hydrogen bond). Furthermore, the use of terminal alkynes in crystal engineering purposes is explored in a number of laboratories; unfortunately, this exciting topic must be outside the scope of the present paper. Significant progress in the fundamental and also in the engineering fields can be expected in the forthcoming years.

ACKNOWLEDGMENTS

The author thanks Prof. W. Saenger for giving him the opportunity to carry out part of this work in his laboratory. All coauthors of the original research work are thanked, in particular G. Koellner, E. B. Starikov and M. Tamm (Berlin), J. A. Kanters, J. Kroon, N. Veldman and B. Lutz (Utrecht), J. J. C. Teixeira-Dias (Coimbra), K. Subramanian (Madras), and G. R. Desiraju (Hyderabad). The concept of this article follows a series of lectures held at the 54th Pittsburgh Diffraction Meeting, Nov. 7–9, 1996, the Université Louis Pasteur, Strasbourg, May 21, 1997, the Heinrich-Heine-Universität, Düsseldorf, June 11, 1997, and the Xth Symposium on Organic Crystal Chemistry, Rydzyna, Poland, Aug. 17–21, 1997. The inviters to these lectures (B. M. Craven, W. Hosseini, W. Kläui, and J. Bernstein, respectively) are thanked.

REFERENCES

1. Allerhand, A.; Schleyer, P. von R. *J. Am. Chem. Soc.* **1963**, *85*, 1715–1723.
2. Green, R. D. *Hydrogen Bonding by C–H Groups*; Macmillan: London, 1974.
3. Pimentel, G. C.; McClellan, A. L. *The Hydrogen Bond*; Freeman: San Francisco, 1960.
4. Joesten, M. D.; Schaad, L. J. *Hydrogen Bonding*; Marcel Dekker: New York, 1974.
5. Jeffrey, G. A.; Saenger, W. *Hydrogen Bonding in Biological Structures*; Springer: Berlin, 1991.
6. Desiraju, G. R. *Acc. Chem. Res.* **1991**, *24*, 290–296.
7. Desiraju, G. R. *Acc. Chem. Res.* **1996**, *29*, 441–448.
8. Steiner, T., *Cryst. Rev.* **1996**, *6*, 1–57.
9. Steiner, T. *Chem. Commun.* **1997**, 727–734.
10. Braga, D.; Grepioni, F. *Acc. Chem. Res.* **1997**, *30*, 81–88.
11. Burley, S. K.; Petsko, G. A. *Adv. Prot. Chem.* **1988**, *39*, 125–189.
12. Mitchell, J. B. O.; Nandi, C. L.; McDonald, I. K.; Thornton, J. M.; Price, S. L. *J. Mol. Biol.* **1994**, *239*, 315–331.
13. Worth, G. A.; Wade, R. C. *J. Phys. Chem.* **1995**, *99*, 17473–17482.
14. Parkinson, G.; Gunasekera, A.; Vojtechovsky, J.; Zhang, X.; Kunkel, T. A.; Berman, H.; Ebright, R. H. *Nature, Struct. Biol.* **1996**, *3*, 837–841.
15. Crisma, M.; Formaggio, F.; Valle, G.; Toniolo, C.; Saviano, M.; Iacovino, R.; Zaccaro, L.; Benedetti, E. *Biopolymers* **1997**, *42*, 1–6.
16. Steiner, T.; Schreurs, A. M. M.; Kanters, J. A.; Kroon, J. *Acta Crystallogr., Sect. D* **1998**, *54*, 25–31.

17. Starikov, E. B.; Steiner, T. *Acta Crystallogr., Sect. B* **1998**, *54*, 94–96.
18. Viswamitra, M. A.; Radhakrishnan, R.; Bandekar, J.; Desiraju, G. R. *J. Am. Chem. Soc.* **1993**, *115*, 4868–4869.
19. Bakshi, P. K.; Linden, A.; Vincent, B. R.; Roe, S. P.; Adhikesavalu, D.; Cameron, T. S.; Knop, O. *Can. J. Chem.* **1994**, *72*, 1273–1293.
20. Rzepa, H. S.; Smith, M. H.; Webb, M. L. *J. Chem. Soc., Perkin Trans. 2* **1994**, 703–707.
21. Allen, F. H.; Hoy, V.; Howard, J. A. K.; Thalladi, V. R.; Desiraju, G. R.; Wilson, C. C.; McIntyre, G. J. *J. Am. Chem. Soc.* **1997**, *119*, 3477–3480.
22. Müller, T. E.; Mingos, D. M. P.; Williams, D. J. *J. Chem. Soc., Chem. Commun.* **1994**, 1787–1788.
23. Brammer, L.; Zhao, D.; Lapido, F. T.; Braddock–Wilking, J. *Acta Crystallogr., Sect. B* **1995**, *51*, 632–640.
24. Braga, D.; Grepioni, F.; Tedesco, E.; Biradha, K.; Desiraju, G. R. *Organometallics* **1996**, *15*, 2692–2699.
25. Crabtree, R. H.; Siegbahn, P. E. M.; Eisenstein, O.; Rheingold, A. L.; Koetzle, T. F. *Acc. Chem. Res.* **1996**, *29*, 348–354.
26. Pedireddi, V. R.; Desiraju, G. R. *J. Chem. Soc., Chem. Commun.* **1992**, 988–990.
27. Steiner, T.; Desiraju, G. R. *Chem. Commun.* **1998**, 891–892.
28. Schödel, H.; Dienelt, R.; Bock, H. *Acta Crystallogr., Sect. C* **1994**, *50*, 1790–1792.
29. Shallcross, F. V.; Carpenter, G. B. *Acta Crystallogr.* **1958**, *11*, 490–496.
30. Steiner, T. *Acta Crystallogr., Sect. C* **1995**, *51*, 1135–1136.
31. Desiraju, G. R.; Murty, B. N. *Chem. Phys. Lett.* **1987**, *139*, 360–362.
32. Lutz, B.; van der Maas, J.; Kanters, J. A. *J. Mol. Struct.* **1994**, *325*, 203–214.
33. Sutor, D. J. *J. Chem. Soc.* **1963**, 1105–1110.
34. Ferguson, G.; Tyrrell, J. *Chem. Commun.* **1965**, 195–197.
35. Ferguson, G.; Islam, K. M. S. *J. Chem. Soc. (B)* **1966**, 593–600.
36. Cameron, A. F.; Cheung, K. K.; Ferguson, G.; Robertson, J. M. *J. Chem. Soc. (B)* **1969**, 559–564.
37. Allen, F. H.; Kennard, O. *Chem. Des. Autom. News* **1993**, *8*, 1–37.
38. Steiner, T.; Kanters, J. A.; Kroon, J. *Chem. Commun.* **1996**, 1277–1278.
39. Steiner, T. *J. Chem. Soc., Chem. Commun.* **1994**, 2341–2342.
40. Steiner, T.; van der Maas, J.; Lutz, B. *J. Chem. Soc., Perkin Trans. 2* **1997**, 1287–1291.
41. Steiner, T.; Lutz, B.; van der Maas, J.; Schreurs, A. M. M.; Kroon, J.; Tamm, M. *Chem. Commun.* **1998**, 171–172.
42. Steiner, T. *J. Chem. Soc., Chem. Commun.* **1994**, 101–102.
43. Dulmage, W. J.; Lipscomb, W. N. *Acta Crystallogr.* **1951**, *4*, 330–334.
44. Kumler, W. D. *J. Am. Chem. Soc.* **1935**, *57*, 600–605.
45. Stevens, R. V.; Beaulieu, N.; Chan, W. H.; Daniewski, A. R.; Takeda, T.; Waldner, A.; Williard, P. G.; Zutter, U. *J. Am. Chem. Soc.* **1986**, *108*, 1039–1049.
46. Mairesse, G.; Boivin, J. C.; Thomas, D. J.; Boute, J. P.; Lesieur, D.; *Acta Crystallogr., Sect. C* **1987**, *43*, 2128–2130.
47. Knolker, H.-J.; Boese, R. *J. Chem. Soc., Perkin Trans. 1* **1990**, 1821–1822.
48. Bernstein, J.; Weismann, W.; Britton, D.; Rippie, E. G.; Hedenstrom, J. *Acta Crystallogr., Sect. C* **1992**, *48*, 214–215.
49. Shalamov, A. E.; Agashkin, O. V.; Yanovskii, A. I.; Struchkov, Yu. T.; Logunov, A. P.; Revenko, G. P.; Bosyakov, Yu. G. *Zh. Strukt. Khim.* **1990**, *31*, 153–3.
50. Allen, F. H.; Bird, C. M.; Rowland, R. S.; Raithby, P. R. *Acta Crystallogr., Sect. B.* **1997**, *53*, 680–695 and 696–701.
51. Steiner, T. *Acta Crystallogr., Sect. B* **1998**, *54*, 456–463.
52. Karle, I. L. *J. Am. Chem. Soc.* **1973**, *95*, 4036–4040.
53. Borzilleri, R. M.; Weinreb, S. M.; Parvez, M. *J. Am. Chem. Soc.* **1994**, *116*, 9789–9790.
54. Gladii, Yu. P.; Buranbaev, M. Zh.; Yalovenko, E. G.; Litvinenko, G. S. *Kristallografija* **1991**, *36*, 395.

55. Steiner, T. *J. Mol. Struct.* **1998**, *443*, 149–153.

56. Steiner, T. *Acta Crystallogr., Sect. C* **1996**, *52*, 2263–2266.

57. Gonzalez, A. G.; Martin, J. D.; Martin, V. S.; Norte, M.; Perez, R.; Ruano, J. Z.; Drexler, S. A.; Clardy, J. *Tetrahedron* **1982**, *38*, 1009–1014.

58. James, S. L.; Verspui, G.; Spek, A. L.; van Koten, G. *Chem. Commun.* **1996**, 1309–1310.

59. Benghiat, V.; Leiserowitz, L. *J. Chem. Soc., Perkin Trans. 2* **1972**, 1772–1778.

60. McMullan, R. K.; Kvick, Å.; Popelier, P. *Acta Crystallogr., Sect. B* **1992**, *48*, 726–731.

61. Desiraju, G. R. *Crystal Engineering: The Design of Organic Solids*; Elsevier: Amsterdam, 1989.

62. Zorky, P. M.; Zorkaya, O. N. In *Advances of Molecular Structure Research*; Hargittai, M.; Hargittai, I. Eds.; JAI Press: Greenwich, CT, 1997, Vol. 3, pp. 147–188.

63. Steiner, T. *J. Chem. Soc., Chem. Commun.* **1995**, 95–96.

64. Steiner, T.; Starikov; E. B.; Amado, A. M.; Teixeira-Dias, J. J. C. *J. Chem. Soc., Perkin Trans. 2* **1995**, 1321–1326.

65. Berkovitch-Yellin, Z.; Leiserowitz, L. *Acta Crystallogr., Sect. B* **1984**, *40*, 149–165.

66. Weiss, H.-C.; Bläser, D.; Boese, R.; Doughan, B. M.; Haley, M. M. *Chem. Commun.* **1997**, 1703–1704.

67. Steiner, T.; Tamm, M.; Grzegorzewski, A.; Schulte, N.; Veldman, N.; Schreurs, A. M. M.; Kanters, J. A.; Kroon, J.; van der Maas, J.; Lutz, B. *J. Chem. Soc., Perkin Trans. 2* **1996**, 2441–2446.

68. Steiner, T.; Tamm, M.; Lutz, B.; van der Maas, J. *Chem. Commun.* **1996**, 1127–1128.

69. Steiner, T.; Starikov, E. B.; Tamm, M. *J. Chem. Soc., Perkin Trans. 2* **1996**, 67–71.

70. Steiner, T.; Mason, S. A.; Tamm, M. *Acta Crystallogr., Sect. B* **1997**, *53*, 843–848.

71. Steiner, T.; Schreurs, A. M. M.; Kanters, J. A.; Kroon., J.; van der Maas, J.; Lutz, B. *J. Mol. Struct.* **1997**, *436–437*, 181–187.

72. Lutz, B.; Kanters, J. A.; van der Maas, J.; Kroon, J.; Steiner, T. *J. Mol. Struct.* **1998**, *440*, 81–87.

73. Rohrer, D. C.; Lauffenburger, J. C.; Duax, W. L.; Zeelen, F. J. *Cryst. Struct. Commun.* **1976**, *5*, 539–542.

74. Lakshmi, S.; Subramanian, K.; Rajagopalan, K.; Koellner, G.; Steiner, T. *Acta Crystallogr., Sect. C.* **1995**, *51*, 2327–2329.

75. Allen, F. H.; Howard, J. A. K.; Hoy, V. J.; Desiraju, G. R.; Reddy, D. S.; Wilson, C. C. *J. Am. Chem. Soc.* **1996**, *118*, 4081–4084.

76. Subramanian, K.; Lakshmi, S.; Rajagopalan, K.; Koellner, G.; Steiner, T. *J. Mol. Struct.* **1996**, *384*, 121–126.

77. Kumar, S.; Subramanian, K.; Srinivasan, R.; Rajagopalan, K.; Schreurs, A. M. M.; Kroon, J.; Koellner, G.; Steiner, T. *Chem. Commun.* Submitted.

78. Steiner, T.; Lutz, B.; van der Maas., J.; Veldman, N.; Schreurs, A. M. M.; Kroon, J.; Kanters, J. *Chem. Commun.* **1997**, 191–192.

HYDROGEN BONDING SYSTEMS IN
ACID METAL SULFATES AND
SELENATES

Erhard Kemnitz and Sergei I. Troyanov

ABSTRACT

This work deals with general structural aspects of crystalline metal hydrogen sulfates and selenates with special emphasis on the hydrogen bonding systems in these

Advances in Molecular Structure Research
Volume 4, pages 79–113
Copyright © 1998 by JAI Press Inc.
All rights of reproduction in any form reserved.
ISBN: 0-7623-0348-4

compounds. The compositions of acid metal sulfates or selenates can be given by the general formula $M_{(2-N)/m}H_NXO_4$, where m represents the stoichiometric valency of the metal M and the value N defines the average number of hydrogen donor functions (OH groups) per XO_4 tetrahedron.

The counterparts of the M^{m+} cations consist of simple HXO_4^- tetrahedra ($N = 1$) or/and of their combinations with XO_4^{2-} anions ($N < 1$, $[H(XO_4)_2]^{3-}$) or H_2XO_4 molecules ($N > 1$, $[H(HXO_4)_2]^-$). The X–O bond lengths correlate with the additional structural functions of O atoms such as hydrogen donor, hydrogen acceptor functions, or coordination by metal atom(s). The X–OH distances are generally 0.1 Å longer than the X–O distances. Both the X–O and the X–OH distances are slightly longer (by 0.02–0.03 Å) in HXO_4^- as compared to those in a H_2XO_4 unit. The X–O bonds in hydrogen selenates are longer than those in hydrogen sulfates, the differences being greater for X–O distances (mean value of the differences, $\Delta = 0.168$ Å) than for X–OH ones (mean value of the differences, $\Delta = 0.150$ Å).

Every H atom forms a hydrogen bond of type O–H$\cdots$O, whereas bifurcated hydrogen bonds were found only in a few cases. The N value defines the main features of the hydrogen bonding systems. There are isolated hydrogen bonds in the case of $N < 1$, rings or infinite chains in the case of $N = 1$, and simple or double layers or columns in the case of $N > 1$. The connectivity pattern of the hydrogen bonding systems for the same N value can vary due to the different distribution of acceptor functions between HXO_4^- and H_2XO_4 units. Some examples of alkali metal hydrogen chalcogenate hydrates as well as structures with disordered hydrogen bonding systems have been included into the discussion.

I. INTRODUCTION

Proton conductivity is a well-known phenomenon especially for acid sulfates and selenates showing superprotonic phase transitions. These include $MHXO_4$ [1], $M_5H_3(XO_4)_4 \cdot H_2O$ (M = Cs, Rb, NH_4, K; X = S, Se [2, 3], $M_3H(XO_4)_2$ [4], $(NH_4)_4H_2(SeO_4)_3$ [5], and $Cs_3(HSO_4)_2(H_2PO_4)$ [6]. For instance, the high-temperature modification of $CsHSO_4$ is one of the best known proton conductors [7, 8]. The mechanistic interpretation of the proton transfer can be basically understood on the basis of either the Grotthus [9, 10] or the Vehicle mechanism [11]. For both models, the arrangement of the hydrogen bonding system is strongly determining the conductivity behavior of the solids under consideration. Consequently, a large series of these compounds has been intensively investigated. Some physical properties of these substances depend on variations in the hydrogen bonding systems. Among sulfates and selenates, alkali metal derivatives are of particular interest due to their nonhygroscopicity and enhanced thermal stability. As a result of widespread spectroscopic, X-ray or neutron diffraction, and solid-state NMR data, there exists a comprehensive knowledge of the hydrogen bonding systems in these compounds (ref. 12 and refs. therein).

Hydrogen chalcogenates which are not or only slightly hygroscopic have been much more intensively investigated than hygroscopic ones. Metal sulfates and selenates with higher acid content usually exhibit strong hygroscopicity. For this reason, they were not that thoroughly studied. Recently, we performed the synthesis of a wide range of compounds in the $M_2^ISO_4/H_2SO_4$ and $M^{II}SO_4/H_2SO_4$ systems and determined their crystal structures which are reviewed in refs. 13, 14, and 15, respectively. We also synthesized and crystallographically characterized a number of new metal selenates. On this basis, it is now possible to extend our knowledge of hydrogen bonding systems in acid oxochalcogenates.

One may expect the respective tellurium compounds to have been included here. However, there are remarkable differences in the chemistry of derivatives of sulfuric and selenic acid, on the one hand, and telluric acid on the other hand. For instance, H_2SO_4 and H_2SeO_4 form stable XO_4 units in their compounds and exhibit nearly the same acidity (the dissociation constants of both acids are very close; $pK_2 = 1.74$). Sulfates and selenates are structurally similar compounds. Contrary to this, the two most important forms of telluric acid, orthotelluric acid, H_6TeO_6, and polymetatelluric acid, H_2TeO_4, exhibit no similarity with H_2SO_4 or H_2SeO_4; the tellurates are not isotypic with sulfates or selenates. Monomeric H_2TeO_4 containing $TeO_2(OH)_2$ molecules is unknown; instead, polymeric metatelluric acid is formed. The crystal structure is unknown but most probably octahedrally coordinated Te is present as has been found in the known structure of orthotelluric acid consisting of octahedral $Te(OH)_6$ molecules whereas tetrahedra are present in crystalline H_2SO_4 and H_2SeO_4. Contrary to the acids of the lighter chalcogens, orthotelluric acid is a very weak one ($pK_1 = 7.7$, $pK_2 = 10.95$, $pK_3 = 14.52$). Consequently, telluric acid forms salts related to the formula of orthotelluric acid, namely MH_5TeO_6, $M_2H_4TeO_6$, $M_4H_2TeO_6$, and M_6TeO_6 which generally crystallize from basic solutions, whereas metal hydrogen sulfates and selenates crystallize from acid solutions. Moreover, in the above mentioned crystalline salts, polymeric anions $[TeO(OH)O_{4/2}]_n^{n-}$ are present forming infinite chains of edge-sharing octahedra. Hydration of such anions results in the formation of chains of vertex-sharing octahedra of $[TeO(OH)_3O_{2/2}]_n^{n-}$ as was found in $K[TeO_2(OH)_3]$ [*16*] or of isolated $[TeO(OH)_5]^-$ units present in the respective ammonium salt [*17*]. Due to the exclusive formation of octahedral coordination of Te(VI) in its oxoacids and oxoanions, it behaves more similar to its neighbors in the same period—tin, antimony, or iodine—which also prefer to form isoelectronic species such as $[Sn(OH)_6]^{2-}$, $[Sb(OH)_6]^-$, and $IO(OH)_5$ with octahedral coordination by oxygen.

Due to quite different chemical and structural properties, the tellurates are not considered in this chapter. Therefore, structural relationships of acid sulfates and acid selenates will be discussed and compared in detail, exhibiting in many cases analogous relationships but also unexpected differences in some cases. Many different types of hydrogen bonding systems were found in structures of acid metal sulfates ranging from isolated units to three-dimensional networks. Some special situations are caused by disordered H atoms.

Special attention will be paid to general principles of structural chemistry, with emphasis laid on structural changes depending on:

1. The H/XO_4 ratio in the formula.
2. The nature of the chalcogen.
3. The nature of the metal.

II. SYNTHESIS AND PROPERTIES

The synthesis of acid alkali metal sulfates with different M_2SO_4/H_2SO_4 ratios is described in detail in [13]. The principal procedure for both sulfates and selenates consists in dissolving a metal salt in 100% sulfuric acid or 95% selenic acid at a temperature above the liquidus curve. The saturated solution was cooled down and left for crystallization at the desired temperature. In the case of sulfates, well-shaped single crystals suitable for X-ray structure determination were formed during some hours to a few days. However, in some systems, crystallization was considerably delayed. In such cases, the crystallization was performed by further cooling the solution down until it had completely solidified and by heating it up just to reach a temperature near the liquidus. As a rule, well-shaped single crystals were obtained after one more cooling process. Some of the investigated acid sulfates were stable below room temperature only.

The situation in selenate systems is somehow complicated since 100% selenic acid shows different properties. In contrast to the respective sulfate systems, the systems $M_2SeO_4/H_2SeO_4/H_2O$ are not stable at low water concentrations and temperatures above 150 °C. This requires working at lower temperatures. However, it takes much longer in comparison to sulfate systems, due to the considerably higher viscosity of selenate systems. In view of this difficulty, it is often a problem to perform the crystallization properly.

Table 1. Compositions of Structurally Characterized Hydrogen Sulfates and Selenates of Mono- and Divalent Metals

$H_2XO_4/M_{2/m}XO_4$ (m = 1 or 2)	Formula[a] M^I Derivatives	Formula M^{II} Derivatives	H/XO_4 Ratio (N)
1/3	$M_3[H(XO_4)_2]$		1/2
3/5	$M_5(XO_4)(HXO_4)_3$		3/4
1/1	$M(HXO_4)$	$M(HXO_4)_2$	1/1
5/3	$M_3[H(HSeO_4)_2](HSeO_4)_2$		5/4
2/1	$M_2(HSO_4)_2(H_2SO_4)$	$M(HSO_4)_2(H_2SO_4)$	4/3
3/1	$M(HXO_4)(H_2XO_4)$	$M(HSO_4)_2(H_2SO_4)_2$	3/2
4/1		$M(HSO_4)_2(H_2SO_4)_3$	8/5
5/1	$M(HSO_4)(H_2SO_4)_2$		5/3
7/1	$M[H(HSO_4)_2](H_2SO_4)_2$		7/4

Note: [a]The presence of S or Se in the formulas at the place of X means that exclusively derivatives of this element are known.

It was generally found that the stability of acid chalcogenates with a higher content of chalcogenic acid decreases with increasing atomic number of the alkali metal. The hygroscopicity of acid chalcogenates changes, depending on the acid content in the formula. The metal-rich compounds are not hygroscopic but their hygroscopicity raises with the increasing content of chalcogenic acid. In some cases, stable metal hydrogen chalcogenate hydrates were obtained when the synthesis was carried out in diluted chalcogenic acid. They are considered in our discussion, too. All hydrogen sulfates and selenates of mono- and divalent metals that have been so far crystallographically characterized are listed in Table 1. The formulas reflect the structural units present in the compounds and are given in order of increasing $H_2XO_4/M_{2/m}XO_4$ ratio (m = stochiometric valency).

III. GENERAL CONSIDERATIONS OF STRUCTURES

The compositions of acid metal sulfates or selenates can be given by the general formula $M_{(2-N)/m}H_NXO_4$, where m represents the stoichiometric valency of the metal M. If the formulas for compounds under consideration are given in a different way, the N value can be calculated using the expression $N = \Sigma H/\Sigma XO_4$, where ΣH represents the number of acid protons. The N value defines the number of OH groups in the above formula which is identical with the average number of hydrogen donor functions per XO_4 tetrahedron. It will be shown that the N value strongly influences the structure of the hydrogen bonding system and can be used for its description and partial prediction. Besides, the hydrogen bonding system is also affected by the nature of the metal and the chalcogen.

In fact, there exist many acid selenates and sulfates exhibiting isotypic relationships, e.g. those having the general composition $M_3H(XO_4)_2$, $MHXO_4$, or $M(HXO_4)(H_2XO_4)$. On the other hand, there exist even some compounds with an analogous stoichiometry adopting different structure types. Moreover, an exchange of H by D in $RbHSeO_4$ results in different structure types for both compounds at room temperature [18, 19]. This is due to a shift of the phase transition temperature as a result of deuteration. At room temperature, $RbDSeO_4$ adopts the structure type that corresponds to the low-temperature phase of $RbHSeO_4$.

As already mentioned, the dependencies of the X–O bond lengths have been described in detail for a series of acid sulfates [13]. On the other hand, there are now much more data available for respective hydrogen selenates providing the possibility to extend the general discussion. One should have in mind that there are some difficulties in determining hydrogen positions in structures of acid selenates using X-ray diffraction techniques due to a large absorption coefficient and fluorescence caused by the Se atoms present.

In such cases, neutron diffractometry provides the best opportunity to determine hydrogen atoms as it has been done where this technique was available [20]. In addition, there are some indirect opportunities to draw conclusions regarding the hydrogen arrangement:

1. Since the structures of many acid sulfates have been already characterized, one can compare the isotypic selenates and sulfates.
2. In case of non-isotypic compounds, one can conclude which O atom is protonated (OH group) and which is not, by comparing the bond lengths X–O.
3. On the basis of the O···O distances, it is also possible to conclude about the hydrogen bonds since O–H···O bonds are significantly shorter than the other O···O contacts.

Some of these aspects will be discussed in the next sections.

IV. X–O DISTANCES

It is well known that a clear distinction between X–O and X–OH is possible on the basis of the respective X–O bond lengths. In hydrogen sulfates, the S–O bond length is generally about 0.1 Å shorter than the S–OH bond. The S–O and S–OH distances in hydrogen sulfates are listed separately for monovalent (Table 2) and divalent metals (Table 3). Here, only the mean values for the S–O and S–OH distances for HSO_4 and H_2SO_4 tetrahedra are given. Generally, both S–O and S–OH distances are shorter in coordinating H_2SO_4 tetrahedra than in HSO_4^- tetrahedra.

Actually, some S–O and S–OH distances can be found within a relatively wide range above and below mean values. Generally, S–O distances depend on such additional functions of the O atoms as their coordination to the metal (as defined by the metal M); their action as hydrogen acceptors (O···H, as defined by acceptor A) or as hydrogen donors (O–H, as defined by donor D). Taking into account all different cases of additional functions of O atoms which can be found in metal hydrogen sulfates, these additional functions can be classified according to their influence on S–O distances. A qualitative ranking of additional coordination influencing S–O distances can be given as follows: M < A < D. The H-donor function affects S–O distances so strongly that it is sometimes considered as an extra type of bonding. The presence of several functions like 2 M, (M + A), or (M + D) weakens the S–O bonds additionally. The fractions, 1/2 A and 1/2 D, correspond to the disordered hydrogen bonds, with O atoms acting half as acceptors and half as donors.

In addition to H_2SO_4 and HSO_4^-, two other structural units, $[H(SO_4)_2]^{3-}$ and $[H(HSO_4)_2]^-$, are present in the structures of $M_3H(SO_4)_2$ and $Li[H(HSO_4)_2](H_2SO_4)_2$, respectively. For all four tetrahedral units, the sequence of coordination functions which increase S–O distances is the same:

$$M < 2M \approx A < (A + M) \approx 3M < 2A < (1/2D + 1/2A)$$
$$< (1/2D + 1/2A + 2M) < D < (D + M)$$

Table 2. Mean Values of S–O and S–OH Distances (Å) in the Crystal Structures of Monovalent Metal Hydrogen Sulfates

Compound	HSO_4^- S–O	S–OH	H_2SO_4 S–O	S–OH	Ref.
Li(HSO$_4$)	1.446	1.546			21
Li$_2$(HSO$_4$)$_2$(H$_2$SO$_4$)	1.442	1.547	1.420	1.521	22
Li[H(HSO$_4$)$_2$](H$_2$SO$_4$)$_2$	1.440	1.542	1.430	1.535	23
α-Na(HSO$_4$)	1.447 1.451*	1.574 1.490*			23, 24
β-Na(HSO$_4$)	1.46	1.58			25
Na$_2$(HSO$_4$)$_2$(H$_2$SO$_4$)	1.444	1.560	1.430	1.524	26
Na(HSO$_4$)(H$_2$SO$_4$)$_2$	1.450	1.556	1.422	1.525	26
K(HSO$_4$)	1.451	1.570			27
K(HSO$_4$)(H$_2$SO$_4$)	1.445	1.563	1.421	1.517	28
Rb(HSO$_4$)	1.45	1.56			29
Rb(HSO$_4$)(H$_2$SO$_4$)	1.448	1.551	1.419	1.533	30
Cs(HSO$_4$)	1.445	1.570			31
Cs(HSO$_4$)(H$_2$SO$_4$)	1.452	1.548	1.419	1.537	30
(NH$_4$)(HSO$_4$)	1.438	1.559			32
(NH$_4$)(HSO$_4$)(H$_2$SO$_4$)	1.451	1.548	1.424	1.531	30
Ag(HSO$_4$)	1.440	1.571			33, 34
Ag$_2$(HSO$_4$)$_2$(H$_2$SO$_4$)	1.450	1.558	1.430	1.540	34
Hg$_2$(HSO$_4$)$_2$	1.450	1.548			34
Mean value	**1.447**	**1.555**	**1.424**	**1.527**	
H$_2$SO$_4$			1.426	1.537	35
(H$_3$O)(HSO$_4$)	1.456	1.558			35

Note: *HSO_4^- tetrahedra with disordered H atoms. There are oxygen atoms with only half an acceptor function (S–O = 1.456 Å); oxygen with half a hydrogen donor function (S–OH = 1.485 Å); and oxygen with half a hydrogen donor and half an acceptor function (S–OH = 1.496 Å).

It is to be noted that some functions are not typical of specific groups. So H-acceptor oxygen atoms are rare in coordinating H$_2$SO$_4$ groups. A more detailed discussion of the influence of additional functions on the S–O distances in metal hydrogen sulfates has been presented in ref. 13.

Basically, such dependencies are also true of hydrogen selenates. However, it seems that the relations can be understood more easily, if the additional functions are subclassified in a not too detailed manner. Therefore, we want to confine our discussion of the Se–O distances to the most important factors.

Table 4 summarizes the Se–O distances in hydrogen selenates. Bond length data are given in columns for different functions of the oxygen atoms which are listed below in the order of their decreasing influence on the Se–O bond length:

Table 3. Mean Values of S–O and S–OH Distances (Å) in the Crystal Structures of Divalent Metal Hydrogen Sulfates

Compound	HSO_4^-		H_2SO_4		Ref.
	S–O	S–OH	S–O	S–OH	
$Mg(HSO_4)_2$	1.447	1.556			36
$Mg(HSO_4)_2(H_2SO_4)_2$	1.451	1.543	1.430	1.523	37
$Ca(HSO_4)_2$	1.444	1.55			38
$Ca(HSO_4)_2(H_2SO_4)_2$	1.450	1.544	1.431	1.522	39
$Sr(HSO_4)_2$	1.444	1.550			40
$Sr(HSO_4)_2(H_2SO_4)$	1.456	1.544	1.425	1.534	41
$Ba(HSO_4)_2$	1.443	1.562			42
$Ba(HSO_4)_2(H_2SO_4)_3$	1.455	1.550	1.434	1.532	41
$Cd(HSO_4)_2$	1.448	1.545			43
$Cd(HSO_4)_2(H_2SO_4)_2$	1.458	1.558	1.435	1.521	44
$Zn(HSO_4)_2$	1.457	1.555			45
$Zn(HSO_4)_2(H_2SO_4)_2$	1.450	1.540	1.431	1.520	43
$Mn(HSO_4)_2$	1.447	1.547			46
$Mn(HSO_4)_2(H_2SO_4)_2$	1.451	1.546	1.435	1.522	46
Mean value	**1.450**	**1.549**	**1.432**	**1.525**	
H_2SO_4			1.426	1.537	35
$(H_3O)(HSO_4)$	1.456	1.558			35

- Hydrogen donor functions, sometimes with additional metal coordination.
- Half hydrogen donor and half hydrogen acceptor functions.
- Acceptor functions only.
- Coordination by metal(s).

It can be clearly seen from the values in Table 4 that if one starts with the "stronger" additional functions (D or D + M) and continue with the "weaker" ones, the Se–O values will decrease. Commonly, a coordination by metal is present in addition to the other functions, showing, however, only small influence on the bond lengths.

The data in Table 4 are given starting with the derivatives of the light alkali metals proceeding via the heavier ones to divalent metal hydrogen selenates. At the bottom, one example for a trivalent metal hydrogen selenate is given. The ammonium derivatives are not discussed here in detail because NH_4^+ offers additional hydrogen acceptor functions for the oxygen atoms instead of metal coordination. However, the acidity of the H atoms of a NH_4 group is much weaker so that these additional hydrogen bonds influence the stronger hydrogen bonds to a small extent only.

In case of the "acid-rich" compounds, besides $HSeO_4$ tetrahedra, there exist coordinated H_2SeO_4 molecules with somewhat shorter Se–O distances for the same

Table 4. Se–O Distances (Å) in Water-Free Metal Hydrogen Selenates Depending on Additional Functions of O Atoms[a]

Compound	nM				A (+M)	2A	(1/2A + 1/2D) (+M)	D	D (+M)	Ref.
	n = 1	2	3	4						
LiHSeO$_4$	1.617				1.633	1.626		1.707	1.710	47
NaHSeO$_4$	1.617				1.627				1.726	48
Na$_3$H$_5$(SeO$_4$)$_4$ [b]		1.611, 1.608			1.623			1.708, 1.692		48
K$_3$H(SeO$_4$)$_2$			1.614–1.623				1.659, 1.694			49
KHSeO$_4$		1.593			1.646				1.724	50
KH$_3$(SeO$_4$)$_2$ [b]		1.597, 1.592			1.626, 1.614			1.685	1.718	51
(NH$_4$)$_3$H(SeO$_4$)$_2$			1.624				1.673			52
(NH$_4$)$_4$H$_2$(SeO$_4$)$_3$ [c]					1.635		1.664		1.713	53
NH$_4$HSeO$_4$		1.615			1.631			1.708		54
Rb$_3$H(SeO$_4$)$_2$ [d]			1.620				1.680			55
Rb$_3$H(SeO$_4$)$_2$			1.621				1.692			56
RbHSeO$_4$ [e]			1.607		1.632		1.671		1.716	57
RbDSeO$_4$			1.591–1.604		1.623				1.714	19

(continued)

87

Table 4. Continued

Compound	nM				A	2 A	(1/2 A + 1/2 D)	D	D	Ref.
	n = 1	2	3	4	(+M)		(+M)		+M	
Cs$_3$H(SeO$_4$)$_2$			1.623				1.698			58
Cs$_3$H(SeO$_4$)$_2$ [f]			1.625				1.691			59
CsHSeO$_4$		1.602			1.623				1.711	60
CsH$_3$(SeO$_4$)$_2$ [b]	1.588				1.616			1.695	1.686	51, 61
Mg(HSeO$_4$)$_2$	1.610				1.620			1.702		61, 62
Zn(HSeO$_4$)$_2$	1.618				1.629			1.701		61, 62
Mn(HSeO$_4$)$_2$	1.619				1.639			1.705		61, 62
EuH(SeO$_4$)$_2$	1.613				1.630			1.757		63
H$_2$SeO$_4$					1.599			1.684		64

Notes: [a]M - coordination by metal; A - acceptor; D - donor functions.

[b]Data for the [H(HSeO$_4$)$_2$]⁻ unit or coordinated H$_2$SeO$_4$, respectively, are given in the second line.

[c]Some Se···O distances are not given due to errors in the ICSD file for this compound.

[d]Neutron powder diffraction data.

[e]Single crystal neutron diffraction data.

[f]High temperature modification at 400 K.

additional oxygen functions (Table 4). Thus, for hydrogen selenates the same relations as in metal hydrogen sulfates have been found. Generally, the X–O distances decrease slightly in the sequence $[H(XO_4)_2]^{3-}$, HXO_4^-, $[H(HXO_4)_2]^-$, H_2XO_4. However, the X–OH bonds, for example, are significantly shorter in the $H(XO_4)_2$ units than in HXO_4 due to the already discussed double functions of oxygen in the former one, acting half as hydrogen donor and half as hydrogen acceptor. In the last rows of Table 2, 3, and 4, the respective values for sulfuric and selenic acid are included demonstrating the good agreement with typical values for acid chalcogenates containing coordinating H_2XO_4.

In the structure of $Na_3H_5(SeO_4)_4$, the complex anion $[H(HSeO_4)_2]^-$ is present so that all anions of the above-mentioned sequence are exemplified also for selenates in Table 4. In such and similar cases, for every kind of tetrahedron, a separate line with the respective values is given. The second column which shows the coordination by metal is subdivided by the number of coordinating metals. The differences are small and depend on various effects the influences of which sometimes contradict each other. The number of coordinating metals will grow, if the radius of the metal increases. In the hydrogen chalcogenates of the lighter alkali metals (Li, Na, partly K), oxygen is usually coordinated by one or two metal atoms. In contrast, the oxygen in hydrogen chalcogenates of the heavier alkali metals adopt coordination by three or four, very seldom by two M atoms only. On the other hand, the effect of influence on the X–O distances is partly compensated due to a smaller M–O distance in the case of light alkali metals. Moreover, the number of coordinating metals M also depends on the stoichiometry of the compound under consideration. For instance, in case of the cesium compounds $Cs_3H(SeO_4)_2$ and $CsH_3(SeO_4)_2$, this effect can be clearly demonstrated. $Cs_3H(SeO_4)_2$ represents a metal-rich compound in comparison to $CsH_3(SeO_4)_2$. As a result, oxygen atoms are coordinated in the first compound by a larger number of metals; all O atoms reveal coordination to three or four cesium atoms [58]. In contrast, in the latter compound all O atoms are involved in hydrogen bonds, some of them also coordinated by one or two metal atoms [51]. In accordance with this, the charge of the metal also strongly influences these relations. If an alkali metal is substituted by di- or trivalent metals, the stoichiometry of the respective hydrogen chalcogenate will change and similar dependencies as discussed above can be found.

$Cs_3H(SeO_4)_2$ forms different modifications at room temperature and at 400 °C [58, 59]. Interestingly, regardless of the differences in the structure type, the additional functions of the oxygen atoms are exactly the same in both modifications. As can be seen from Table 4, the Se–O distances are virtually unchanged clearly demonstrating the dominating effect of additional functions.

In Table 4, data on compounds are given which have been also investigated by neutron diffraction. Although the hydrogen positions were poorly determined in case of X-ray diffraction measurements, the X–O distances originating from neutron diffraction and X-ray diffraction for the same compounds are consistent in a satisfying manner. For the europium compound, $EuH(SeO_4)_2$ [63], the Se–O value

Table 5. Comparison of X–O Distances (Å) in Isotypic Metal Hydrogen Sulfates and Selenates

Formula	M	Unit	S–O	S–OH	Se–O	Se–OH	Δ(Se–S)	
							X–O	X–OH
$M_3H(XO_4)_2$	K, Rb	$H(XO_4)_2$	1.460	1.539	1.612	1.693	0.152	0.154
$MHXO_4$	K	HXO_4	1.451	1.570	1.611	1.724	0.160	0.154
$MHXO_4$	Cs	HXO_4	1.445	1.570	1.609	1.711	0.164	0.141
$MH_3(XO_4)_2$	K	HXO_4	1.445	1.563	1.617	1.718	0.172	0.155
		H_2XO_4	1.421	1.517	1.603	1.685	0.182	0.168
$MH_3(XO_4)_2$	Cs	HXO_4	1.452	1.548	1.616	1.695	0.164	0.147
		H_2XO_4	1.419	1.537	1.588	1.686	0.169	0.149
$M(HXO_4)_2$	Mg, Zn	HXO_4	1.452	1.555	1.618	1.702	0.166	0.147

for the oxygen atom with D function (1.757 Å) deviates remarkably from all the other distances in the same column (about 1.70 Å). This is the only case where the situation is somehow unclear, whereas all the other values are consistent with the expected ones. Possibly, the accuracy of this value has to be checked before any further discussion of this discrepancy.

Only the values for isotypic acid sulfates and selenates are summarized in Table 5 providing a reliable comparison between sulfur and selenium derivatives. On the basis of isotypic pairs with the same metal, the influence of the chalcogen on the X–O distances can be best compared. If the replacement of a metal in a compound with the same stoichiometry results in a change of the structure type, such cases are treated in separate lines. Summarizing these relations, one can state that the differences between sulfur and selenium are stronger in cases of X–O bonds as compared with X–OH. The mean values for the differences in the lengths of X–O bonds in hydrogen sulfates and selenates are 0.168 Å for X–O and 0.150 Å for X–OH bonds. There is only one compound class, $M_3H(XO_4)_2$, where these relationships are not strictly obeyed and the differences are minor, being 0.152 and 0.154 Å for X–O and X–OH, respectively.

V. HYDROGEN BONDING SYSTEMS IN WATER-FREE ACID SALTS

As has been mentioned above, the type of the hydrogen bonding system is essentially dependent on the H/XO_4 ratio in the general formula $M_{(2-N)/m}H_NXO_4$ for the water-free acid salts, with N representing the average number of H atoms per XO_4 tetrahedron. If water (or ammonium) is present, the hydrogen bonding system will become more complicated and is not so easy to predict. Some examples of hydrogen bonding in hydrates will be given in the next section. Simple structural principles for hydrogen bonding systems in the structures of acid salts have been formulated by Wells [*65*]. These principles have been further developed in our review concerning the structural chemistry of alkali metal hydrogen sulfates [*14*]. In most structures of this compound class, every H atom participates in a hydrogen bond of type D–H···A where D and A symbolize oxygen atoms with hydrogen donor and hydrogen acceptor functions, respectively. The bifurcated (three-center) hydrogen bond is a rare case for this class of compounds, found in few structures only, e.g. in $NH_4(HSO_4)(H_2SO_4)$ [*30*].

It is obvious that, for the structure with the formula $M_{(2-N)/m}H_NXO_4$, N donors (D) and N acceptors (A) should be present (on an average) per one XO_4 tetrahedron. When discussing the structures of hydrogen sulfates and selenates, we consequently start with lower N values and finish with the maximum of $N = 2$ which corresponds to crystalline sulfuric and selenic acids.

If we compare H_2XO_4 and HXO_4^- units the general tendency is that H_2XO_4 only seldom acts as hydrogen acceptor. In most cases it is included in a hydrogen bonding system using its D functions exclusively whereas HSO_4^- mostly adopts both D and A functions.

For compounds with the same N value, special attention has to be paid to the distribution of hydrogen donor and acceptor functions between different units which causes sometimes quite different patterns of the hydrogen bonding system. Even in cases of the same D and A distributions, different structures can be formed for different metals. Here, the connectivity pattern of the hydrogen bonding systems is influenced also by the coordination ability of the metal cations.

A large number of alkali metal hydrogen sulfates and selenates of the composition $M_3H(XO_4)_2$ have been synthesized and studied using X-ray and neutron diffraction. Not only monometallic, but also bimetallic as well as the deuterated derivatives are known. For these compounds with the general formula $M_{1.5}H_{0.5}XO_4$, N is equal to 0.5 that corresponds to 0.5 D and 0.5 A functions per XO_4 tetrahedron. The formation of H-bridged dimers has been found for all structures with an H atom situated between two oxygen atoms from two XO_4 groups. The position of H can be asymmetrical as in $Na_3H(SO_4)_2$ [*66*] or $EuH(SeO_4)_2$ [*63*], with 1 D function of one XO_4 tetrahedron and 1 A function of another one, giving (1/2 D + 1/2 A) on an average (Figure 1).*

An alternative possibility is a symmetrical (one or two minima) arrangement of the H atom between two O atoms with the same functions (1/2 D + 1/2 A) for both. Examples of such an H arrangement include, for example, $K_3H(XO_4)_2$ [*67, 49*], $Rb_3H(XO_4)_2$ [*68, 55*], and $Cs_3H(SeO_4)_2$ [*58*].

The value $N = 3/4$ is realized in compounds with the general formula $M_4LiH_3(XO_4)_4$: M = Rb, K, NH_4; X = S [*69–71*] and M = Rb, X = Se [*72*]. All these compounds are isotypic, therefore their structural features will be discussed on the example of $Rb_4LiH_3(SO_4)_4$ [*69*]. The hydrogen atoms were not localized but a distinct description of the hydrogen bonding system was possible on the basis of differences in the S–O and O–O distances. As shown in Figure 2, three hydrogen bonds connect four SO_4 tetrahedra to zigzag chain fragments from which the

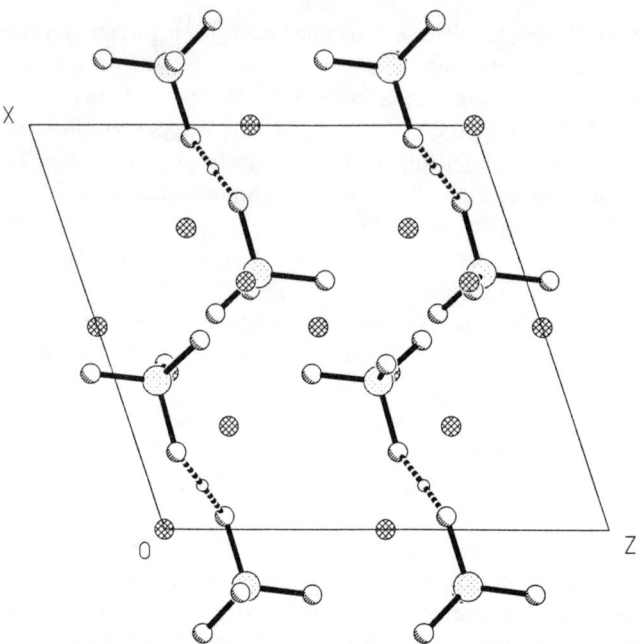

Figure 1. Nonsymmetrical hydrogen-bonded dimers, $H(SO_4)_2$, in the structure of $Na_3H(SO_4)_2$ [66].

complete ones are indicated by arrows. The SO_4 chain fragments are connected to each other by coordination to the metal atoms (Rb - empty circles, Li - hatched circles). In each chain fragment, three H donor and three H acceptor functions are distributed between the four SO_4 tetrahedra in the following way. The two SO_4 tetrahedra at the ends of the chain possess only 1 D function each. One of the two middle SO_4 groups provides 1 D and 1 A function whereas the other one exhibits only 2 A functions. Such distribution of H-acceptor and H-donor functions within the chain fragments has been confirmed later on the basis of a more precise structure determination for the potassium and ammonium compounds [70, 71]. A phase transition at elevated temperatures results in a change of the hydrogen bonding system (probably disordering), however, without change of the space group.

The case of $N = 1$ is known for monovalent, $M^{I}HXO_4$, and divalent metals, $M^{II}(HXO_4)_2$. This value implies the presence (on an average) of one hydrogen donor (1 D) and one hydrogen acceptor (1 A) function per XO_4 tetrahedron. These structures can be divided into two groups. In the structures of the first group, all XO_4 tetrahedra have the same distribution of donor and acceptor functions, D + A. This situation has been found in most alkali and alkali earth derivatives. The availability of one donor and one acceptor function for each XO_4 group results in

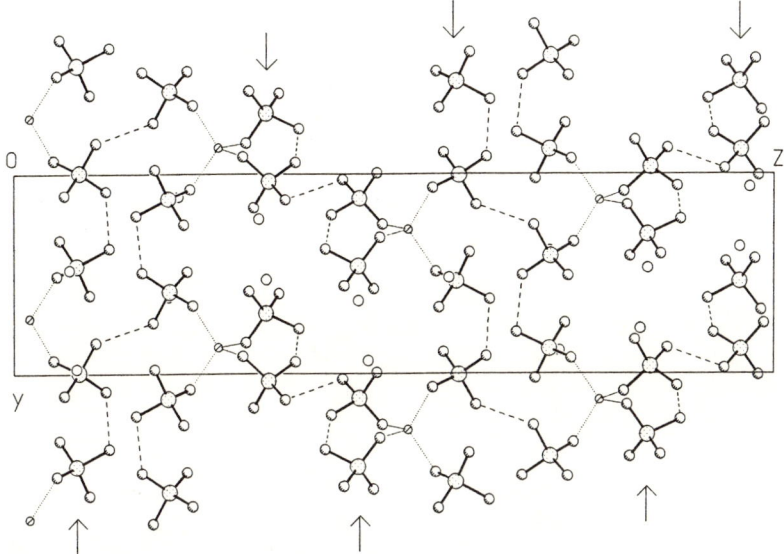

Figure 2. Structure of $Rb_4LiH_3(SO_4)_4$. Isolated zigzag chain fragments are formed from four SO_4 tetrahedra connected by three hydrogen bonds. Rb - *empty circles*, Li - *hatched circles*. The complete fragments are indicated by arrows [69].

the formation of infinite hydrogen-bonded chains of tetrahedra or cyclic dimers or combinations of both. The infinite zigzag chains of XO_4 with a typical $X\cdots X\cdots X$ angle of $109 \pm 1°$ have been found in the structure of recently synthesized $LiHSeO_4$ [47] (Figure 3). The analogous chains are present in $LiHSO_4$ [21], however, with completely different Li coordination by O atoms (distorted 5 + 1 octahedral for the former and tetrahedral for the latter example). The similar chain geometry occurs in many structures such as $CsHXO_4$ [31, 60], $RbHSO_4$ [73, 57], $Mg(HXO_4)_2$ [36, 62], $Zn(HXO_4)_2$ [45, 62] and $Hg_2(HSO_4)_2$ [34].

A different type of zigzag chain with smaller $X\cdots X\cdots X$ angle is present in the unique structure of $Cd(HSO_4)_2$ [43], probably·due to the larger ionic radius of Cd^{2+}. An example for a combination of both zigzag patterns has been found in the structure of $Mn(HSeO_4)_2$ with screwed chains of H-bonded SeO_4 tetrahedra [62] (Figure 4).

Structures of compounds with $N = 1$, containing cyclic H-bonded dimers, are rare. It has been derived from powder diffraction data on the basis of S–O and O$\cdots$O distances that β-$NaHSO_4$ contains exclusively such dimers [25]. Recently, this conclusion was confirmed by synthesis and structure determination of $NaHSeO_4$ by X-ray single-crystal diffraction [48]. Both compounds are isotypic and charac-

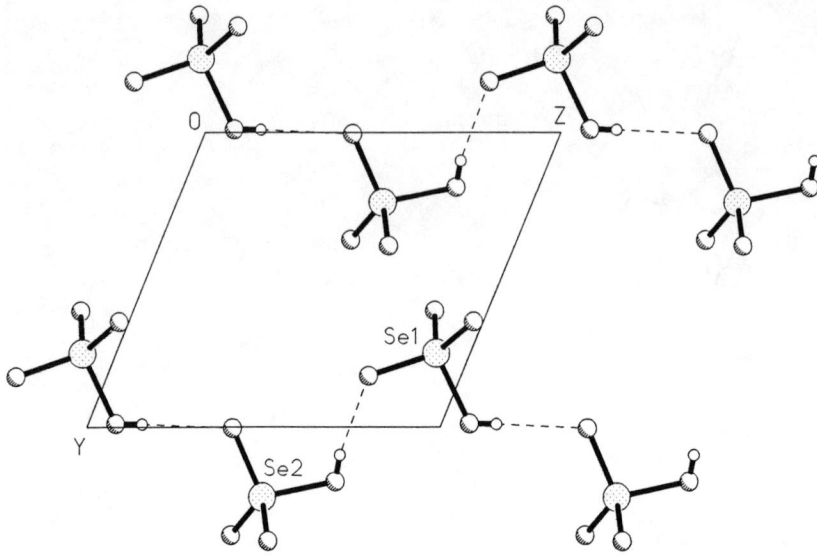

Figure 3. Zigzag chains of HSeO$_4$ tetrahedra in the structure of LiHSeO$_4$ [47].

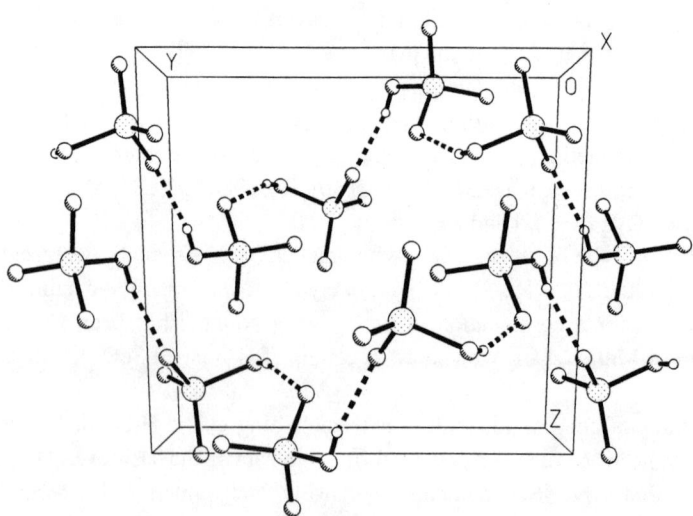

Figure 4. Screw-like chains of HSeO$_4$ tetrahedra in the structure of Mn(HSeO$_4$)$_2$ [62].

terized by a hydrogen bonding system in the form of cyclic dimers as shown in Figure 5 for the Se derivative.

At last, the structures containing both zigzag chains and dimers are known. The isotypic structures $KHXO_4$ contain half of the tetrahedra connected to the dimers, whereas the second half forms chains with a geometry shown in Figure 3 [27, 50].

A new example of tetrahedra connectivity was found recently in the structure of $AgHSO_4$ [33, 34] where puckered tetrameric rings of HSO_4 tetrahedra are formed beside flat wave-like infinite chains (Figure 6). Although hydrogen atoms were not localized, the hydrogen bonding system was derived from the analysis of the S–O and O⋯O distances [34]. Weak hydrogen bonds between rings and chains due to the presence of bifurcated hydrogen bonds in the structure are not shown in Figure 6.

For the second group of compounds with $N = 1$, the distribution of D and A functions is not the same for each XO_4 tetrahedron which is present in the structure. The average value $N = 1$ (D + A) results from the presence of tetrahedra exhibiting D + 2 A functions and an equal amount of tetrahedra with 1 D function only. This situation occurs in some hydrogen sulfates of divalent metals, $M^{II}(HSO_4)_2$, in the structures of which the hydrogen bonding system is described as branched chains. The tetrahedra of the former type (D + 2 A) are H-connected to infinite chains to

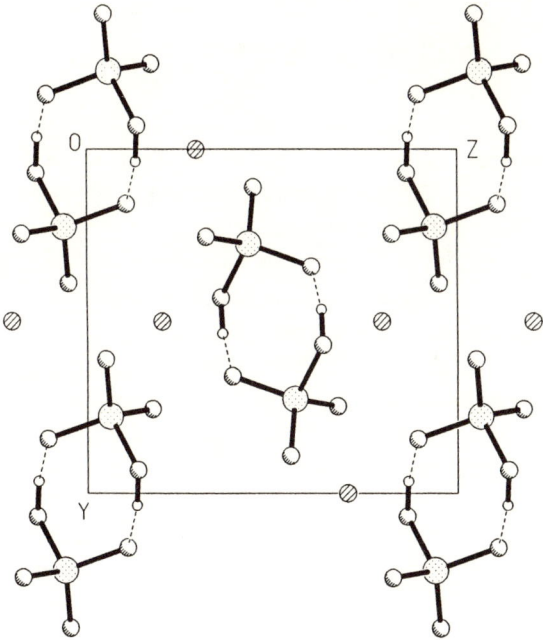

Figure 5. Closed dimers, $[HSeO_4]_2$, in the structure of $NaHSeO_4$ [48].

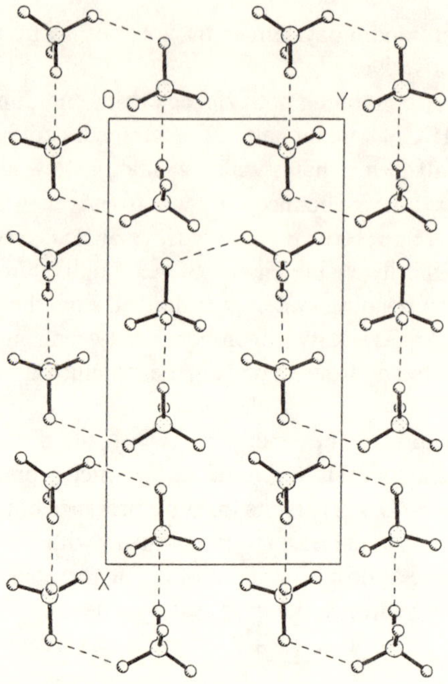

Figure 6. Hydrogen bonding system in AgHSO$_4$ consisting of puckered HSO$_4$ tetramers and screwed chains of HSO$_4$ tetrahedra [34].

which the tetrahedra of the latter type are attached additionally. The attached tetrahedra use their D functions, whereas the chain tetrahedra exhibit A functions additionally. The branched chains found in the structures of Ca(HSO$_4$)$_2$ [38] (and the isotypic Sr(HSO$_4$)$_2$ [40]), on the one hand, and Ba(HSO$_4$)$_2$ [42], on the other hand, differ in the form of the chain (zigzag and linear, respectively) and in the position of the branched tetrahedra on the alternative or on the same sides along the chain (Figure 7). The corresponding selenates would probably adopt the same or similar structure types. However, the difficulties in the synthesis of selenates (low solubility of M^{II}SeO$_4$ combined with a low thermal stability of selenic acid) prevent the systematic investigation of the whole series of divalent metal hydrogen selenates.

If the H/XO$_4$ ratio increases to values of $N > 1$, more complex hydrogen bonding systems can be expected. In general, only a few compounds with $N > 1$ have been synthesized for X = Se (compared to the larger number of sulfate derivatives) due to the special difficulties in working with highly concentrated selenic acid. Nevertheless, some selenium derivatives of M$_{2-N}^I$H$_N$XO$_4$ with $N > 1$ were obtained and structurally investigated recently.

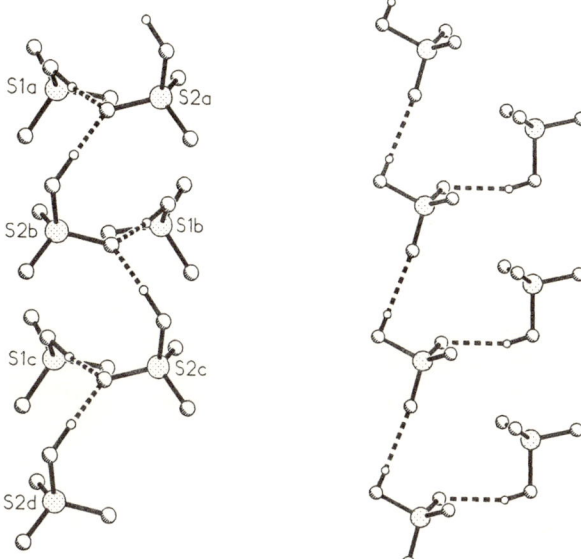

Figure 7. Two different geometries of branched chains of HSO_4 tetrahedra in the structures of $Ca(HSO_4)_2$ (zigzag; *left*) [38] and $Ba(HSO_4)_2$ (linear; *right*) [42].

The compound $Na_3H_5(SeO_4)_4$, properly formulated as $Na_3[H(HSeO_4)_2]$ $(HSeO_4)_2$, has a general formula $Na_{0.75}H_{1.25}SeO_4$ with $N = 5/4$ [48]. This value implies the availability of 5 D and 5 A functions per four SeO_4 groups (content of the unit cell). It should be mentioned that (at first) the positions of H atoms in this structure could not be obtained from the structure solution. The hydrogen bonding system has been derived exclusively from a comparison of Se–O and O⋯O distances based on the regularities presented in the previous section. Each of both $HSeO_4$ groups possesses a (D + 2 A) functionality giving altogether 2 D and 4 A functions. The situation for a complex anion $[H(HSeO_4)_2]^-$ can be described as consisting of two $HSeO_4$ tetrahedra (with one ordinary D function in each) connected via a symmetrical hydrogen bond with additional (1/2 D + 1/2 A) functions for each tetrahedra. Consequently, this ion provides 2(D + 1/2 D + 1/2 A) = 3 D + A functions. In such a way, four SeO_4 groups have altogether 5 D and 5 A functions. The hydrogen bonding system can be interpreted as cyclic dimers $[HSeO_4]_2$ connected by hydrogen bonds with $[H(HSeO_4)_2]^-$ ions into infinite chains going in the [221] direction (Figure 8). On the basis of later investigations, using a crystal of higher quality, the hydrogen positions could be determined confirming the description of the hydrogen bonding system given above.

There exist some compounds with $N = 4/3$ among hydrogen sulfates containing two HSO_4 units and one H_2SO_4 molecule. It is worth noting that $Li_2(HSO_4)_2$

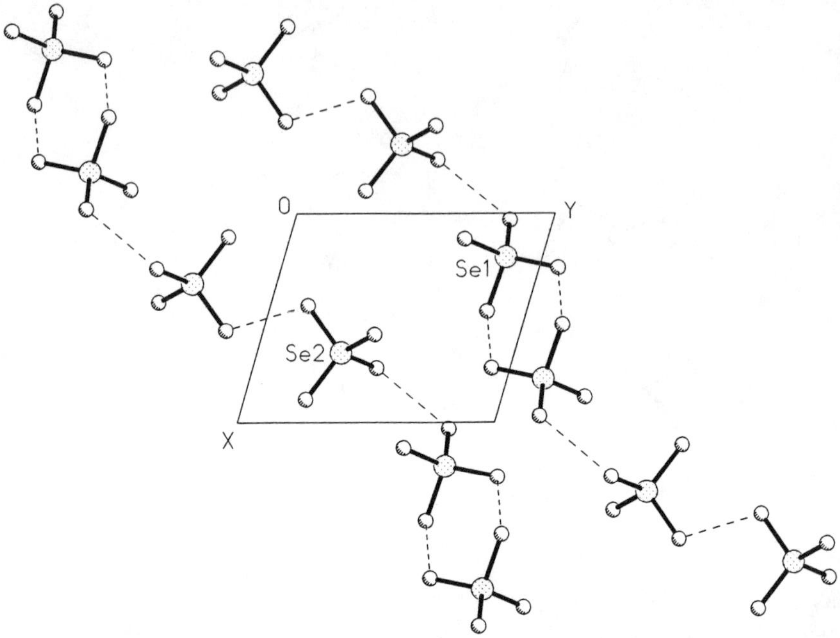

Figure 8. Hydrogen-bonded chains formed by $[HSeO_4]_2$ dimers and $[H(HSeO_4)_2]$ units in the structure of $Na_3[H(HSeO_4)_2](HSeO_4)_2$ [48].

(H_2SO_4) [23], $Na_2(HSO_4)_2(H_2SO_4)$ [26], and $Sr(HSO_4)_2(H_2SO_4)$ [41] have the same distribution of the D and A function between units, namely D + 2 A for both HSO_4 units and 2 D for H_2SO_4 (altogether 4 D + 4 A for three SO_4 tetrahedra). However, the hydrogen bonding systems are essentially different due to the change of the size of the cations (Li, Na) and the number of cations in the formula (Sr). In the Li structure, HSO_4 tetrahedra form infinite chains using their 1 D and 1 A functions (as in $LiHSO_4$), but additionally they act as acceptors for donor functions of a H_2SO_4 molecule which connects two HSO_4 tetrahedra from different chains (Figure 9 and Scheme 1, left).

In the structures of $Na_2(HSO_4)_2(H_2SO_4)$ and the isotypic $Ag_2(HSO_4)_2(H_2SO_4)$ [34], the geometry of the hydrogen bonding system differs essentially from that in the Li structure. As can be seen in Figure 10, both HSO_4 groups (S1 and S2, both with D + 2 A functions) form an infinite chain, whereas the H_2SO_4 molecule (S3) is connected by its two donor functions to the neighboring S1 and S2 tetrahedra of the same chain (Scheme 1, right). The whole hydrogen bonding system can be considered as consisting of SO_4 trimers connected with each other by hydrogen bonds.

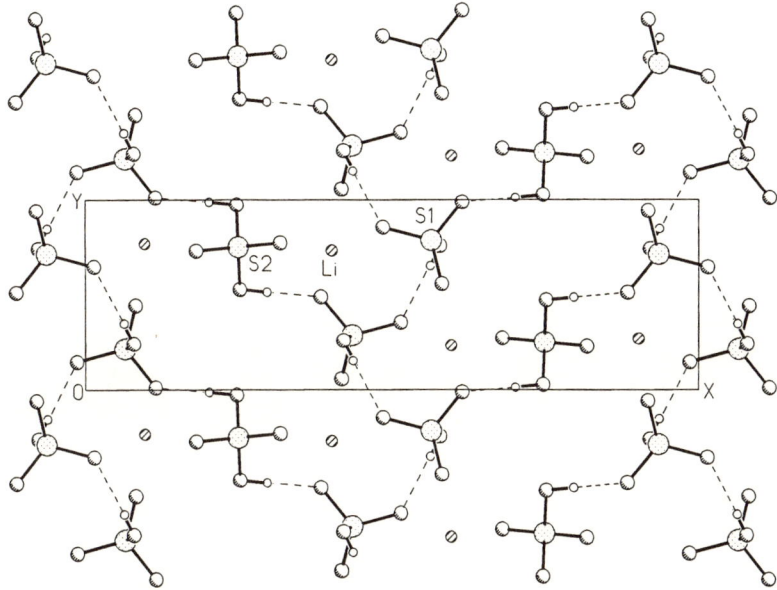

Figure 9. Crystal structure of $Li_2(HSO_4)_2(H_2SO_4)$. S1 and S2 designate sulfur atoms from HSO_4^- and H_2SO_4 units, respectively [22].

In addition to the two different hydrogen bonding systems described above, a third type occurs in the structure of $Sr(HSO_4)_2(H_2SO_4)$ [41] being somehow analogous to that in the Li structure. The principle of the sulfate tetrahedra connectivity via hydrogen bonds is the same in the Sr structure as in the Li structure, but whereas a flat layer of hydrogen-bonded SO_4 tetrahedra is characteristic of the Li compound (Figure 9 and Scheme 1, left), these layers are rolled up to flattened cylinders in the Sr compound (Figure 11). Such a rolling up of layers represents a two-dimensional analogue of the relationships between the one-dimensional infinite chains and rings, discussed above for examples with $N = 1$. Thus, the variation of the connectivity pattern depends on the nature of the metal (stochiometric valence and size) being present in the structure with the same N value and the same distribution of D and A functions.

There are two structure types for compounds $M^I(HXO_4)(H_2XO_4)$ with $N = 3/2$. The difference in the hydrogen bonding systems is manifested by the function distribution for HXO_4 and H_2XO_4 tetrahedra. In the structure type $Cs(HXO_4)(H_2XO_4)$, the distribution of D and A functions is very different, being D + 3 A for HXO_4 and 2 D for H_2XO_4 [30, 51]. The infinite chains of HXO_4 tetrahedra are connected via H_2XO_4 tetrahedra forming puckered layers shown in Figure 12 for $Cs(HSeO_4)(H_2SeO_4)$. The corresponding rubidium hydrogen sulfate adopts the same structure type [30].

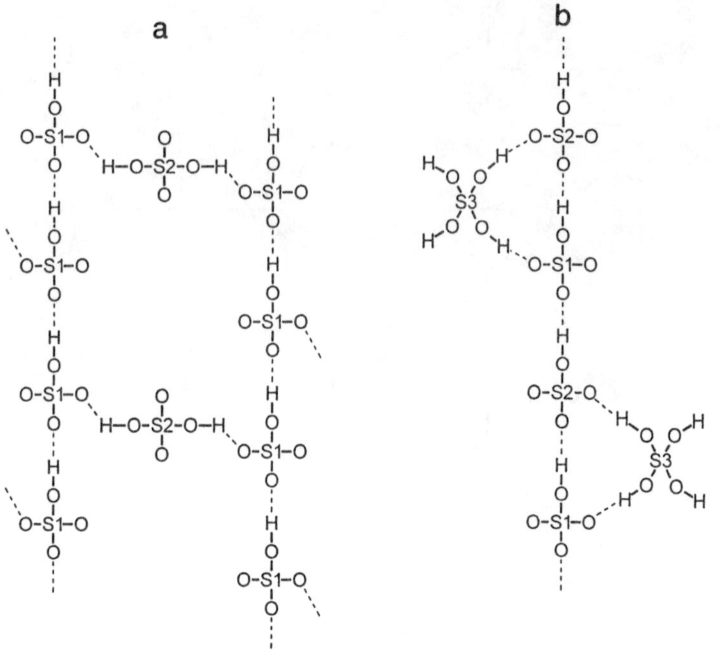

Scheme 1. Different H-bond connectivity patterns in two structures of $M_2(HSO_4)_2(H_2SO_4)$. (a) M = Li, every S2 (H_2SO_4) tetrahedron links two S1 (HSO_4) chains forming an infinite layer. (b) M = Na, S3 (H_2SO_4) tetrahedra are attached via hydrogen bonds to the same chain of S1 + S2 (both HSO_4) tetrahedra forming a composite chain consisting of S1 + S2 + S3 tetrahedra. For clarity, in both cases, zigzag HSO_4 chains are shown as linear ones.

More uniformly distributed functions for HXO_4 and H_2XO_4 tetrahedra have been found in structures of $K(HXO_4)(H_2XO_4)$ [28, 51]. Due to the availability of three functions for each group, D + 2 A for HXO_4 and 2 D + A for H_2XO_4, three-connected 4.8^2 nets are produced in the form of strongly puckered layers (Figure 13). Similar D/A distribution occurs in the structures of divalent metal hydrogen sulfates having the formula $M^{II}(HSO_4)_2(H_2SO_4)_2$: M = Mg [37], Ca [39], Zn [43], and Cd [44]. Although there are two different structure types for Cd, Mg, and Zn, on the one hand, and Ca, on the other hand, they are similar in that the HSO_4 groups with D + 2 A functions and the H_2SO_4 groups with 2 D + A functions are combined to flat three-connected 6^3 nets (Figure 14). Differences in their structure types concern only the kind of linkage of the flat hydrogen bonded H_2SO_4/HSO_4 layers by the metal. In the calcium compound every second space between layers of the hydrogen-bonded tetrahedra is occupied by the metal, whereas every space between such layers is occupied in the other compounds.

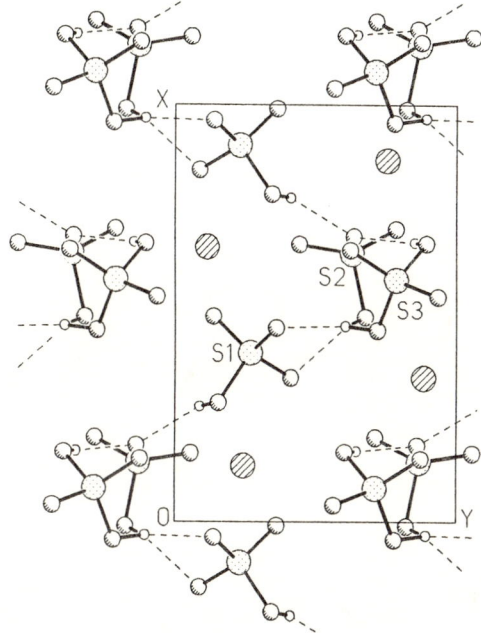

Figure 10. Crystal structure of $Na_2(HSO_4)_2(H_2SO_4)$. Zigzag chains of trimers are formed by two HSO_4 (S1, S2) and one H_2SO_4 tetrahedron (S3) [26].

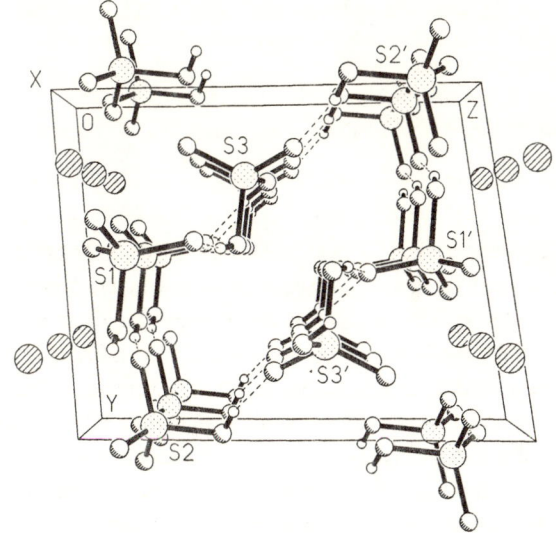

Figure 11. Crystal structure of $Sr(HSO_4)_2(H_2SO_4)$. S1 and S3' (as well as S1' and S3) tetrahedra, (all HSO_4^-) form infinite chains in the *z* direction which are further connected via S2 and S2' tetrahedra (both H_2SO_4) forming flattened cylinders [41].

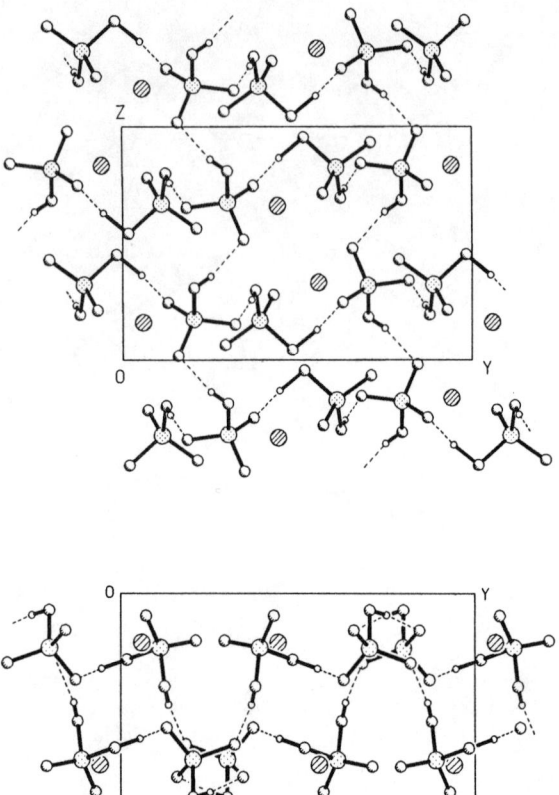

Figure 12. Puckered layers of hydrogen-bonded $HSeO_4$ and H_2SeO_4 tetrahedra in the structure of $Cs(HSeO_4)(H_2SeO_4)$ [51].

A further increase of the H/XO_4 ratio to $N = 8/5$, as in case of $Ba(HSO_4)_2(H_2SO_4)_3$ [41] leads to the formation of infinite columns consisting of HSO_4^- and H_2SO_4 tetrahedra with a high degree of cross linking (Figure 15). Eight hydrogen donor and acceptor functions are distributed between five SO_4 tetrahedra in the following way: S1 and S2 (both HSO_4) possess 1 D and 3 A functions each, S3 and S4 (both H_2SO_4) provide 2 D and 1 A functions each, whereas S5 (H_2SO_4) acts twice as donor (2 D). The complex hydrogen bonding system can be described as consisting of columns formed by hydrogen-bonded chains in x-direction. These columns are formed by linear S2, S3, and S4 chains and S1 zigzag chains by using their 1 D + 1 A functions. Additional A functions (S1 and S2) or D functions (S3 and S4) as well as 2 D functions of S5 are used for cross linking of the chains. The preferential avoiding of hydrogen acceptor functions for H_2SO_4 is again exemplified in this

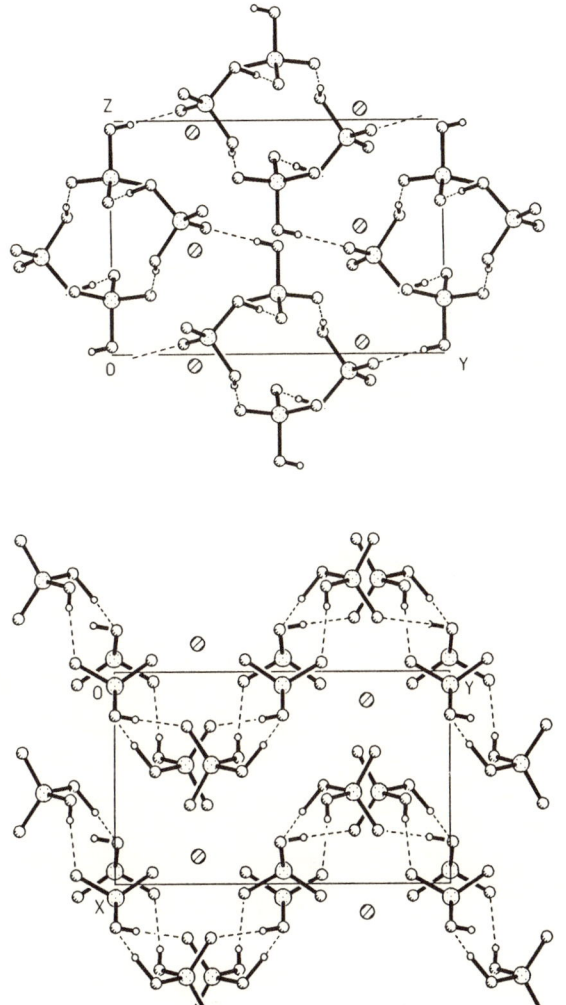

Figure 13. Puckered layers of hydrogen-bonded $HSeO_4$ and H_2SeO_4 tetrahedra in the structure of $K(HSeO_4)(H_2SeO_4)$. Within the layer the tetrahedra connectivity corresponds to a 4.8^2 net [51].

structure. On the other hand, the enhanced acceptor ability of an HSO_4^- tetrahedron can be understood on the basis of its negative charge.

A similar distribution of D and A functions was found in the structure of $Na(HSO_4)(H_2SO_4)_2$ with a slightly increased H/XO_4 ratio of $N = 5/3$ [26]. That means the three SO_4 tetrahedra have to perform 5 D and 5 A functions which are

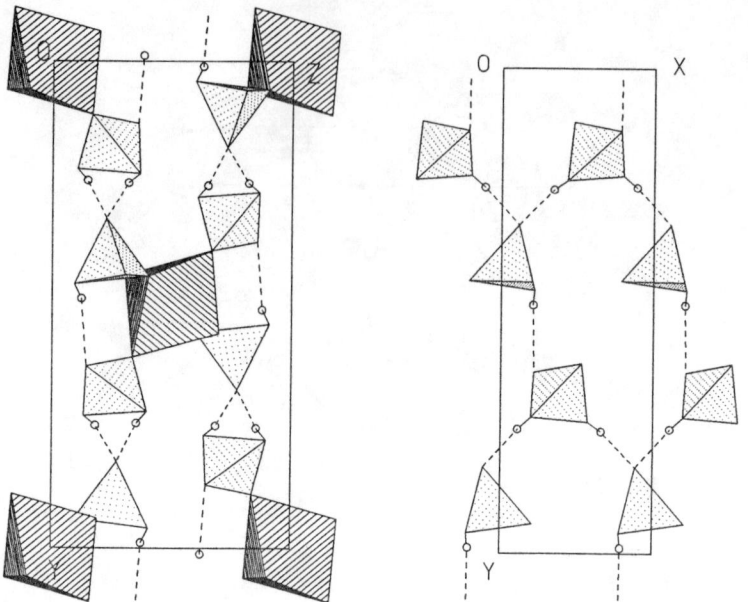

Figure 14. Flat three-connected 6^3 nets of hydrogen-bonded HSO_4 and H_2SO_4 tetrahedra in the structure of $Zn(HSO_4)_2(H_2SO_4)_2$ (*right*). View of the structure in x direction (*left*) [*43*].

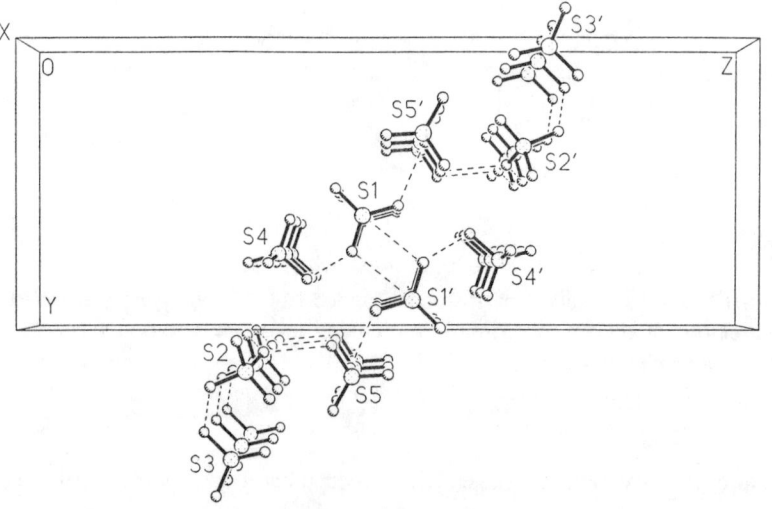

Figure 15. Infinite columns consisting of HSO_4^- and H_2SO_4 tetrahedra with a high degree of cross-linking in the structure of $Ba(HSO_4)_2(H_2SO_4)_3$. The columns are formed by linear S2, S3, and S4 chains and S1 zigzag chains in x-direction [*41*].

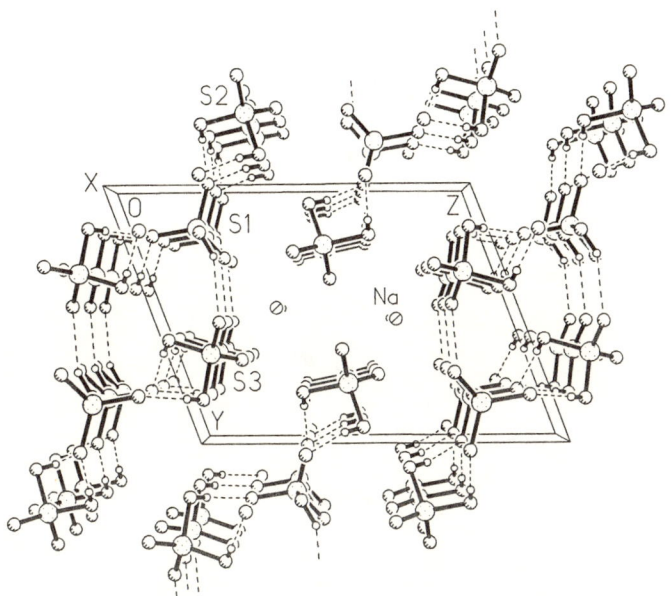

Figure 16. Hydrogen bonding system in the crystal structure of Na(HSO$_4$)(H$_2$SO$_4$)$_2$ [26].

distributed between them in such a way that a major part of acceptor functions are adopted by HSO$_4^-$ group (D + 4 A), while two H$_2$SO$_4$ molecules possess mainly donor functions, (2 D + A) for 1 and 2 D for the other molecule. The hydrogen bonding system consists of infinite columns of cross-linked HSO$_4^-$ and H$_2$SO$_4$ tetrahedra (Figure 16). In contrast to Ba(HSO$_4$)$_2$(H$_2$SO$_4$)$_3$, linear chains of SO$_4$ tetrahedra are not present.

High values of N make it possible to form a three-dimensional network of hydrogen bonds. However, in most structures where N is less than or equal to 2, the systems prefer to be separated by MO$_n$ polyhedra and form infinite structural motifs in one or two directions (chains/columns or single/double layers). This can also be illustrated on the example of the Li[H(HSO$_4$)$_2$](H$_2$SO$_4$)$_2$ structure [23] having a very high H/SO$_4$ ratio of N = 7/4. The system of hydrogen bonds can be described as consisting of double layers which are parallel to the xy0 plane (Figure 17). These layers (6^3 nets) are formed by hydrogen bridges and are situated at z –0.25 and 0.25. Two single layers are connected to each other via symmetrical O···H···O' bridges. In principle, the preparation of the metal hydrogen selenates with such high N values would be possible if some problems in working with 100% H$_2$SeO$_4$ were overcome.

The maximum value of N = 2 is achieved in pure sulfuric and selenic acids. The structure of the former one was recently reinvestigated and refined [35]. Two of

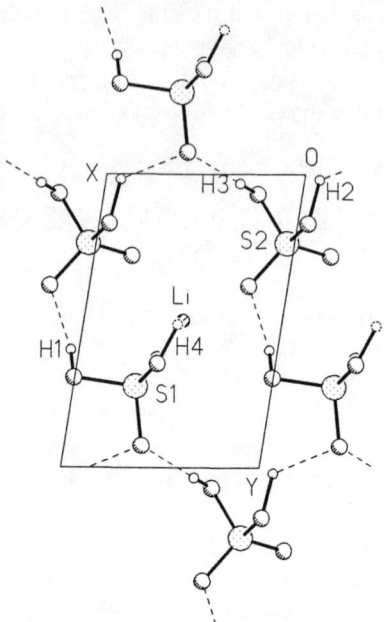

Figure 17. 6^3 connected layer formed by hydrogen-bonded S1 and S2 tetrahedra in the crystal structure of $Li[H(HSO_4)_2](H_2SO_4)_2$ [23].

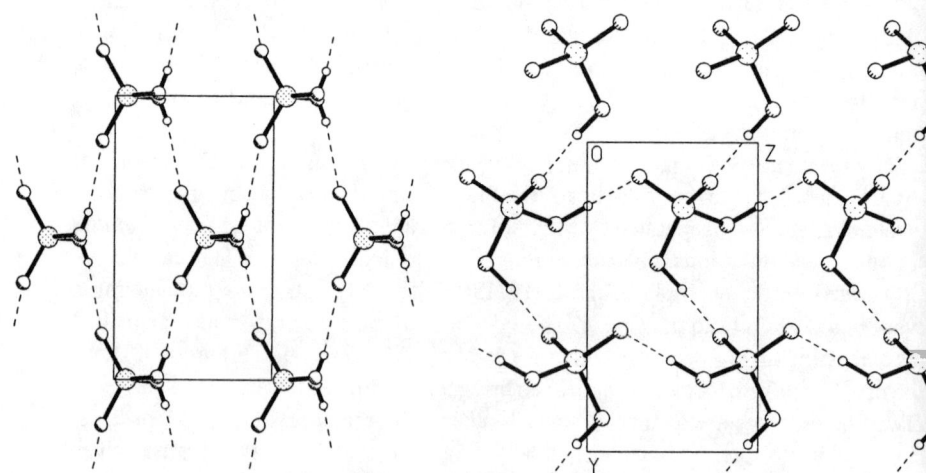

Figure 18. Two different 4^4 nets of hydrogen-bonded H_2XO_4 tetrahedra in the structures of crystalline H_2SO_4 (*left*) [35] and H_2SeO_4 (*right*) [64].

four O atoms are acting as donors and the other two as acceptors. Consequently, the layer structure with four-connected 4^4 nets is formed in crystalline H_2SO_4. Solid selenic acid adopts a different crystal structure. Although the connectivity patterns within the layers are similar, the puckering of the layers is different in these two structures (Figure 18).

VI. HYDROGEN BONDING SYSTEMS IN ACID SALT HYDRATES

If water is present in the system from which the metal hydrogen sulfate or selenate crystallizes, it is sometimes incorporated into the crystal structure, though its structural function can be different. The defining factor is the presence of other species in the structure which can dissociate and originate H^+ ions. Water is incorporated into the structure as a molecule if dissociation constants of other species are not high. For example, all known structures of $M^{II}(HXO_4)_2 \cdot H_2O$ (M^{II} = Mn, Mg, Cd; X = S [46, 74, 75] or M^{II} = Mn, Cd; X = Se [62]) contain water molecules connected to the other part of the structure via hydrogen bonds acting twice as hydrogen donor and twice as hydrogen acceptor (Figure 19). The water molecule can "coexist" with the HXO_4^- ion due to a not so high dissociation constant of the latter. The difference in the acidity of the two species H_2O and

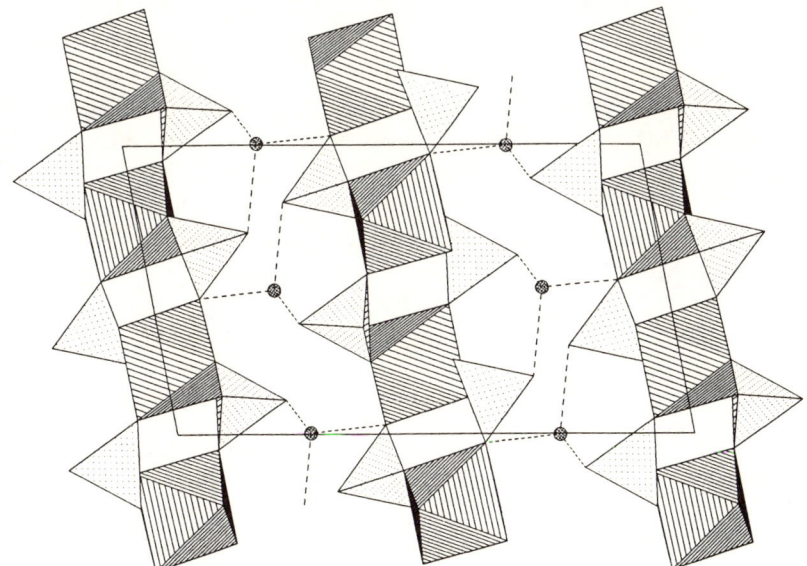

Figure 19. Water molecules connecting $Mn(HSeO_4)_2$ layers via hydrogen bonds in the structure of $Mn(HSeO_4)_2 \cdot H_2O$ [46].

HXO_4^- is reflected by the different lengths of hydrogen bonds. Those of type $O-H \cdots O_w$ are considerably shorter (2.71–2.72 Å) compared to the hydrogen bonds of type $O_w-H \cdots O$ (2.98–3.11 Å for the structure of $Mn(HSeO_4)_2 \cdot H_2O$ [62]). A similar situation occurs in the hydrates of the compositions $Cs_5H_3(SeO_4)_4 \cdot H_2O$ [76] and $Na_5H_3(SeO_4)_4 \cdot 2H_2O$ [77]. The availability of only three protons per four SeO_4 groups further decreases the acidic dissociation. In the sodium derivative, the water molecules are connected via hydrogen bonds of type $O_w-H \cdots O$ (2.75 and 2.86 Å) and are additionally coordinated by sodium cations.

The value $N > 1$ in the general formula means that the structures contain some excess hydrogen in respect of HSO_4^- ions, e.g. as H_2SO_4 or $[H(HSO_4)_2]^-$. Such species have a larger dissociation constant and can not "coexist" with water molecules. As a result, hydroxonium cations are formed. A good example is the compound of the composition $K(HSO_4)(H_2SO_4) \cdot H_2O$. The corresponding water-free compound $K(HSO_4)(H_2SO_4)$ is also known (see the previous section) [28]. It contains a sulfuric acid as a ligand incorporated in the structure via hydrogen bonds and coordination by potassium. Its hydrate, however, does not contain H_2SO_4 and H_2O. According to the relations discussed above, two ions, H_3O^+ and HSO_4^-, are formed instead, correctly represented by the formula $K(H_3O)(HSO_4)_2$ [78]. The corresponding sodium [26] and silver [34] compounds are isotypic to that of potassium. Also the monohydrates of $H_2SO_4 \cdot H_2O$ or $H_2SeO_4 \cdot H_2O$ are proper examples for this behavior and should correctly be described as hydroxonium hydrogen sulfates or selenates, $(H_3O)(HXO_4)$ [35, 79]. Comparing hydroxonium salts with the corresponding water-free hydrogen sulfates structural similarities can be noticed. Although, the half of metal ions is replaced by hydroxonium cations, the original hydrogen bonding system remains nearly unchanged. The same arrangement of HSO_4 chains and dimers can be found; even the coordination of the remaining metal is the same. Of course, due to H donor functions of H_3O^+ ions, there are additional weaker hydrogen bonds which, however, do not influence the original structure arrangement too strongly (Figure 20). Ammonium derivatives fit well in this consideration, too. For example, $M(HSO_4)(H_2SO_4)$ (M = Rb, Cs) is isotypic with $(NH_4)(HSO_4)(H_2SO_4)$, the latter showing only small changes of the original hydrogen bonding system by full replacement of the metal by ammonium [30].

A similar situation can be found in hydrogen sulfates and selenates of higher valent metals. Apparently, the coordination of HSO_4^- by such cations results in the formation of a stronger complex. Water molecules attract a proton and, consequently, form hydroxonium ions in such systems. Therefore, it is not surprising that the hydrate $Zr(SO_4)(HSO_4)_2 \cdot 2H_2O$ exists as a hydroxonium sulfate zirconate(IV), $(H_3O)_2[Zr(SO_4)_3]$ [80]. A similar process leads to the formation of twice hydrated protons, $(H_5O_2)^+$, in the structures of $(H_5O_2)M^{III}(SO_4)_2 \cdot 2H_2O$ (M = Fe [81], Ti [80] (Figure 21), Cr [82]) and $(H_5O_2)Sc(SeO_4)_2$ [83].

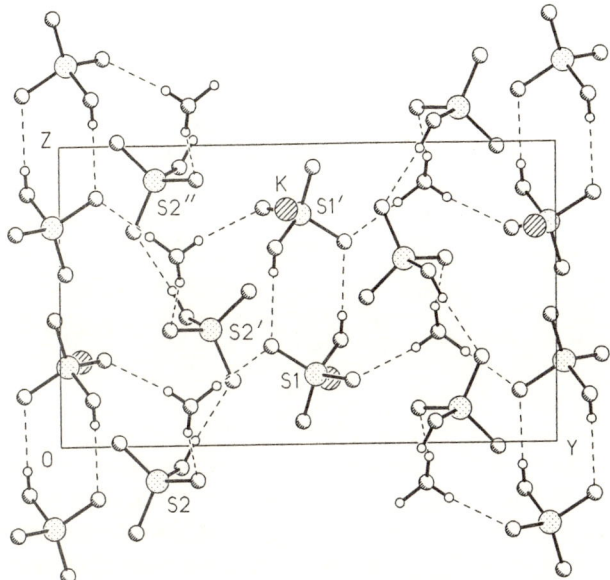

Figure 20. The hydrogen bonded dimers and chains in the structure of K(H₃O)(HSO₄)₂ [78] which are very similar to those in the structure of KHSO₄.

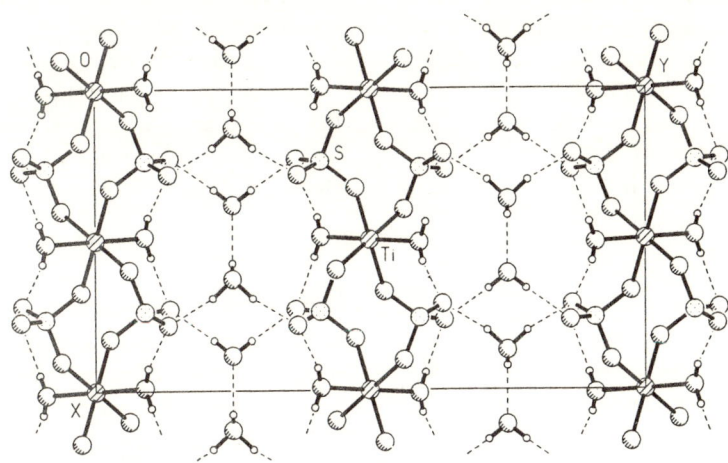

Figure 21. The $(H_5O_2)^+$ cations connected to $[Ti(SO_4)_2(H_2O)]_n$ infinite layers by hydrogen bonds in the structure of TiH(SO₄)₂·4H₂O [80].

VII. SYSTEMS WITH DISORDERED HYDROGEN BONDS

Different types of disorder can be found in the structures of hydrogen chalcogenates. The most common case is a proton disorder within a hydrogen bond. It occurs in the strong hydrogen bonds with two minima and $O\cdots O$ separation of 2.45–2.55 Å. The best known examples include strongly bonded, one-proton-containing dimers in the structures of hydrogen disulfates and diselenates, $M_3H(XO_4)_2$ (e.g. M = K, Rb, X = S, Se). Recently, the structure of $Rb_3H(SeO_4)_2$ was refined using neutron powder diffraction data collected at 4 K [55]. The $O\cdots O$ distance of 2.473 Å represents a short hydrogen bridge. A hydrogen atom was found to be disordered between two positions of half occupancy at a $H\cdots H$ distance of 0.34 Å. Symmetrical hydrogen bonds have been also found in the acid-rich compounds such as $LiH_7(SO_4)_4$ and $Na_3H_5(SeO_4)_4$. In both cases, there are short symmetrical hydrogen bonds (2.44 and 2.51 Å, respectively). The disordered proton connects two HXO_4^- anions thus forming $[H(HXO_4)_2]^-$ units. Therefore, the formulas of these compounds should be better formulated as $Li[H(HSO_4)_2](H_2SO_4)_2$ [23] and $Na_3[H(HSeO_4)_2](HSeO_4)_2$ [48]. A more complicated hydrogen disorder was found in the structure of α-NaHSO$_4$ [23]. In addition to the disordered H atom of the type discussed above, one H atom occupies two positions connected by symmetry but taking part in two different hydrogen bonds.

An orthorhombic phase of $Cs_5H_3(SeO_4)_4\cdot H_2O$ undergoes a phase transition at 345 K to hexagonal phase with partially disordered SeO_4 tetrahedra and hydrogen bonding system [84]. A similar structure possesses $Cs_5H_3(SO_4)_4\cdot xH_2O$ at room temperature [85]. At elevated temperatures (400–500 K), some compounds of the formulas $M_3H(XO_4)_2$ and $MHXO_4$ undergo a phase transition to superionic modifications with high protonic conductivity. Such phases are characterized by higher crystallographic symmetry and increased orientational disorder of the XO_4 tetrahedra. Consequently, the whole of hydrogen bonding systems are strongly disordered which leads to a weakening and even partial breaking of hydrogen bonds [86]. For example, such a disorder in $Cs_3H(SeO_4)_2$ at 483 K has been investigated by neutron diffraction [87]. Within the hydrogen bonds $O\cdots O$ (2.66 Å), the H atoms with a site occupancy of 1/3 have been found to be symmetrically disordered between two minima at an $H\cdots H$ separation of 0.77 Å.

ACKNOWLEDGMENTS

This work is the result of a long lasting cooperation within the joint research program between the Humboldt-University Berlin and the Moscow State University. It is a pleasure to acknowledge colleagues who have contributed to this work over the last few years. They include C. Werner (Berlin), I. Morozov (Moscow), A. Stiewe (Berlin) and V. Rybakov (Moscow). We kindly acknowledge financial support by the Deutsche Forschungsgemeinschaft, the Volkswagenstiftung and the Russian Foundation for Basic Research. E. Kemnitz thanks the Fonds der Chemischen Industrie for financial assistance.

NOTE

*All figures of crystal structures presented in this work were newly created by taking the coordinates from the Inorganic Crystal Structure Database (ICSD), Fachinformationszentrum Karlsruhe, and using the graphic program SHAKAL-92.

REFERENCES

1. Baranov, A. I.; Fedosyuk, R. M.; Schagina, N. M.; Shuvalov, L. A. *Ferroelect. Lett.* **1984**, *2*, 25.
2. Merinov, B. V.; Baranov, A. I.; Shuvalov, L. A.; Schagina, N. M. *Sov. Phys. Cryst.* **1991**, *36*, 321.
3. Belushkin, A. V.; Carlile, C. J.; Shuvalov, L. A. *Ferroelectrics* **1995**, *167*, 21.
4. Pawlowski, A.; Pawlaczyk, Cz.; Hilczer, B. *Solid State Ionics* **1990**, *44*, 17.
5. Pawlaczyk, Cz.; Salman, F. E.; Pawlowski, A.; Czapla, Z.; Pietraszko, A. *Phase Transit.* **1986**, *8*, 9.
6. Haile, S. M.; Kreuer, K.-D.; Maier, J. *Acta Cryst.* **1995**, *B51*, 680.
7. Baranov, A. I.; Shuvalov L. A.; Schagina, N. M. *JETP Lett.* **1982**, *36*, 459.
8. Hainovsky, N. G.; Pavlukhin, Yu. T.; Hairetdinov, E. *Solid State Ionics* **1986**, *20*, 249.
9. Howe, A. T.; Shilton, M. G. *J. Solid State Chem.* **1979**, *23*, 345.
10. Bernard, L.; Fitch, A.; Wright, A. F.; Fender, B. E. F.; Howe, A. T. *Solid State Ionics* **1981**, *5*, 459.
11. Kreuer, K.-D.; Rabenau, A.; Weppner, W. *Angew. Chem.* **1982**, *94*, 224.
12. Kreuer, K.-D. *Chem. Mater.* **1996**, *8*, 610.
13. Kemnitz, E.; Werner, C.; Troyanov, S. I. *Eur. J. Solid State Inorg. Chem.* **1996**, *33*, 563.
14. Kemnitz, E.; Werner, C.; Troyanov, S. I. *Eur. J. Solid State Inorg. Chem.* **1996**, *33*, 581.
15. Kemnitz, E.; Troyanov, S. I.; Worzala, H. *Eur. J. Solid State Inorg. Chem.* **1993**, *30*, 629.
16. Lammers, P.; Zemann, J. *Z. Anorg. Allg. Chem.* **1965**, *334*, 225.
17. Johansson, G. B.; Lindquist, O.; Moret, J. *Acta Cryst.* **1975**, *B35*, 1684.
18. Makarova, I. P. *Acta Cryst.* **1993**, *B49*, 11.
19. Waskowska, A.; Czapla, Z. *Acta Cryst.* **1982**, *B38*, 2017.
20. Modenbaugh, A. R.; Hartt, J. E.; Hurst, J. J. *Phys. Rev.* **1983**, *28 B*, 3501.
21. Kemnitz, E.; Werner, C.; Worzala, H.; Troyanov, S.; Strutschkov, Yu. T. *Z. Anorg. Allg. Chem.* **1995**, *621*, 675.
22. Werner, C.; Kemnitz, E.; Troyanov, S.; Strutschkov, Yu.T.; Worzala, H. *Z. Anorg. Allg. Chem.* **1995**, *621*, 1266.
23. Werner, C.; Troyanov, S.; Kemnitz, E.; Worzala, H. *Z. Anorg. Allg. Chem.* **1996**, *622*, 337.
24. Sonneveld, E. J.; Visser, J. W. *Acta Cryst.* **1979**, *B35*, 1975.
25. Sonneveld, E. J.; Visser, J. W. *Acta Cryst.* **1978**, *B34*, 643.
26. Troyanov, S.; Werner, C.; Kemnitz, E.; Worzala, H. *Z. Anorg. Allg. Chem.* **1995**, *621*, 1617.
27. Cotton, F. A.; Frenz, B. A.; Hunter, D. L. *Acta Cryst.* **1975**, *B31*, 302.
28. Kemnitz, E.; Werner, C.; Worzala, H.; Troyanov, S. *Z. Anorg. Allg. Chem.* **1995**, *621*, 1075.
29. Ashmore, J.; Petch, H. *Can. J. Phys.* **1975**, *53*, 2694.
30. Kemnitz, E.; Werner, C.; Worzala, H.; Troyanov, S. *Z. Anorg. Allg. Chem.* **1996**, *622*, 380.
31. Itoh, K.; Ukeda, T.; Ozaki, T.; Nakamura, E. *Acta Cryst.* **1990**, *C46*, 358.
32. Nelmes, R. J. *Acta Cryst.* **1971**, *B27*, 272.
33. DellÀmico, D. B.; Calderazzo, F.; Maarchetti, F.; Merlino, St. *Chem. Mat.* **1998**. In press.
34. Stiewe, A.; Troyanov, S.; Kemnitz, E. *Z. Anorg. Allg. Chem.* **1998**. Submitted.
35. Kemnitz, E.; Werner, C.; Troyanov, S. *Acta Cryst.* **1996**, *C52*, 2665.
36. Simonov, M. A.; Troyanov, S.; Kemnitz, E.; Hass, D.; Kammler, M. *Kristallografiya* **1986**, *31*, 1220.
37. Troyanov, S.; Merinov, B. V.; Verin, I. P.; Kemnitz, E.; Hass, D. *Kristallografiya* **1990**, *35*, 852.

38. Troyanov, S.; Simonov, M. A.; Kemnitz, E.; Hass, D.; Grunze, B. *Dokl. Akad. Nauk SSSR* **1985**, *283*, 1241.

39. Kemnitz, E.; Hass, D.; Worzala, H.; Troyanov, S.; Rybakov, V. B. *Z. Anorg. Allg. Chem.* **1989**, *576*, 179.

40. Simonov, M. A.; Troyanov, S.; Kemnitz, E.; Hass, D.; Kammler, M. *Kristallografiya* **1988**, *33*, 245.

41. Werner, C.; Kemnitz, E.; Worzala, H.; Troyanov, S. *Z. Naturforsch.* **1996**, *51b*, 952.

42. Troyanov, S. I.; Simonov, M. A.; Kemnitz, E.; Hass, D.; Kammler, M. *Dokl. Akad. Nauk SSSR* **1986**, *288*, 1376.

43. Kemnitz, E.; Werner, C.; Stiewe, A.; Worzala, H.; Troyanov, S. *Z. Naturforsch.* **1996**, *51b*, 14.

44. Simonov, M. A.; Troyanov, S.; Gauk, V. Yu. *Kristallografiya* **1988**, *33*, 1520.

45. Troyanov, S.; Simonov, M. A. *Kristallografiya* **1989**, *34*, 233.

46. Stiewe, A.; Troyanov, S.; Kemnitz, E. *Acta Cryst.* **1998**. In press.

47. Zakharov, M.; Troyanov, S.; Kemnitz, E. *Zhurn. Neorg. Khim.* Submitted.

48. Zakharov, M. A.; Troyanov, S.; Rybakov, V. B.; Aslanov, L. A.; Kemnitz, E. *Kristallografia*, **1998**. In press.

49. Ichikawa, M.; Sato, S.; Komukae, M.; Osaka, T. *Acta Cryst.* **1992**, *C48*, 1569.

50. Baran, J.; Lis, T. *Acta Cryst.* **1986**, *C42*, 270.

51. Troyanov, S.; Morozov, I.; Zakharov, M.; Kemnitz, E. *Kristallografia*, **1998**. In press.

52. Pietraszko, A.; Lukaszewicz, K.; Augustyniak, M. A. *Acta Cryst.* **1992**, *C48*, 2069.

53. Kruglik, A. I.; Simonov, M. A. *Kristallografiya* **1977**, *22*, 1082.

54. Aleksandrov, K. S.; Kruglik, A. I.; Misyul', S.V. *Kristallografiya* **1980**, *25*, 1142.

55. Melzer, R.; Sonntag, R.; Knight, K. S. *Acta Cryst.* **1996**, *C52*, 1061.

56. Makarova, I. P.; Verin, I. A.; Shchagina, N. M. *Kristallografiya* **1986**, *31*, 178.

57. Makarova, I. P.; Rider, E. E.; Sarin, V. A.; Aleksandrova, I. P.; Simonov, V. I. *Kristallografiya* **1989**, *34*, 853.

58. Ichikawa, M.; Gustafsson, T.; Olovsson, I. *Acta Cryst.* **1992**, *B48*, 633.

59. Merinov, B. V.; Bolotina, N. B.; Baranov, A. I.; Shuvalov, L. A. *Kristallografiya* **1991**, *36*, 1131.

60. Baran, J.; Lis, T. *Acta Cryst.* **1987**, *C43*, 811.

61. Troyanov, S.; Morozov, I.; Zakharov, M.; Stiewe, A.; Kemnitz, E. *17th Eur. Cryst. Meeting*; Lisboa, Aug. 1997, Abstracts, P3.2-18, 138.

62. Morozov, I.; Troyanov, S.; Stiewe, A.; Kemnitz, E. *Z. Anorg. Allg. Chem.* **1998**. In press.

63. Iskhakova, L. D.; Ovanisyan, S. M.; Trunov, V. K. *Zhurn. Struct. Khim.* **1991**, *32*, 30.

64. Erfany-Far, H.; Fuess, H.; Gregson, D. *Acta Cryst.* **1987**, *C43*, 395.

65. Wells, A. F. *Structural Inorganic Chemistry*; 5th ed., Oxford University Press: Oxford, 1984, p. 690.

66. Joswig, W.; Fuess, H.; Ferraris, G. *Acta Cryst.* **1982**, *B38*, 2798.

67. Noda, Y.; Kasatani, H.; Watanabe, Y.; Terauchi, H.; Gesi, K. *J. Phys. Soc. Jpn.* **1990**, *59*, 3243.

68. Fortier, S.; Fraser, M. E.; Heyding, R. D. *Acta Cryst.* **1985**, *C41*, 1139.

69. Zuniga, F. J.; Etxebarria, J.; Madariaga, G.; Breczewski, T. *Acta Cryst.* **1990**, *C46*, 1199.

70. Haznar, A.; Pietraszko, A. *17th Eur. Cryst. Meeting*; Lisboa, Aug. 1997, Abstracts, P3.3-3, 140.

71. Pietraszko, A.; Haznar, A. *17th Eur. Cryst. Meeting*; Lisboa, Aug. 1997, Abstracts, P5.7-20, 191.

72. Pietraszko, A.; Lukaszewicz, K. *Z. Krist.* **1988**, *158*, 564.

73. Itoh, K.; Ohno, H.; Kuragaki, S. *J. Phys. Soc. Jpn.* **1995**, *64*, 479.

74. Worzala, H.; Schneider, M.; Kemnitz, E.; Troyanov, S. *Z. Anorg. Allg. Chem.* **1991**, *596*, 167.

75. Kemnitz, E.; Troyanov, S.; Worzala, H. *GDCH-Hauptversamlung*; Munich, Sept. 1991, Abstracts.

76. Merinov, B. V.; Baranov, A. I.; Shuvalov, L. A.; Shchagina, N. M. *Kristallografia* **1991**, *36*, 584.

77. Kozlova, N. N.; Iskhakova, L. D.; Marugin, V. V.; Zhadanov, B. V.; Polyakova, I. A. *Zhurn. Neorg. Khim.* **1990**, *35*, 1363.

78. Kemnitz, E.; Werner, C.; Troyanov, S.; Worzala, H. *Z. Anorg. Allg. Chem.* **1994**, *620*, 1921.

79. Lundgren, J. O.; Teasler, I. *Acta Cryst.* **1979**, *B35*, 2384.

80. Troyanov, S.; Stiewe, A.; Kemnitz, E. *Z. Naturforsch.* **1996**, *51b*, 19.
81. Mereiter, K. *Tschermaks Min. Petr. Mitt.* **1974**, *21*, 216.
82. Tudo, J.; Jolibois, B.; Laplace, G.; Nowogrocki, G. *Acta Cryst.* **1979**, *B35*, 1580.
83. Valkonen, J. *Acta Cryst.* **1978**, *B34*, 3064.
84. Merinov, B. V.; Baranov, A. I.; Shuvalov, L. A.; Schneider, J.; Schulz, H. *Solid State Ionics* **1994**, *69*, 153.
85. Merinov, B. V.; Baranov, A. I.; Shuvalov, L. A.; Schneider, J.; Schulz, H. *Solid State Ionics* **1994**, *74*, 53.
86. Colomban, Ph.; Novak, A. In *Proton Conductors: Solids, Membranes and Devices*, Colomban, Ph., Ed.; Cambridge Univ. Press: Cambridge, 1992, Chap. 11, p. 165.
87. Sonntag, R.; Melzer, R.; Knight, K. S. *Physica* **1997**, *B234–236*, 89.

A CRYSTALLOGRAPHIC STRUCTURE REFINEMENT APPROACH USING *AB INITIO* QUALITY ADDITIVE, FUZZY DENSITY FRAGMENTS

Paul G. Mezey

Advances in Molecular Structure Research
Volume 4, pages 115–149
Copyright © 1998 by JAI Press Inc.
All rights of reproduction in any form reserved.
ISBN: 0-7623-0348-4

ABSTRACT

Recent developments in electron density computations of large molecules and in quantum crystallography provide the basis for a new approach to the X-ray structure refinement process as well as to the extension of conventional quantum chemical computational approaches to macromolecules. Some of the relevant, fundamental aspects of density functional theory are reviewed, providing the foundation to new theoretical developments and to new computational approaches in macromolecular quantum chemistry which until recently were assumed impractical. Some of the new methods are based on an additive, fuzzy partitioning of the electron density. Using such density fragments, some of the chemical properties of local molecular regions can be studied. Since the fuzzy density fragments incorporate all the short-range interfragment interactions, in subsequent steps the density fragments can be used to construct electron densities of large systems. This method requires computer time that grows only linearly with the number of density fragments. As a consequence of the linear size dependence, the computational properties of the new approach are favorable as compared to earlier approaches. Besides practical applications in crystallography, the ready availability of Hartree–Fock-type ab initio quality electron densities raise the possibilities for a systematic application of some fundamental energy and correlation energy relations of density functionals to macromolecules. In this contribution, the fundamental ideas of a new approach are described, and the basic relations of density functional theory as well as the interrelations between global and local density descriptors are discussed from a new perspective.

I. INTRODUCTION

According to the fundamental relations of crystallography (see e.g. refs. 1–5), appropriate Fourier transforms of the electron density $\rho(\mathbf{r})$ define a set of structure factors, $\mathbf{F_h}$. These quantities are the coefficients of a Fourier series representing the electronic density distribution $\rho(\mathbf{r})$ in a crystal, where the three-dimensional periodicity of the crystal is exploited:

$$\rho(\mathbf{r}) = V^{-1} \sum_{\mathbf{h}} \mathbf{F_h} \exp(-2\pi i \mathbf{h} \cdot \mathbf{r}) \tag{1}$$

Here $\mathbf{h}$ is a vector of integral components h, k, and l of values inversely proportional to the intercepts of the axes defining the unit cell with the imaginary plane $P_\mathbf{h}$, cutting through the crystal, and V is the volume of the unit cell of the crystal. The $\mathbf{F_h}$ structure factors are complex numbers representing the essential information of the X-ray scattering associated with the planes $P_\mathbf{h}$. These structure factors $\mathbf{F_h}$ can be expressed as,

$$\mathbf{F_h} = |\mathbf{F_h}| \exp(i\phi_\mathbf{h}) \tag{2}$$

where the angle $\phi_\mathbf{h}$ is the formal *phase* associated with $\mathbf{F_h}$.

In most X-ray crystallographic experiments the goal is molecular structure determination. The standard approaches are based on phases, where the local electron density distributions about individual nuclei are initially approximated by spherical or elliptical functions; for example, by spherical or elliptical Gaussian functions. Based on accumulated bond length and bond angle information of similar molecules, and on the initial analysis of the X-ray scattering results, an initial estimate of the nuclear coordinates and the associated Gaussian electron densities is compared to the actual diffraction data. By gradually readjusting the assumed nuclear positions and the associated Gaussian electron density distributions in an iterative process, called *structure refinement*, the assumed nuclear geometry is improved until satisfactory agreement is obtained with the actual diffraction data.

It has been recognized early that the electronic density $\rho(\mathbf{r})$ is a natural link between crystallography and quantum chemistry. Electronic density, an observable, can be measured by crystallographic (or other) experimental methods, and electronic density can also be computed by quantum chemical methods of increasing sophistication.

A combination of crystallographic and quantum chemical approaches is quantum crystallography [6]. The origins of quantum crystallography can be found in the early techniques of fitting N-representable density matrices and molecular wavefunctions to experimental diffraction data [7–25]. The primary goal of quantum crystallography is the determination of molecular density matrices or wavefunctions from crystallographic diffraction data. If such density matrices and wavefunctions are generated, these quantum chemical representations can be used to compute a variety of important molecular properties, such as electrostatic potentials and molecular and bond energies.

In a parallel development, a new approach has been proposed for the computation of quantum chemical electron densities of large molecules using fuzzy electron density fragments. The general scheme of the additive fuzzy density fragmentation approach [26, 27], implemented in its simplest Mulliken–Mezey form within the MEDLA program of Walker and Mezey [28–33], and in a more advanced form within the ALDA and ADMA methods [34–37], is suitable for the generation of ab initio quality electron densities for proteins and other macromolecules. The simplest of the additive fuzzy density fragmentation (AFDF) schemes can be regarded as analogous to a Mulliken population analysis [38, 39] without integration, and the motivating influence of Mulliken's approach is also recognizable in the more advanced AFDF approaches [26, 27, 34–37]. The computer time requirements of these methods scale linearly with the molecular size, an important advantage over conventional implementations of the *ab initio* Hartree–Fock–Roothaan–Hall method [40–43].

Within the framework of quantum crystallography, the AFDF-based electron density computation methods provide a novel alternative to a combined, experimental–theoretical analysis of electron densities of large molecules. In this chapter a framework is presented for the utilization of AFDF methods within the crystallographic structure refinement process.

II. A QUANTUM CRYSTALLOGRAPHIC ROUTE FROM QUANTUM CHEMICAL ELECTRON DENSITIES OF MACROMOLECULES TO CRYSTALLOGRAPHIC STRUCTURE DETERMINATION

One may consider a formal reverse of the process employed in quantum crystallography: using quantum chemical electron densities as tools for a more complete, and possibly more accurate interpretation of crystallographic scattering experiments and experimental electron densities. In principle, sufficiently accurate theoretical electron densities might enhance, as well as simplify, the interpretation of scattering experiments of X-ray crystallography. There seems to be little to gain in the case of small molecules, where both experimental techniques and conventional computational methods perform with remarkable accuracy. However, the new approach is likely to have advantages in the case of macromolecules, where the linear size dependence of AFDF methods permits the computation of *ab initio* quality macromolecular electron densities, and where good quality theoretical densities can provide useful guides for the interpretation of complex scattering results.

A fundamental component in the interpretation of diffraction data is the structure refinement process that usually involves comparisons and least square fits between experimental data and electron density models for free atoms that are locally spherically symmetric quantum chemical representations; for example, those based on simple Gaussian density distributions. The locally spherical electron density models approximate the formal atomic contributions to the molecular electron densities, and the fit between these models and the experiment can be improved by appropriately changing the nuclear locations within the locally spherical model. In principle, the structure refinement process determines the nuclear arrangement that provides the best fit. However, such locally spherical distributions, and even locally elliptic distributions are somewhat oversimplified representations of the actual electron densities, where bonding, lone pair distributions, and conjugated π-bond systems may cause severe shape deviations from locally spherical or elliptical distributions, such as Gaussian distributions assigned to atoms. One can expect improvements in the efficiency and, possibly, in the accuracy, of the structure refinement process if more accurate electron density models are used. Some of the recent advances in the quantum chemical computation of electron densities of large molecules suggest that using these more accurate theoretical electron densities might be advantageous at the initial and intermediate stages of the X-ray structure refinement process.

In terms of a set of atomic orbitals $\varphi_i(\mathbf{r})$ $(i = 1, 2, \ldots, n)$, and the self-consistent SCF LCAO *ab initio* density matrix $\mathbf{P}$ of dimensions $n \times n$, the molecular electronic density $\rho(\mathbf{r})$ of a fixed nuclear geometry K is calculated as:

$$\rho(\mathbf{r}) = \sum_{i=1}^{n} \sum_{j=1}^{n} P_{ij} \, \varphi_i(\mathbf{r}) \, \varphi_j(\mathbf{r}) \tag{3}$$

The AFDF family of electron density computation methods involves a formal assignment of fuzzy, additive electron density contributions to subsets of nuclei. One divides the set of nuclei of the molecule into m mutually exclusive families, $f_1, f_2, \ldots, f_k, \ldots f_m$, corresponding to fuzzy electron density fragments $F_1, F_2, \ldots, F_k, \ldots F_m$, described by fragment density functions $\rho^1(\mathbf{r}), \rho^2(\mathbf{r}), \ldots, \rho^k(\mathbf{r}), \ldots \rho^m(\mathbf{r})$, respectively. The general AFDF scheme proposed in refs. 26, 27 can be given in terms of an AO membership function $m_k(i)$ defined as.

$$m_k(i) = 1 \text{ if AO } \varphi_i(\mathbf{r}) \text{ is centered on one of the nuclei of set } f_k,$$

$$0 \text{ otherwise} \tag{4}$$

In terms of these membership functions $m_k(i)$, and weighting factors w_{ij}, w_{ji},

$$w_{ij} + w_{ji} = 1, \qquad w_{ij}, w_{ji} > 0 \tag{5}$$

the elements P_{ij}^k of the $n \times n$ fragment density matrix $\mathbf{P}^k$ of the k-th fragment F_k is defined as:

$$P_{ij}^k = [m_k(i) \, w_{ij} + m_k(j) \, w_{ji}] \, P_{ij} \tag{6}$$

Note that the simplest, Mulliken–Mezey fragmentation scheme employed in the MEDLA method, and in the more advanced macromolecular density matrix approach of the ADMA method [34–37], corresponds to the special choice of $w_{ij} = w_{ji} = 0.5$.

As can be verified by simple substitution, the fragment density matrices $\mathbf{P}^k$ add up to the density matrix $\mathbf{P}$ of the molecule:

$$P_{ij} \dot{=} \sum_{k=1}^{m} P_{ij}^k \tag{7}$$

The k-th density fragment $\rho^k(\mathbf{r})$ is defined as:

$$\rho^k(\mathbf{r}) = \sum_{i=1}^{n} \sum_{j=1}^{n} P_{ij}^k \, \varphi_i(\mathbf{r}) \, \varphi_j(\mathbf{r}) \tag{8}$$

As implied by Eq. 7, the $\rho^k(\mathbf{r})$ fragment densities add up to the density $\rho(\mathbf{r})$ of the molecule:

$$\rho(\mathbf{r}) = \sum_{k=1}^{m} \rho^k(\mathbf{r}) \tag{9}$$

The general AFDF rules (Eqs. 4–9) are exact within the given *ab initio* LCAO framework. If the fuzzy fragment densities are obtained from small molecules where the nuclear geometry and the environment of the fragment match well the geometry and local environment of the fragment in the target macromolecule, then the AFDF scheme is suitable for the generation of *ab initio* quality macromolecular electronic densities [*29, 33*].

The linear size dependence of the AFDF family of methods is an advantage that can be exploited in the rapid generation of macromolecular electron density representations which are of better quality than those obtained by locally spherical distributions in the usual structure refinement process. The proposed approach, a quantum crystallographic application of the AFDF method (QCR–AFDF) is outlined below.

For each assumed nuclear arrangement K_i of the iterative structure refinement process, an AFDF electron density distribution $\rho(\mathbf{r}, K_i)$ can be calculated, replacing the conventional Gaussian density representations. The least square fit process using AFDF densities consists of the following formal steps, carried out iteratively:

Step A   Initial estimate of nuclear geometry $K_i = K_0$
Step B   Generation of AFDF density $\rho(\mathbf{r}, K_i)$
Step C   Comparison of scattering results and AFDF density $\rho(\mathbf{r}, K_i)$
Step D   Estimation of nuclear geometry correction ΔK_i; test for convergence
Step E   Generation of new approximation to the nuclear geometry:
$$K_{i+1} = K_i + \Delta K_i; \quad \text{set } i = i + 1; \quad \text{return to step B.} \tag{10}$$

In the above scheme, step B can follow the established methods of computing macromolecular electronic densities based on the ADMA technique.

Steps C and D determine the success of this quantum crystallographic approach. In step C one may take advantage of some aspects of the currently used fitting techniques of X-ray crystallography, with appropriate modifications to account for the new functional representation of the electronic density. In particular, one may follow the approach given below:

C1.   For an optimal placement (position and orientation) of the calculated AFDF electron density in the unit cell standard translation–rotation programs can be used, that involves optimization of three translational and three orientational parameters.

C2. For the current, i-th approximation K_i of the theoretical nuclear configuration, take the AFDF electron density $\rho(\mathbf{r}, K_i)$ of the entire unit cell in numerical form and numerically Fourier transform it. This gives a set of AFDF structure factor magnitudes and phases compatible with the computed *ab initio* quality electron density.

C3. One may expect various degrees of discrepancies between the calculated structure factor magnitudes, $|F_{calc}|$, and the experimental ones, $|F_{obs}|$. In particular, a scaling problem may arise, that can be corrected by a simple adjustment. One may adjust the scale of the observed structure factor magnitudes by taking the average of the whole set and scaling it to fit the average of the calculated structure factor magnitudes (or *vice versa*). Furthermore, for a valid comparison between computed and observed structure factor magnitudes, the experimental data require adjustments in order to minimize the effect of vibrational motion and general positional disorder within the unit cells on the experimental structure factor magnitudes, $|F_{obs}|$.

C4. Using the quantities $|F_{obs}| - |F_{calc}|$ as coefficients and the *calculated phases*, a density difference map is constructed. This map indicates where adjustments of the current approximation K_i of the theoretical nuclear configuration is required. One potential problem involves the effects of cut-off errors related to limitations of $|F_{obs}|$ with regard to scattering angles. Another potential problem is that the phases may need adjustments. Nevertheless, the difference density map so obtained is expected to indicate the likely components of the nuclear geometry correction ΔK_i for step D.

C5. Another density difference map that provides input for estimating the nuclear geometry correction ΔK_i is obtained by using the quantities $|2F_{obs}| - |F_{calc}|$ as coefficients and the *calculated phases*, where the cut-off errors are no longer relevant. This map is expected to indicate the *dominant* components in the nuclear geometry correction ΔK_i for step D.

In the above scheme, step B is expected to be the slowest. However, the increased accuracy of the theoretical electronic density is likely to reduce the number of iterations within the structure refinement process substantially. Whereas the estimated overall computer time requirement of the QCR–AFDF approach is comparable to that of the standard structure refinement method, some improvements in speed as well as in accuracy are expected.

The crystallographic structure refinement approach suggested is based on the viability of a fuzzy and additive density fragmentation technique. Ultimately, the fundamental quantum mechanical properties of density functionals justify the fuzzy fragmentation procedure, and the very same fundamental properties also suggest new approaches to local electron density shape analysis. These possibilities have relevance both to crystallographic structure refinement as well as to a density-based interpretation of the chemical properties of functional groups and other local moieties within molecules. In the following sections some of the consequences of quantum mechanical and information theoretical features of molecular electron densities are discussed.

III. SOME RELEVANT PROPERTIES OF DENSITY FUNCTIONALS

A. Electron Density, Molecular Properties, and the Hohenberg–Kohn Theorem

Modern density functional theory has provided a very natural framework for molecular physics and quantum chemistry, especially since the formulation of the famous theorem of Hohenberg and Kohn [44], that gives the justification of many of the contemporary computational approaches in electron density modeling.

This theorem itself is a very natural statement on the relations between the ground-state electron density $\rho(\mathbf{r})$, assumed in the form of spin-averaged density,

$$\rho(\mathbf{r}) = n \sum_{s_1} \cdots \sum_{s_n} \int d^3 \mathbf{r}_2 \cdots \int d^3 \mathbf{r}_n \mid \Psi(\mathbf{r}, s_1, \mathbf{r}_2, s_2, \ldots \mathbf{r}_n, s_n)\mid^2 \tag{11}$$

of a system of n electrons in a local spin-independent external potential V, and the Hamiltonian H given as:

$$H = \sum_{i=1}^{n} V(\mathbf{r}_i) + T + V_{ee} \tag{12}$$

Here T is the kinetic energy operator,

$$T = -(1/2) \sum_{i=1}^{n} \Delta_i \tag{13}$$

V_{ee} is the electron–electron repulsion operator,

$$V_{ee} = \sum_{i=1}^{n-1} \sum_{j=i+1}^{n} (\mid \mathbf{r}_i - \mathbf{r}_j \mid)^{-1} \tag{14}$$

and,

$$V(\mathbf{r}) = \sum_{i=1}^{n} V(\mathbf{r}_i) \tag{15}$$

is the external potential, where in the usual molecular case the term $V(\mathbf{r}_i)$ is the electron–nuclear attraction operator for the interaction of the i-th electron with the nuclear system of the molecule. In practical terms, the theorem states that a nondegenerate ground-state electron density $\rho(\mathbf{r})$ determines H within an additive constant; that is, $\rho(\mathbf{r})$ determines all ground-state and all excited-state properties of H.

This statement of the theorem of Hohenberg and Kohn is the justification of many of the modern density functional approaches to chemistry. Here we shall outline the generalized and very concise "constrained search" proof as described by Levy [45–48]. This proof, as we shall see, lends itself to further generalizations that are of special advantage in the case of large molecules, such as proteins.

B. The Constrained Search Approach

First choose a fixed electron density $\rho(\mathbf{r})$ and consider the family,

$$\Phi_\rho = \{\Psi_{\rho,\alpha}\}_{(\rho\ \text{fixed})} \tag{16}$$

of all antisymmetrized wavefunctions $\Psi_{\rho,\alpha}$ which yield the given density $\rho(\mathbf{r})$, where the index α, taken from a continuum, distinguishes each wavefunction $\Psi_{\rho,\alpha}$ in the family.

Since all the wavefunctions in family Φ_ρ yield the same electron density $\rho(\mathbf{r})$, they must also have the same expectation value for any local multiplicative operator; for example, for the $V(\mathbf{r}) = \Sigma_{i=1,n} V(\mathbf{r}_i)$ electron–nuclear attraction operator, as implied by the relation,

$$\int d^3\mathbf{r}\, V(\mathbf{r})\, \rho(\mathbf{r}) = \langle\Psi_{\rho,\alpha}\, |\, \sum_{i=1}^{n} V(\mathbf{r}_i)\, |\, \Psi_{\rho,\alpha}\rangle \tag{17}$$

for any index α.

From this family Φ_ρ, a wavefunction that yields the lowest energy is denoted by $\Psi_{\rho,\text{min}}$, then:

$$\langle\Psi_{\rho,\text{min}}\, |H|\, \Psi_{\rho,\text{min}}\rangle = \min_{\alpha,(\rho\ \text{fixed})} \{\langle\Psi_{\rho,\alpha}\, |H|\, \Psi_{\rho,\alpha}\rangle\} \tag{18}$$

Of course, this wavefunction $\Psi_{\rho,\text{min}}$ also gives the same $\rho(\mathbf{r})$-determined value of the expectation value for the electron–nuclear attraction operator as that given by Eq. 17:

$$\int d^3\mathbf{r}\, V(\mathbf{r})\, \rho(\mathbf{r}) = \langle\Psi_{\rho,\text{min}}\, |\, \sum_{i=1}^{n} V(\mathbf{r}_i)\, |\, \Psi_{\rho,\text{min}}\rangle \tag{19}$$

Next, release the constraint of a fixed electron density $\rho(\mathbf{r})$ and take the family,

$$\Phi_{min} = \{\Psi_{\rho,min}\} \tag{20}$$

of energy-minimizing wavefunctions $\Psi_{\rho,min}$ for every electron density $\rho(\mathbf{r})$. Within this family, a wavefunction $\Psi_{\rho_o,min}$ that belongs to a special density $\rho_o(\mathbf{r})$ that minimizes energy is in fact a variationally optimum wavefunction. Of course, this follows directly from the actual construction, since within the family $\Phi_{min} = \{\Psi_{\rho,min}\}$ of wavefunctions which correspond to the minimum energy for each possible electron density $\rho(\mathbf{r})$, there must exist a wavefunction for some density $\rho_o(\mathbf{r})$ that corresponds to the lowest possible energy expectation value E_o,

$$<\Psi_{\rho_o,min} |H| \Psi_{\rho_o,min} > = \min_\rho \{< \Psi_{\rho,min} |H| \Psi_{\rho,min} >\} \tag{21}$$

that is:

$$< \Psi_{\rho_o,min} |H| \Psi_{\rho_o,min} > = \min_\rho \{\min_{\alpha,(\rho\ fixed)} \{< \Psi_{\rho,\alpha} |H| \Psi_{\rho,\alpha} >\}\}$$

$$= \min_\Psi \{<\Psi|H|\Psi>\} = E_o \tag{22}$$

Consequently, $\rho_o(\mathbf{r})$ is the ground-state density.

Equation 21 can be rewritten as,

$$< \Psi_{\rho_o,min} |H| \Psi_{\rho_o,min} > = < \Psi_{\rho_o,min} | \sum_{i=1}^{n} V(\mathbf{r_i}) | \Psi_{\rho_o,min} >$$

$$+ < \Psi_{\rho_o,min} |T + V_{ee}| \Psi_{\rho_o,min} > \tag{23}$$

that is, as

$$< \Psi_{\rho_o,min} |H| \Psi_{\rho_o,min} > = \int d^3\mathbf{r}\, V(\mathbf{r})\, \rho_o(\mathbf{r})$$

$$+ < \Psi_{\rho_o,min} |T + V_{ee}| \Psi_{\rho_o,min} > \tag{24}$$

where in the first term on the right hand side we utilized the fact that this term depends only on the actual density, as described in Eqs. 17 and 19.

According to the actual formulation of the Hohenberg–Kohn theorem [44], there exists a universal variational functional $F(\rho)$ of trial electron densities ρ, such that if E_o is the ground-state energy and $\rho_o(\mathbf{r})$ is the ground-state electron density that belongs to the given external potential $V(\mathbf{r})$ specified in Eq. 15, then,

$$E_o = \min_\rho \{\int d^3\mathbf{r}\, V(\mathbf{r})\, \rho(\mathbf{r}) + F(\rho)\} \tag{25}$$

and:

$$E_o = \int d^3\mathbf{r}\, V(\mathbf{r})\, \rho_o(\mathbf{r}) + F(\rho_o)\} \tag{26}$$

Since the left-hand side of Eq. 24 is E_o, and since $\Psi_{\rho,\text{min}}$ must be the ground-state wavefunction that belongs to the external potential $V(\mathbf{r})$, by comparing Eqs. 24 and 26, we conclude that the Hohenberg–Kohn universal functional $F(\rho)$ can be identified with the second term on the right-hand side of Eq. 24, taken for any general trial electron density ρ,

$$F(\rho) = \; < \Psi_{\rho,\text{min}} \, | T + V_{ee} | \, \Psi_{\rho,\text{min}} > \tag{27}$$

that in the special case of $\rho(\mathbf{r}) = \rho_o(\mathbf{r})$ gives:

$$F(\rho_o) = \; < \Psi_{\rho_o,\text{min}} \, | T + V_{ee} | \, \Psi_{\rho_o,\text{min}} > \tag{28}$$

Furthermore, one should note that the quantity $\int d^3\mathbf{r} \, V(\mathbf{r}) \, \rho(\mathbf{r})$ is the same for any wavefunction $\Psi_{\rho,\alpha}$ of family Φ_ρ defined in Eq. 16—specifically, the quantity $\int d^3\mathbf{r} \, V(\mathbf{r}) \, \rho_o(\mathbf{r})$ is the same for any wavefunction $\Psi_{\rho_o,\alpha}$ of family Φ_{ρ_o}. Consequently, a constrained search for the energy-minimizing wavefunction $\Psi_{\rho_o,\text{min}}$ within family Φ_{ρ_o} can be restricted to a minimization of the expectation value of $T + V_{ee}$, instead of the minimization of the expectation value of the full Hamiltonian $H = \Sigma \, V(\mathbf{r}_i) + T + V_{ee}$:

$$\text{min} < \Psi_{\rho_o,\alpha} \, |T + V_{ee}| \, \Psi_{\rho_o,\alpha} > \; = \; < \Psi_{\rho_o,\text{min}} \, |T + V_{ee}| \, \Psi_{\rho_o,\text{min}} > \tag{29}$$

The ground-state electron density fully determines the first term in energy Eqs. 24 and 26 and it also determines the family Φ_{ρ_o}. In turn, the family Φ_{ρ_o} determines the second term of Eqs. 24 and 26, since the ground-state wavefunction $\Psi_{\rho_o,\text{min}}$ is specified as the wavefunction that gives the ground-state electron density $\rho_o(\mathbf{r})$ (i.e. it is a member of family Φ_{ρ_o}) and also minimizes the expectation value of the operator $T + V_{ee}$. Consequently, the second term of energy Eqs. 24 and 26 is also determined by the ground-state electron density $\rho_o(\mathbf{r})$. That is, the ground-state density $\rho_o(\mathbf{r})$ determines the ground-state energy E_o.

One should also note that the operator $T + V_{ee}$ is independent of the external potential $V(\mathbf{r})$; consequently, the functional $F(\rho)$ is truly universal since it depends only on the electron density ρ; specifically, $F(\rho)$ does not depend on the actual external potential $V(\mathbf{r})$.

However, the external potential $V(\mathbf{r})$ is also a unique functional of the ground-state electron density, up to an arbitrary additive constant. One can show this easily by obtaining an expression from Eq. 26 where all terms appear under a single integral. The functional derivative of the Hohenberg–Kohn universal functional $F(\rho)$, denoted by,

$$\partial \, F(\rho) \, / \, \partial \, \rho(\mathbf{r}) \tag{30}$$

taken at some density ρ', is defined by the relation,

$$\partial F(\rho' + \varepsilon g)/\partial \varepsilon|_{\varepsilon=0} = \int d^3 \mathbf{r} \, \partial \, F(\rho)/\partial \, \rho(\mathbf{r}) \, |_{\rho=\rho'} \, g(\mathbf{r}) \tag{31}$$

where $g(\mathbf{r})$ is an arbitrary function and ε is a scalar.

For any function $g(\mathbf{r})$, such that,

$$\int d^3 \mathbf{r} \, g(\mathbf{r}) = 0 \tag{32}$$

Eq. 26 can be rewritten as:

$$E_o = \min_\varepsilon \{ \int d^3 \mathbf{r} \, V(\mathbf{r}) \, [\rho_o(\mathbf{r}) + \varepsilon \, g(\mathbf{r})] + F(\rho_o + \varepsilon \, g) \} \tag{33}$$

We do know that the choice of $\varepsilon = 0$ minimizes the right hand side of Eq. 33, consequently,

$$\partial \{ \int d^3 \mathbf{r} \, V(\mathbf{r}) \, [\rho_o(\mathbf{r}) + \varepsilon \, g(\mathbf{r})] + F(\rho_o + \varepsilon \, g) \} / \partial \varepsilon \, |_{\varepsilon=0} = 0 \tag{34}$$

that is,

$$\int d^3 \mathbf{r} \, V(\mathbf{r}) \, g(\mathbf{r}) + \partial F(\rho_o + \varepsilon \, g) \} / \partial \varepsilon \, |_{\varepsilon=0} = 0 \tag{35}$$

where, after recognizing the functional derivative of the Hohenberg–Kohn universal functional $F(\rho)$, one obtains an expression of a single integral:

$$\int d^3 \mathbf{r} \, g(\mathbf{r}) \, [V(\mathbf{r}) + \partial F(\rho)/\partial \, \rho(\mathbf{r}) \, |_{\rho=\rho_o}] = 0 \tag{36}$$

Since the function $g(\mathbf{r})$ is arbitrary, except for the integral condition (Eq. 32), Eq. 36 can hold only if the expression $V(\mathbf{r}) + \partial \, F(\rho)/\partial \, \rho(\mathbf{r}) \, |_{\rho=\rho_o}$ in the square brackets is zero, that is,

$$V(\mathbf{r}) = - \partial \, F(\rho)/\partial \, \rho(\mathbf{r}) \, |_{\rho=\rho_o} \tag{38}$$

a relation that is the fundamental Euler equation of density functional theory.

Indeed, the external potential $V(\mathbf{r})$ is also determined by the ground-state electron density $\rho_o(\mathbf{r})$.

In fact, the ground-state electron density $\rho_o(\mathbf{r})$ determines the Hamiltonian $H = \Sigma \, V(\mathbf{r}_i) + T + V_{ee}$; consequently, the ground-state electron density also implies excited-state information.

C. An Information-Theoretical Proof of the Hohenberg–Kohn Theorem

The general result that the ground-state electron density determines the molecular properties is not really surprising if one takes an information-theoretical approach to the Hohenberg–Kohn density functional model.

When we are in the position of applying the Hohenberg–Kohn theorem, what else are we supposed to know besides the density? We have two pieces of information: the density itself as a three-dimensional function given in some form, and the additional knowledge that it is the ground-state density! In fact, this is the key idea that has lead to the constrained search methods, which provide perhaps the clearest picture of the fundamental relations of density functional theory. We must emphasize and also utilize the fact that the Hohenberg–Kohn theorem refers to the ground-state density.

The same two pieces of information is the starting point in the information-theoretical approach. For simplicity, we shall restrict the following argument to molecules; however, the treatment is equally valid for any physical system built from nuclei and electrons.

In fact, a simple proof of the Hohenberg–Kohn theorem for molecules can be given as follows:

1. A molecule contains nothing else than a family of nuclei and an electron density cloud. All information concerning the static properties of the molecule must be contained in the nuclear and electron distributions; there is simply no other material present to encode information.
2. If the ground-state electron density $\rho_o(\mathbf{r})$ is known, than the location and atomic numbers of the nuclei are fully determined.
3. Consequently, the ground-state electron density $\rho_o(\mathbf{r})$ contains all information concerning all static properties of the molecule, including its ground-state energy, and any other molecular properties.

In the above argument we have relied on the assumption that the information concerning a molecule is localized in the following sense: the molecule, as an entity distinguishable from the rest of the universe, has its own identity, and the molecule fully characterizes all of its own properties. In fact, a similar argument is the basis of the "holographic electron density theorem," discussed in a later section.

D. Subsystems of Finite-Bounded Systems

The problem of the application of the Hohenberg–Kohn theorem to a subsystem and the associated information content of the subsystem has been first addressed by Riess and Münch [49] using analyticity arguments. One of their assumptions that was essential for the utilization of analyticity, namely that the entire electron density of a system (such as a molecule) is confined to a finite region of the space, is not strictly valid. Here we shall review their argument and also present a technique that circumvents the difficulty, leading to the four-dimensional holographic electron density theorem [50]. This theorem applies to quantum mechanically valid, boundaryless molecular electron densities [50a], with some important consequences for the interrelations between local and global symmetries and local and global chirality properties of molecules [50b].

The nonrelativistic, spin-free, n-electron Born–Oppenheimer Hamiltonian of Eq. 12 is self-adjoint, bounded from below, and below some energy has a discrete energy spectrum; it belongs to an elliptic differential equation with coefficients analytic almost everywhere within its domain G, $R^{3n} \supset G$ (that is, analytic everywhere in G except a set of measure zero) that leaves the rest G_0 of the domain G-connected.

Here G is the n-fold direct sum of an arbitrary, simply connected region g in the ordinary 3-space R^3, $R^3 \supset g$, the physical region of the system where at this stage there is no boundedness requirement for g. Note, however, that in the strict quantum mechanical sense g should be taken as the entire 3-space R^3, a fact that will require special attention in the proof of the holographic electron density theorem.

An important auxiliary result discussed in ref. 49 is the following. Assume nondegenerate, variationally determined ground states for two Hamiltonians, H_1 and H_2,

$$H_1 = V_1(\mathbf{r}) + T + V_{ee} \tag{39}$$

and

$$H_2 = V_2(\mathbf{r}) + T + V_{ee} \tag{40}$$

The corresponding normalized ground-state eigenfunctions Ψ_1 and Ψ_2 are related to each other by,

$$\Psi_1 = \alpha \, \Psi_2 \tag{41}$$

where α is a complex phase factor,

$$|\alpha| = 1 \tag{42}$$

if and only if the two Hamiltonians H_1 and H_2 differ only in an additive real constant β in their external potentials:

$$V_1(\mathbf{r}) = V_2(\mathbf{r}) + \beta \tag{43}$$

Indeed, if $\Psi_1 = \alpha\Psi_2$ then,

$$H_1 \, \Psi_1 = E_1 \, \Psi_1 \tag{44}$$

and,

$$H_2 \, \Psi_1 = E_2 \, \Psi_1 \tag{45}$$

implying:

$$[V_1(\mathbf{r}) - V_2(\mathbf{r})] \, \Psi_1 = [E_1 - E_2] \, \Psi_1 \tag{46}$$

Since the coefficients of the elliptic Hamiltonians H_1 and H_2 are analytic in G_0 of G, the wavefunction Ψ_1 must also be analytic in G_0, hence,

$$\Psi_1 \neq 0 \tag{47}$$

except on a set of measure zero, consequently:

$$V_1(\mathbf{r}) - V_2(\mathbf{r}) = E_1 - E_2 = \beta \tag{48}$$

Conversely, if $V_1(\mathbf{r}) - V_2(\mathbf{r}) = \beta$, a real scalar, then the two Hamiltonians H_1 and H_2, where,

$$H_2 = H_1 - \beta \tag{49}$$

must have the same spectrum, and by nondegeneracy of the lowest eigenvalue, the two ground-state eigenfunctions Ψ_1 and Ψ_2 must be related by,

$$\Psi_1 = \alpha\, \Psi_2 \tag{50}$$

for some complex factor α, where $|\alpha| = 1$.

The second important auxiliary result discussed in ref. 49 concerns the strong relation between the electron density $\rho(\mathbf{r})$ and the external potential $V(\mathbf{r})$. A substantial difference in two external potentials, $V_1(\mathbf{r})$ and $V_2(\mathbf{r})$, implies that the corresponding two ground-state electron densities, $\rho_1(\mathbf{r})$ and $\rho_2(\mathbf{r})$, must also be different, and conversely, a difference in the ground-state electron densities implies a substantial difference in the external potentials.

More precisely, if the two external potentials differ by more than a simple additive constant, that is, if,

$$V_1(\mathbf{r}) \neq V_2(\mathbf{r}) + \beta \tag{51}$$

for every additive real constant β, this is a sufficient and necessary condition for the two electron densities being different:

$$\rho_1(\mathbf{r}) \neq \rho_2(\mathbf{r}) \tag{52}$$

Indeed, if $V_1(\mathbf{r}) \neq V_2(\mathbf{r}) + \beta$, then $\Psi_1 \neq \alpha\, \Psi_2$ follows, according to the previous result. Invoking the variational theorem, and the nondegeneracy of the ground-states,

$$E_1 = \;<\Psi_1|H_1|\Psi_1> \;<\; <\Psi_2|H_1|\Psi_2> \;=\; E_2 + <\Psi_2|V_1(\mathbf{r}) - V_2(\mathbf{r})|\Psi_2> \tag{53}$$

that is,

$$E_1 < E_2 + \int_g d^3\mathbf{r}\, \rho_2(\mathbf{r})\, [V_1(\mathbf{r}) - V_2(\mathbf{r})] \tag{54}$$

where one should emphasize the fact that the strict inequality applies.

By a similar treatment one obtains,

$$E_2 < E_1 + \int_g d^3\mathbf{r}\rho_1(\mathbf{r})\,[V_2(\mathbf{r}) - V_1(\mathbf{r})] \tag{55}$$

another strict inequality.

Combining inequalities 54 and 55, one obtains:

$$0 < \int_g d^3\mathbf{r}\,[\rho_2(\mathbf{r}) - \rho_1(\mathbf{r})]\,[V_1(\mathbf{r}) - V_2(\mathbf{r})] \tag{56}$$

The strict inequality in 56 is possible only if $\rho_1(\mathbf{r}) \neq \rho_2(\mathbf{r})$, as claimed; otherwise the integral on the right-hand side would be zero.

Conversely, if $\Psi_1 = \alpha\,\Psi_2$ for some complex factor α, $|\alpha| = 1$ then $\rho_1(\mathbf{r}) = \rho_2(\mathbf{r})$; consequently, if $\rho_1(\mathbf{r}) \neq \rho_2(\mathbf{r})$, then necessarily $\Psi_1 \neq \alpha\,\Psi_2$ follows, that in turn implies that $V_1(\mathbf{r}) \neq V_2(\mathbf{r}) + \beta$, as claimed.

These two auxiliary results, in fact, state all of the conclusions of the original Hohenberg–Kohn theorem: the ground-state energy E of a molecule, as well as the ground-state wavefunction Ψ; consequently, the expectation values of all spin-free observables are unique functionals of the ground-state electron density $\rho(\mathbf{r})$.

The proofs of these auxiliary results described above would not hold if the wavefunctions Ψ_1 and Ψ_2 were allowed to become zero on a subset of nonzero measure; also, the connectedness of the domain G_0 of G is essential, underlining the special role of nodeless wavefunctions.

In the possession of the two auxiliary results, we shall consider a subdomain of the 3D space R^3 and the associated electron density restricted to the subdomain, and investigate the problem of analytic continuation of the subdomain electron density beyond the subdomain.

First we shall describe the argument followed in ref. 49 describing a treatment where the complete physical system is confined to a finite domain of the space. Next, we shall extend the treatment to infinite systems, leading to the "holographic electron density theorem."

Consider now a set d of nonzero volume, such that d is a subset of the physical region g in the ordinary 3-space R^3, where the physical domain g is assumed to be bounded and is assumed to contain the entire physical system, for example, the entire molecule:

$$R^3 \supset g \supset d \tag{57}$$

The complete electron density $\rho_g(\mathbf{r})$ of the physical system is given by the relation,

$$\rho_g(\mathbf{r}) = n \sum_{s_1} \cdots \sum_{s_n} \int_g d^3\mathbf{r}_2 \cdots \int_g d^3\mathbf{r}_n\,|\,\Psi(\mathbf{r}, s_1, \mathbf{r}_2, s_2, \ldots \mathbf{r}_n, s_n)|^2 \tag{58}$$

where the $n - 1$ integrations are carried out over the finite domain g of the 3D space R^3.

One may note that both the wavefunction $\Psi(\mathbf{r}, s_1, \mathbf{r}_2, s_2, \ldots \mathbf{r}_n, s_n)$ and the n-electron density,

$$\rho_g(\mathbf{r}_1, \mathbf{r}_2, \ldots \mathbf{r}_n) = n \sum_{s_1} \cdots \sum_{s_n} | \Psi(\mathbf{r}_1, s_1, \mathbf{r}_2, s_2, \ldots \mathbf{r}_n, s_n)|^2 \tag{59}$$

are analytic in G almost everywhere, except on a set of singular points including the sets of Coulomb singularities of coincident electronic locations where,

$$\mathbf{r}_i = \mathbf{r}_j \tag{60}$$

for pairs of electrons $i, j = 1, 2, \ldots n$, $i \neq j$, and the nucleus–electron Coulomb singularities,

$$\mathbf{R}_\alpha = \mathbf{r}_i \tag{61}$$

for nuclei $a = 1, 2, \ldots k$, and electrons $i = 1, 2, \ldots n$. These points form a set of measure zero in G, and this set of Coulomb singularities does not cut any of the $(3n - 3)$-dimensional subspaces of G into two disjoint parts with non-analytic common boundary.

Hence, the function $\rho_g(\mathbf{r}_1, \mathbf{r}_2, \ldots \mathbf{r}_n)$ is integrable in G; consequently, it is integrable on the subset $(\mathbf{r}_2, \ldots \mathbf{r}_n)$ of its $n - 1$ variables, leading to the actual density $\rho_g(\mathbf{r})$ within the finite domain g:

$$\rho_g(\mathbf{r}) = n \int_g d^3\mathbf{r}_2 \cdots \int_g d^3\mathbf{r}_n \, \rho_g(\mathbf{r}, \mathbf{r}_2, \ldots \mathbf{r}_n) \tag{62}$$

Furthermore, we recall that $\rho_g(\mathbf{r}_1, \mathbf{r}_2, \ldots \mathbf{r}_n)$ is an almost everywhere analytic function of the variables $\mathbf{r} = \mathbf{r}_1, \mathbf{r}_2, \ldots \mathbf{r}_n$; consequently, the density function $\rho_g(\mathbf{r})$ obtained in Eq. 62 is also almost everywhere analytic over the three-dimensional domain g, since it was obtained by integrating an almost everywhere analytic function according to a subset $\mathbf{r}_2, \ldots \mathbf{r}_n$ of its variables over a finite, $(3n - 3)$-dimensional domain G^{3n-3}, where G^{3n-3} is the $(n - 1)$-fold direct sum of the finite, 3-dimensional domain g, and where $\mathbf{r}$ is the variable not involved in the integration.

Consequently, if $\rho_d(\mathbf{r})$ denotes the electron density over a subset d of g, $g \supset d$, where d has nonzero volume, then the conditions of a general theorem of analytic continuation are fulfilled, and $\rho_g(\mathbf{r})$ over the entire domain g is uniquely determined by $\rho_d(\mathbf{r})$ over the subdomain d, that is, if the restrictions $\rho_d^a(\mathbf{r})$ and $\rho_d^b(\mathbf{r})$ of densities $\rho_g^a(\mathbf{r})$ and $\rho_g^b(\mathbf{r})$, respectively, agree on the subdomain d,

$$\rho_d^a(\mathbf{r}) = \rho_d^b(\mathbf{r}) \tag{63}$$

for every point $\mathbf{r}$ of d, then necessarily,

$$\rho_g^a(\mathbf{r}) = \rho_g^b(\mathbf{r}) \tag{64}$$

for every point **r** of g.

The assumption that the physical system, for example a molecule, is bounded within a finite region g of the ordinary space R^3 is essentially classical in nature. By contrast, in rigorous quantum mechanics the localizability of a molecular system within any finite domain of space is not strictly valid, hence, the above proof is limited to the finite model case, even though the result appears rather plausible.

Nevertheless, by choosing an alternative approach to the proof of a more general problem, based on the natural convergence properties of electron densities, this difficulty can be circumvented [50], leading to the holographic electron density theorem on quantum mechanically correct, boundaryless molecular electron densities [50a]. This also implies some fundamental relations between local and global symmetries and local and global chirality properties of electron densities [50b] of molecules.

It will be shown in the next section that by using a four-dimensional electron density model and the Alexandrov one-point compactification of the ordinary three-dimensional space R^3, it will be possible to use analyticity arguments on compact sets to establish the claim that the electron density of any finite subsystem of nonzero volume determines the electron density of the rest of the system.

E. The Holographic Electron Density Theorem

It is a natural expectation that a fragment of the molecular electron density contains some information about the rest of the molecular electron density. In fact, a much stronger statement can be made: fragments of molecular electron densities have properties analogous with the fragments of a holographic plate. Just as a piece of a holographic plate is sufficient to reconstruct the complete image; a piece of the molecular electron density is sufficient to reconstruct the entire molecular electron density.

This result will be formulated as a four-dimensional holographic electron density theorem [50], where the reference to holographic properties is justified by the analogy with holographic plates, and where a mere fragment of the holographic plate contains, in principle, the full information about the entire holographic image, in a nice illustration of the power of matching cardinalities of any continua of a fixed dimension.

In order to introduce a compactification of a three-dimensional, quantum chemical electron density that, in principle, is neither finite nor bounded, we first shall consider a two-dimensional example, illustrated in Figure 1. Note that compactness is a generalization of the properties "finite" and "bounded." Consider a plane P, supplied with a coordinate system and a sphere S^2 of center c and of finite radius that is placed on the plane with a tangential contact with the plane at the origin of the coordinate system. Furthermore, assume that an almost everywhere continuous and infinitely differentiable (smooth) function $F(x)$ is specified over this plane that converges exponentially to zero at infinity. Whereas the plane is neither bounded

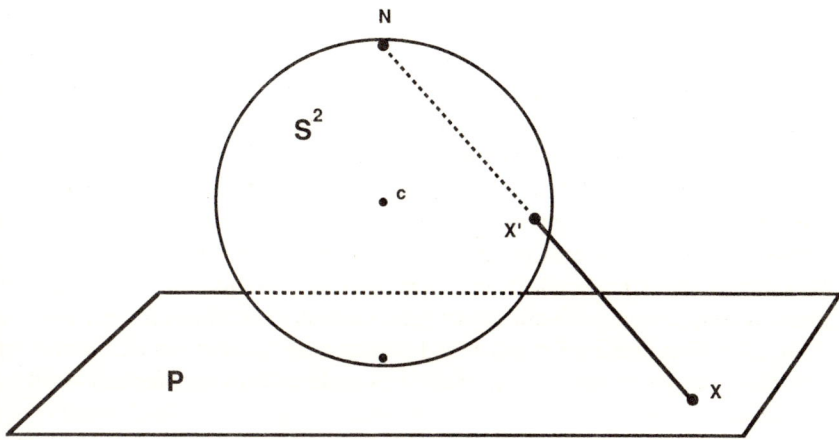

Figure 1. An illustration of the Alexandrov one-point compactification, as applied to a two-dimensional plane. See text for details.

nor finite, it can be replaced by a compact set using the technique called the "Alexandrov one-point compactification." To each point x of the plane P issue a line from the north pole N of the sphere. This line will pierce the sphere at a point x'. This establishes a one-to-one assignment of the points x of the plane P to points x' of the sphere S^2. By adding the north pole N to all the points x' obtained by the above procedure, the sphere is completed. Since the sphere S^2 of finite radius is compact, the plane P, with the addition of a single point N, is now replaced with a compact set. This technique has also been used to study the topological properties of various domains of multidimensional potential energy hypersurfaces [51].

On the sphere S^2, the function F can be redefined as function $F'(x')$, using the assignment

$$F'(x') = F(x) \tag{65}$$

Here the exponential convergence of $F(x)$ to zero as x becomes infinitely distant from the origin c ensures that by assigning the value,

$$F'(N) = 0 \tag{66}$$

to the north pole, the function $F'(x')$ on the sphere S^2 is a continuous and infinitely differentiable (smooth) function.

Consider now the electron density $\rho(\mathbf{r})$ of a molecular system in a formal four-dimensional space where the first three dimensions correspond to the ordinary three-space R^3, and the fourth dimension corresponds to the function values of the molecular electron density $\rho(\mathbf{r})$. A point $\mathbf{Y}$ of this four dimensional space R^4 is characterized by four coordinates, given as,

$$\mathbf{Y} = [r_x, r_y, r_z, \rho(\mathbf{r})]' \tag{67}$$

where the prime symbol over the square bracket stands for "transpose." Note that a similar, four-dimensional approach has been introduced for the shape characterization of electron densities in terms of higher dimensional curvatures, leading to an eventual characterization in terms of three-dimensional Hessian matrices [52, 53]. The same technique was used subsequently for a similarity analysis and a differential geometrical study of electron densities [54, 55].

The three-dimensional molecular electron density $\rho(\mathbf{r})$ fulfills all the continuity, differentiability, and exponential convergence conditions specified for the function $F(x)$ of the two-dimensional example. Consequently, a four-dimensional variant of the Alexandrov one-point compactification is feasible, and the three-dimensional space R^3 can be replaced by a three-dimensional sphere S^3 embedded in a four-dimensional space R^4, using a one-to-one assignment of points $\mathbf{r}$ of space R^3 to the points $\mathbf{r}'$ of the 3-sphere S^3. In addition, a single point, a formal "north pole" $\mathbf{n}$ of the sphere S^3 corresponds to all formal points of infinite displacement from the center of mass of the molecule.

The molecular electron density can also be redefined on the sphere S^3 by simply taking,

$$\rho'(\mathbf{r}') = \rho(\mathbf{r}) \tag{68}$$

for any corresponding point pair $\mathbf{r}$ and $\mathbf{r}'$. Furthermore, the

$$\rho'(\mathbf{n}) = 0 \tag{69}$$

assignment completes the definition of electron density on the sphere S^3. The differentiability and the uniform exponential convergence of $\rho(\mathbf{r})$ to zero as $|\mathbf{r}|$ approaches infinity ensures that the function $\rho'(\mathbf{r}')$ on the sphere S^3 is continuous and infinitely differentiable (smooth) [50].

Analyticity of $\rho'(\mathbf{r}')$ almost everywhere on the sphere S^3 is simply inherited from $\rho(\mathbf{r})$. On the closed and bounded sphere S^3 there exists a unique correspondence between the electron density of a subsystem and the electron density of the complete system. This proves the following holographic electron density theorem [50]:

If $\rho'_d(\mathbf{r}')$ denotes the ground-state electron density over a subset d of three-dimensional sphere S^3, $S^3 \supset d$, where d has nonzero volume, then the conditions of the general theorem of analytic continuation are fulfilled, and $\rho'(\mathbf{r}')$ over the entire sphere S^3 is uniquely determined by $\rho'_d(\mathbf{r}')$ over the subdomain d; that is, if the restrictions $\rho'^a_d(\mathbf{r}')$ and $\rho'^b_d(\mathbf{r}')$ of densities $\rho'^a(\mathbf{r}')$ and $\rho'^b(\mathbf{r}')$, respectively, agree on the subdomain d,

$$\rho'^a_d(\mathbf{r}') = \rho'^b_d(\mathbf{r}') \tag{70}$$

for every point $\mathbf{r}'$ of d, then necessarily,

$$\rho'^{a}(\mathbf{r}') = \rho'^{b}(\mathbf{r}') \tag{71}$$

for every point $\mathbf{r}'$ of S^3.

That is, a piece of the ground-state electron density fully determines the ground state electron density of the entire system. As also follows from the Hohenberg–Kohn theorem, the ground-state energy E, and the ground-state wavefunction Ψ are uniquely determined by $\rho'(\mathbf{r}')$, hence they are also uniquely determined by the electron density $\rho'_d(\mathbf{r}')$ of the subsystem d. Consequently, the expectation values of all spin-free operators defined by the ground-state wavefunction Ψ are uniquely determined by the electron density $\rho'_d(\mathbf{r}')$ of any nonzero volume subsystem d.

The proof of the holographic electron density theorem on quantum mechanically correct, boundaryless molecular electron densities [50a] provides a justification of using local electron density shape information in various correlations with experimental data which often depend on global chemical properties of molecules, such as those employed in quantitative shape activity relations (QShAR) approaches to drug design and toxicological risk assessment [26, 27, 53]. It is expected that in practical applications the computational accessibility of global molecular information present in a larger molecular fragment is more favorable than that in a small fragment; nevertheless, the theoretical guarantee provided by the theorem implies that by improving accuracy and information-detection methods, the correlations with biochemical properties will also improve, even if based on small molecular fragments. The holographic electron density theorem also provides proof of rigorous relations between local and global symmetries, local and global chirality properties, as well as of other local and global symmetry deficiencies of molecular electron densities [50b].

IV. AN APPROACH TO LOW-DENSITY ANALYSIS BASED ON THE ADMA METHOD

The contributions from the low-density ranges of space have not received sufficient attention until recently, yet many of the interactions which determine the optimum nuclear arrangements and hence the shape of the overall electron density cloud are dependent on these low-density ranges. In small molecules, most of the low electron density can be found in the peripheral regions of the molecule, hence their role is somewhat limited in the actual bonding pattern and molecular conformation. By contrast, in large molecules, for example, in proteins, extensive low-density ranges occur well within the interior of the nuclear framework, and their role in the folding pattern is far from negligible. In an earlier study the contribution of the low electron density regions to chemical bonding was compared to a formal, low-viscosity "glue," motivating the terminology "low-density glue" (LDG) bonding [56].

For the study of LDG bonding in macromolecules, reasonable quality macromolecular electron densities are required, which were not available until recently. The main purpose of quantum crystallography methods is the construction of quantum chemical representations of molecular electron densities based on experimental electron density information. Within the framework of the quantum crystallography approaches, such as those based on kernel projection matrices [6], the accuracy of the information transferred from experiments to a quantum crystallographic description is ultimately determined by the accuracy of experimental electron density data. This accuracy has been improved considerably in recent years, although the accurate detection of local electron density regions corresponding to H nuclei is not always satisfactory in large molecules, such as proteins. In a certain sense the fully theoretical approach, such as the AFDF approach, has some advantages in this regard since the AFDF methods, in particular the adjustable density matrix assembler (ADMA) method, provide effective computational tools for the analysis of macromolecular electron densities, even in the ranges of low densities usually not available by experimental techniques [26–37].

The AFDF fragment density matrices given by Eq. 6 can be computed for each fragment of a macromolecule M, based on a high-quality *ab initio* computation of a small molecule. For each nuclear family f_k contained within an actual electron density fragment F_k of the macromolecule M (called the *target molecule*), a small *parent molecule* M_k is designed, where M_k contains the same nuclear family f_k as well as the complete local arrangement of other nuclei and the associated electron density which surrounds the nuclear family f_k within the target macromolecule M. In fact, a "custom-made" small parent molecule M_k with a suitably chosen "coordination shell" about the actual nuclei of the fragment is used to compute the fragment density matrix of fuzzy electron density fragment F_k of the macromolecule M. Note that, due to the density matrix decomposition scheme (Eqs. 4–6) approximately half of the local interactions of this fragment with its surroundings is included in the fragment density F_k, hence the same information is present in the fragment density matrix $\mathbf{P}^k$ if the basis set is specified. The other half of interactions is supplied by those fragments $F_{k'}$ and their fragment density matrices $\mathbf{P}^{k'}$, which are computed using other parent molecules $M_{k'}$ with central regions of the actual other fragments $F_{k'}$, where the current density fragment F_k and the associated nuclear family f_k appear in the peripheral region—that is, in the "coordination shell" merely as a part of the surroundings for nuclei $f_{k'}$.

The fragment density matrix $\mathbf{P}^k$ is determined by the fuzzy density fragmentation of the density matrix of the small parent molecule M_k, resulting in the matrix $\mathbf{P}^k$ of the fuzzy density fragment $\rho^k(\mathbf{r}, K)$ of nuclear configuration K corresponding to the nuclear set f_k. By repeating this procedure for each nuclear family f_k of M, a series of fragment density matrices,

$$\mathbf{P}^1(\mathbf{r}), \mathbf{P}^2(\mathbf{r}), \dots, \mathbf{P}^k(\mathbf{r}), \dots \mathbf{P}^m(\mathbf{r}) \tag{72}$$

as well as a set of the fuzzy density fragments,

$$\rho^1(\mathbf{r}, K), \rho^2(\mathbf{r}, K), \dots, \rho^k(\mathbf{r}, K), \dots, \rho^m(\mathbf{r}, K) \tag{73}$$

are obtained from a set of m small "parent" molecules:

$$M_1, M_2, \ldots, M_k, \ldots, M_m \tag{74}$$

Fragment density matrices $\mathbf{P}^k(\mathbf{r})$, $k = 1, 2, \ldots m$, can be combined and used to construct the density matrix $\mathbf{P}(\mathbf{r}, K)$ of the large target molecule M. At the same time, the fragment density matrices $\mathbf{P}^k(\mathbf{r})$ and the fragment densities $\rho^k(\mathbf{r}, K)$ can be used to study the local features of the macromolecule M, using one of the electron density shape analysis methods [57].

The MEDLA (molecular electron density "loge" assembler, or molecular electron density "Lego" assembler) method is a numerical technique, based on the AFDF principle, on a numerical electron density fragment databank, and on the direct assembly of numerical electron density fragments into a macromolecular electron density.

The ADMA method is the more advanced of the AFDF methods; it does not require a numerical electron density databank. Instead various sets of compatible fragment density matrices are generated. Since the ADMA method generates a macromolecular density matrix, all quantum chemical property calculations which are based on density matrices (e.g. approximate force calculations for the study of conformational rearrangements) are now extended via the ADMA approach to macromolecules, such as proteins. The additive fuzzy density fragmentation approach has led to the computation of the first, *ab initio* quality electron densities for proteins, including crambin, bovine insulin, the gene-5 protein (g5p) of bacteriophage M13, the HIV-1 protease monomer of 1564 atoms, and the proto-oncogene tyrosine kinase protein 1ABL containing 873 atoms [30–32, 58].

Besides the possibilities of the computation of all molecular properties expressible in terms of density matrices, the ADMA method has other advantages. For example, if electron density computations are compared, then the accuracy of the electron density obtained using the ADMA macromolecular density matrix $\mathbf{P}(\varphi(K))$ corresponds to the ideal MEDLA result that could be obtained using an infinite resolution numerical grid. The memory requirements of the ADMA method is also substantially lower than that of the numerical MEDLA method since it takes much less memory to store density matrices than three-dimensional numerical grids of electron densities, especially if reasonably detailed electron densities are required.

In order to simplify the construction of the macromolecular density matrix $\mathbf{P}(\varphi(K))$, it is useful if the fragment density matrices $\mathbf{P}^k(\varphi(K_k))$ obtained from small parent molecules M_k fulfill a set of mutual compatibility requirements, as follows:

(1)   The AO basis sets of all the fragment density matrices $\mathbf{P}^k(\varphi(K_k))$ are defined in local coordinate systems which have axes that are parallel and have matching orientations with the axes of the reference coordinate system defined for the macromolecule M.

(2) In the fragmentation scheme the nuclear families of both the target and the parent molecules are compatible in the following sense: each parent molecule M_k may contain only complete nuclear families from the sets of nuclear families $f_1, f_2, \ldots, f_k, \ldots, f_m$, present in the target macromolecule M. Note that, within each parent molecule additional nuclei may be involved which link to dangling bonds at the peripheries of the parent molecules M_k.

The local coordinate systems always can be reoriented using a simple similarity transformation of the fragment density matrix $\mathbf{P}^k(\varphi(K_k))$, based on a suitable orthogonal transformation matrix $\mathbf{T}^{(k)}$ of the AO sets. The nuclear families f_k for the various fragments within the macromolecule M and within the "coordination shells" of parent molecules M_k can always be chosen so that the second condition is also fulfilled.

If these two mutual compatibility conditions are fulfilled, then the approach is referred to as the mutually compatible AFDF approach, or in short, the MC–AFDF approach.

For the actual implementation of the MC–AFDF ADMA approach, extensive index manipulations are required. A brief review of the index conventions is listed below, following the original notations and terminology in refs. 35, 37.

A quantity $c_{k'k}$ is defined for each pair $(f_k, f_{k'})$ of nuclear families:

$$c_{k'k} = \begin{cases} 1, \text{ if nuclear family } f_{k'} \text{ is present in parent molecule } M_k \\ 0 \text{ otherwise,} \end{cases} \quad (75)$$

where the number of AOs in the nuclear family f_k of the target macromolecule M denoted by n_k.

If the serial number b of an AO $\varphi(\mathbf{r})$ in the AO set,

$$\{\varphi_{a,k'}(\mathbf{r})\}_{a=1}^{n_{k'}} \quad (76)$$

of nuclear a family $f_{k'}$ is relevant, then this AO basis function is referred to as $\varphi_{b,k'}(\mathbf{r})$.

If the serial index j of the same AO $\varphi(\mathbf{r})$ in the basis set,

$$\{\varphi_i^k(\mathbf{r})\}_{i=1}^{n_{pk}} \quad (77)$$

of the k-th fragment density matrix $\mathbf{P}^k(\varphi(K_k))$ is relevant, where the total number of these AOs is n_{p^k},

$$n_{p^k} = \sum_{k'=1}^{m} c_{k'k}\, n_{k'} \quad (78)$$

then the notation $\varphi_j^k(\mathbf{r})$ is used.

If the serial index y of the same AO $\varphi(\mathbf{r})$ in the AO set,

$$\{\varphi_x(\mathbf{r})\}_{x=1}^{n} \tag{79}$$

of the density matrix $\mathbf{P}(K)$ of the target macromolecule M is relevant, then the AO is denoted by $\varphi_y(\mathbf{r})$, where for each AO $\varphi_{a,k'}(\mathbf{r}) = \varphi_i^k(\mathbf{r}) = \varphi_x(\mathbf{r})$ the index x is determined by the index a in family k' as follows:

$$x = x(k', a, f) = a + \sum_{b=1}^{k'-1} n_b \tag{80}$$

In this notation the last entry f in $x(k',a,f)$ indicates that k' and a refer to a family of nuclei.

Furthermore, three quantities are introduced for each index k and nuclear family $f_{k''}$ for which $c_{k''k} \neq 0$:

$$a_k'(k'', i) = i - \sum_{b=1}^{k''} n_b c_{bk} \tag{81}$$

$$k' = k'(i, k) = \min \{k'': a_k'(k'', i) \leq 0\} \tag{82}$$

and

$$a_k(i) = a_k'(k', i) + n_{k'} \tag{83}$$

These auxiliary quantities facilitate the determination of the index x from the element index i and serial index k of fragment density matrix $\mathbf{P}^k(\varphi(K_k))$. In fact, the index $x = x(k, i, P)$ of an AO basis function in the density matrix $\mathbf{P}(K)$ of target molecule M depends on indices i and k and can be expressed using index k' and the function $x(k', a, f)$:

$$x = x(k, i, P) = x(k', a_k(i), f) \tag{84}$$

The last entry P in the index function $x(k, i, P)$ indicates that indices k and i refer to the fragment density matrix $\mathbf{P}^k(\varphi(K_k))$.

These index assignments also help to identify the nonzero elements of the rather sparse macromolecular density matrix, based on the also rather sparse fragment density matrices. By looping over only the nonzero elements of each fragment density matrix $\mathbf{P}^k(\varphi(K_k))$, the macromolecular density matrix $\mathbf{P}(K)$ is assembled iteratively:

$$P_{x(k,i,P),y(k,j,P)}(K) = P_{x(k,i,P),y(k,j,P)}(K) + P_{ij}^k(K_k) \tag{85}$$

If the parent molecules M_k are confined to some limited size, then the entire iterative procedure depends linearly on the number of fragments and on the size of the target macromolecule M, a feature that provides considerable savings of computer time when compared to any of the more conventional *ab initio* type methods which have computer time requirements that grows with the third or fourth power of the number of electrons.

The sparse nature of the macromolecular density matrix $P(K)$ simplifies its determination, its storage, and all the subsequent computations. If the macromolecular AO basis is stored using a list of appropriate indices referring to a standard list of AO basis functions, then the macromolecular electron density can be computed according to Eq. 3.

One advantage of the ADMA approach is the possibility of calculating those low-density parts of a macromolecular electron distribution that is not well described experimentally. Most of the experimental data are relevant to the high-density regions; for example, X-rays scatter predominantly by the high electron density atomic neighborhoods. As long as our interest is limited to the determination of the locations of the nuclei, especially the heavy nuclei, this is not a serious disadvantage. However, for an assessment of chemical reactivity, or for the analysis of local shape features of the electron density where fine details may influence the feasibility of an enzyme–substrate interaction, the low-density ranges are of importance. Whereas the shape of the high-density ranges is commonly considered the shape of the molecule, since these ranges are usually associated with individual atomic regions and with the internuclear regions commonly regarded the chemical bonds, nevertheless, the shapes of the low-density ranges often influence the initial stages of chemical interactions and eventual chemical reactions between molecules. These latter ranges also describe interactions that are of lower energy; however, if there are large low-density regions of the space with low-energy interactions, their cumulative effect can be significant. Within a folded macromolecule, such as a globular protein, there are extensive regions with electron density lower than those typically found in the regions of formal chemical bonds; nevertheless, these electron density contributions collectively have an important influence on the folding pattern, hence on the global shape of the macromolecule.

In biopolymers and other, large, folded chain molecules the low-density components of self-interactions contribute to the structural features of the molecule. These interactions affect the molecular shape in a fundamentally different way than formal chemical bonds which are usually assigned to pairs of atomic nuclei, hence they have well-defined directional properties. The low-density contributions are less directional, they are more widely distributed within the macromolecule, and they cannot be assigned to individual atom pairs. These low-density contributions are better described as interactions between larger structural elements; for example, as an interaction between an aromatic ring and another functional group, or one between two helical fragments. The mutual interpenetration of the fuzzy, peripheral electron density clouds can be used to detect these low-density interactions. In this context, one may consider the low-density cloud as a dilute "glue," with a non-negligible role in mediating the interactions between the various molecular structural entities, and in the stabilization of a mutual arrangement of larger structural elements of a macromolecule.

Some of the fundamental relations of fuzzy set theory and the actual formulation of fuzzy set methods [59–63] appear ideally suitable for the description of the fuzzy, low-density electron distributions [64]. A molecular electron density exhibits a natural fuzziness, in part due to the quantum mechanical uncertainty relation. Evidently, a molecular electron distribution, when considered as a formal molecular body, is inherently fuzzy without any well-defined boundaries. A molecule does not end abruptly, since the electron density is gradually decreasing with distance from the nearest nucleus, and becomes zero, in the strict sense, at infinity. In fact, the same consideration has been the motivation for the compactification technique described in the proof of the holographic electron density theorem. For a correct description of molecules, the models used for electron densities must exhibit the natural fuzziness of quantum chemical electron distributions [64].

In ordinary set theory a point either does or does not belong to a given set. That is, value of the membership function of a point x in the set is either 1 (when the point belongs to the set) or it is zero (if the point does not belong to the set). In fuzzy set theory, there is usually a continuous range of the possible "degree" of belonging to a fuzzy set, for example, the membership of a point x in a fuzzy set A may well be 50%, indicating that this point is not a full-fledged member of the set, but it is not completely excluded either. This fact is expressed by defining the value of the membership function of point x in fuzzy set A as 0.5.

By assigning a formal membership function to each point $\mathbf{r}$ of the ordinary three-dimensional space R^3 describing the "degree" of belonging of this point $\mathbf{r}$ to any one of the molecular electronic charge clouds, a fuzzy set description is obtained.

We may select a spatial domain D containing the nuclei of molecule M, and choose the value ρ_{max} as the maximum value of the electron density within this domain D:

$$\rho_{max} = \max \{\rho(\mathbf{r}), \mathbf{r} \in D\} \tag{86}$$

It is natural to decide the relative importance of various density values with reference to this maximum value ρ_{max}. With respect to ρ_{max}, a fuzzy membership function $\mu_M(\mathbf{r})$ is defined for all points $\mathbf{r}$ of the space R^3, expressing the "degree" of $\mathbf{r}$ belonging to the molecule M:

$$\mu_M(\mathbf{r}) = \rho(\mathbf{r})/\rho_{max} \tag{87}$$

Of course, if the molecule M is not isolated, that is other molecules also have electron density contribution to a selected point $\mathbf{r}$, then this point $\mathbf{r}$ may partially belong to several different molecules to various degrees. This fact can also be expressed easily using fuzzy membership functions, as follows.

In the general case, if the electron densities of m molecules,

$$M_1, M_2, \ldots M_i, \ldots M_m \tag{88}$$

of some molecular family L have non-negligible contributions to the total electron density at some point $\mathbf{r}$, and if the individual contributions are,

$$\rho_{M_1}(\mathbf{r}), \rho_{M_2}(\mathbf{r}), \ldots \rho_{M_i}(\mathbf{r}), \ldots \rho_{M_m}(\mathbf{r}) \tag{89}$$

respectively, then these individual electron densities $\rho_{M_i}(\mathbf{r})$ can be taken to represent the "share" of each molecule M_i in the total electron density generated by the entire molecular family L. These individual electron densities also define appropriate fuzzy membership functions for each point $\mathbf{r}$ with respect to each molecule of family L, as determined by the relative magnitudes of the individual electron densities.

Of course, the "share" $\rho_{M_i}(\mathbf{r})$ of an individual molecule M_i, being a part of the total electron density can also be considered separately, in the absence of all other molecules, as an individual fuzzy electron density cloud.

Take a domain D_{M_i} of the space containing all the nuclei of molecule M_i and no additional nuclei, and take the maximum value $\rho_{max,i}$ of the electron density $\rho_{M_i}(\mathbf{r})$ within this domain, that is, define:

$$\rho_{max,i} = \max\{\rho_{M_i}(\mathbf{r}), \mathbf{r} \in D_{M_i}\} \tag{90}$$

If D_{M_i} is chosen as a closed set, then there must exist a point $\mathbf{r}_{max,i}$ within D_{M_i} where this maximum density value $\rho_{max,i}$ is realized for the given molecule M_i:

$$\rho_{M_i}(\mathbf{r}_{max,i}) = \rho_{max,i} \tag{91}$$

If we consider only a single molecule M_i in the absence of all other molecules, then a fuzzy membership function for points $\mathbf{r}$ of the space R^3 belonging to molecule M_i is defined as:

$$\mu_{M_i}(\mathbf{r}) = \rho_{M_i}(\mathbf{r})/\rho_{max,i} \tag{92}$$

An entirely different case is obtained if all other molecules of the family L are also considered to be present.

In this case each molecule M_i, $i = 1, 2, \ldots i, \ldots m$, may have some partial claim for each point $\mathbf{r}$ of the space R^3, consequently, the degree of belonging of each point $\mathbf{r}$ to a given molecule M_i is determined collectively by the electron density contributions $\rho_{M_1}(\mathbf{r}), \rho_{M_2}(\mathbf{r}), \ldots \rho_{M_i}(\mathbf{r}), \ldots \rho_{M_m}(\mathbf{r})$ of all m molecules.

In the nonisolated case, the total electron density $\rho_L(\mathbf{r})$ of the entire molecular family $M_1, M_2, \ldots M_j, \ldots M_m$ taken at point $\mathbf{r}$ can be taken as a local reference:

$$\rho_L(\mathbf{r}) = \Sigma_j \rho_{M_j}(\mathbf{r}) \tag{93}$$

The individual fuzzy membership functions $\mu_{M_i,L}(\mathbf{r})$ for each molecule are defined relative to this total density. In this case the fuzzy membership function $\mu_{M_i,L}(\mathbf{r})$ for points $\mathbf{r}$ of the space belonging to molecule M_i of the family L is defined as:

$$\mu_{M_i,L}(\mathbf{r}) = \mu_{M_i}(\mathbf{r}) \, [\rho_{max,i}/\rho_L(\mathbf{r}_{max,i})] \qquad (94)$$

The scaling factor $[\rho_{max,i}/\rho_L(\mathbf{r}_{max,i})]$ is included in this definition in order to provide proportionality among the actual density contributions from various molecules M_i.

The same fuzzy membership function $\mu_{M_i,L}(\mathbf{r})$ can also be defined with direct reference to the maximum total density of the combined family:

$$\mu_{M_i,L}(\mathbf{r}) = \mu_{M_i}(\mathbf{r}) \, [\rho_{max,i}/\rho_L(\mathbf{r}_{max,i})]$$

$$= [\rho_{M_i}(\mathbf{r})/\rho_{max,i}] \, [\rho_{max,i}/\rho_L(\mathbf{r}_{max,i})]$$

$$= \rho_{M_i}(\mathbf{r})/\rho_L(\mathbf{r}_{max,i}) \qquad (95)$$

The fuzzy electron density membership functions described above are useful for the characterization of the interactions and mutual interpenetration of individual, fuzzy electron density clouds within a molecular family.

If several molecules share some common regions of the space R^3, then the fuzzy membership functions $\mu_{M_i,L}(\mathbf{r})$ provide measures of their relative importance at each point $\mathbf{r}$.

Two additional concepts of electron density modeling and shape analysis are of special relevance to the study of low-density ranges: the molecular isodensity contours, MIDCOs, denoted by $G(K, a)$, and the associated density domains $DD(K, a)$ [57]. In the parentheses, an electron density threshold a is specified, besides the actual nuclear configuration K. A MIDCO $G(K, a)$ is the collection of all those points $\mathbf{r}$ of the three-dimensional space where the electron density $\rho(K, \mathbf{r})$ of the molecule M of conformation K is equal to the threshold a:

$$G(K, a) = \{\mathbf{r}: \rho(K, \mathbf{r}) = a\} \qquad (96)$$

The corresponding density domain $DD(K, a)$ is the collection of all points $\mathbf{r}$ where the electron density $\rho(K, \mathbf{r})$ is greater than or equal to the threshold a:

$$DD(K, a) = \{\mathbf{r}: \rho(K, \mathbf{r}) \geq a\} \qquad (97)$$

The boundary surface of the density domain $DD(K, a)$ is the MIDCO $G(K, a)$.

The following formalism was first proposed for the study of electron density ranges including those density values a_m at which some individual density domains of a molecule (initially taken with a high density threshold a) merge into a common domain (at some lower threshold a').

For example, one may select a threshold value a_m that corresponds to the onset of some merger of MIDCOs $G(K, a_m)$. One may, in fact, generate three MIDCOs,

$G(K, a_m + \Delta a)$, $G(K, a_m)$, and $G(K, a_m - \Delta a)$ for the selected threshold value a_m and density increment Δa for a family R of nuclear configurations K. In several actual test calculations a value of the threshold a_m falling within the range of [0.003 a.u., 0.005 a.u.], and a density increment of $\Delta a \sim 0.001$ a.u. were found suitable.

The corresponding "low-density glue" (LDG) part of the electron distribution can be identified with the object:

$$LDG(K, a_m, \Delta a) = DD(K, a_m - \Delta a) \backslash DD(K, a_m + \Delta a) \qquad (98)$$

Here the density domains $DD(K, a_m - \Delta a)$ and $DD(K, a_m + \Delta a)$ correspond to the MIDCOs $G(K, a_m - \Delta a)$ and $G(K, a_m + \Delta a)$, respectively.

Note that, the low-density glue formalism and the objects $LDG(K, a_m, \Delta a)$ are not restricted to the choices of thresholds a_m and density ranges Δa discussed above, in fact, in most instances the more interesting structural features are exhibited by LDG sets of lower a_m values.

In the terminology of fuzzy set theory [60–63], the object $LDG(K, a_m, \Delta a)$ can be generated in terms of two α-cuts, using the two values $\alpha = a_m - \Delta a$ and $\alpha = a_m + \Delta a$, respectively. For large molecules, the set $LDG(K, a_m, \Delta a)$ usually contains at least one hollow interior cavity, corresponding to the high-density ranges of the molecule near the nuclear neighborhoods and the high-density chemical bonds.

These $LDG(K, a_m, \Delta a)$ objects are usually complicated three-dimensional bodies with intricate shape features which seem to be well suited for a topological shape analysis. Most $LDG(K, a_m, \Delta a)$ objects obtained from a globular protein are multiply connected, as a consequence of the gradual mergers of electron density clouds due to nonbonded interactions, as the density threshold a_m is gradually lowered.

In an earlier study [56] the focus was placed on the simplest type of connectedness property which appeared the most relevant: arcwise multiple connectedness, or 1-connectedness. For the analysis of 1-connectedness the one-dimensional homotopy group, also called the fundamental group of the object, provides a concise algebraic–topological description. A summary of the mathematical background of homotopy groups and a discussion of their chemical applications in the study of potential energy hypersurfaces of chemical reactions and conformational rearrangements can be found in ref. 51.

The fundamental group $\Pi_1(LDG(K, a_m, \Delta a))$ of the "low-density glue" part of the macromolecular electron density is the one-dimensional homotopy group of the object $LDG(K, a_m, \Delta a)$, describing the relations between equivalence classes of loops within $LDG(K, a_m, \Delta a)$, where within each equivalence class all loops are continuously deformable into one another without any part of a loop leaving the object $LDG(K, a_m, \Delta a)$. The product operation between loops is the continuation of one loop by another, whereas the product operation between any two equivalence classes results in an equivalence class that contains a product loop of two loops, one from each of the two equivalence classes.

The presence of internal cavities within the object LDG(K, a_m, Δa) influences the 2-connectedness properties of LDG(K, a_m, Δa) that is manifested in the contractibility of (possibly distorted) spherical surfaces into a single point. The two-dimensional homotopy group Π_2(LDG(K, a_m, Δa)) of the LDG(K, a_m, Δa) low-density glue region of the macromolecular electron density describes the relations between equivalence classes of topological spheres within LDG(K, a_m, Δa), where within each equivalence class all topological spheres are continuously deformable into one another without any part of a topological sphere leaving the object LDG(K, a_m, Δa). Consider a simple example of an LDG(K, a_m, Δa) low-density glue region where there is a single internal cavity. Evidently, some but not all of the topological spheres within LDG(K, a_m, Δa) can be contracted to a point. One such topological sphere that is not contractible to a point is the inner wall of the cavity: clearly, this topological sphere cannot be contracted to a single point without "leaving" the body of the LDG(K, a_m, Δa) low-density glue region of the macromolecular electron density. In this example the two-dimensional homotopy group Π_2(LDG(K, a_m, Δa)) of the object LDG(K, a_m, Δa) is a relatively simple group, a free group with a single generator; note, however, that for more complicated LDG(K, a_m, Δa) objects of more intricate internal structure the two-dimensional homotopy groups Π_2(LDG(K, a_m, Δa)) are also more complicated.

Similarly, in actual LDG(K, a_m, Δa) bodies obtained from globular proteins there are many different equivalence classes of loops with homotopically different contractibility properties within LDG(K, a_m, Δa), hence the fundamental groups Π_1(LDG(K, a_m, Δa)) usually have a large number of generators.

When comparing the electron densities of globular proteins (e.g. if proteins of a given biological function in several different animal species are compared) then the similarities in the LDG(K, a_m, Δa) contributions to the macromolecular structures can be expressed in terms of the algebraic–topological descriptors of the LDG(K, a_m, Δa) bodies. Alternatively, it is possible to use the same algebraic–topological method for comparing different conformations of the same macromolecule.

For example, one may compare the fundamental groups Π_1(LDG(K, a_m, Δa)) of two proteins or of two different folding patterns K_a and K_b of the same protein. If the two fundamental groups are isomorphic,

$$\Pi_1(\text{LDG}(K_a, a_m, \Delta a)) = \Pi_1(\text{LDG}(K_b, a_m, \Delta a)) \tag{99}$$

then the corresponding low-density bonding contributions, LDG(K_a, a_m, Δa) and LDG(K_b, a_m, Δa) are similar in some of their essential shape features. We may formalize this observation by introducing the concept of LDG Π_1-similarity, expressed by the fact that the two conformations, K_a and K_b are equivalent by their 1-homotopical LDG structure, expressed in the notation:

$$K_a \sim [\Pi_1 - \text{LDG}(K, a_m, \Delta a)] \sim K_b \tag{100}$$

By a similar argument, if for two proteins or for two different folding patterns K_a and K_b of the same protein the two two-dimensional homotopy groups of their respective low-density bonding contributions, $LDG(K_a, a_m, \Delta a)$ and $LDG(K_b, a_m, \Delta a)$ are isomorphic,

$$\Pi_2(LDG(K_a, a_m, \Delta a)) = \Pi_2(LDG(K_b, a_m, \Delta a)) \qquad (101)$$

then the corresponding low-density bonding contributions, $LDG(K_a, a_m, \Delta a)$ and $LDG(K_b, a_m, \Delta a)$ are similar according to a different aspect of their shapes, expressing the presence and topological distribution of their internal cavities. By introducing the concept of LDG Π_2-similarity, the fact that the two conformations, K_a and K_b are equivalent by their 2-homotopical LDG structure is expressed in the notation:

$$K_a \sim [\Pi_2 - LDG(K, a_m, \Delta a)] \sim K_b \qquad (102)$$

If the two homotopy groups are not isomorphic, then the group–subgroup relations among all possible LDG homotopy groups still provides a rich characterization.

Analogous characterization of the similarities of the low-density glue bodies is possible using the homology groups of algebraic topology instead of the homotopy groups. A brief mathematical introduction to homology groups of algebraic topology and some details of their chemical applications in the study of molecular shape, in particular, of the shape of fuzzy electron density clouds can be found in ref. 57. The one- and two-dimensional homology groups, $H_1(LDG(K, a_m, \Delta a))$ and $H_2(LDG(K, a_m, \Delta a))$, of the low-density glue region $LDG(K, a_m, \Delta a)$ of a macromolecule express the relations between the possible subdivisions and the boundaries of subdivisions of these objects, in the case of one- and two-dimensional boundaries, respectively.

If for two different protein conformations K_a and K_b the two one-dimensional homology groups are isomorphic,

$$H_1(LDG(K_a, a_m, \Delta a)) = H_1(LDG(K_b, a_m, \Delta a)) \qquad (103)$$

then the corresponding low-density bonding contributions, $LDG(K_a, a_m, \Delta a)$ and $LDG(K_b, a_m, \Delta a)$ have 1-homologically equivalent shape features. This similarity is expressed as a homological equivalence of the two $LDG(K_a, a_m, \Delta a)$ and $LDG(K_b, a_m, \Delta a)$ objects, manifested in the LDG H_1-similarity of the two conformations, K_a and K_b, as expressed in the notation:

$$K_a \sim [H_1 - LDG(K, a_m, \Delta a)] \sim K_b \qquad (104)$$

Similarly, if for two different protein conformations K_a and K_b the two two-dimensional homology groups are isomorphic, that is, if,

$$H_2(LDG(K_a, a_m, \Delta a)) = H_2(LDG(K_b, a_m, \Delta a)) \tag{105}$$

then the corresponding low-density glue contributions of the macromolecules, $LDG(K_a, a_m, \Delta a)$ and $LDG(K_b, a_m, \Delta a)$ have 2-homologically equivalent shape features. Again, this similarity can be expressed as a 2-homological equivalence of the two $LDG(K_a, a_m, \Delta a)$ and $LDG(K_b, a_m, \Delta a)$ low-density glue objects. This LDG H2-similarity of the two macromolecular conformations, K_a and K_b, is expressed by the notation:

$$K_a \sim [H_2 - LDG(K, a_m, \Delta a)] \sim K_b \tag{106}$$

Additional, more detailed characterization is obtained by the group–subgroup relations among all possible LDG homology groups of macromolecular electron densities.

V. CLOSING COMMENTS AND SUMMARY

We anticipate two advantages of using the more realistic electron densities obtained by the AFDF methods. More reliable theoretical electronic charge densities calculated for each assumed nuclear geometry in the course of the iterative structure refinement process will improve the reliability of comparisons with the experimental diffraction pattern. In particular, AFDF electron densities are expected to serve as more sensitive and more reliable criteria for accepting or rejecting an assumed structure than the locally spherical or possibly elliptical electron density models used in the conventional approach. We also expect that the more accurate density representations within the QCR-AFDF framework will facilitate a more complete utilization and interpretation of the structural information contained in the observed X-ray diffraction pattern.

The density functional approach is used for a study of the fundamental properties of electron densities, and the role of subsystems in determining the electron density of a complete system. The "holographic electron density theorem" is discussed that provides a more general justification of the determining role of subsystems than earlier approaches.

Some of the most essential shape features of the low-density regions of macromolecules can be captured by the homotopy groups and the homology groups of algebraic topology, without any recourse to truncations of the objects according to some geometric features, such as local curvature and relative convexity properties. A topological analysis of the "low-density glue" contributions to macromolecular structure provides a tool for the evaluation of a special aspect of similarities among macromolecules.

ACKNOWLEDGMENTS

The author is grateful for stimulating discussions with Prof. Jerome Karle, and Prof. Lou Massa, of the Laboratory for the Structure of Matter, Naval Research Laboratory, Washington, D.C. 20375-5341 USA. This work was supported in part by NSERC of Canada.

REFERENCES AND NOTES

1. Sands, D. E. *Introduction to Crystallography*; Benjamin: New York, 1969.
2. Buerger, M. J. *Contemporary Crystallography*; McGraw-Hill: New York, 1970.
3. Woolfson, M. M. *An Introduction to X-ray Crystallography*; Cambridge University Press: Cambridge, 1970.
4. Glusker, J. P.; Trueblood, K. N. *Crystal Structure Analysis*; Oxford University Press: New York, 1972.
5. Karle, J. *Proc. Natl. Acad. Sci. USA* **1991**, *88*, 10099.
6. (a) Massa, L.; Huang, L.; Karle, J. *Int. J. Quantum Chem., Quant. Chem. Symp.* **1995**, *29*, 371.
   (b) Huang, L.; Massa, L.; Karle, J. *Int. J. Quantum Chem., Quant. Chem. Symp.* **1996**, *30*, 1691.
7. Löwdin, P.-O. *Phys. Rev.* **1955**, *97*, 1474.
8. McWeeny, R. *Rev. Mod. Phys.* **1960**, *32*, 335.
9. Coleman, A. J. *Rev. Mod. Phys.* **1963**, *35*, 668.
10. Clinton, W. L.; Galli, A. J.; Massa, L. J. *Phys. Rev.* **1969**, *177*, 7.
11. Clinton, W. L.; Galli, A. J.; Henderson, G. A.; Lamers, G. B.; Massa, L. J.; Zarur, J. *Phys. Rev.* **1969**, *177*, 27.
12. Clinton, W. L.; Massa, L. J. *Int. J. Quantum Chem.* **1972**, *6*, 519.
13. Clinton, W. L.; Massa, L. J. *Phys. Rev. Lett.* **1972**, *29*, 1363.
14. Clinton, W. L.; Frishberg, C.; Massa, L. J.; Oldfield, P. A. *Int. J. Quantum Chem. Quantum Chem. Symp.* **1973**, *7*, 505.
15. Henderson, G. A.; Zimmermann, R. K. *J. Chem. Phys.* **1976**, *65*, 619.
16. Tsirel'son, V. G.; Zavodnik, V. E.; Fonichev, E. B.; Ozerov, R. P.; Kuznetsolirez, I. S. *Kristallogr.* **1980**, *25*, 735.
17. Frishberg, C.; Massa, L. J. *Phys. Rev. B* **1981**, *24*, 7018.
18. Frishberg, C.; Massa, L. J. *Acta Cryst.* **1982**, *A38*, 93.
19. Massa, L. J.; Goldberg, M.; Frishberg, C.; Boehme, R. F.; LaPlaca, S. J. *Phys. Rev. Lett.* **1985**, *55*, 622.
20. Frishberg, C. *Int. J. Quantum Chem.* **1986**, *30*, 1.
21. Cohn, L.; Frishberg, C.; Lee, C.; Massa, L. J. *Int. J. Quantum Chem., Quantum Chem. Symp.* **1986**, *19*, 525.
22. Massa, L. J. *Chemica Scripta* **1986**, *26*, 469.
23. Boehme, R. F.; LaPlaca, S. J. *Phys. Rev. Lett.* **1987**, *59*, 985.
24. Tanaka, K. *Acta Cryst.* **1988**, *A44*, 1002.
25. Aleksandrov, Y. Y.; Tsirel'son, V. G.; Resnik, I. M.; Ozerov, R. P. *Phys. Status Solidi* **1989**, *B155*, 201.
26. Mezey, P. G. In *Molecular Similarity in Drug Design*; Dean, P. M., Ed.; Chapman & Hall-Blackie: Glasgow, U.K., 1995.
27. Mezey, P. G. In *Topics in Current Chemistry*; Sen, K., Ed.; Springer-Verlag: Heidelberg, 1995.
28. Walker, P. D.; Mezey, P. G. *Program MEDLA 93*; Mathematical Chemistry Research Unit, University of Saskatchewan: Saskatoon, Canada, 1993.
29. Walker, P. D.; Mezey, P. G. *J. Amer. Chem. Soc.* **1993**, *115*, 12423.
30. Walker, P. D.; Mezey, P. G. *J. Amer. Chem. Soc.* **1994**, *116*, 12022.
31. Walker, P. D.; Mezey, P. G. *Canad. J. Chem.* **1994**, *72*, 2531.

32. Walker, P. D.; Mezey, P. G. *J. Math. Chem.* **1995**, *17*, 203.
33. Walker, P. D.; Mezey, P. G. *J. Comput. Chem.* **1995**, *16*, 1238.
34. Mezey, P. G. In *Computational Chemistry: Reviews and Current Trends*; Leszczynski, J., Ed.; World Scientific: Singapore, 1996, Vol. 1, pp. 109–137.
35. Mezey, P. G. *J. Math. Chem.* **1995**, *18*, 221.
36. Mezey, P. G. *Program ALDA 95*; Mathematical Chemistry Research Unit, University of Saskatchewan: Saskatoon, Canada, 1995.
37. Mezey, P. G. *Program ADMA 95*; Mathematical Chemistry Research Unit, University of Saskatchewan: Saskatoon, Canada, 1995.
38. (a) Mulliken, R. S. *J. Chem. Phys.* **1955**, *23*, 1833, 1841, 2338, 2343. (b) Mulliken, R. S. *J. Chem. Phys.* **1962**, *36*, 3428.
39. Mulliken, R. S. Unpublished comments. Topic of the discussions held at the crater of the volcano Teide, Tenerife, Canary Islands, June 20, 1976, between Prof. Mulliken and the author, on the fundamentals of local components of molecular wavefunctions, Mulliken's population analysis, and various atomic charge models based on overlap integrals. These discussions had a motivating role in the later development of the AFDF methods MEDLA and ADMA, both based on the realization by the author that Mulliken's population analysis *without integration* provides a simple, additive fuzzy electron density fragmentation (AFDF) scheme.
40. Hartree, D. R. *Proc. Cambridge Phil. Soc.* **1928**, *24*, 111, 426, *ibid.* **1929**, *25*, 225, 310.
41. Fock, V. Z. *Physik* **1930**, *61*, 126.
42. Roothaan, C. C. *Rev. Mod. Phys.* **1951**, *23*, 69, *ibid.* **1960**, *32*, 179.
43. Hall, G. G. *Proc. Roy. Soc. London* **1951**, *A205*, 541.
44. Hohenberg, P.; Kohn, W. *Phys. Rev.* **1964**, *136*, B864.
45. Levy, M. *Proc. Natl. Acad. Sci. USA* **1979**, *76*, 6062.
46. Levy, M. *Bull. Amer. Phys. Soc.* **1979**, *24*, 626.
47. Levy, M. *Phys. Rev. A* **1982**, *26*, 1200.
48. Levy, M. *Adv. Quant. Chem.* **1990**, *21*, 69.
49. Riess, J.; Münch, W. *Theor. Chim. Acta* **1981**, *58*, 295.
50. (a) Mezey, P. G. *Molec. Phys.* In press. (b) Mezey, P. G. *J. Math. Chem.* **1998**, *23*, 65.
51. Mezey, P. G. *Potential Energy Hypersurfaces*; Elsevier: Amsterdam, 1987.
52. Mezey, P. G. In *Reports in Molecular Theory*; Weinstein, H.; Náray-Szabó, G., Eds.; CRC Press: Boca Raton, 1990, Vol. 1, pp. 165–183.
53. Mezey, P. G. In *Concepts and Applications of Molecular Similarity*; Johnson, M. A.; Maggiora, G. M., Eds.; Wiley: New York, 1990.
54. Zimpel, Z.; Mezey, P. G. *Int. J. Quant. Chem.* **1996**, *59*, 379.
55. Zimpel, Z.; Mezey, P. G. *Int. J. Quantum Chem.* **1997**, *64*, 669.
56. Mezey, P. G. In *Pauling's Legacy: Modern Modelling of the Chemical Bond*; Maksic, Z.; Orville-Thomas, W. J., Eds.; Elsevier Science: Amsterdam, The Netherlands, 1998.
57. Mezey, P. G. *Shape in Chemistry: An Introduction to Molecular Shape and Topology*; VCH Publishers: New York, 1993.
58. Mezey, P. G.; Walker, P. D. *Drug Discovery Today (Elsevier Trend Journal)* **1997**, *2*, 6.
59. Zadeh, L. A. *Inform. Control* **1965**, *8*, 338.
60. Zadeh, L. A. *J. Math. Anal. Appl.* **1968**, *23*, 421.
61. Kaufmann, A. *Introduction à la Théorie des Sous-Ensembles Flous*; Masson: Paris, 1973.
62. Zadeh, L. A. In *Encyclopedia of Computer Science and Technology*; Marcel Dekker: New York, 1977.
63. Klir, G. J.; Yuan, B. *Fuzzy Sets and Fuzzy Logic, Theory and Applications*; Prentice-Hall: Englewood Cliffs, NJ, 1995.
64. Mezey, P. G. In *Fuzzy Logic in Chemistry*; Rouvray, D. H., Ed.; Academic Press: San Diego, 1997, pp. 139–223.

NOVEL INCLUSION COMPOUNDS WITH UREA/THIOUREA/ SELENOUREA–ANION HOST LATTICES*

Thomas C. W. Mak and Qi Li

*Dedicated to Prof. George A. Jeffrey on the occasion of his 83rd birthday.

Advances in Molecular Structure Research
Volume 4, pages 151–225
Copyright © 1998 by JAI Press Inc.

ABSTRACT

This chapter covers currently available information on the crystal structures of urea, thiourea, selenourea, and their inclusion compounds. We concentrate here on the systematics of novel host lattices generated by the combined use of urea, thiourea, or selenourea plus selected anionic species as building blocks and hydrophobic organic cations as the enclosed guests. Different kinds of anions, such as halides, simple trigonal and square planar anions, monocarboxylate and dicarboxylate anions, as well as larger planar or nonplanar anions were used, together with symmetric tetraalkyl-ammonium or phosphonium ions and unsymmetrical quaternary ammonium cations. The results have demonstrated that the classical urea/thiourea/selenourea channel-type host lattice can be modified in interesting ways by the incorporation of various anionic moieties, with or without cocrystallized water or other uncharged hydrophilic molecules, and that novel hydrogen-bonded frameworks bearing different urea, thiourea, or selenourea/guest molar ratios are generated by variation in size of the hydrophobic, pseudo-spherical R_4N^+ guest species that act as templates. The versatility of urea, thiourea, or selenourea as a versatile component in the construction of novel host lattices is exemplified by the occurrence of many new types of linkage modes, which include various discrete units, chains, ribbons, and composite ribbons with incorporation of the corresponding anions. As envisaged, the urea/thiourea/selenourea–anion host lattices exhibit a rich variety of inclusion topologies besides the typical unidirectional channel structure, such as a channel with widened cross section accommodating two parallel columns of guest species, corrugated-layer or double-layer sandwich structure, and a three-dimensional network containing isolated cages, intersecting tunnels, or dual channel systems.

I. INTRODUCTION

The chemistry of inclusion compounds has a long history and is nowadays a subject of wide-ranging and intense study. Early advances on inclusion complexes that contain layer-, channel-, or cage-type (clathrate) host structures were summarized in a monograph in 1964 [1]. Specialized literature on in-depth investigations in "molecular inclusion and molecular recognition" began to appear 15 years ago, including the *Journal of Inclusion Phenomena and Molecular Recognition in Chemistry* [2], a two-volume set in *Topics in Current Chemistry* [3], and a series of five books entitled *Inclusion Compounds* in which developments in all principal aspects of inclusion chemistry were described [4]. Structural modification of inclusion compounds, study of the properties of existing "host–guest" systems, and design of new host molecules have since been pursued with ever-increasing vigor [5–7]. With the award of the 1987 Nobel Prize in Chemistry to Donald J. Cram [8], Jean-Marie Lehn [9] and Charles J. Pedersen [10] for their fundamental work on host–guest or "supramolecular" systems, inclusion chemistry has come to the fore in contemporary research. Increasing varieties of novel inclusion compounds and new host molecules have been synthesized recently [11–13]. A comprehensive overview of the forefronts of modern supramolecular chemical research has appeared as an 11-volume set in 1996 [14].

The term "crystal engineering" was coined by Schmidt [15] to describe the rational design and control of molecular packing arrangements in the solid state, and the structural study of clathrates has contributed substantially to our understanding of the problem [16, 17]. Generalizations concerning the preferred structural motifs generated by hydrogen bonding and weaker noncovalent interactions between specific functional groups or molecular fragments have led to the realization of some impressive predictions about the construction of supramolecular networks [17–20]. Recently, there has been growing interest in "nanoporous" molecular crystals, which in principle can provide molecular-scale voids with controlled size, shape, and chemical environment that may be exploited for separation of mixtures, shape-selective catalysis, and optoelectronic applications [21].

Urea was the first organic compound to be synthesized in the laboratory, a feat accomplished by Wöhler in 1828. Since it is a small molecule with high symmetry and crystals of good quality can be easily obtained, the structure of urea was one of the earliest to be determined by X-ray crystallography [22]. The structure was later refined several times using improved X-ray and neutron diffraction data [23]. The coefficients of thermal expansion of urea were measured, and changes in atomic coordinates, bond lengths, and angles with temperature were computed by Natalie et al. [24]. In the most precise study to date, detailed comparisons were made of the charge density distribution in urea based on accurate X-ray and neutron diffraction data along with ab initio quantum chemical calculations [25]. Thiourea is a well-known ferroelectric which exhibits many interesting higher order commensurate and incommensurate phases between the paraelectric and ferroelectric phases. Since the first report on ferroelectricity in thiourea over four decades ago [26], the successive phase transitions have been thoroughly investigated [27].

Urea, thiourea, and selenourea are good host molecules because they have a well-defined trigonal planar geometry (see Figure 1) and can form at least six hydrogen bonds. The cocrystallization behavior of urea and thiourea with straight- and branched-chain aliphatic compounds, respectively, was first reported over a half-century ago, and detailed reviews on the resulting channel inclusion compounds have appeared from time to time [28]. New concepts about various structural aspects of urea inclusion compounds, such as temperature-dependent structural properties, phase transition and structure change [29, 30], disordered crystal structure [31–33], and migration of molecules into the channel structure [34], as well as unusual conformational behavior in thiourea inclusion compounds

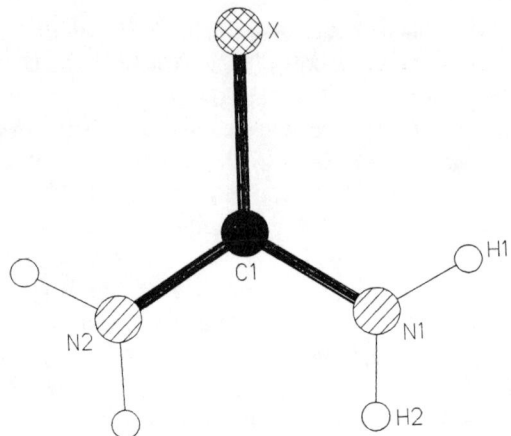

Figure 1. Molecular geometry of urea and its group VI analogues (X = O, urea; X = S, thiourea; X = Se, selenourea). H(1) and H(2) are *syn* and *anti*, respectively, to the X atom.

[35] have been reported. Recent work by Thomas and Harris [36–38] and Hollingsworth [38] have shown that the inclusion phenomenon is much more complex, varied, and interesting than had been realized all along, and the latest review on urea/thiourea/selenourea inclusion compounds was published by them in 1996 [38b]. The formation, characterization, and segmental mobilities of block copolymers in crystals of their urea inclusion compounds have been reported recently [39].

Meanwhile many new host lattices generated by the combined use of urea [40], thiourea [41], or selenourea [42] and other molecular species as the building blocks have led to the discovery of different inclusion topologies, such as networks of intersecting tunnels, two-dimensional interlamellar regions within a layered structure, isolated cages, and systems of interconnected cages by incorporating additional components into the construction of the hydrogen-bonded host lattice.

The purpose of this review is to assemble and assess currently available information about structural modification of urea/thiourea inclusion compounds. We concentrate here on the structural aspects of new inclusion compounds with urea/thiourea/selenourea–anion host lattices, most of which were prepared and structurally analyzed in our laboratory. The versatility of urea or thiourea as a key component in the construction of novel anionic host lattices is clearly demonstrated by occurrence of many new types of linkage modes. The results show that co-molecular aggregates of urea or thiourea with other neutral molecules or anionic moieties can be considered as fundamental building blocks for the constructions of various types of novel host lattices. Comments on structure–property relationship for these inclusion compounds are made wherever appropriate.

II. CLASSICAL CHANNEL INCLUSION COMPOUNDS OF UREA, THIOUREA, AND SELENOUREA

A. Crystal Structures of Urea, Thiourea, and Selenourea

A brief review on the crystal structures of urea, thiourea, and selenourea is included for understanding the properties of hydrogen-bonded channel frameworks in their inclusion compounds. The N–H⋯E (E = O, S, Se) hydrogen bonds which play an instrumental role in the structures of urea/thiourea/selenourea inclusion compounds are discussed in this section.

Crystal Structure of Urea

The crystal and molecular structure of urea, which crystallizes in the tetragonal space group $P\bar{4}2_1m$ with $Z = 2$, has been studied repeatedly since the first report dated 1923 [22a]. As part of a study of the charge density distribution in amide groups, the bond electron density distribution in urea was determined from X-ray data collected at 123 K and analyzed with two different structure models [23f]. In 1984, the experimental charge density distribution was reinvestigated with more extensive X-ray and neutron diffraction data collected at 123 K and found to be in satisfactory agreement with the SCF theoretical charge distribution calculated using a 6-31G basis set [25a]. The lattice parameters of urea were measured at seven temperatures in the range 12 to 173 K, as shown in Tables 1 and 2. The crystal structure was subsequently determined at 12, 60, and 123 K from neutron diffraction data, and the molecular thermal motion of urea was analyzed [25b].

In these highly accurate studies on the crystal structure of urea, precise oxygen, nitrogen, and carbon positional parameters were determined together with reliable location of the hydrogen atoms in positions coplanar with the rest of the molecule [25]. This indication of planarity corroborated similar findings from infrared [43] and nuclear magnetic resonance studies [44].

In the crystalline state, the urea molecule occupies a special position where its molecular symmetry (*mm*2) is fully utilized. The two crystallographically distinct types of hydrogen atoms in the urea molecule are labeled H(1) and H(2) as shown

Table 1. Lattice Parameters (Å) for Urea: X-ray Diffraction Method

	RT^a	RT^b	$133\ K^b$	$123\ K^c$	$123\ K^a$
a	5.661	5.662(2)	5.582(2)	5.576(3)	5.578(1)
c	4.712	4.716(2)	4.686(2)	4.686(3)	4.686(1)

Notes: [a]Ref. 23a.
  [b]Ref. 23c.
  [c]Ref. 23f.

Table 2. Latice Parameters (Å) for Urea: Neutron Diffraction Method

	12 K[a]	30 K[a]	60 K[a]	90 K[a]	123 K[a]	150 K[a]	173 K[a]	RT[b]
a	5.565	5.565	5.570	5.576	5.584	5.590	5.598	5.662(3)
c	4.684	4.685	4.688	4.689	4.689	4.692	4.694	4.716(3)

Notes: [a]Ref. 25b, e.s.d. is 0.001 Å for all entries.
        [b]Ref. 23e.

in Figure 1, where H(2) is the one *anti* to the carbonyl oxygen. Interatomic distances
and bond angles obtained from diffraction methods, including electron diffraction
[45], at different temperatures are listed in Table 3. The bond angles of approximately 120° found for C–N–H(1), C–N–H(2), and H(1)–N–H(2) are in fair
agreement with the nuclear magnetic resonance results [44]. These bond angles and
the planar structure of urea are consistent with the assignment of sp^2 hybridization
to the C and N atoms.

The molecular arrangement and three-dimensional hydrogen bonding network
in the urea crystal structure are illustrated in Figure 2. Every hydrogen atom is
involved in a N–H⋯O hydrogen bond, so that the urea molecules are linked in a
head-to-tail mode to generate a straight chain. Such chains run parallel to the c axis,
but neighboring chains are arranged in an antiparallel fashion and individually lie
on almost vertical planes. In any particular urea molecule, the carbonyl oxygen is
tethered to both *anti*-hydrogen atoms H(2) of an adjacent molecule in the same
head-to-tail chain, and also to *syn*-hydrogen atoms H(1)′ of two other molecules
belonging to adjacent chains lying on both sides (Figure 2a). The resulting three-
dimensional framework provides a rare example of a carbonyl O atom forming four
acceptor N–H⋯O hydrogen bonds [46].

As seen in the projection down c (Figure 2b), the structure is quite open with the
molecules and the hydrogen bonds confined to planes parallel to (110) and (1$\bar{1}$0),
which intersect to form channels having a square cross section of 3.94 × 3.94 Å
[25b].

The closest contact of 2.57 Å between two adjacent chains, that between H(1)
and H(2)′, is slightly greater than the expected van der Waals separation. At each
temperature and for each atom in the urea molecule, the mean-square amplitude of
nuclear thermal motion is considerably less along the tetragonal c-axis direction
than in directions normal to c; also, the a and c lengths are almost proportional to
the temperature in the range 30 to 173 K but the expansion is smaller for c than a
(Table 1). These observations were explained qualitatively by Swaminathan et al.
[25b] in terms of the molecular arrangement and the three-dimensional hydrogen-
bonding network in the crystal structure, as shown in Figure 2. The largest
amplitudes of atomic vibration are normal to c because of the relatively greater

Table 3. Interatomic Distances (Å) and Angles (°) for Urea

	X-ray Diffraction Method				Neutron Diffraction Method				Electron
	RT^a	RT^b	$133\ K^b$	$123\ K^c$	$12\ K^d$	$60\ K^d$	$123\ K^d$	RT^e	RT^f
C=O	1.262(16)	1.264(4)	1.262(3)	1.257(1)	1.265(1)	1.265(1)	1.264(1)	1.260(3)	1.28
C–N	1.335(13)	1.336(7)	1.341(3)	1.340(2)	1.349(1)	1.349(1)	1.351(1)	1.352(2)	1.35
N–H(1)	—	0.918*	1.080*	1.06(8)	1.022(3)	1.021(3)	1.015(3)	0.998(5)	—
N–H(2)	—	0.739*	0.975*	1.04(8)	1.018(3)	1.015(3)	1.005(3)	1.003(4)	—
N–C–O	121.0(7)	120.9(3)	120.9(2)	121.6*	121.4(1)	121.4(1)	121.4(1)	121.7(1)	120.9
N–C–N	118.0(13)	118.2(3)	118.2(2)	116.9*	117.2(1)	117.3(1)	117.2(1)	116.7*	—
C–N–H(1)	—	119.5*	123.6*	118.8	119.1(1)	119.2(1)	119.2(1)	119.0(3)	—
C–N–H(2)	—	120.0*	119.4*	114.7*	120.5(1)	120.6(1)	120.7(1)	120.2(3)	—
H(1)–N–H(2)	—	120.4*	117.1*	126.5*	120.4(2)	120.2(2)	120.1(2)	120.8*	—
N···O	2.989	2.998(5)	2.968(3)	2.957*	2.985(1)	2.989(1)	2.998(1)	2.978(8)	2.97
N···O'	3.035	3.036(7)	2.963(4)	2.992*	2.955(1)	2.958(1)	2.960(1)	3.035(5)	3.02
N–H(1)···O	—	165.9*	172.3*	166.0*	167.2(2)	167.0(2)	166.8(2)	166.2*	—
N–H(2)···O'	—	151.5*	149.1*	155.9*	147.4(2)	147.4(2)	147.6(2)	148.6*	—

Notes: [a]Ref. 23a.
[b]Ref. 23c.
[c]Ref. 23f.
[d]Ref. 25b.
[e]Ref. 23e.
[f]Ref. 45.
*Calculated from crystal data and atomic coordinates in the original references.

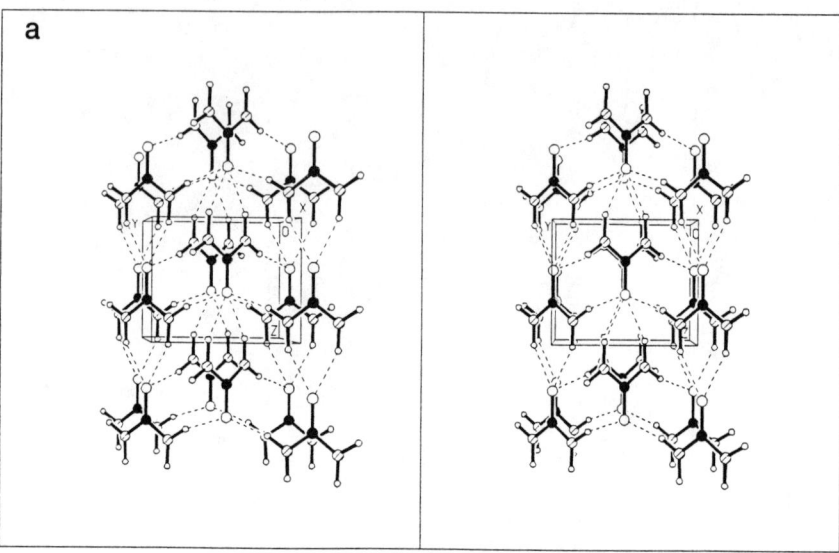

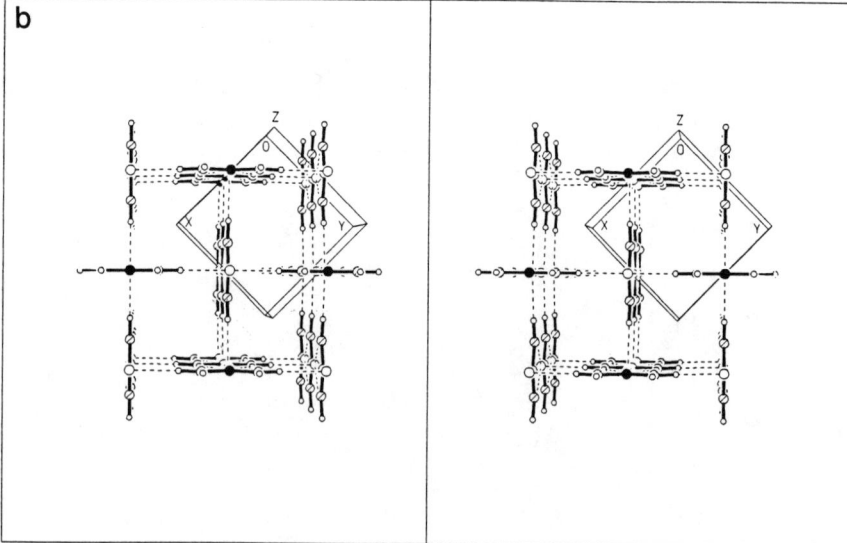

Figure 2. (a) Stereoview of the crystal structure of urea. Note that the hydrogen-bonded head-to-tail chains run parallel to the c axis and each O atom forms four N–H⋯O acceptor hydrogen bonds. (b) The hydrogen-bond framework in crystalline urea viewed approximately along the c axis. (Data from ref. 25b).

freedom for motion of the NH_2 groups into the channels. Motion parallel to c is hindered because it occurs within the walls of the channels where the atoms are densely packed.

Crystal Structure of Thiourea

Thiourea has been the subject of several structural investigations. After the first structural report by Wyckoff and Corey [47], X-ray diffraction studies of the room-temperature structure of thiourea were made by Kunchur and Truter [48] using photographic data, and then by Truter [49] using counter data. Subsequently the crystal structure of thiourea was refined at different temperatures using X-ray and neutron diffraction methods [27]. Crystalline thiourea at room temperature is orthorhombic with $Z = 4$ in space group *Pnma*, and the lattice parameters measured at different temperatures are shown in Table 4. The thiourea molecules occupies a site of symmetry m such that its carbon and sulfur atoms lie on a crystallographic mirror plane with the NH_2 groups related by reflection symmetry. The molecule of thiourea is very flat as both independent hydrogen atoms whose positions were determined by neutron diffraction are coplanar with the rest of the molecule. Interatomic distances and bond angles obtained from diffraction methods, including electron diffraction, at different temperatures are listed in Table 5.

In the crystal structure adjacent thiourea molecules related by a 2_1 screw axis are linked by a pair of N–H⋯S hydrogen bonds in the usual shoulder-to-shoulder manner to form a zigzag ribbon running parallel to the b axis (Figure 3a). Each thiourea molecule is further linked to another molecule belonging to a neighboring

Table 4. Lattice Parameters (Å) for Thiourea (X-ray and Electron Diffraction Method)

	Space Group Pnma			Space Group P2₁ma			Electron
	$RT^{a,b}$	$295\ K^c$	$170\ K^d$	$120\ K^e$	$153\ K^d$	$119\ K^e$	RT^f
a	7.655(7)	7.663(4)	7.545(1)	7.516(7)	7.531(4)	7.500(5)	7.65(1)
b	8.537(7)	8.564(4)	76.867(11)*	8.519(10)	8.541(3)	8.535(5)	8.53(1)
c	5.520(7)	5.488(3)	5.467(1)	5.494(5)	5.476(2)	5.464(4)	5.52(1)

Notes: [a]Ref. 48.
[b]Ref. 49.
[c]Ref. 27a.
[d]Ref. 27b.
[e]Ref. 27e.
[f]Ref. 53b.
*Ninefold superstructure

Table 5. Interatomic Distances (Å) and Angles (°) for Thiourea from Diffraction
Methods

	X-ray		Neutron			Electron (Solid)
	RT^a	RT^b	170 $K^{c,f}$	RT^d	110 K^d	RT^e
S=C	1.720(9)	1.713(12)	1.721(4)	1.746(9)	1.742(9)	1.680*
C–N	1.340(6)	1.329(12)	1.322(8)	1.350(4)	1.341(4)	1.336*
N–H(1)	0.96(6)	—	0.97(9)	1.022(7)	1.011(9)	0.94
N–H(2)	0.80(9)	—	0.83(3)	1.012(11)	1.006(9)	1.01
N–C–S	120.5(5)	122.2(6)	120.9(2)	121.7(2)	121.4(2)	123.9*
N–C–N	119.0(5)	115.6(11)	118.6(4)			112.2*
C–N–H(1)	118(4)	—	116(2)	120.1(4)	120.4(5)	123.4*
C–N–H(2)	114(5)	—	124(2)	118.3(5)	119.8(6)	126.5*
H(1)–N–H(2)	109(6)	—	117(4)	121.3(6)	119.7(7)	108.6*
N···S	3.39*	3.42	—	3.394(4)	3.384(4)	3.437*
N···S′	3.53*	3.51	—	3.526(8)	3.580(11)	3.483*
N···S″	3.72*	3.70*	—	3.696(7)	3.452(10)	3.700*
N–H(1)···S	163.7*	—	—	168.9(5)	169.5(6)	151.1*
N–H(2)···S′	134.9*	—	—	133.7(7)	115.2(6)	134.4*
N–H(2)···S″	102.9*	—	—	123.5(7)	135.3(7)	118.0*

Notes: [a]Ref. 49.
            [b]Ref. 48.
            [c]Ref. 27b.
            [d]Ref. 27d.
            [e]Ref. 53b.
            [f]Ninefold superstructure.
            *Calculated from crystal data and atomic coordinates in their original references.

ribbon via two additional N–H···S acceptor hydrogen bonds, so that each S atom
forms four acceptor hydrogen bonds, as shown in Figure 3.

Upon cooling from room temperature, thiourea passes through at least five
different phases, including a commensurate high-temperature or disordered phase,
an incommensurate phase, and a commensurate low-temperature phase. The crystal
structures of the room-temperature paraelectric phase (phase V; space group $Pnma$,
$Z = 4$) and the lowest temperature ferroelectric phase (phase I; space group $P2_1ma$,
$Z = 4$), have been determined at different temperatures [27a,b,e]. According to these
studies, the molecule in phase V at room temperature has almost the same shape as
that in phase I at 110 K. In the b direction there are alternating sheets consisting of

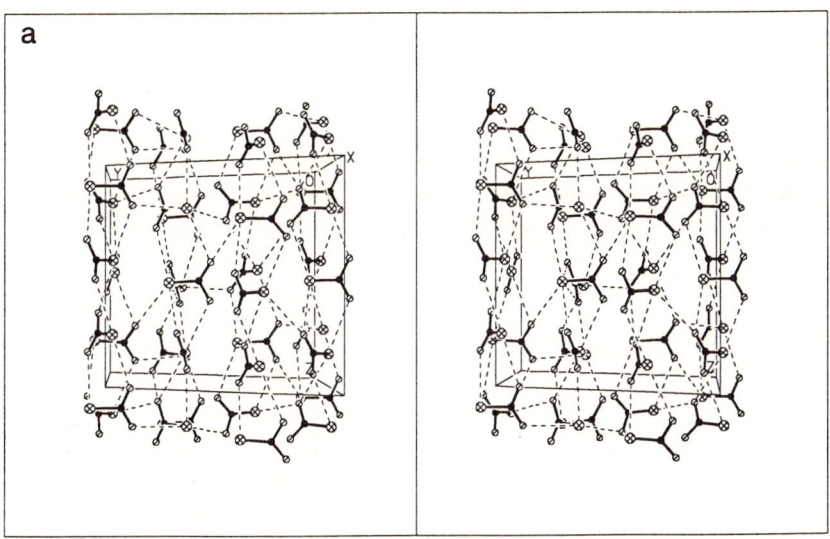

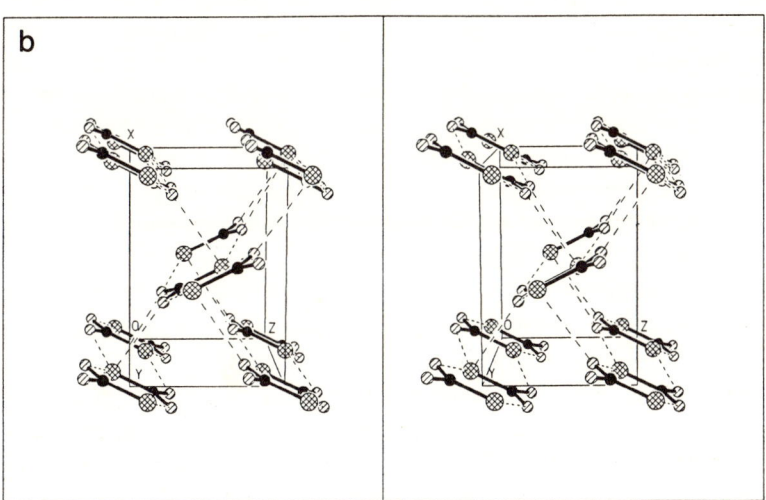

Figure 3. (a) Stereoview of the crystal structure of thiourea. Note that the hydrogen-bonded shoulder-to-shoulder ribbons run parallel to the *c* axis and each S atom forms four N–H⋯S acceptor hydrogen bonds. (b) The hydrogen-bond framework in crystalline thiourea viewed along the *c* axis. (Data from ref. 48).

similar polar molecules having opposite "average" orientation for the disordered phase V where the two subsystems correspond to each other by a n glide, which disappears in the low-temperature phase I [27c]. In this phase, the four molecules in the unit cell constitute two nonequivalent pairs with different inclinations to the ferroelectric a axis. Goldsmith and White [27e] concluded that the ferroelectric reversal was caused by the action of the field on the resultant of the dipole moments of the molecules (in the [100] direction), causing the molecules to rotate and interchange their tilts with respect to the [100] direction.

The three intermediate phases II, III, and IV of thiourea are believed to be incommensurate, and the incommensurately modulated structures in phase II and IV have been analyzed [50]. Another commensurate phase which is stable between phase I and phase II in a narrow temperature region of about 2 K has been recognized many years ago [27c,f], and was refined at 170 K by Tanisaki et al. [27b]. This ninefold superstructure of thiourea is characterized by a rotation of the $(NH_2)_2CS$ molecule along the c axis coupled with a displacement of the center of mass in a plane perpendicular to the axis. Around mirror planes ($y = 1/4$ and $3/4$) the local structure is isostructural with phase I. Therefore the superstructure is constructed of alternately polarized layers which are sandwiched by domain walls (or discommensurations) whose local structure is that of the paraelectric room temperature phase (V) around $y = 0$ and $1/2$.

Crystal Structure of Selenourea

Reports on the crystal structure of selenourea appeared in the early 1960s, which gave preliminary accounts of the unit-cell dimensions of selenourea [51a,b] and information about the carbon–selenium double bond in the crystal structure of N-phenyl-N'-benzoylselenourea [51c]. More accurate determination of the lattice parameters and atomic positions of selenourea was carried out at room temperature and at 173 K by Rutherford and Calvo in 1969 [52]. All of these reports assigned the structure to one of the enantiomorphous trigonal space groups $P3_1$ and $P3_2$ with $Z = 27$, which differs from an earlier electron diffraction study based on space group $Pnma$ with $a = 6.48$, $b = 8.75$, and $c = 7.04$ Å and $Z = 4$ [53a] that can be compared with the room-temperature phase of thiourea. The lattice parameters of selenourea from these studies are listed in Table 6.

There are no significant differences in the molecular geometry between the nine independent selenourea molecules in the asymmetric unit. These planar molecules, which have the average dimensions: C=Se 1.86(3) Å, C–N 1.37(2) Å, Se–C–N and N–C–N both 120°, are linked by hydrogen bonds to generate nine spiral ribbons per unit cell, which are essentially similar in form and run parallel to the c axis, as shown in Figure 4a. The nine independent selenourea molecules contribute differently to the construction of the spiral ribbons. Molecules M1, M2, and M3, each lying close to a 3_1 screw axis, form three separate spiral ribbons (type rib1, rib2, and rib3, respectively), but the remaining molecules constitute two sets, namely (M4, M5, and M6) and (M7, M8, and M9), which generate spiral ribbons of type

Table 6. Lattice Parameters (Å) for Selenourea (X-ray Diffraction Method)

	RT^a	RT^b	RT^c	$173\ K^c$
a	15.37	15.34	15.285(5)	15.201(5)
c	13.08	12.99	13.007(5)	12.950(5)

Notes: [a]Ref. 51a.
       [b]Ref. 51b.
       [c]Ref. 52.

rib4 and rib5, respectively. These relationships can be seen in projection along the c axis (Figure 4b).

The ribbons have distinctly twisted configurations as shown by their C–N···Se=C torsion angles (Table 7). The N–H···Se hydrogen bonds between adjacent intra-ribbon selenourea molecules have an average length of 3.51Å. Rutherford and Calvo [52] concluded that "not all nitrogen atoms are involved in inter-ribbon hydrogen bonds, and because the hydrogen bonds are so weak that some have been sacrificed in favor of compactness, that is, to gain energy by van der Waals interaction."

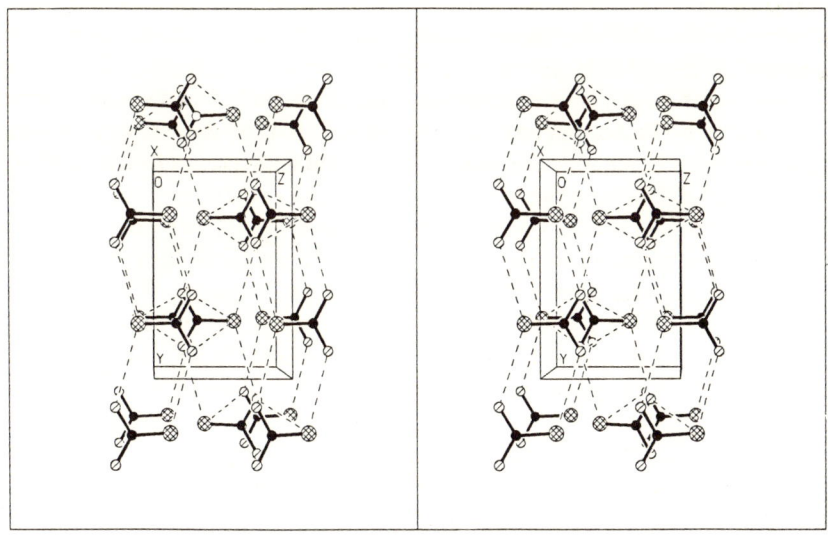

Figure 4. (a) Stereoview of the crystal structure of selenourea. Note that the hydrogen-bonded spiral ribbons run parallel to the c axis.

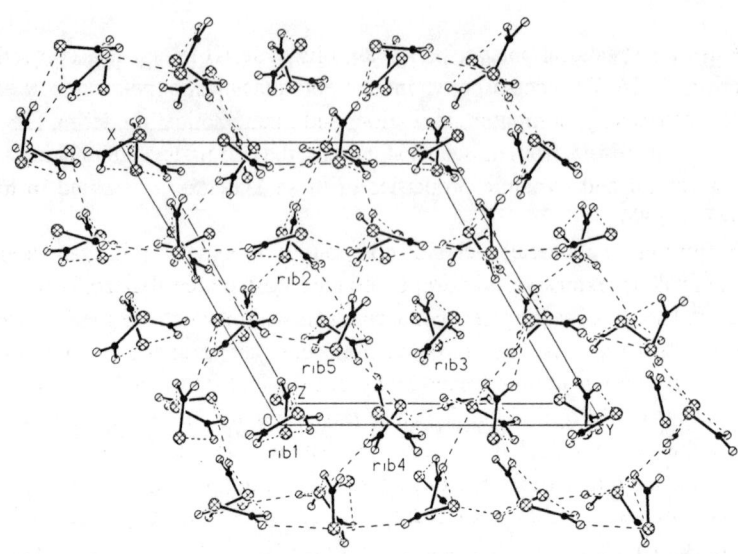

Figure 4. (b) The hydrogen-bond framework in crystalline selenourea viewed along the *c* axis. The relationships between five different types of spiral ribbon are shown. (Data from ref. 52).

B. Channel Inclusion Compounds of Urea, Thiourea, and Selenourea

The chemistry of urea and thiourea inclusion compounds has received much attention and is still a subject of continuing interest since their accidental discovery in 1940 [*54*] and 1947 [*55*], respectively.

Full reviews on the formation, structure, and properties of a variety of inclusion compounds of urea and thiourea were published about 10 years ago [*28*], but very significant advances have been made in the present decade with the aid of modern

Table 7. C–N···Se═C Torsion Angles for Spiral Ribbons in Selenourea

Ribbon	Molecules[a]	Torsion Angle (°)
rib1	M1′–M1–M1″	48.1; 50.6
rib2	M2′–M2–M2″	53.5; 69.6
rib3	M3′–M3–M3″	41.4; 59.1
rib4	M4–M5′–M6″	35.6; 36.2; 41.2; 64.5
rib5	M7–M8″–M9′	43.2; 49.0; 52.1; 74.9

Note: [a]Equivalent positions generated by the symmetry transformations of space group $P3_1$ are denoted by superscripts: $'y - x, -x, z + 2/3;\ ''-y,$ $x - y, z + 1/3$.

techniques for structural analysis, such as solid-state NMR spectroscopy, X-ray diffraction, EXAFS spectroscopy, incoherent quasielastic neutron scattering, Raman spectroscopy, computer simulation, and mathematical modeling [56]. On the basis of new results from these investigations, the current level of understanding of the structural and dynamic properties of these material is assessed in recent reviews [37, 38b].

In the normal (or classical) inclusion compounds of both urea [36a] and thiourea [36b] there is an extensively hydrogen-bonded host lattice that contains linear, parallel, and non-intersecting hexagonal channels within which the guest molecules are densely packed. The urea/thiourea molecules are arranged in three-twined helical spirals, which are interlinked by hydrogen bonds to form the walls of each channel see (Figures 5a and 6a). The host molecules are almost coplanar with the walls of the hexagonal channel, and the channels are packed in a distinctive honeycomb-like manner (see Figures 5b and 6b). The host lattice is stabilized by the maximum possible number of hydrogen bonds that lie virtually within the walls of the hexagonal channels: each NH_2 group forms two donor bonds, and each O or S atom four acceptor bonds, with their neighbors.

Inclusion Compounds of Urea

Urea forms an isomorphous series of crystalline non-stoichiometric inclusion compounds with n-alkanes and their derivatives (including alcohols, esters, ethers, aldehydes, ketones, carboxylic acids, amines, nitriles, thioalcohols, and thioethers)

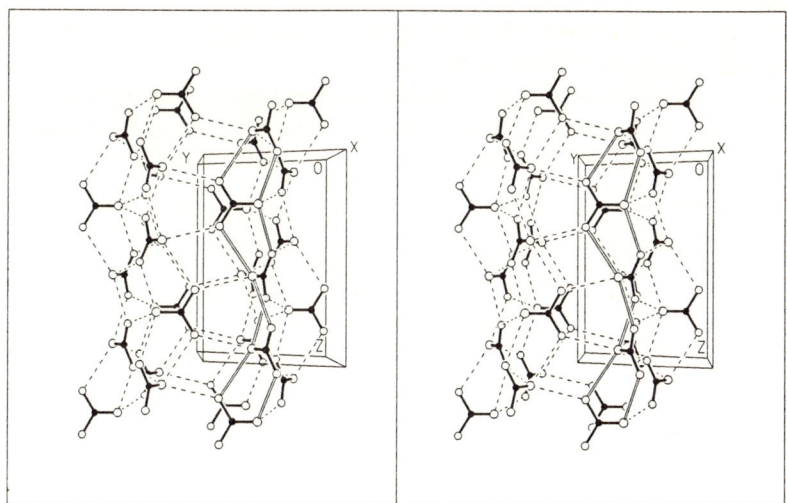

Figure 5. (a) Stereodrawing of the hydrogen-bonded urea host lattice showing an empty channel extending parallel to the *c* axis. A shoulder-to-shoulder urea ribbon is highlighted by showing its N–H···O hydrogen bonds as open lines.

Figure 5. (b) The channel-type structure of inclusion compounds of urea viewed along the *c* axis. For clarity the guest molecules are represented by large circles. (Data from ref. 36a).

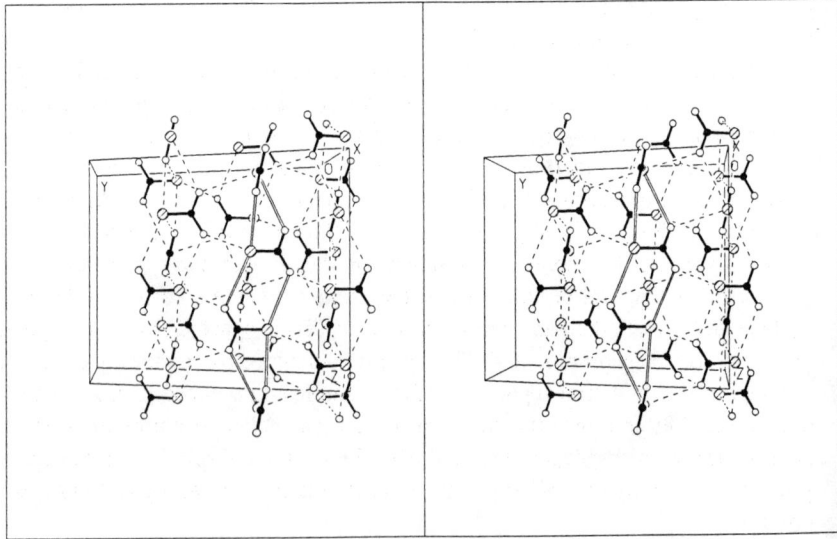

Figure 6. (a) Stereodrawing of the hydrogen-bonded thiourea host lattice showing an empty channel extending parallel to the *c* axis. A shoulder-to-shoulder thiourea ribbon is highlighted by showing N–H⋯S hydrogen bonds as open lines.

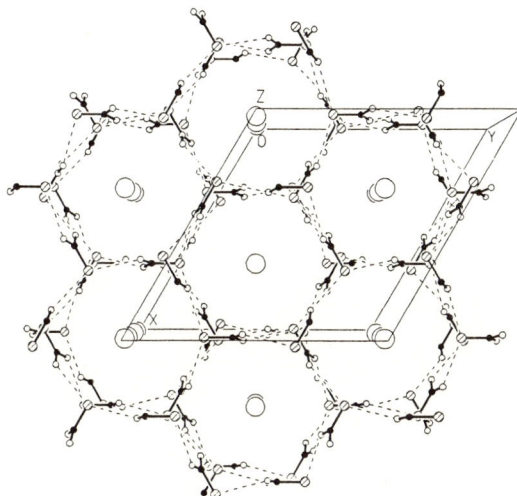

Figure 6. (b) The channel-type structure of inclusion compounds of thiourea viewed along the *c* axis. For clarity the guest molecules are represented by large circles. (Data from ref. 36b).

provided that their main chain contains of six or more carbon atoms. The unit cell is hexagonal, space group $P6_122$ or $P6_522$ with $a = 8.230(4)$ and $c = 11.005(5)$ Å, and contains six urea molecules [57a, 58].

In the urea-n-hydrocarbon complex, the urea molecules are connected by hydrogen bonds to form helical ribbons, the relevant torsion angle (C–N⋯X=C) being given in Table 8. The guest molecules occupy the hollow channels of the host lattice (Figure 5b) in an extended planar zigzag configuration, generally with positional disorder of the atoms. Since the minimum channel diameter ranging between 5.50 and 5.80 Å [38b] is only slightly larger than the van der Waals envelope of an aliphatic hydrocarbon, there are limitations on the amount and type of branching or substitution that can be tolerated in the guest molecules. For example, while n-octane and 3-methyheptane can fit into the available space, 2,2,4-trimethylpentane cannot be so accommodated. The selectivity dictated by channel size has been exploited in the industrial separation of n-alkane from branched-chain and cyclic compounds. Furthermore, the occurrence of urea inclusion compounds in two enantiomorphic forms, depending on the handedness of the spirals, can be used for optical resolution by crystallizing part of a recemic mixture as a guest in the urea complex.

The composition of the nonstoichiometric urea-n-hydrocarbon adduct can be calculated from a formula given by Smith [57b]: Urea/hydrocarbon mole ratio $= 0.684 (n-1) + 2.175$, where n is the number of carbon atoms in the extended zigzag chain (by taking C–C = 1.54 Å, C–C–C = 109.5°, and radius of CH_3 group = 2.0 Å).

Table 8. Typical Linkage Modes of Self-Assembled Hydrogen-Bonded Chains and Ribbons of Urea, Thiourea or Selenourea Molecules in Their Pure Crystals and Channel Inclusion Compounds

		$C-N\cdots X{=}C$ Torsion Angle (°)	Ref.
head-to-tail chain		180.0	25b
shoulder-to-shoulder ribbon		8.7	48
helical spiral in urea inclusion compounds		56.1	36a
helical spiral in thiourea inclusion compounds		52.2	36b
helical spiral in selenourea inclusion compounds		51.8	79

In the disordered crystal structure of the urea inclusion compounds of C_nH_{2n+2} ($n = 12, 16$), the orientation of the included paraffin chain was deduced and the hydrogen (deuterium) position determined for the hexadecane complex ($n = 16$) using neutron diffraction data [33]. A wide variety of guest molecules accommodated within urea channels were reported recently, including the synthesis and structural characterizations of a dialkylamine adduct [59] and a series of polymer–urea inclusion compounds [39e]. Interchannel ordering of n-alkane guest molecules have been experimentally and theoretically studied [31].

A considerable amount of recent work has been directed towards the study of detailed molecular orientations and motions of guest molecules in urea channel inclusion compounds, as well as the generation of crystalline modifications in different space groups, such as $P2_12_12_1$ [30a,b] $R3c$ [60] and $Pbcn$ [61], but the principal structural characteristics of the channel-type host lattices remain virtually unaltered. Structural properties of the 1,10-dibromodecane/urea and 1,12-dibromodecane/urea inclusion compounds have been determined by single-crystal X-ray analysis for both the room-temperature ($P6_122$) and low-temperature ($P2_12_12_1$) phases [62]. In the low-temperature phase, the guest molecules have a narrow distribution of preferred orientations that correlates well with the observed distortion of the host tunnel.

Direct binding of guest molecules to the channel wall via hydrogen bonding occurs, however, only in very exceptional cases. Recently, very interesting stress-induced domain reorientation has been observed in commensurate ($3c_h = 2c_g$, where c_h and c_g are the host and guest repeat distances along the channel axis) 2,10-undecanedione/urea (1:9), whose lattice symmetry is lowered from $P6_122$ to $C222_1$ by hydrogen-bonding interaction between the guest molecules and every third urea molecule along the channel wall [38a].

Both experimental evidence [63] and Monte Carlo simulation [64] have shown that the removal of guest molecules from urea inclusion compounds leads to collapse of the channel host framework to produce the tetragonal crystal structure of pure urea. Thus there is always a dense packing of guest molecules within the channel structure of urea inclusion compounds [36a]. The migration of molecules into the channel structure of urea inclusion compounds has been probed by high-resolution solid-state ^{13}C NMR spectroscopy [34].

Inclusion Compounds of Thiourea

Thiourea inclusion compounds crystallize in several space groups in two crystal systems: rhombohedral $R\overline{3}2/c$ [65], $R3c$ [66], $R\overline{3}c$ [36b, 67], $R\overline{3}$ [68], and monoclinic $P2_1/a$ [69–71]. In the last space group, the adducts exist only at reduced temperatures in the range $-170-25°$ C except for those containing squalene or aromatic guests [72]. In the most common type of thiourea inclusion compounds (space group $R\overline{3}c$; $a \approx 15.50-16.20$ Å, $c = 12.50-12.50$ Å in the hexagonal setting with $Z = 6$) the host structure is built of thiourea molecules in an extensively hydrogen-bonded lattice that contains parallel channels of non-uniform cross section (Figure 6b), showing constrictions (minimum diameter = 5.8 Å) at $z = 0$, 1/2 and bulges (minimum diameter = 7.1 Å) at $z = 1/4$, 3/4 [73]. In contract to the urea host lattice, a wider variety of guests, such as branched-chain alkanes and their derivatives, haloalkanes, 5-, 6- and 8-membered ring compounds, condensed aromatic ring systems, and even ferrocene and other metallocenes [74–76], can be enclosed in the more spacious thiourea host lattice. Recent examples of appropriate guest molecules that have been investigated include cyclohexane and its derivatives such as chlorocyclohexane and bromocyclohexane [35, 77], as well as some organometallics such as (benzene)$Cr(CO)_3$ and (1,3-cyclohexadiene)$Fe(CO)_3$ [78].

Inclusion Compounds of Selenourea

Selenourea, like urea and thiourea, can form inclusion compounds with a variety of hydrocarbons. Its inclusion properties were studied by van Bekkum and coworkers in 1969 [79], well over 20 years after the discovery of the analogous urea [54] and thiourea [55] inclusion compounds. The 1:3 adduct of adamantane with selenourea $[(CH_2)_6(CH)_4 \cdot 3(NH_2)_2CSe$, space group $R\bar{3}c$ with $a = 16.548(11)$, $c = 12.830(5)$ Å, $Z = 6$] was the first selenourea inclusion compound to be subjected to single crystal X-ray analysis [80]. Though the difference in channel diameter between thiourea and selenourea inclusion compounds is very small, as can be seen by comparing the above unit-cell dimensions with those of the adduct of the same adamantane guest with thiourea $[(CH_2)_6(CH)_4 \cdot 3(NH_2)_2CS$, space group $R\bar{3}c$ with $a = 16.187(7)$, $c = 12.578(7)$ Å, $Z = 6$], selenourea seems to be much more selective in its choice of guest molecules [28a].

There are more striking differences in the unit-cell dimensions of the selenourea inclusion compounds than are observed for the thiourea inclusion compounds, as adaptation of the selenourea lattice to the shape and size of the included molecules is apparently relatively easy [81–83]. A theoretical study of the conformations and vibrational frequencies in urea, thiourea, and selenourea compounds has been reported [84]. Studies on the complexation of Tl^+ with I^- and urea, thiourea, or selenourea (and their Me derivetives) using a horizontal zone electromigration method have indicated that the complexing capacities of the urea derivatives increase in the order urea < thiourea < N,N,N',N'-tetramethylthiourea < selenourea [85].

III. CRYSTAL STRUCTURES OF SOME UREA/THIOUREA/SELENOUREA–ANION INCLUSION COMPOUNDS

The strategy for the generation of inclusion compounds by the use of various organic host materials with appropriate guest molecules as templates has been shown to be an important approach in the structural design or modification of the host lattice, and has the potential of yielding fruitful results in crystal engineering.

Our interest in urea and thiourea adducts stems from an attempt to generate different inclusion topologies by incorporating additional components into the construction of the hydrogen-bonded host lattice. We consider the addition of a second anionic component, for example: (a) halide and pseudohalide ions that can play a bridging role in the cross-linkage of zigzag chains of hydrogen-bonded urea or thiourea molecules; (b) simple trigonal and square planar oxo-anions such as NO_3^-, CO_3^{2-}, $C_4O_4^{2-}$, HCO_3^-, and $BO(OH)_2^-$ which can easily form O···H–N acceptor hydrogen bonds (the last two anions also function as donors in hydrogen bonding), so that they can link together with urea or thiourea molecules to form novel building blocks; (c) the novel planar allophanate ion $NH_2CONHCO_2^-$; (d) anions of some monocarboxylic and dicarboxylic acids, such as HCO^{2-}, CH_3CO^{2-}, $HC_2O_4^-$ and $C_4H_2O_4^{2-}$; as well as (e) some nonplanar anions such as the pentaborate ion $[B_5O_6(OH)_4]^-$, which have functional O atoms or hydroxyl groups that can act as proton donors or acceptors in hydrogen bonding, as shown in Figure 7a.

Several symmetric tetraalkylammonium ions (R_4N^+, Figure 7b, **1**) have been explored as template guest moieties for the generation of new urea/thiourea/selenourea-anion host lattices. As a logical extension of the urea/thiourea-peralkylated ammonium salt family, we considered some unsymmetrical quaternary ammonium cations $R_3R'N^+$ (**2** and **3**, with smaller and larger disparity in size of R and R', respectively; and **4**, in which R and R' differ greatly in both size and shape), in order to study the influence of change in size and shape of the enclosed guests. Following this idea of altering the quaternary organic cation, we subsequently prepared analogous inclusion compounds containing quaternary phosphonium cations R_4P^+ (**5**). The choline ion $(CH_3)_3N^+(CH_2)_2OH$ (**6**) was also used in an attempt to investigate possible hydrogen bonding interaction between the host and guest components.

In our research program a wide variety of inclusion compounds of urea, thiourea, or selenourea with peralkylated ammonium salts have been prepared and unambi-

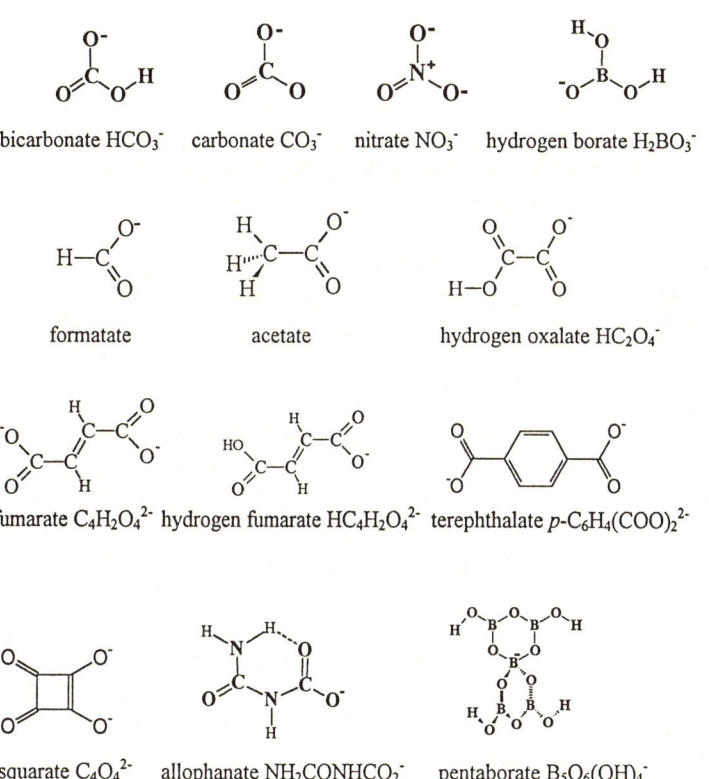

bicarbonate HCO_3^- carbonate CO_3^- nitrate NO_3^- hydrogen borate $H_2BO_3^-$

formatate   acetate   hydrogen oxalate $HC_2O_4^-$

fumarate $C_4H_2O_4^{2-}$ hydrogen fumarate $HC_4H_2O_4^{2-}$ terephthalate $p\text{-}C_6H_4(COO)_2^{2-}$

squarate $C_4O_4^{2-}$ allophanate $NH_2CONHCO_2^-$ pentaborate $B_5O_6(OH)_4^-$

Figure 7. (a) Anions used together with urea/thiourea/selenourea in the construction of new host lattices.

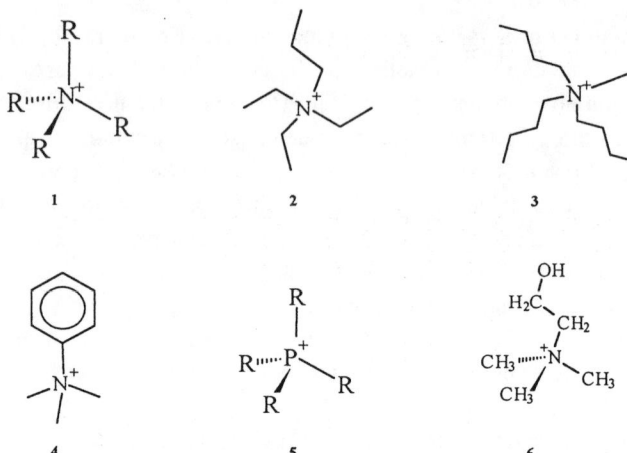

Figure 7. (b) Cationic guest moieties used in the generation of urea/thiourea/se-lenourea-anion host lattices.

guously characterized by single-crystal X-ray analysis. The crystallographic data of these compounds are summarized in Table 9. Some selected host lattices with novel structural features are presented in this section.

A. Urea–Anion Host Lattices

Open-Channel System in $2(n\text{-}C_3H_7)_4N^+F^-\cdot7(NH_2)_2CO\cdot3H_2O$ (1.1)

In complex **1.1** [*40f*], the urea molecules, water molecules, and fluoride ions constitute a three-dimensional host lattice containing an open-channel system running in the [010] direction, as shown in Figure 8. This urea–water–anion lattice is built from twisted urea ribbons running parallel to the [110] and [$\bar{1}$10] directions and cross-bridged by water molecules and fluoride anions via hydrogen bonds. Features of and relations between these ribbons and the bridging species may be conveniently described with reference to the hydrogen bonding scheme shown in Figure 9.

Five independent urea molecules **C(1)** [for simplicity the urea molecule composed of atoms C(1), O(1), N(1), and N(2) is designated as **C(1)**, other molecules are designated in the same manner], **C(2)**, **C(3)**, **C(4)**, and **C(5)** are arranged sequentially and connected by pairs of N–H···O hydrogen bonds in the usual shoulder-to-shoulder manner to form a pentamer, which serves as a repeating structural unit in generating an infinite urea ribbon. The torsion angles in the range 12.8° to 72.4° between adjacent urea molecules indicate that this ribbon has a highly twisted configuration. Two urea ribbons related by inversion symmetry are extended along the [110] direction, and at a location which is one-half translation

Table 9. Crystallographic Data of Some Inclusion Compounds of Urea/Thiourea/Selenourea with Peralkylated Ammonium Salts

No.	Formula	Space Group	Z	a	b	c (Å)	α	β	γ(°)	Ref.
1.1	$2(n\text{-}C_3H_7)_4N^+F^-\cdot 7(NH_2)_2CO\cdot 3H_2O$	$P2_1/c$	4	8.560(2)	16.301(3)	37.004(7)		92.31(3)		40f
1.2	$(n\text{-}C_3H_7)_4N^+Cl^-\cdot 2(NH_2)_2CO$	$P2_1/n$	4	9.839(2)	15.160(3)	14.583(3)		108.82(3)		40f
1.3	$(n\text{-}C_3H_7)_4N^+Cl^-\cdot 3(NH_2)_2CO$	$P2_1/n$	4	9.866(2)	16.274(3)	15.277(3)		103.36(3)		40f
1.4	$(n\text{-}C_3H_7)_4N^+Br^-\cdot 3(NH_2)_2CO\cdot H_2O$	$P\bar{1}$	2	8.857(2)	10.639(2)	15.115(3)	88.01(3)	75.02(3)	66.72(3)	40f
1.5	$(n\text{-}C_3H_7)_4N^+I^-\cdot 3(NH_2)_2CO\cdot H_2O$	$P\bar{1}$	2	9.045(2)	10.781(2)	15.160(3)	87.98(3)	76.00(3)	65.73(3)	40f
1.6	$(C_2H_5)_4N^+HCO_3^-\cdot (NH_2)_2CO\cdot 2H_2O$	$P2_1/n$	8	9.356(1)	29.156(4)	12.161(1)		90.03(1)		40b
1.7	$(n\text{-}C_4H_9)_4N^+HCO_3^-\cdot 3(NH_2)_2CO$	$P\bar{1}$	2	8.404(2)	12.352(2)	14.377(3)	88.20(2)	89.56(2)	71.68(1)	40b
1.8	$(CH_3)_4N^+NH_2CONHCO_2^-\cdot 5(NH_2)_2CO$	$P2_1/n$	4	9.553(4)	16.715(7)	15.576(5)		94.14(1)		40a
1.9	$(n\text{-}C_3H_7)_4N^+NH_2CONHCO_2^-\cdot 3(NH_2)_2CO$	Cc	4	8.811(2)	18.156(3)	16.394(8)		97.15(1)		40a
1.10	$(CH_3)_4N^+BO(OH)_2^-\cdot 2(NH_2)_2CO\cdot H_2O$	$Pnma$	4	16.671(2)	6.940(1)	13.180(2)				40g
1.11	$[(C_2H_5)_4N^+]_2CO_3^{2-}\cdot (NH_2)_2CO\cdot 2B(OH)_3\cdot H_2O$	$P2_1/n$	4	14.165(1)	13.753(1)	15.612(1)		108.87(1)		40g
1.12	$(n\text{-}C_3H_7)_4N^+[B_5O_6(OH)_4]^-\cdot 4(NH_2)_2CO\cdot H_2O$	$P2_1$	2	8.343(2)	16.037(3)	13.343(3)		104.75(3)		40d
1.13	$(n\text{-}C_4H_9)_4N^+[B_5O_6(OH)_4]^-\cdot 2(NH_2)_2CO\cdot B(OH)_3$	$P2_1/n$	4	11.582(3)	17.270(4)	17.819(5)		96.85(2)		40d
1.14	$(CH_3)_3N^+(CH_2)_2OH\cdot NH_2CONHCO_2^-\cdot (NH_2)_2CO$	$Pna2_1$	4	8.549(4)	12.583(2)	12.587(2)				40c
1.15	$(C_2H_5)_4N^+Cl^-\cdot 2(NH_2)_2CO$	$C2/m$	4	14.455(3)	6.920(1)	7.097(1)		115.80(3)		40e
1.16	$(C_2H_5)_4P^+Cl^-\cdot 2(NH_2)_2CO$	$P2_1/c$	4	10.492(6)	14.954(8)	10.335(6)		91.02(5)		40e
1.17	$[(n\text{-}C_4H_9)_4N^+]_2[p\text{-}C_6H_4(COO)_2]^{2-}\cdot 6(NH_2)_2CO\cdot 2H_2O$	$P\bar{1}$	1	8.390(2)	9.894(2)	18.908(3)	105.06(2)	94.31(1)	93.82(2)	40h
2.1	$(n\text{-}C_4H_9)_4N^+Cl^-\cdot 2(NH_2)_2CS$	$P2_1$	2	8.754(2)	8.857(2)	16.748(3)		92.00(3)		41c
2.2	$(C_2H_5)_4N^+HCO_3^-\cdot (NH_2)_2CS\cdot H_2O$	$Pbca$	8	8.839(2)	14.930(3)	24.852(5)				41a
2.3	$(n\text{-}C_3H_7)_4N^+HCO_3^-\cdot 2(NH_2)_2CS$	$C222_1$	8	8.521(3)	16.941(4)	32.022(7)				41a

(continued)

173

Table 9. Continued

No.	Formula	Space Group	Z	a	b	c (Å)	α	β	γ (°)	Ref.
2.4	$(n\text{-}C_3H_7)_4N^+HCO_3^-\cdot3(NH_2)_2CS$	$P\bar{1}$	2	9.553(2)	12.313(3)	14.228(4)	90.44(2)	103.11(2)	110.12(2)	41a
2.5	$(C_2H_5)_4N^+NO_3^-\cdot3(NH_2)_2CS$	$P\bar{1}$	4	10.300(2)	14.704(3)	15.784(4)	75.30(3)	86.98(3)	72.25(3)	41b
2.6	$(n\text{-}C_3H_7)_4N^+NO_3^-\cdot3(NH_2)_2CS\cdot H_2O$	$P2_1/n$	4	8.433(2)	9.369(2)	34.361(7)		91.01(3)		41b
2.7	$(CH_3)_4N^+NO_3^-\cdot(NH_2)_2CS$	$Pnma$	4	15.720(3)	8.218(2)	8.709(2)				41b
2.8	$(n\text{-}C_3H_7)_4N^+NO_3^-\cdot(NH_2)_2CS$	$P2_1/n$	4	8.784(2)	14.421(3)	15.078(3)		92.31(3)		41b
2.9	$(n\text{-}C_4H_9)_4N^+NO_3^-\cdot(NH_2)_2CS$	$Pna2_1$	4	19.934(3)	12.680(2)	9.092(3)				41b
2.10	$(C_2H_5)_4N^+HCO_2^-\cdot(NH_2)_2CS\cdot H_2O$	$P2_1/c$	4	7.199(2)	16.851(2)	13.044(2)		100.13(2)		41e
2.11	$2(C_2H_5)_4N^+HCO_2^-\cdot2(NH_2)_2CS\cdot HCO_2H$	$Pca2_1$	4	25.803(5)	7.190(2)	17.394(2)				41e
2.12	$(n\text{-}C_3H_7)_4N^+HCO_3^-\cdot3(NH_2)_2CS\cdot H_2O$	$P2_1/n$	4	8.533(2)	9.423(5)	33.517(7)		90.44(2)		41e
2.13	$(n\text{-}C_4H_9)_4N^+[(HCO_2)_2H]^-\cdot2(NH_2)_2CS$	$Pbca$	8	17.389(3)	16.622(2)	20.199(3)				41e
2.14	$(C_2H_5)_4N^+CH_3CO_2^-\cdot4(NH_2)_2CS$	$C2/c$	8	28.702(4)	8.457(1)	22.906(7)		98.91(1)		41f
2.15	$(n\text{-}C_3H_7)_4N^+[(CH_3CO_2)_2H]^-\cdot2(NH_2)_2CS$	$P2/n$	2	8.536(2)	8.613(1)	18.360(2)		90.66(2)		41f
2.16	$(n\text{-}C_4H_9)_4N^+[(CH_3CO_2)_2H]^-\cdot2(NH_2)_2CS$	$P\bar{1}$	2	8.771(3)	10.720(1)	16.742(2)	99.08(6)	94.07(2)	95.25(2)	41f
2.17	$(CH_3)_4N^+CH_3CO_2^-\cdot(NH_2)_2CS$	$P2_1/n$	4	8.421(2)	16.532(3)	8.628(4)		90.25(3)		41f

174

			Z	a	b	c	α	β	γ	
2.18	$(n\text{-}C_4H_9)_4N^+HC_2O_4^-\cdot2(NH_2)_2CS$	$P2_1/n$	4	8.854(6)	9.992(3)	32.04(2)		97.34(3)		41d
2.19	$(CH_3)_4N^+HC_4H_2O_4^-\cdot(NH_2)_2CS$	$P\bar{1}$	2	6.269(2)	8.118(4)	14.562(8)	104.79(4)	91.72(4)	101.30(4)	41d
2.20	$[(C_2H_5)_4N^+]_2C_4H_2O_4^{2-}\cdot2(NH_2)_2CS$	$P2_1/n$	2	11.340(2)	9.293(6)	14.619(2)		102.41(1)		41d
2.21	$(n\text{-}C_3H_7)_4N^+HC_4H_2O_4^-\cdot(NH_2)_2CS\cdot2H_2O$	$P2/n$	4	16.866(4)	8.311(1)	17.603(2)		104.94(1)		41d
2.22	$(n\text{-}C_4H_9)_3(CH_3)N^+HC_2O_4^-\cdot(NH_2)_2CS\cdot1/2H_2C_2O_4$	$Fdd2$	16	18.603(3)	63.661(8)	7.830(1)				41g
2.23	$[(C_2H_5)_3(n\text{-}C_3H_7)N^+]_2CO_3^{2-}\cdot6(NH_2)_2CS$	$P\bar{1}$	4	12.593(2)	13.075(2)	16.238(2)	76.16(1)	71.17(1)	66.32(1)	41g
2.24	$2[(CH_3)_3(C_6H_5)N^+]_2CO_3^{2-}\cdot11(NH_2)_2CS\cdot H_2O$	$P\bar{1}$	2	11.605(2)	17.059(3)	22.779(5)	109.46(3)	92.72(3)	107.07(3)	41g
2.25	$2[(C_2H_5)_4N^+]_2C_4O_4^{2-}\cdot4(NH_2)_2CS\cdot2H_2O$	$P2_1/n$	2	8.070(1)	28.365(5)	8.622(1)		91.28(1)		41h
3.1	$(C_2H_5)_4N^+Cl^-\cdot2(NH_2)_2CSe$	$P2_1/n$	4	8.768(5)	11.036(6)	19.79(1)		96.92(4)		42
3.2	$(n\text{-}C_3H_7)_4N^+Cl^-\cdot3(NH_2)_2CSe$	Cc	4	18.091(4)	13.719(3)	11.539(2)		111.93(3)		42
3.3	$(n\text{-}C_3H_7)_4N^+Br^-\cdot3(NH_2)_2CSe$	Cc	4	18.309(4)	13.807(3)	11.577(2)		112.45(3)		42
3.4	$(n\text{-}C_3H_7)_4N^+I^-\cdot(NH_2)_2CSe$	$P2_1/n$	4	8.976(1)	14.455(2)	15.377(3)		94.16(1)		42

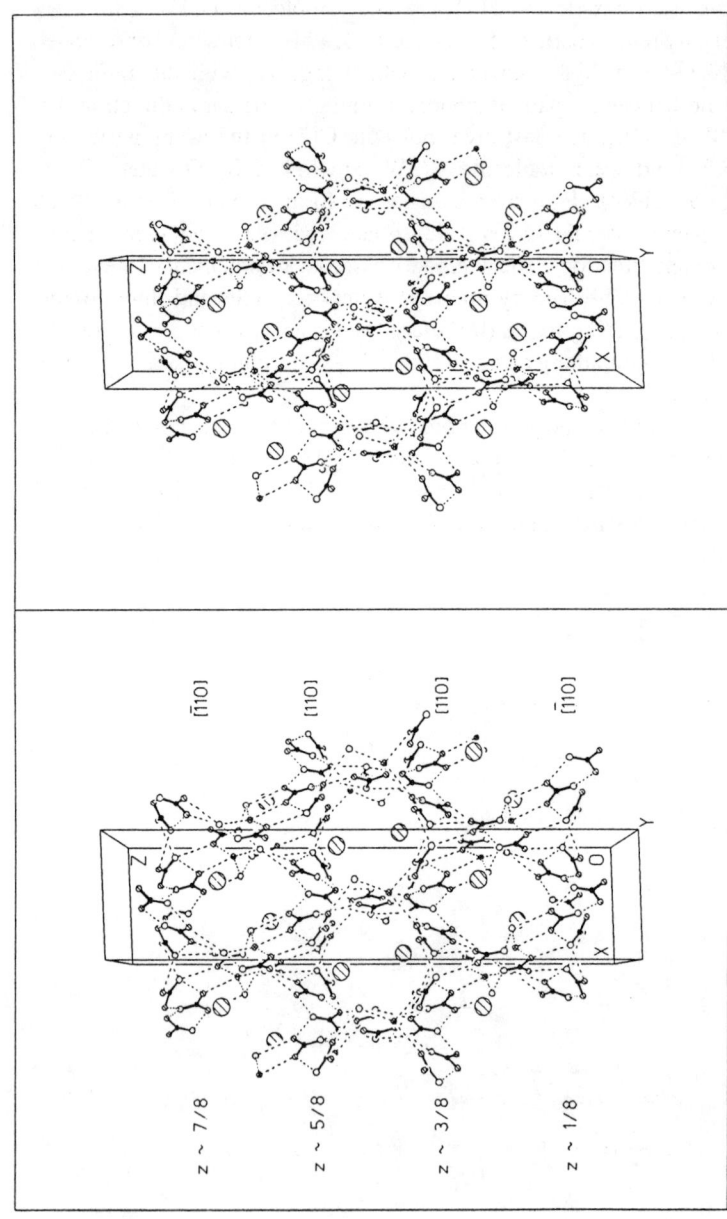

Figure 8. Stereodrawing showing the channel structure of $2(n\text{-}C_3H_7)_4N^+F^-\cdot7(NH_2)_2CO\cdot3H_2O$ **(1.1)**. The origin of the unit cell lies at the upper left corner, with a pointing from left to right, b towards the reader, and c downwards. The directions of the urea ribbons at different levels along the c axis are indicated. Broken lines represent hydrogen bonds. For clarity the enclosed $(n\text{-}C_3H_7)_4N^+$ ions are represented by large shaded circles. Note that urea ribbons at $z \sim 1/8$ and 7/8 extend in the [$\bar{1}10$] direction, whereas urea ribbons at $z \sim 3/8$ and 5/8 extend in the [110] direction.

176

away along the *c* axis, there are two similar ribbons which extend in the [$\bar{1}$10] direction (Figure 8). Fluoride ion F(1) and water molecule O(1W) plus their centrosymmetrically related partners form a cyclic $(H_2O \cdot F^-)_2$ tetramer consolidated by two pairs of O–H···F hydrogen bonds, which together with the sixth urea molecule **C(6)** lie between a pair of ribbons running in the same direction, i.e. parallel to [110] or [$\bar{1}$10]. The last urea molecule **C(7)** in the asymmetric unit, fluoride ion F(2), and water molecule O(3W) are linked by O(water)–H···O, O(water)–H···F, and N–H···F hydrogen bonds to form a cyclic trimer, which together with water molecule O(2W) are situated between two urea ribbons extending in different directions. The alternate crisscross arrangement of these urea ribbons, which are cross-bridged by urea and water molecules and fluoride ions lying in layers corresponding to the (004) family of planes, generates a three-dimensional host lattice containing an open-channel system running parallel to the *b* axis, as illustrated in Figure 8.

The basic component of the host lattice of **1.1** may also be described as a composite ribbon constructed from urea molecules **C(1)**–**C(6)** and $(H_2O \cdot F^-)_2$ units (Figure 9). These composite ribbons point alternately in the [110] and [$\bar{1}$10] directions, and their cross-linkage by **C(7)**, F(2), O(2W), and O(3W) generates the channel-type host network.

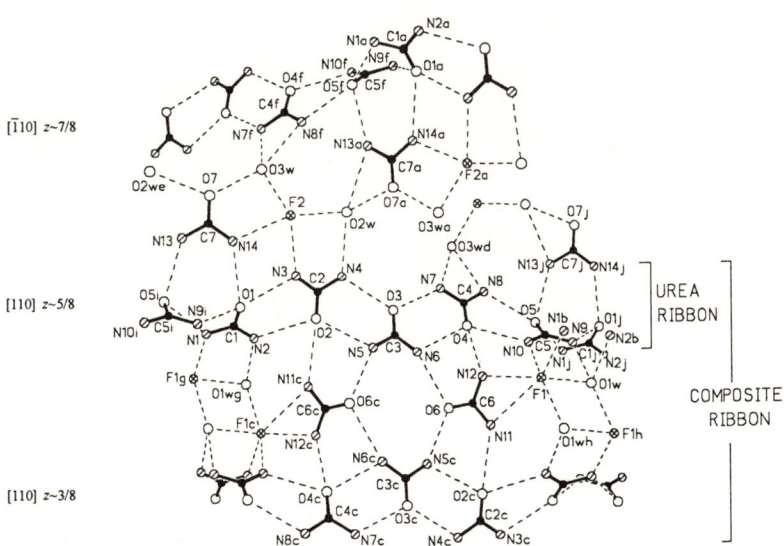

Figure 9. Projection drawing of the hydrogen-bonded urea ribbons in **1.1**. Portions of a urea ribbon and a composite ribbon are indicated. Broken lines represent hydrogen bonds, and atom types are distinguished by size and shading.

Cage-Type Host Lattice in $(CH_3)_4N^+NH_2CONHCO_2^-\cdot 5(NH_2)_2CO$ *(1.8)*

In the three-dimensional network structure of **1.8** [*40a*] (Figure 10), each tetra-methylammonium cation is enclosed in a box-like cage, the separation between individual pairs of opposite walls being 8.3 (*b*/2), 7.8 (*c*/2), and 7.3 Å, respectively. The allophanate ion and one independent urea molecule **C(1)** are joined together by a pair of N–H···O hydrogen bonds to form a structural unit. Repetition of this unit in the *a* direction generates a hydrogen-bonded zigzag ribbon, and a pair of adjacent ribbons related by an inversion center are cross-linked by additional N–H···O hydrogen bonds to form a nearly planar double ribbon (Figure 11). Two such double ribbons which are arranged parallel to the (010) plane constitute a pair of opposite walls of the hexahedral cage. Another pair of walls is formed by straight chains of urea molecules extending along the [101] direction, which are derived from two independent urea molecules **C(2)** and **C(3)** that are alternately linked by hydrogen bonds in a head-to-tail mode (Figure 12). The third pair of walls of the cage is constructed from urea molecules **C(4)** and **C(5)** joined in the usual shoulder-to-shoulder fashion via a pair of hydrogen bonds, as shown in Figure 12.

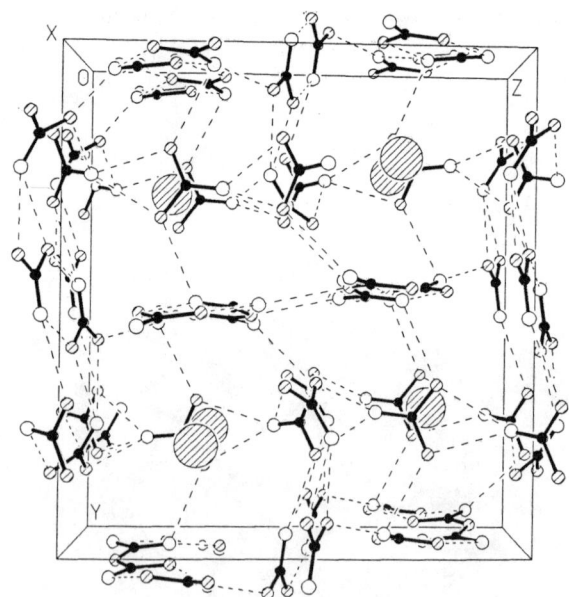

Figure 10. Crystal structure of $(CH_3)_4N^+NH_2CONHCO_2^-\cdot 5(NH_2)_2CO$ (**1.8**) along the *a* axial direction. Broken lines represent hydrogen bonds, and atom types are distinguished by size and shading. For clarity the enclosed $(CH_3)_4N^+$ ions are represented by large shaded circles.

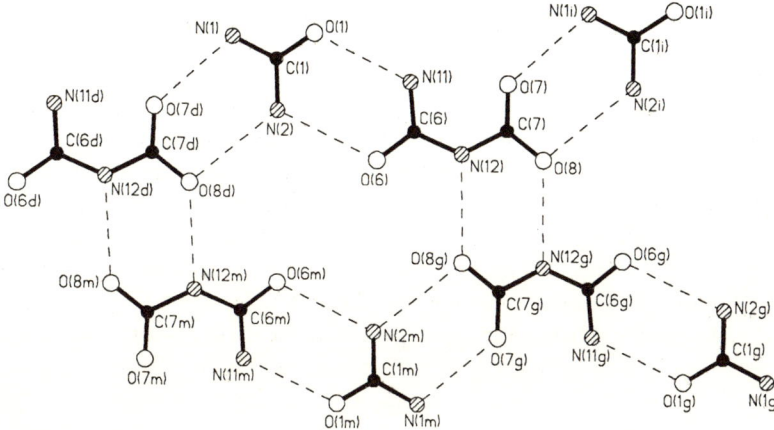

Figure 11. Perspective view of a portion of the planar double ribbon built by urea molecules and allophanate ions in **1.8**.

Channel with Peanut-Shaped Cross Section in $(CH_3)_4N^+BO(OH)_2^-\cdot 2(NH_2)_2CO\cdot H_2O$ (1.10)

In compound **1.10** [40g], both independent urea molecules and the $BO(OH)_2^-$ ion (dihydrogen borate, hereafter abbreviated as DHOB) in the asymmetric unit occupy special positions of symmetry m such that only the C–O or B–O double

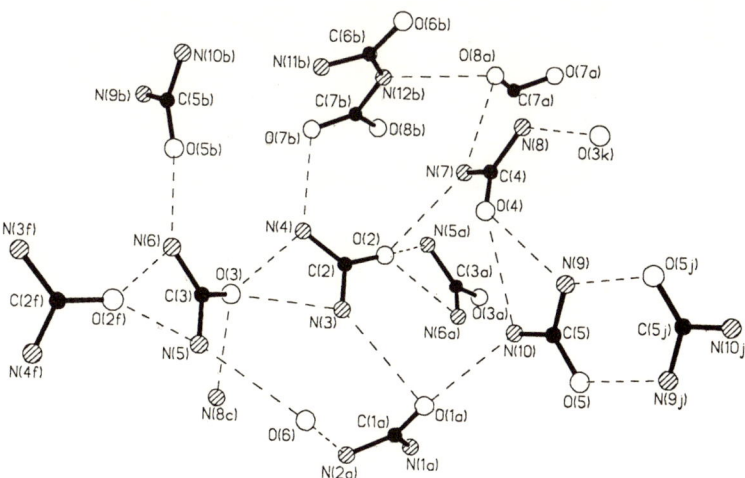

Figure 12. The head-to-tail hydrogen-bonded chains extending along the [101] direction, which constitute another pair of walls of the cage, and the shoulder-to-shoulder dimer of urea in the crystal structure of **1.8**.

bond lies on the mirror plane. A perspective view of the crystal structure of **1.10** along the [010] direction is presented in Figure 13. The host lattice consists of a parallel arrangement of unidirectional channels whose cross section has the shape of a peanut. The diameter of each spheroidal half is about 7.04 Å, and the separation between two opposite walls at the waist of the channel is about 5.85 Å. The well-ordered tetramethylammonium cations are accommodated in double columns within each channel.

As shown in Figure 14, the two independent urea molecules are alternately connected by hydrogen bonds to form a zigzag ribbon running parallel to b, which has a highly twisted configuration as shown by the torsion angles $C(1)$–$N(1a)\cdots O(2)$–$C(2) = 55.0(4)$ and $C(2)$–$N(2)\cdots O(1b)$–$C(1b) = 48.3(3)°$. The DHOB ions generated from successive inversion centers aligned parallel to the c direction are linked into a zigzag ribbon by O–H$\cdots$O hydrogen bonds, but the torsion angle between neighboring ions, $B(1)$–$O(4)\cdots O(3d)$–$B(1d) = -9.1(3)°$, shows that this ribbon is almost planar. Each DHOB anion is further connected to two urea molecules belonging to different neighboring urea ribbons (mutually related by a

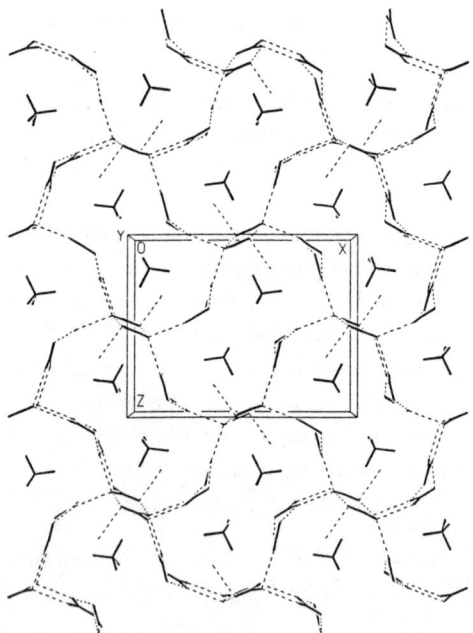

Figure 13. Crystal structure of $(CH_3)_4N^+BO(OH)_2^-\cdot 2(NH_2)_2CO\cdot H_2O$ (**1.10**) showing the channels extending parallel to the b axis and the enclosed cations. Note the peanut shape of the cross section of each channel and the hydrogen-bonded Y-shaped junction. Broken lines represent hydrogen bonds, and the atoms are shown as points for clarity.

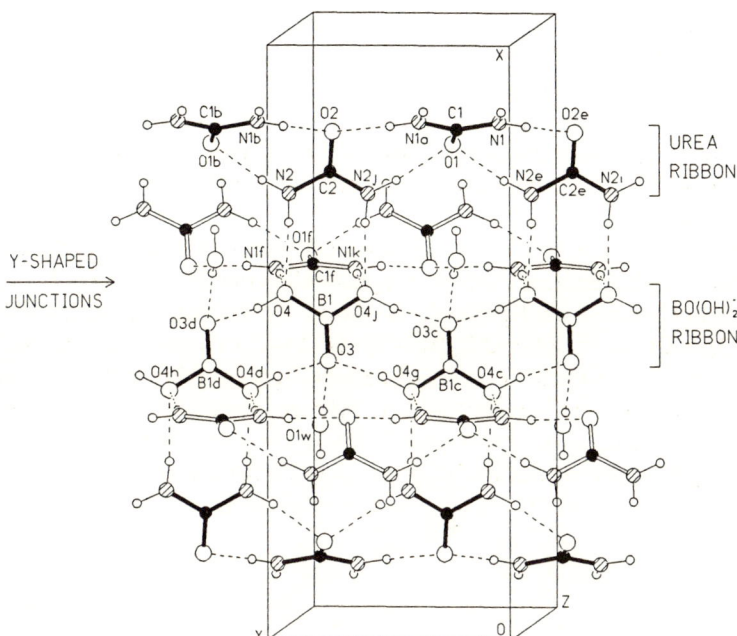

Figure 14. The hydrogen-bonding scheme in the urea-BO(OH)$_2^-$ host lattice of complex **1.10**. Broken lines represent hydrogen bonds.

2_1 axis) by four N–H···O hydrogen bonds that extend outward on the same side of the ribbon (Figure 14). The dihedral angles between the planar moieties involved in the cross-linkage are 106.7(3)° for urea with urea and 126.7(3)° for urea with DHOB, so that a cross section of each junction is shaped like the capital letter "Y." With these Y-shaped junctions oriented normal to (010) and concentrated at the layers $y = 1/4$ and 3/4, a unidirectional channel-like host lattice is formed (Figure 13). Notably, the water molecule makes no contribution to the construction of the host network, and its role is limited to that of an additional guest that forms a donor O–H···O hydrogen bond to a DHOB oxygen host atom.

Pentaborate Ion Taking the Unusual Hybrid Role of Both Host and Guest in
(n-C$_4$H$_9$)$_4$N$^+$[B$_5$O$_6$(OH)$_4$]$^-$ ·2(NH$_2$)$_2$CO·B(OH)$_3$ (1.13)

As shown in Figure 15, the host lattice in the crystal structure of **1.13** [40d] comprises a set of two-dimensional infinite layers extending parallel to the (202) plane with interlayer spacing d_{101}. One type of guest species, the tetra-n-butylam-monium cations (represented by large shaded circles in Figure 15), are sandwiched between adjacent layers, while another most unusual guest component, namely a hydrogen-bonded ribbon formed by B$_3$O$_3$(OH)$_2$ fragments belonging to different

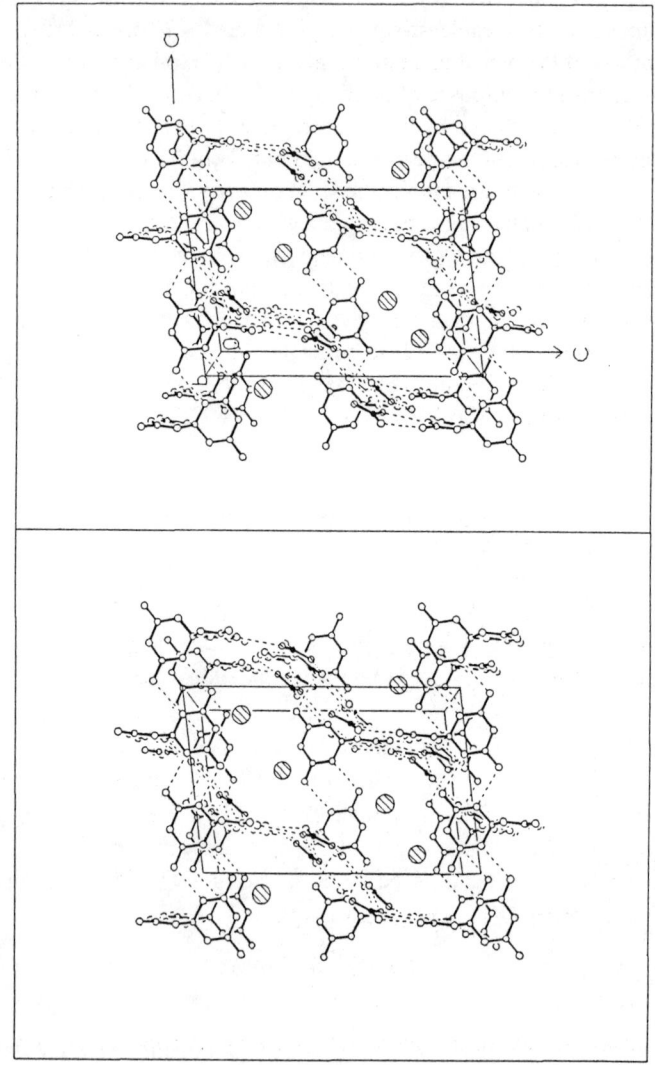

Figure 15. Stereodrawing of the crystal structure of $(n\text{-}C_4H_9)_4N^+[B_5O_6(OH)_4]^-\cdot 2(NH_2)_2CO\cdot B(OH)_3$ (**1.13**) showing the puckered layers parallel to (202) with interlayer spacing d_{101}, and planar $B_3O_3(OH)_2^-$ ribbons extending in the [010] direction. The origin of the unit cell lies at the upper left corner, with a pointing from left to right, b towards the reader and c downwards. Broken lines represent hydrogen bonds, and atom types are distinguished by size and shading. For clarity the enclosed $(n\text{-}C_4H_9)_4N^+$ ions are represented by large shaded circles.

pentaborate ions, threads a row of central holes of the macrocycles in a stacking of the host layers. The host architecture may be conveniently described with reference to the hydrogen bonding scheme shown in Figure 16. All atoms of the pentaborate group lie approximately in two planes each passing through an almost planar six-membered B_3O_3 ring and two oxygen atoms of the attached hydroxyl groups; the mean deviations from the best least-squares planes are 0.022 and 0.035 Å. The two planar fragments of the pentaborate group makes a dihedral angle of 88.0(4)°. One $B_3O_3(OH)_2$ fragment [composed of atoms B(1), B(2), B(3), O(1), O(2), O(5), O(7), and O(8); B(1) is also shared by the other fragment] lies in the main plane of the host layer and interacts with its nearest neighbors, forming a pair of N–H···O acceptor hydrogen bonds with urea molecule **C(2)** on one side, and both O···H–O acceptor and O–H···O donor hydrogen bonds with a boric acid molecule [composed of atoms B(6), O(13), O(14), and O(15), and hereafter abbreviated as **B(6)**] on the other side, to yield a trimeric aggregate. This trimer unit together with its centro-symmetrically related companion are linked by two pairs of hydrogen bonds involving boric acid **B(6)** and urea **C(2)** to generate a doughnut-like macrocyclic ring with a central hollow (Figure 16). The relevant torsion angles, C(2b)–N(3b)···O(14)–B(6) = 12.0(4)° and B(6)–O(15)···O(12b)–C(2b) = –4.5(4)°, show

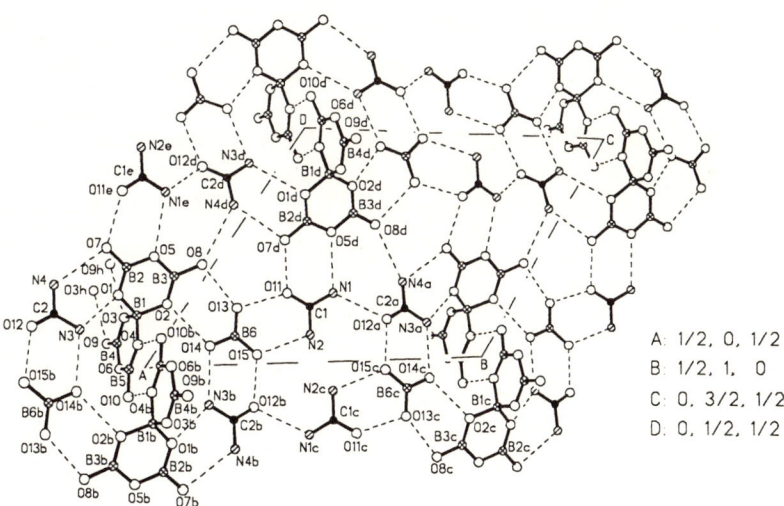

A: 1/2, 0, 1/2
B: 1/2, 1, 0
C: 0, 3/2, 1/2
D: 0, 1/2, 1/2

Figure 16. Hydrogen-bonded host layer in **1.13** showing an approximately planar array of macrocyclic rings formed by urea and boric acid molecules and pentaborate $B_3O_3(OH)_2$ fragments, plus the bridging urea molecules that weave them together. The corresponding pseudo-hexagonal "two-dimensional unit cell" is outlined and the coordinates of its corners listed. Note that a portion of a planar $B_3O_3(OH)_2$ ribbon passes obliquely through the central hole of each macrocyclic ring. Broken lines represent hydrogen bonds.

that boric acid **B(6)** and urea **C(2)** are nearly coplanar, but the $B_3O_3(OH)_2$ fragment is inclined to them at $C(2)-N(4)\cdots O(7)-B(2) = 34.2(4)°$ and $B(3)-O(8)\cdots O(13)-B(6) = -38.7(4)°$. The hydrogen-bonded macrocyclic rings constitute a pseudo-hexagonal array: the distances between the centers of adjacent macrocycles are $AD = 13.694(4)$ and $DC = 17.270(4)$ ($= b$) Å, and the angle ADC between these two repeat distances is $127.7(4)°$ (Figure 16). The urea molecule **C(1)** functions as a bridge in linking these macrocycles into a hydrogen-bonded puckered layer that is oriented parallel to the (202) plane (Figure 15), and the well-ordered tetrahedral $(n\text{-}C_4H_9)_4N^+$ cations are sandwiched between adjacent anionic urea–pentaborate–boric acid layers.

The remaining $B_3O_3(OH)_2$ fragment of the pentaborate ion [composed of atoms $B(1)$, $B(4)$, $B(5)$, $O(3)$, $O(4)$, $O(6)$, $O(9)$, and $O(10)$], together with identical units generated from a row of inversion centers, form an essentially planar hydrogen-bonded ribbon that is oriented parallel to (010). These ribbons extend parallel to the a axis and pass through the central holes of the macrocycles of stacked layers, so that the pentaborate ion plays a dual role as both host and guest.

The spacing between two adjacent urea–pentaborate–boric acid layers is 9.12 Å, which is considerably larger than the corresponding value (a = 8.40 Å) for tetra-n-butylammonium cations accommodated in the urea–bicarbonate layer structure [$40b$]. These parameters are consistent with the difference in size between the bicarbonate and pentaborate ions and the particular modes of molecular packing.

Compound **1.13** is of interest as the pentaborate ion takes the unusual hybrid role of both host and guest. From an alternative point of view, the pentaborate ion may be considered to participate fully in the construction of the host framework. In that case the resulting architecture resembles a multistoried building supported by leaning slab-like pillars.

Host–Guest Hydrogen Bonding in the Channel Inclusion Compound $(CH_3)_3N^+(CH_2)_2OH \cdot NH_2CONHCO_2^- \cdot (NH_2)_2CO$ (1.14)

The three-dimensional, open-channel host framework of **1.14** [$40c$] viewed along the c direction is presented in Figure 17. It can be seen that the choline cations are arranged in a single column about a 2_1 axis so that their hydroxyl groups point toward opposite channel walls to form donor hydrogen bonds with $O(4)$ atoms of the allophanate ions. This hydrogen bond is rather strong, as its length of 2.69(1) Å lies closer to the short side of general $O-H\cdots O$ hydrogen bonds in the range 2.40–3.10 Å [86].

A pair of hydrogen-bonded zigzag chains, composed of allophanate ions related by a 2_1 axis, constitute two opposite walls of a hexagonal channel extending along the [001] direction. The allophanate ions are connected to form a nearly planar ribbon, as illustrated by the value of the torsion angle $C(2)-N(3)\cdots O(3c)-C(3c) = -175.6(2)°$. Urea molecules related by the a glide are alternately linked by hydrogen bonds in a head-to-tail mode to generate an undulated chain extending along the

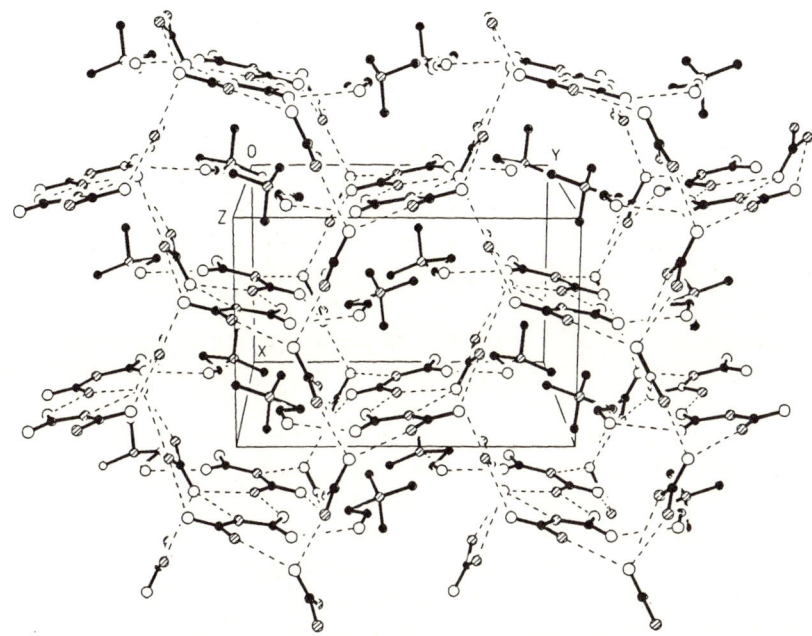

Figure 17. Three-dimensional host framework containing open channels in the crystal structure of $(CH_3)_3N^+(CH_2)_2OH\cdot NH_2CONHCO_2^-\cdot(NH_2)_2CO$ (**1.14**), viewed in a direction that makes a small inclined angle with respect to the *c* axis. Broken lines represent hydrogen bonds, and atom types are distinguished by size and shading (open circles for O, shaded circles for N, and filled circles for C).

[100] direction. Since the angle between two successive urea molecular planes along this chain is 129.0(2)°, two families of parallel urea chains directed along [001] cross-link the chains of allophanate ions to form a three-dimensional host framework containing open channels whose cross section is a distorted hexagon, whose three independent interior angles are 103.0(2), 128.0(2), and 129.0(2)° respectively (Figure 17).

The channel-type host structure of **1.14** can be compared with those of the classical urea inclusion compounds, in which the separation between the centers of two adjacent channels is 8.2 Å, leading to an effective cross-sectional diameter of 5.2 Å for the inclusion of aliphatic guest molecules. On the other hand, these two corresponding values in the host lattice of **1.14** are 8.5 and 5.5 Å, respectively, which are consistent with the greater bulk of the quaternary ammonium group in the choline cation. In the related urea–allophanate host lattice in $(n\text{-}C_3H_7)_4N^+\cdot NH_2CONHCO_2^-\cdot 3(NH_2)_2CO$ (**1.9**), the channel has an approximately elliptical shape with effective major and minor axes of 5.2/2 and 6.0/2 Å, respectively, and neighboring channels are separated by about 8.2 and 10.1 Å [*40a*].

Elongated Hexagonal Channel in
[(n-C₄H₉)₄N⁺]₂[p-C₆H₄(COO)₂]²⁻·6(NH₂)₂CO·2H₂O (1.17)

The asymmetric unit of the complex **1.17** [*40h*] consists of one-half of a centrosymmetric terephthalate anion, a tetra-*n*-butylammonium cation, three independent urea molecules, and one water molecule. Urea molecules **C(1)** and **C(2)** are connected by a pair of N–H···O hydrogen bonds to form a twisted cyclic dimer, the conformation of which can be described by the dihedral angle of 64.5(2)° between the urea molecules, and the torsion angles C(1)–N(1)–O(2)–C(2) = 43.8(2)° and C(2)–N(3)–O(1)–C(1) = 50.6(2)°. Pairs of centrosymmetrically related N–H···O hydrogen bonds centered at (0, 0, 0) and (0, 1/2, 0), respectively, link the urea dimers into a zigzag ribbon running parallel to the *b* axis. Note that the urea molecules are arranged about the inversion center at (0, 1/2, 0) in an unusual head-to-head fashion, so that each urea molecule of type **C(2)** is in a favorable orientation to form a pair of donor N–H···O hydrogen bonds to a neighboring carboxylate group. The terephthalate anions are orientated parallel to the *c* axial direction and bridge the urea ribbons to generate a puckered layer normal to the *a* axis (Figure 18).

As shown in the molecular packing diagram viewed along the *b* axis (Figure 19), the puckered layer has a stairs-like profile, in which the rigid terephthalate anions constitute the level portion with step length ca. 17.0 Å, and urea ribbons, the vertical portion with step height ca. 6.0 Å. The third independent urea molecule **C(3)** in the asymmetric unit connects adjacent layers via two donor and one acceptor N–H···O

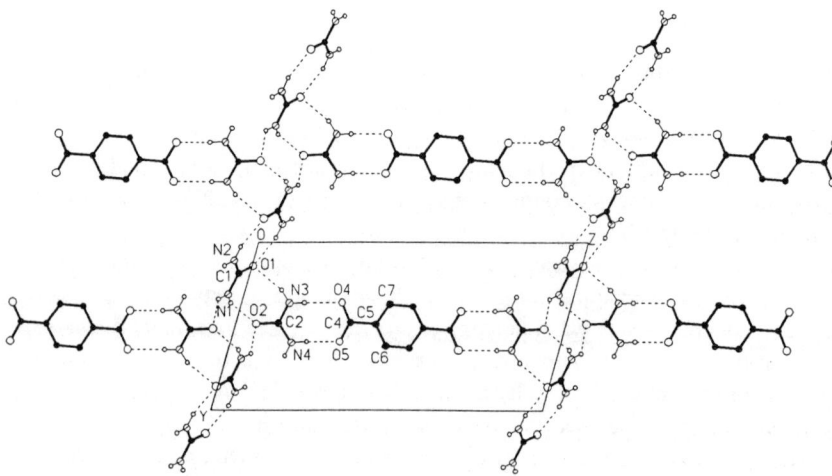

Figure 18. Hydrogen-bonded layer constructed from urea dimers and terephthalate anions in [(n-C₄H₉)₄N⁺]₂[p-C₆H₄(COO)₂]²⁻·6(NH₂)₂CO·2H₂O (**1.17**).

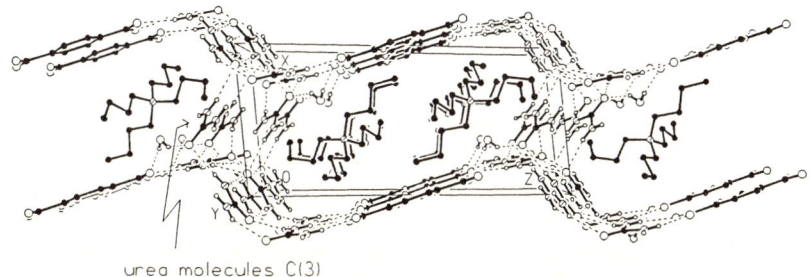

urea molecules C(3)

Figure 19. Perspective view down the *b* axis showing inclusion of the cationic guest molecules in the channel-type anionic host lattice of **1.17**.

hydrogen bonds with adjacent urea ribbons. The cross-linkage is further consolidated by a bridging water molecule which forms two O–H···O donor hydrogen bonds with separate O atoms of the terephthalate anion and urea molecule **C(3)**. A three-dimensional hydrogen-bonded framework is thus formed with large channels extending along the *b* axis with the dimension of the elongated hexagonal cross section of each channel being ca. 8.0 × 17.0 Å. The tetra-*n*-butylammonium ions are arranged in two columns within each channel in a well-ordered pattern (Figure 19).

B.   Thiourea–Anion Host Lattices

Composite Double Layers in the Host Lattice of
(n-C₃H₇)₄N⁺HCO₃⁻·2(NH₂)₂CS (2.3)

In the crystal structure of **2.3** [*41a*], both independent thiourea molecules are involved in forming zigzag, puckered ribbons (torsion angles C(2)–N(4)···S(1c)–C(1c) = 48.4(6)°, C(1)–N(2)···S(2a)–C(2a) = –47.3(6)°) running parallel to *a*, and these ribbons are aligned in such a way that they lie approximately in layers that are normal to the *c* axis. The bicarbonate ions form cyclic (HCO₃⁻)₂ dimers of symmetry 2 in space group *C*222₁, and each exocyclic O atom forms four acceptor hydrogen bonds with two adjacent thiourea ribbons belonging to the same layer. As a consequence of this unusual type of orthogonal cross-linkage, a composite double layer whose thickness equals the longest dimension of the (HCO₃⁻)₂ dimer emerges as the principal component of the organized hydrogen-bonded host lattice (Figure 20).

Figure 21 shows the crisscross arrangement of molecules and details of the hydrogen bonding between them. The four acceptor hydrogen bonds formed by O(3) with two molecules of thiourea resemble those formed by the carbonyl O atom in crystalline urea [*25b*] and its normal inclusion compounds [*28b*]. In the present structure the four N–H···O hydrogen bonds have an average N···O bond length of about 3.01 Å and are approximately coplanar; the angles between them are:

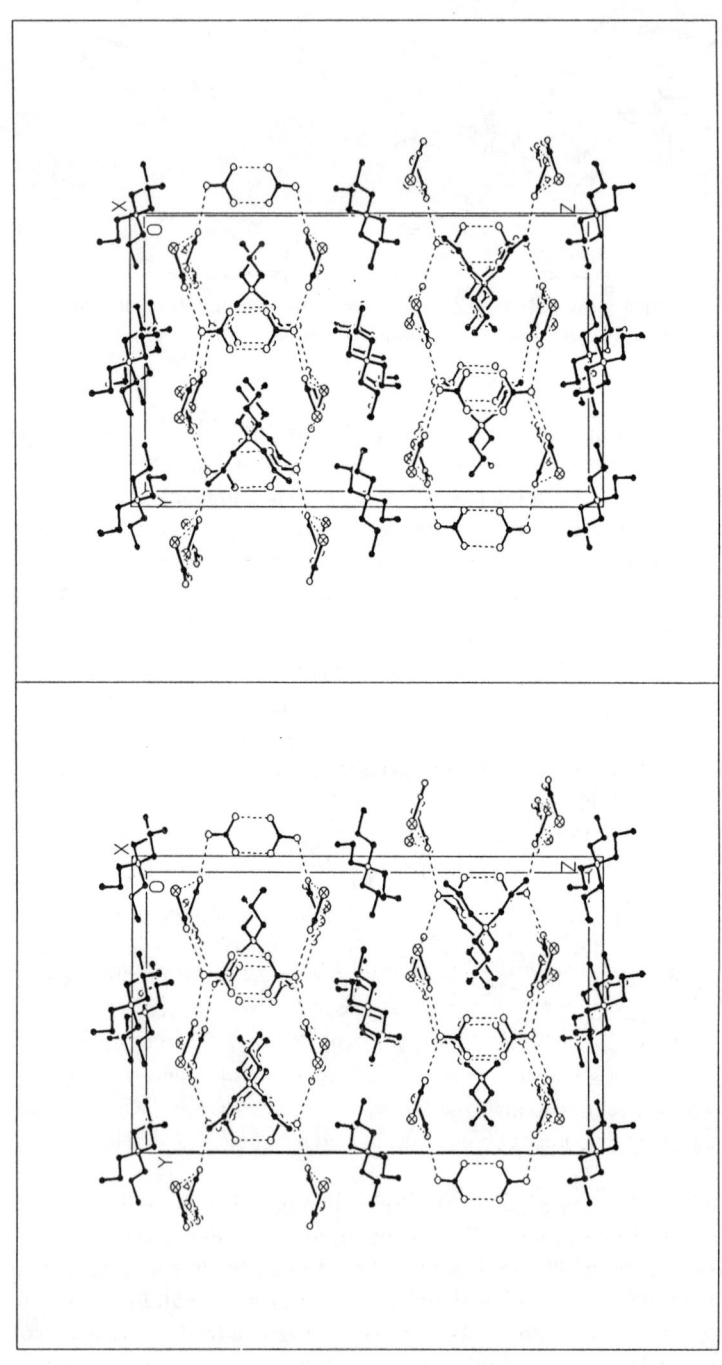

Figure 20. Stereodrawing of the crystal structure of **2.3** showing the double layers and the two different types of enclosed cations. The origin of the unit cell lies at the upper left corner, with *a* towards the reader, *b* from right to left, and *c* downwards. Broken lines represent hydrogen bonds.

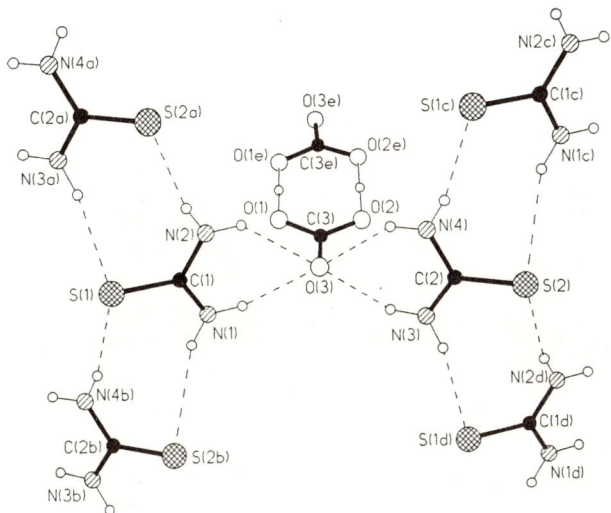

Figure 21. Perspective view of a portion of the thiourea-bicarbonate lattice in **2.3**.

N(1)···O(3)···N(2) = 43.7(6)°, N(2)···O(3)···N(4) = 116.5(6)°, N(3)···O(3)···N(4) = 43.4(6)°, and N(1)···O(3)···N(3) = 126.4(6)°, respectively, with a sum of 330.0°. The values of the torsion angles between the bridging bicarbonate dimer and thiourea ribbons, C(3)–O(3)···N(1)–C(1) = –91.7(6) and C(3)–O(3)···N(3)–C(2) = 80.7(6), clearly show their nearly orthogonal relationship in the construction of the double layer.

Both well-ordered tetrapropylammonium cations in the asymmetric unit occupy special positions of site symmetry 2 in space group $C222_1$: atoms N(5) and N(6) are situated at Wyckoff positions 4(b) at (1/2, y, 1/4) and 4(a) at (x, 0, 0), respectively. Figure 20 shows a honeycomb-like double layer at $z = 1/4$ with large octagonal windows and (n-$C_3H_7)_4$N(5)$^+$ cations trapped within it. When the crystal structure of **2.3** is viewed parallel to the a axis, the (n-$C_3H_7)_4$N(6)$^+$ cations are seen to be concentrated about the (002) planes and sandwiched between adjacent double layers.

Intersecting Channels in the Host Lattice of $(C_2H_5)_4N^+CH_3CO_2^-\cdot4(NH_2)_2CS$ (2.14)

In complex **2.14** [*41f*], the thiourea molecules and acetate ions build a three-dimensional host lattice in a 4:1 ratio, generating two channel systems that are arranged alternately along the [110] and [$\overline{1}$10] directions, as shown in Figures 22 and 23, respectively. The tetraethylammonium cations are accommodated in a single column within each channel. The thiourea–anion lattice comprises zigzag ribbons running parallel to the [110] and [$\overline{1}$10] directions, and parallel undulate

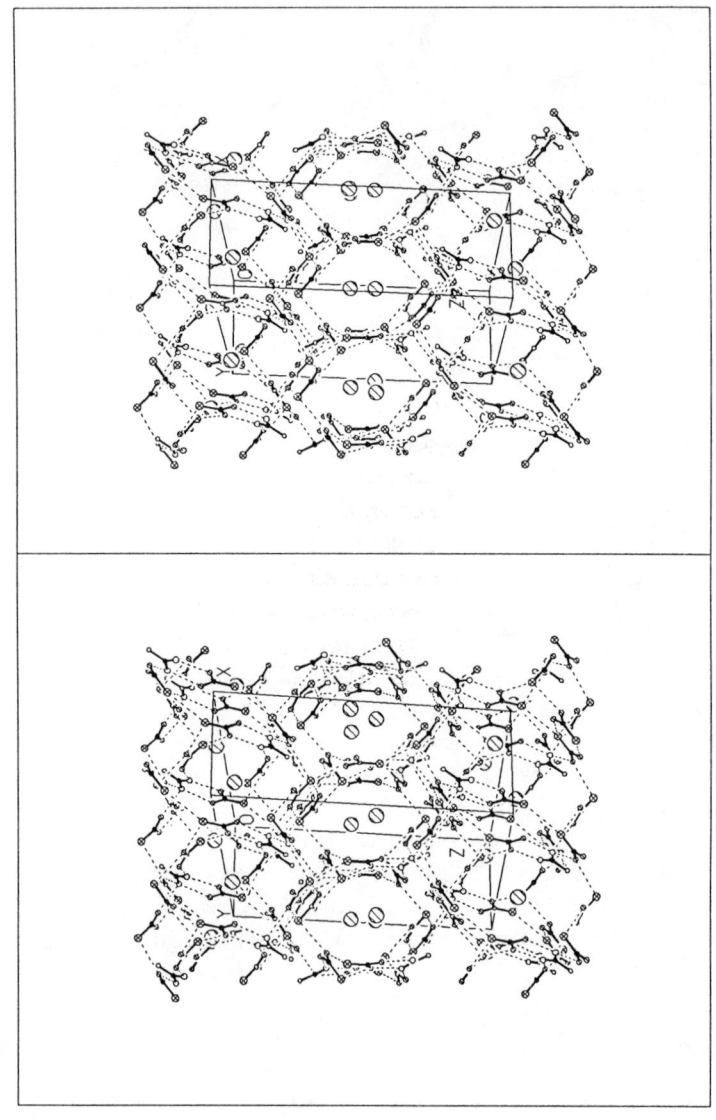

Figure 22. Stereodrawing of the crystal structure of $(C_2H_5)_4N^+CH_3CO_2^-\cdot4(NH_2)_2CS$ (**2.14**) showing the channels extending parallel to the [110] direction and the enclosed cations. Broken lines represent hydrogen bonds, and atom types are distinguished by size and shading. For clarity the enclosed $(C_2H_5)_4N^+$ ions are represented by large shaded circles.

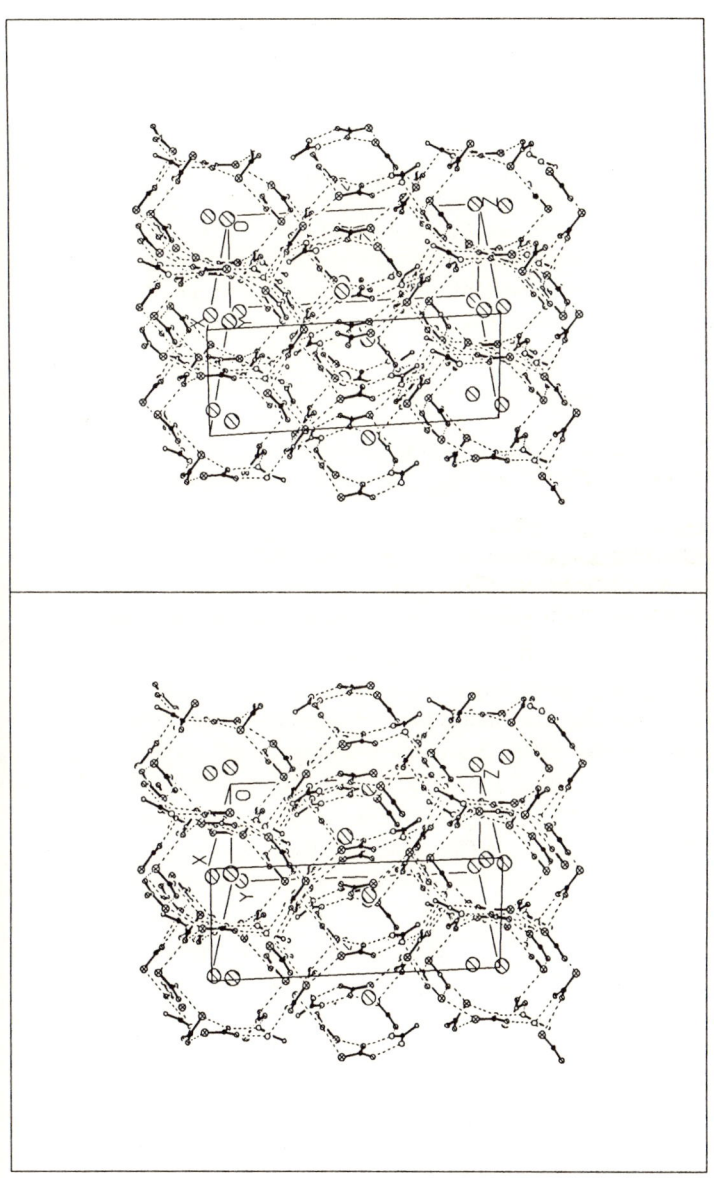

Figure 23. Stereodrawing of the crystal structure of **2.14** showing the channels extending parallel to the [$\overline{1}10$] direction and the enclosed cations. Broken lines represent hydrogen bonds, and atom types are distinguished by size and shading. For clarity the enclosed $(C_2H_5)_4N^+$ ions are represented by large shaded circles.

layers connected by hydrogen bonds which are oriented almost perpendicular to the ribbons and parallel to the (110) family of planes. As shown in Figure 22, the mean planes of these highly undulate layers are positioned at $z = 1/4$ and $3/4$. Features of the ribbons and relation between ribbons and layers may be conveniently described with reference to the hydrogen bonding scheme shown in Figure 24a.

The infinite chains running through the structure in the $[110]$ and $[\bar{1}10]$ directions, which form a pair of opposite walls of the corresponding channels, are generated by two independent thiourea molecules, $C(1)$ and $C(2)$, together with an acetate anion $C(6)$. Thiourea molecules $C(1)$ and $C(2)$ are linked by a pair of N–H$\cdots$S hydrogen bonds in a shoulder-to-shoulder fashion while $C(2)$ is further linked to an acetate ion via two N–H$\cdots$O donor hydrogen bonds to generate a trimer (Figure 24a). Two centrosymmetrically related trimers constitute a hexamer consolidated by a pair of N–H$\cdots$O hydrogen bonds between $C(2)$ and its symmetry equivalent partner $C(2d)$. With this hexamer as a repeating unit, formation of hydrogen bonds of the type between $C(1)$ and $C(6c)$ generates a wide ribbon running through the structure in the direction of $[110]$. At a location which is one-half translation away along the c axis, there is a similar ribbon which extends in the $[\bar{1}10]$ direction. In this ribbon the molecules originating from thiourea $C(2)$ and acetate $C(6)$ of the hexamer are essentially coplanar, as shown by the torsion angle $C(2)$–$N(4)\cdots O(2)$–$C(6) = -1.4(4)°$ and $C(2)$–$N(3)\cdots O(1)$–$C(6) = -13.1(4)°$. However, those derived from thiourea $C(1)$ are inclined with respect to the main plane of the ribbon, and the relevant torsion angles are $C(1)$–$N(2)\cdots S(2)$–$C(2) = 53.1(4)°$ and $C(1)$–$N(1)\cdots O(1c)$–$C(6c) = 121.1(4)°$, respectively.

The acetate ion $C(6)$ also contributes to the construction of a layer built of the other two thiourea molecules: thiourea molecules of type $C(3)$ related by a 2_1 axis (space group $C2/c$) are alternately linked by hydrogen bonds nearly in a head-to-tail mode to generate a highly twisted zigzag ribbon extending along the $[010]$ direction, as shown by the torsion angle $C(3)$–$N(5)\cdots S(3i)$–$C(3i) = -63.7(4)°$ (Figure 24b). Only one N–H$\cdots$S hydrogen bond is formed between adjacent molecules along the ribbon, so that each pair of thiourea molecules are also bridged by N–H$\cdots$O hydrogen bonds with the $O(2)$ atom belonging to an acetate ion. Two thiourea molecules $C(4)$ related by a 2 axis (space group $C2/c$) are linked by a pair of hydrogen bonds in the shoulder-to-shoulder fashion to form a rare dimer, in which they are almost mutually orthogonal, as shown by the torsion angle $C(4)$–$N(8)\cdots S(4e)$–$C(4e) = -83.0(4)°$. The twisted zigzag ribbons lying side by side in an alternate arrangement are cross-linked by the thiourea dimers via N–H$\cdots$S hydrogen bonds of the type between $C(3)$ and $C(4)$ to form a puckered layer as shown in Figure 24b. Cross-linkage of the ribbons and layers positioned at $z = 1/4$ and $3/4$ generates a three-dimensional host lattice containing two open-channel systems running parallel to the $[110]$ and $[\bar{1}10]$ directions, as illustrated in Figures 22 and 23, respectively.

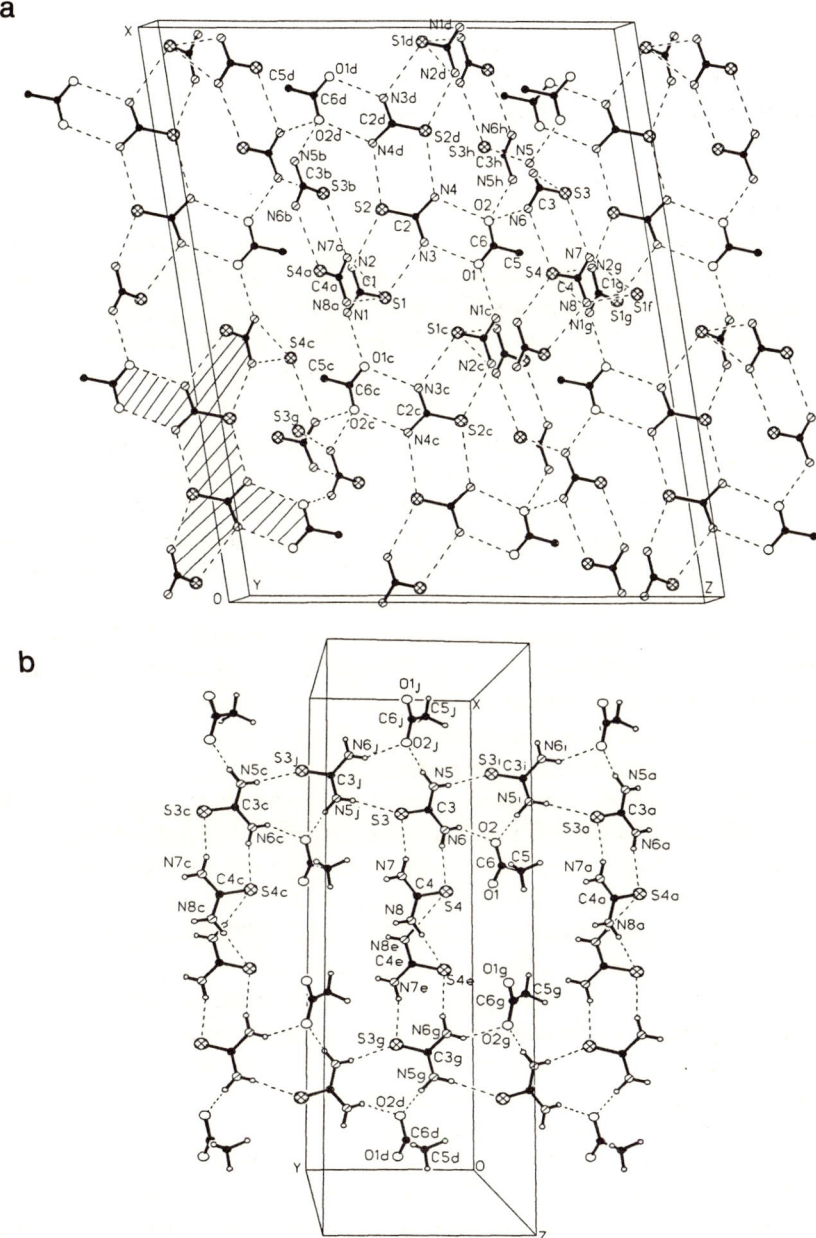

Figure 24. (a) The hydrogen-bonding scheme in the thiourea–anion host lattice of **2.14**, in which a centrosymmetric hexametric structure unit is highlighted by shading. (b) Projection drawing of the hydrogen-bonded thiourea–acetate anionic layer parallel to the (001) family of planes in **2.14**.

Two Open-Channel Systems in the Host Lattice of $[(C_2H_5)_3(n\text{-}C_3H_7)N^+]_2CO_3^{2-}\cdot6(NH_2)_2CS$ (2.23)

As shown in Figure 25 and Figure 26, compound **2.23** [*41g*] has two open-channel systems extending parallel to the [100] and [010] directions. This three-dimensional host framework is built of an alternate arrangement of layers of two types of thiourea double ribbons running in the corresponding channel directions, which are bridged by cabonate ions via N–H···O hydrogen bonds.

The two types of double ribbons lie in layers at $z = 1/2$ (type I) and $z = 0$ (type II) and extend parallel to the [100] and [010] directions, respectively. As illustrated in Figure 27a, the type I thiourea chain is generated by a trimeric unit comprising thiourea molecules **C(1)**, **C(2)**, and **C(3)**, and the C–N···S–C torsion angles between consecutive thiourea molecules are −41.7(9)°, −49.8(9)°, and −67.4(9)°. Two adjacent, parallel chains related by the inverse operation are cross-linked by pairs of N–H···S hydrogen bonds involving donor N atoms belonging to **C(1)** in one chain and acceptor S atoms belonging to **C(2)** and **C(3)** of the other chain, so that a double ribbon is formed. The torsion angles between two chains are C(1)–N(1)···S(2b)–C(2b) = −110.1(9)° and C(1)–N(2)···S(3b)–C(3b) = −96.8(9)°. These

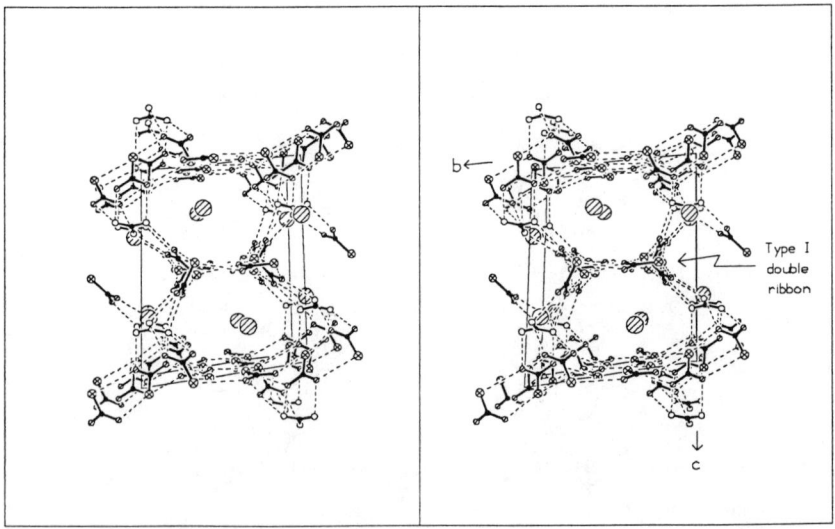

Figure 25. Stereodrawing of the crystal structure of $[(C_2H_5)_3(n\text{-}C_3H_7)N^+]_2$ $CO_3^{2-}\cdot6(NH_2)_2CS$ (**2.23**) showing the channels extending parallel to the [100] direction and the enclosed cations. The origin of the unit cell lies at the upper right corner, with *a* towards the reader, *b* pointing from right to left and *c* downwards. Broken lines represent hydrogen bonds, and atom types are distinguished by size and shading. For clarity the enclosed $(C_2H_5)_3(n\text{-}C_3H_7)N^+$ ions are represented by large shaded circles.

twisted thiourea double ribbons, lying side by side and extending parallel to the *a* axis, are located in the (002) plane as shown in Figure 25 and Figure 27a.

The type II double ribbon (Figure 26 and Figure 27b) is constructed from thiourea molecules **C(4)**, **C(5)**, and **C(6)** and has essentially the same structure as type I, except that the torsion angles between **C(4)** and adjacent molecules have much larger values, namely C(6g)–N(12g)···S(4)–C(4) = −71.5(9)° and C(4)–N(7)···S(5)–C(5) = −68.6(9)°.

The carbonate ion plays an important role in generating the host lattice. The O(1) and O(2) oxygen atoms each forms three acceptor hydrogen bonds in which two donor N atoms belong to adjacent double ribbons of the same type and one donor N atom belongs to a double ribbon of the other type: e.g. in the O(1)···N(8), O(1)···N(10c), and O(1)···N(3) hydrogen bonds, N(8) and N(10c) belong to Type II but N(3) belongs to Type I (Figure 27a). Thus a three-dimensional framework containing two open-channel systems is built from two different types of thiourea double ribbons that are bridged by carbonate anions. Stacked columns of well-ordered **N(13)** $(C_2H_5)_3(n\text{-}C_3H_7)N^+$ cations are accommodated in channels extending in the [100] direction (Figure 25), and likewise single columns of **N(14)** cations occupy channels extending in the [010] direction (Figure 26).

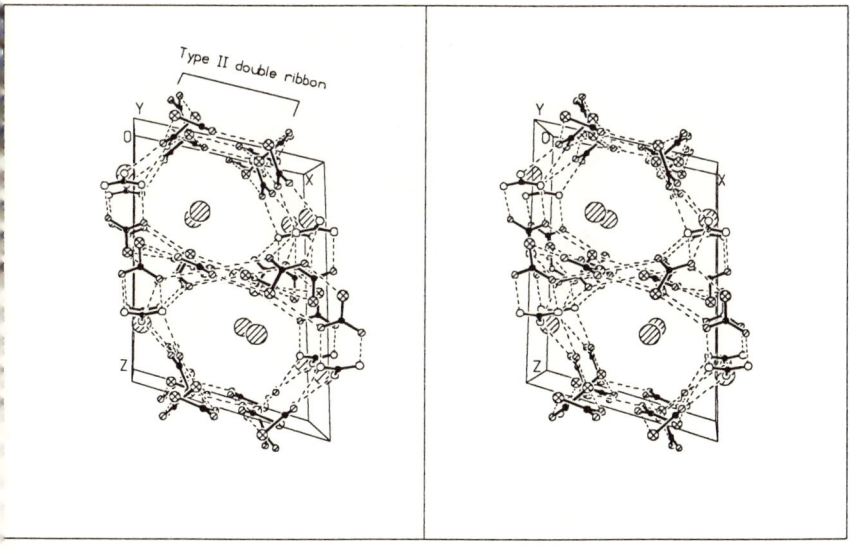

Figure 26. Stereodrawing of the crystal structure of **2.23** showing the channels extending parallel to the [010] direction and the enclosed cations. Broken lines represent hydrogen bonds, and atom types are distinguished by size and shading. For clarity the enclosed $(C_2H_5)_3(n\text{-}C_3H_7)N^+$ ions are represented by large shaded circles.

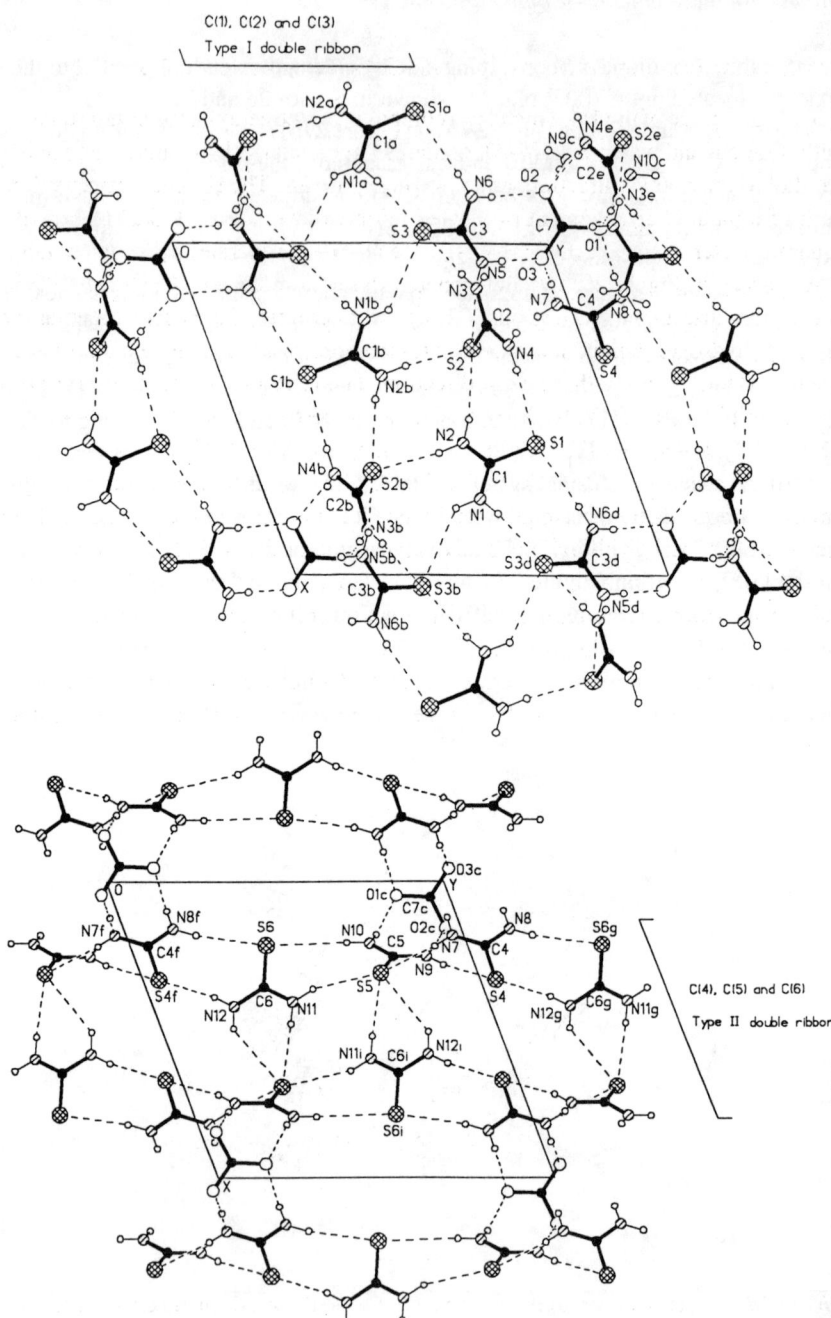

Figure 27. Hydrogen-bonded thiourea double ribbons in **2.23**. (a) Type I double ribbon running parallel to the *a* axis at $z = 1/2$; (b) Type II double ribbon running parallel to the *b* axis at $z = 0$. Broken lines represent hydrogen bonds.

Rectangular Channel in $[(C_2H_5)_4N^+]_2C_4O_4^{2-}\cdot4(NH_2)_2CS\cdot2H_2O$ (2.25)

The host lattice of the 1:4:2 inclusion compound of tetraethylammonium squarate with thiourea and water, **2.25** [*41h*], contains large rectangular channels extending parallel to the *a*-axial direction, as shown in Figure 28. The channel framework is built a parallel arrangement of pleated thiourea layers cross-linked by straight squarate-water chains, $[C_4O_4^{2-}\cdot(H_2O)_2]_\infty$, via N–H···O(squarate) hydrogen bonds.

As shown in Figure 29, two independent thiourea molecules **C(1)** and **C(2)** in the asymmetric unit (space group $P2_1/n$) are alternately connected together by N–H···S hydrogen bonds in a shoulder-to-shoulder fashion to generate a twisted ribbon extending along the *a*-axial direction. The relevant torsion angles in each thiourea ribbon are: C(1)–N(1)···S(2)–C(2) = –45.0(6)°, C(2)–N(3)···S(1)–C(1) = 52.4(6)°, C(1)–N(2)···S(2b)–C(2b) = 23.7(6)°, and C(2b)–N(4b)···S(1)–C(1) = 48.0(6)°, respectively. Atom S(2) of molecule **C(2)** forms two acceptor hydrogen bonds in the head-to-tail fashion with a type **C(1)** thiourea molecule belonging to an adjacent thiourea ribbon. This kind of cross-linkage gives pleated (or puckered) layers which are normal to the *b* axis and located at *b* = 1/4 and 3/4. The extent of puckering is quite large, as can be seen from the relevant torsion angles: C(1)–N(1)···S(2a)–C(2a) = –83.8(6)° and C(1)–N(2)···S(2a)–C(2a) = 98.9(6)°.

Pairs of centrosymmetrically related squarate dianions and bridging water molecules are connected by O1W···O1(squarate) hydrogen bonds to yield a seat belt-like

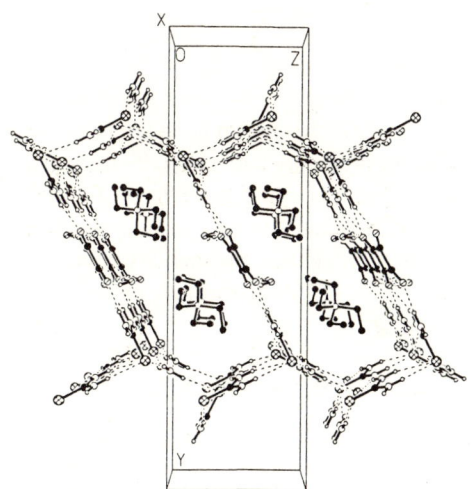

Figure 28. Rectangular channels running parallel to the *a* axis in the crystal structure of $[(C_2H_5)_4N^+]_2C_4O_4^{2-}\cdot4(NH_2)_2CS\cdot2H_2O$ (**2.25**). The origin of the unit cell lies at the upper left corner, with *a* pointing towards the reader, *b* downwards, and *c* from left to right. Note that only thiourea molecules of type **C(2)** interact with the squerate-water chains.

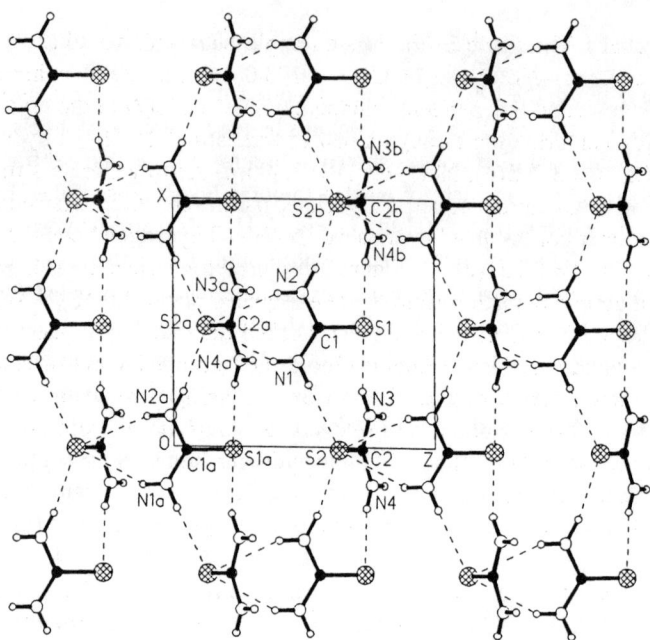

Figure 29. Hydrogen-bonded pleated thiourea layer in compound **2.25**. Broken lines represent hydrogen bonds.

chain running parallel to the a axis. These $[C_4O_4^{2-}\cdot(H_2O)_2]_\infty$ chains link two adjacent thiourea layers together by forming N–H···O (squarate) hydrogen bonds with thiourea molecules of type **C(2)** that protrude on both sides of each layer. Thus the space between the thiourea layers are partitioned by the squarate-water chains and a channel system is generated. In each channel the $[C_4O_4^{2-}\cdot(H_2O)_2]_\infty$ chains constitute two sidewalls which are parallel to $(0\ 2\ \bar{1})$, and the thiourea layers constitute the other two walls (Figure 28). Each channel has a rectangular cross-section of approximate dimensions 7.7×17.0 Å and accommodates two rows of stacked tetraethylammonium ions.

C. Selenourea–Anion Host Lattices

Selenourea–Chloride Puckered Layers in $(C_2H_5)_4N^+Cl^-\cdot2(NH_2)_2CSe$ (3.1)

In compound **3.1** [42], the cations are separated by selenourea–chloride puckered layers to generate a sandwich-like packing mode, and the layer structure may be conveniently described with reference to the hydrogen bonding scheme shown in Figure 30. An independent selenourea molecule **C(1)** is repeated by a 2_1 screw axis (space group $P2_1/n$), and pairs of adjacent molecules are interconnected by a single N–H···Se hydrogen bond of length N(2e)···Se(1) = 3.629(6) Å, to generate an infinite chain. The value of the torsion angle between two adjacent selenourea molecules, C(1e)–N(2e)···Se(1)–C(1) = –173.0(4)°, indicates that this chain is near planar. The second independent selenourea molecule **C(2)** in the asymmetric unit links with a chloride ion through a pair of "chelating" N–H···Cl hydrogen bonds [N(3)···Cl(1) = 3.205(6), N(4)···Cl(1) = 3.405(6) Å] to generate a dimer, which together with its centrosymmetrically related partner constitute a (selenourea–chloride)₂ tetramer consolidated by a pair of N–H···Cl hydrogen bonds [N(3)···Cl(1d) = N(3d)···Cl(1) = 3.251(6) Å] (Figure 30). This tetrameric unit is almost planar as the mean atomic deviation from the least-squares plane is 0.122 Å (including all non-hydrogen atoms in the tetramer). Linked alternately by pairs of N–H···Se hydrogen bonds between selenourea molecules arranged about inversion centers, these tetrameric units form a zigzag ribbon running parallel to the *b* axis. The selenourea–chloride ribbons and selenourea chains are alternately arranged and cross-linked by N–H···Cl and N–H···Se hydrogen bonds to form a puckered layer matching the (101) plane.

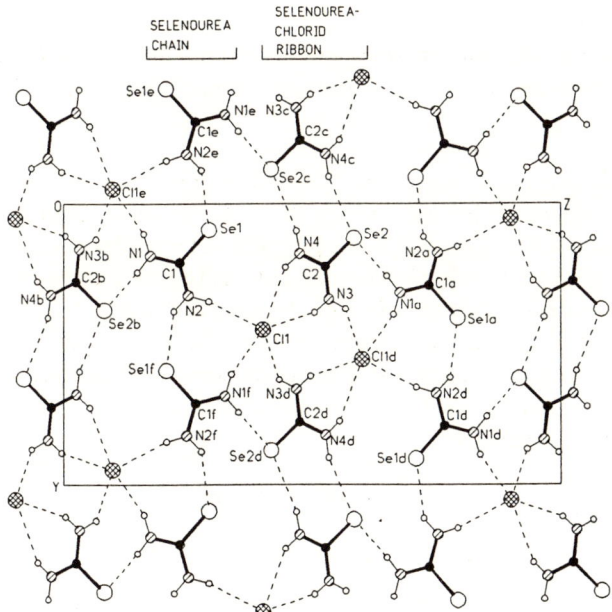

Figure 30. Projection drawing of the hydrogen-bonded layer in (C₂H₅)₄N⁺Cl⁻ ·2(NH₂)₂CSe (**3.1**) formed by the cross-linkage of selenourea chains and selenourea–chloride ribbons. Broken lines represent hydrogen bonds, and atom types are distinguished by size and shading.

One-Dimensional Open Channel System in $(n-C_3H_7)_4N^+Cl^-\cdot3(NH_2)_2CSe$
(3.2) and $(n-C_3H_7)_4N^+Br^-\cdot3(NH_2)_2CSe$ (3.3)

Complexes **3.2** and **3.3** [*42*] are virtually isostructural with the same basic
skeleton and only differ in the halide ions. The hydrogen bonding scheme of **3.2** is
shown in Figure 31. Two independent selenourea molecules of an asymmetric unit
(space group *Cc*) are connected by a pair of N–H⋯Se hydrogen bonds in a
shoulder-to-shoulder fashion to form a dimer, which is approximately planar as
shown by the C(1)–N(1)⋯Se(2)–C(2) torsion angle of –22.7° for **3.2** and –13.1°
for **3.3**. These selenourea dimers are interlinked laterally by additional N–H⋯Se
hydrogen bonds and bridging chloride ions on one side, each via three N–H⋯X
hydrogen bonds, with N(3a)⋯Cl(1) = 3.342(8), N(3)⋯Cl(1) = 3.434(8), and
N(4)⋯Cl(1) = 3.647(8) Å for **3.2**, and N(3a)⋯Br(1) = 3.522(8), N(3)⋯Br(1) =
3.403(8), and N(4)⋯Br(1) = 3.829(8) Å for **3.3**, respectively. On the other side, the
Se atom of the third independent selenourea molecule in the asymmetric unit
bridges adjacent selenourea dimers via three N–H⋯Se hydrogen bonds in the same
manner as a halide ion. Thus an anionic double ribbon running parallel to the [001]
direction is generated (Figure 31). For this buckled double ribbon the deviation of
its molecular components from coplanarity can be judged by the angle between the
least-square planes of a pair of adjacent dimers, which is 44.1(4)° for **3.2** and
42.9(4)° for **3.3**. The double ribbons are concentrated in layers that match the (020)
family of planes (Figure 32). The N atoms of the third bridging selenourea molecule

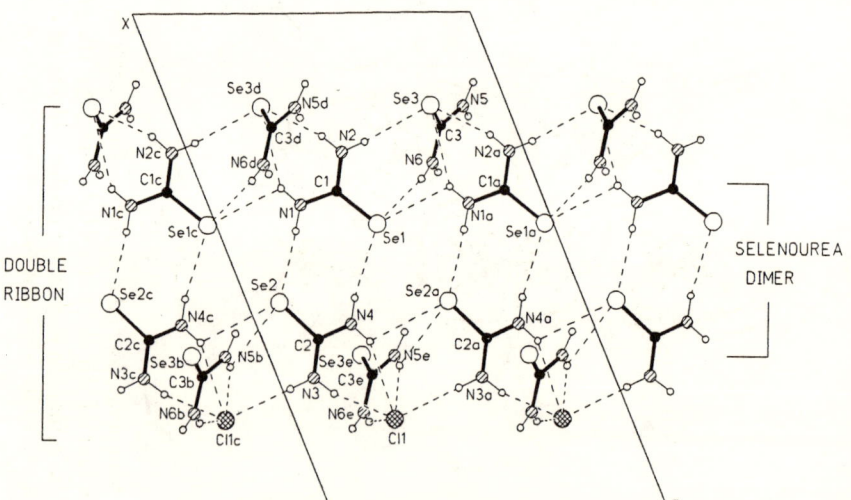

Figure 31. Selenourea-chloride double ribbon in $(n-C_3H_7)_4N^+Cl^-\cdot3(NH_2)_2CSe$ (**3.2**),
which is isostructural to the selenourea-bromide double ribbon in $(n-C_3H_7)_4N^+Br^-$
$\cdot3(NH_2)_2CSe$ (**3.3**). Broken lines represent hydrogen bonds.

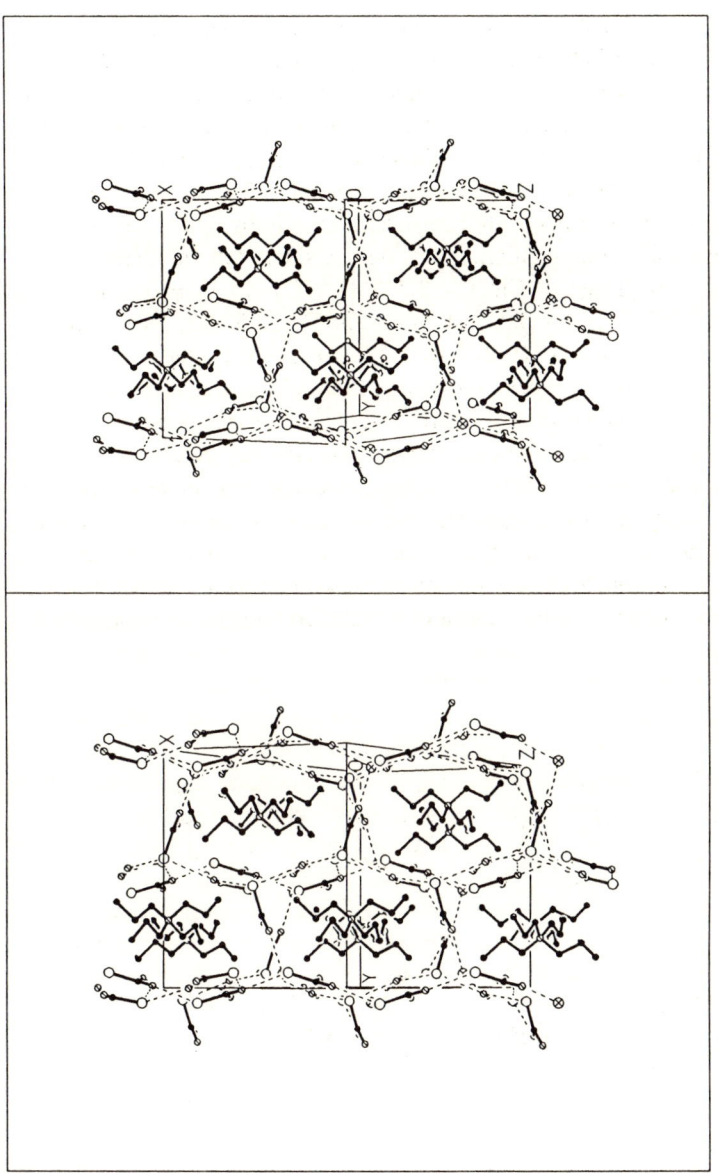

Figure 32. Stereodrawing of the crystal structure of **3.2** (or **3.3**) showing the accommodation of a single column of (n-C$_3$H$_7$)$_4$N$^+$ ions in a channel running parallel to the [101] direction. The host structure is built from selenourea–halide double ribbons and bridging selenourea molecules. Note that the double ribbons lie close to the (020) planes and are cross-linked by the bridging selenourea molecules. Broken lines represent hydrogen bonds, and atom types are distinguished by size and shading.

201

form a pair of N–H···X hydrogen bonds to a halide ion in the "chelating" fashion [N(5e)···Cl(1) = 3.219(8), N(6e)···Cl(1) = 3.127(8), N(5e)···Br(1) = 3.460(7), and N(6e)···Br(1) = 3.288(7) Å]. Since the Se atom of this selenourea molecule also forms three N–H···Se hydrogen bonds with two selenourea molecules in the ribbon that are related by a n glide, it plays a very similar and complementary role as a halide ion. Thus the wide selenourea–halide double ribbons are cross-linked by bridging selenourea molecules to generate a three-dimensional network containing an open-channel system running in the [101] direction (Figure 32).

IV. GENERAL STRUCTURAL FEATURES AND RELATIONSHIPS OF UREA/THIOUREA/SELENOUREA–ANION INCLUSION COMPOUNDS

A.   General Structural Features and Relationships

The results of our studies on urea/thiourea/selenourea–anion inclusion compounds have demonstrated that the classical urea or thiourea hydrogen-bonded host lattice can be modified in interesting ways by the incorporation of various anionic moieties, with or without cocrystallized water or other uncharged molecules, and that novel host frameworks bearing different urea, thiourea, or selenourea/guest molar ratios are generated by variation in size of the hydrophobic, pseudo-spherical R_4N^+ guest species. The stoichiometric formulas of 46 inclusion compounds and their structural details are listed in Table 10. For convenient description of stoichiometric ratios, the letters u, a, and c are used to denote the urea/thiourea/selenourea molecule, the anion, and the cocrystallized neutral molecule, respectively.

As envisaged, the inclusion complexes of urea, thiourea, or selenourea with quaternary ammonium salts exhibit many different inclusion topologies, such as a network of isolated cages (**1.8**, Figure 10), intersecting tunnels (**2.23**, Figures 25 and 26), two-channel systems (**2.14**, Figures 22 and 23), a channel with a peanut-shaped cross section in which two parallel columns of guest species can be accommodated (**1.10**, Figure 13), a double-layer system (**2.3**, Figure 20), wave-like layers (**1.16**, Figure 33), as well as the typical unidirectional channels and sandwich-like layers.

It is noted that the types of host lattices formed are dependent on the stoichiometric ratio of urea derivatives to anions and cocrystallized solvent molecules (water in most cases) or neutral molecules as additional host components.

The sole cage-type complex **1.8** has almost the largest u/a ratio (5:1) in the present series of inclusion compounds except **2.23** (6:1) and **2.24** (11:2), but clearly the guest species in the latter cases are larger and the incorporated anionic moieties are smaller than that in the first one.

The u/a ratios of all the channel-type compounds are higher or equal to 3:1, and if it is 2:1 then at least one cocrystallized solvent molecule is usually present, as in

Table 10. Types of Host Lattices in Inclusion Compounds of Urea Derivatives and Peralkylated Ammonium Salts

No.	Cation	Urea Derivative, u	Anion, a	Cocrystallized Neutral Molecule, c	u/a/c	Size Parameters[e]
Cage-Type						
1.8	$(CH_3)_4N^+$	$(NH_2)_2CO$	$NH_2CONHCO_2^-$		5/1	$7.3 \times 7.8 \times 8.3$
Channel-Type						
1.1	$(n\text{-}C_3H_7)_4N^+$	$(NH_2)_2CO$	F^-	H_2O	7/2/3	5.6
1.3	$(n\text{-}C_3H_7)_4N^+$	$(NH_2)_2CO$	Cl^-		3/1	5.2
1.4[a]	$(n\text{-}C_3H_7)_4N^+$	$(NH_2)_2CO$	Br^-	H_2O	3/1/1	7.5
1.5[a]	$(n\text{-}C_3H_7)_4N^+$	$(NH_2)_2CO$	I^-	H_2O	3/1/1	7.6
1.9	$(n\text{-}C_3H_7)_4N^+$	$(NH_2)_2CO$	$NH_2CONHCO_2^-$		3/1	5.2×6.1
1.10[a]	$(CH_3)_4N^+$	$(NH_2)_2CO$	$BO(OH)_2^-$	H_2O	2/1/1	7.04
1.12	$(C_2H_5)_4N^+$	$(NH_2)_2CO$	$[B_5O_6(OH)_4]^-$	H_2O	4/1/1	5.3×9.3
1.17	$(n\text{-}C_4H_9)_4N^+$	$(NH_2)_2CO$	$[p\text{-}C_6H_4(COO)_2]^{2-}$	H_2O	6/1/2	8.0×17.0
2.4[c]	$(n\text{-}C_4H_9)_4N^+$	$(NH_2)_2CS$	HCO_3^-		3/1	6.1×9.7
2.5	$(CH_3)_4N^+$	$(NH_2)_2CS$	NO_3^-		3/1	6.5

(continued)

203

Table 10. Continued

No.	Cation	Urea Derivative, u	Anion, a	Cocrystallized Neutral Molecule, c	u/a/c	Size Parameters[e]
2.6	$(n\text{-}C_3H_7)_4N^+$	$(NH_2)_2CS$	NO_3^-	H_2O	3/1/1	5.4×6.2
2.12	$(n\text{-}C_3H_7)_4N^+$	$(NH_2)_2CS$	HCO_2^-	H_2O	3/1/1	5.5×6.3
2.14[b]	$(C_2H_5)_4N^+$	$(NH_2)_2CS$	$CH_3CO_2^-$		4/1	5.5×8.5
2.21	$(n\text{-}C_3H_7)_4N^+$	$(NH_2)_2CS$	$HC_4H_2O_4^-$	H_2O	2/1/2	5.3×8.6
2.22[b]	$(n\text{-}C_4H_9)_3(CH_3)N^+$	$(NH_2)_2CS$	$HC_2O_4^-$	$H_2C_2O_4$	2/2/1	7.2×31.8
2.23[c]	$(C_2H_5)_3(n\text{-}C_3H_7)N^+$	$(NH_2)_2CS$	CO_3^{2-}		6/1	7.4×11.4
2.24	$(CH_3)_3(C_6H_5)N^+$	$(NH_2)_2CS$	CO_3^-	H_2O	11/2/1	6.7
2.25	$(C_2H_5)_4N^+$	$(NH_2)_2CS$	$C_4O_4^{2-}$	H_2O	4/1/2	7.7×17.0
3.2	$(n\text{-}C_3H_7)_4N^+$	$(NH_2)_2CSe$	Cl^-		3/1	6.8×7.3
3.3	$(n\text{-}C_3H_7)_4N^+$	$(NH_2)_2CSe$	Br^-		3/1	6.9×7.4
1.14	$(CH_3)_3N^+(CH_2)_2OH$	$(NH_2)_2CO$	$NH_2CONHCO_2^-$		1/1/1	5.2
Layer-Type						
1.2	$(n\text{-}C_3H_7)_4N^+$	$(NH_2)_2CO$	Cl^-		2/1	7.58
1.6	$(C_2H_5)_4N^+$	$(NH_2)_2CO$	HCO_3^-	H_2O	1/1/2	7.29
1.7	$(n\text{-}C_4H_9)_4N^+$	$(NH_2)_2CO$	HCO_3^-		3/1	8.40
1.11	$(C_2H_5)_4N^+$	$(NH_2)_2CO$	CO_3^{2-}	$B(OH)_3, H_2O$	1/1/3	6.88
1.13	$(n\text{-}C_4H_9)_4N^+$	$(NH_2)_2CO$	$[B_5O_6(OH)_4]^-$	$B(OH)_3$	2/1/1	9.64
1.16	$(C_2H_5)_4N^+$	$(NH_2)_2CO$	Cl^-		2/1	7.48
1.17[d]	$(C_2H_5)_4P^+$	$(NH_2)_2CO$	Cl^-		2/1	9.20
2.1	$(n\text{-}C_4H_9)_4N^+$	$(NH_2)_2CS$	Cl^-		2/1	9.31
2.2	$(C_2H_5)_4N^+$	$(NH_2)_2CS$	HCO_3^-	H_2O	1/1/1	7.47

2.3	$(n\text{-}C_3H_7)_4N^+$	$(NH_2)_2CS$	HCO_3^-	2/1		8.01
2.10	$(C_2H_5)_4N^+$	$(NH_2)_2CS$	HCO_2^-	1/1/1	H_2O	7.20
2.11	$(C_2H_5)_4N^+$	$(NH_2)_2CS$	HCO_2^-	2/1/1	HCO_2H	7.19
2.13	$(n\text{-}C_4H_9)_4N^+$	$(NH_2)_2CS$	$[(HCO_2)_2H]^-$	2/1		8.70
2.15[d]	$(n\text{-}C_3H_7)_4N^+$	$(NH_2)_2CS$	$[(CH_3CO_2)_2H]^-$	2/1		8.61
2.16	$(n\text{-}C_4H_9)_4N^+$	$(NH_2)_2CS$	$[(CH_3CO_2)_2H]^-$	2/1		8.37
2.18	$(n\text{-}C_4H_9)_4N^+$	$(NH_2)_2CS$	$HC_2O_4^-$	2/1		7.70
2.20	$(C_2H_5)_4N^+$	$(NH_2)_2CS$	$C_4H_2O_4^{2-}$	2/1		7.31
3.1	$(C_2H_5)_4N^+$	$(NH_2)_2CSe$	Cl^-	2/1		8.77

Separated Ribbons

2.7	$(CH_3)_4N^+$	$(NH_2)_2CS$	NO_3^-	1/1
2.8	$(n\text{-}C_3H_7)_4N^+$	$(NH_2)_2CS$	NO_3^-	1/1
2.9	$(n\text{-}C_4H_9)_4N^+$	$(NH_2)_2CS$	NO_3^-	1/1
2.17	$(CH_3)_4N^+$	$(NH_2)_2CS$	$CH_3CO_2^-$	1/1
2.19	$(CH_3)_4N^+$	$(NH_2)_2CS$	$HC_4H_2O_4^-$	1/1
3.4	$(n\text{-}C_3H_7)_4N^+$	$(NH_2)_2CSe$	I^-	1/1

Notes: [a]Cross-section has the shape of a peanut in which two parallel columns of guests can be accommodated.
[b]Dual channel systems.
[c]Intersecting tunnels.
[d]Wave-like layer.
[e]Cavity dimensions for cage type, dimensions of cross-section for channel type, or interlayer spacing for layer type.
[f]Stoichiometric ratio of urea : anion : cocrystallized solvent molecule or additional neutral host molecule.

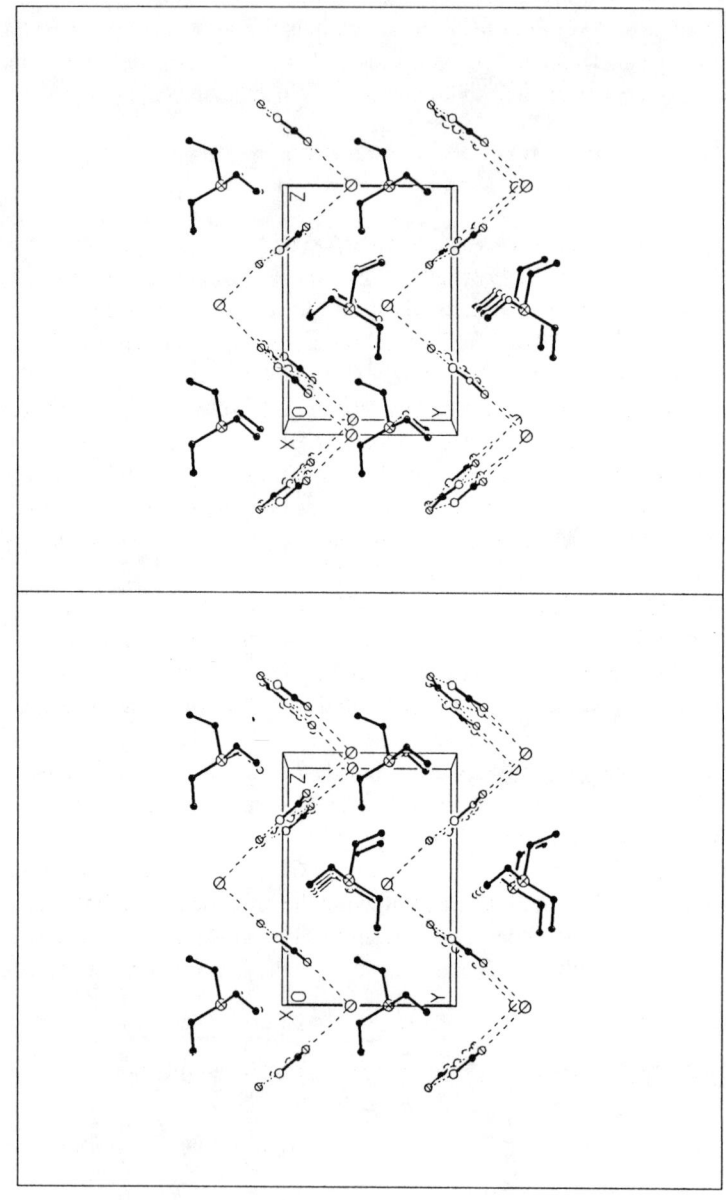

Figure 33. Stereodrawing of the crystal structure of **1.16**. The origin of the unit cell lies at the upper left corner, with *a* pointing towards the reader, *b* downwards, and *c* from left to right. Broken lines represent hydrogen bonds, and atom types are distinguished by size and shading.

examples **1.10** (2:1:1) and **2.21** (2:1:2). In the two exceptions, **2.22** (2:2:1) and **1.14** (1:1:1), it can be considered that the larger anionic or neutral moieties play important roles in the construction of the host lattices.

When the complexes have a low u/a ratio and without the presence of other neutral molecules, only a parallel arrangement of separate ribbons can be constructed from the thiourea or selenourea with anionic building blocks, as in **2.7** and **3.4**.

B. Hydrogen Bonding in Urea/Thiourea/Selenourea–Anion Inclusion Compounds

Only van der Waals contacts are made between the included guest molecules and the molecules or anions of the host lattices in almost all inclusion compounds, except a complex consolidated by host–host and host–guest hydrogen bonding [urea–choline allophanate (**1.14**)]. The fundamental role of hydrogen bonding in stabilizing the host lattices is so outstandingly important that regiospectific hydrogen bonds control the way in which small molecular units, such as the present urea derivatives and various anions, form aggregate host lattices. It can be seen that the requirement for strong host–host hydrogen bonding in the maintenance of a host lattice need not be restrictive on the shape or dimensions of the lattice. Hydrogen bonding dimensions, particularly angles, are variable, and host–host hydrogen bonding networks can possess a substantial flexibility.

We now present results illustrating that the N–H···O and N–H···S hydrogen bonds play an instrumental role in the structures of a novel class of urea/thiourea inclusion compounds. The average lengths of 84 measured N–H···O hydrogen bonds in 16 urea–anion inclusion complexes and 114 measured N–H···S hydrogen bonds in 23 thiourea–anion host lattices are listed in Table 11.

Although these bond lengths vary over a wide range, the average values of N–H···O and N–H···S are in good agreement with those found for several types of hydrogen bonds, N···O = 2.93 ± 0.11 and N···S = 3.40 ± 0.20 Å [87].

Generally, the hydrogen bond length is related to the environment and configuration at both donor and acceptor atoms, following the order head-to-tail ("chelating") mode < shoulder-to-shoulder fashion < singly H-bond. An excellent example is provided by the box-like cage-type urea–tetramethylammonium allophanate

Table 11. Average Lengths N···O and N···S (Å) in Urea/Thiourea–Anion Host Lattices

	Min.	*Max.*	*Average*	*Number of H Bonds*	*Number of Complexes*
N–H···O	2.828(5)	3.194(8)	2.964 ± 0.071	84	16
N–H···S	3.281(9)	3.990(3)	3.481 ± 0.096	114	23

compound (**1.8**, Figures 11 and 12), in which these three types of hydrogen bonds all exist and the average values of each type, 2.898(9), 2.990(9), and 3.043(9) Å respectively, are in accord with this order.

The N–H$\cdots$Se hydrogen bond is of interest as few reported data are available. The measured N$\cdots$Se distances [ranging from 3.386(8) to 3.741(8) Å, with an average value of 3.527(7) Å] for **3.1–3.4** are comparable to those found in the selenourea–adamantane inclusion complex, 3.51(2) and 3.65(2) Å, for N–H$\cdots$Se hydrogen bonds between chains and within chains, respectively [*80*]. The weakest N–H$\cdots$Se hydrogen bond in the present series of inclusion compounds occurs in the singly hydrogen-bonded selenourea chain in complex **3.1**, and the bridging selenourea molecule between double ribbons in complexes **3.2** and **3.3**. There is a distinct difference between the average value of the present N$\cdots$Se distances and above-mentioned mean N$\cdots$S bond length (3.481 Å with standard deviation 0.096 Å).

It is noted that the N–H$\cdots$X hydrogen bonds in the present series of selenourea–halide complexes play an important role to consolidate the host lattices. In compounds **3.1**, **3.2**, and **3.3** there are five N–H$\cdots$X hydrogen bonds per chloride or bromide ion. The 10 measured N$\cdots$Cl distances range from 3.127(8) to 3.647(8) Å with an average value of 3.304(8) Å, and five N$\cdots$Br distances from 3.288(7) to 3.829(7) Å with an average value of 3.502(7) Å. The respective averages lie within the range of N–H$\cdots$Cl [N$\cdots$Cl 3.206(7) to 3.393(9) Å] and N–H$\cdots$Br [N$\cdots$Br 3.353(9) to 3.53(2) Å] bond lengths compiled from 29 N–H$\cdots$Cl hydrogen bonds in eight complexes and 6 N–H$\cdots$Br hydrogen bonds in three complexes possessing urea or thiourea–halide host lattices (Table 12). In inclusion compound **3.4** the measured "chelating" N–H$\cdots$I hydrogen bonds [N$\cdots$I 3.565(8) and 3.651(8) Å, with an average value of 3.608(8) Å] and the N$\cdots$I$\cdots$N angle [36.8(6)°] are in good agreement with those [3.622(5) Å and 36.4(3)°] determined for the related thiourea inclusion compound (n-C$_3$H$_7$)$_4$N$^+$I$^-$·(NH$_2$)$_2$CS [*88*].

Notably the cyclic hydrogen-bonded water–halide dimers (H$_2$O–X$^-$)$_2$ (X = F, Br, I) make their presence in the inclusion compounds of urea with tetra-n-propylammonium halides. As shown in Table 13, the dimensions of these centrosymmetric cyclic moieties can be compared with those reported in a variety of complexes, such as [Zn(en)$_3$]F$_2$·2H$_2$O (en = ethylenediamine) (**1a**) [*89*], (C$_2$H$_5$)$_4$N$^+$Cl$^-$·(NH$_2$)$_2$CO·2H$_2$O (**2a**), (C$_2$H$_5$)$_4$N$^+$Cl$^-$·H$_2$O (**2b**) [*90*], 2[Mo(S$_2$CNEt$_2$)$_4$]$^+$·(H$_2$O·Cl$^-$)$_2$· xCHCl$_3$ ($x \cong$ 1.76, **2c**) [*91*], (C$_2$H$_5$)$_4$N$^+$Br$^-$·(NH$_2$)$_2$CO·2H$_2$O (**3a**) [*92*], and [(CH$_2$)$_6$N$_4$CH$_3$]Br·H$_2$O (**3b**) [*93*].

The C–H$\cdots$O hydrogen bond exists in only one case (**2.11**, Figure 34), in which the formyl atom **C(5)** forms an C–H$\cdots$O hydrogen bond with the carboxy O atom of **C(3b)** with H$\cdots$O = 2.303(6) Å, C$\cdots$O = 3.213(6) Å, and C–H$\cdots$O = 157.9(4)°. The H$\cdots$O distance is shorter than the sum of the relevant van der Waals radii (W$_H$ + W$_O$ = 2.4 Å) and lies closer to the lower end of the H$\cdots$O range of 1.5 to 3.0 Å for C–H$\cdots$O hydrogen bonding established by accurate X-ray and neutron diffraction analyses [*94, 95*], indicating that the formyl CH group has a relatively strong

Table 12. N–H⋯X and O–H⋯X Hydrogen Bonds (Å) in Some
Urea/Thiourea/Water–Halide Host Lattices

	Compound	Min.	Max.	Number of H Bonds/X⁻	Ref.
N–H⋯X	**In urea-halide host lattices**				
	$(C_2H_5)_4N^+Cl^-\cdot2(NH_2)_2CO$	3.254(5)	3.368(5)	4	40e
	$(C_2H_5)_4P^+Cl^-\cdot2(NH_2)_2CO$	3.331(9)	3.402(9)	4	40e
	$(n\text{-}C_3H_7)_4N^+Cl^-\cdot2(NH_2)_2CO$	3.317(7)	3.366(7)	4	40f
	$(n\text{-}C_3H_7)_4N^+Cl^-\cdot3(NH_2)_2CO$	3.303(6)	3.424(6)	6	40f
	In urea-halide-water host lattices				
	$2(n\text{-}C_3H_7)_4N^+F^-\cdot7(NH_2)_2CO\cdot3H_2O$	2.737(8)	3.187(8)	4, F(1)	40f
		2.805(8)	2.914(8)	2, F(2)	
	$(C_2H_5)_4N^+Cl^-\cdot(NH_2)_2CO\cdot2H_2O$	3.379(5)		1	92
	$(C_2H_5)_4N^+Br^-\cdot(NH_2)_2CO\cdot2H_2O$	3.53(2)		1	92
	$(n\text{-}C_3H_7)_4N^+Br^-\cdot3(NH_2)_2CO\cdot H_2O$	3.491(3)		1	40f
	$(n\text{-}C_3H_7)_4N^+I^-\cdot3(NH_2)_2CO\cdot H_2O$	3.722(4)		1	40f
	In thiourea-halide host lattices				
	$(n\text{-}C_4H_9)_4N^+F^-\cdot3(NH_2)_2CS$	2.819(5)	2.994(7)	6	88
	$(n\text{-}C_4H_9)_3N(CH_3)^+Cl^-\cdot2(NH_2)_2CS$	3.252(3)	3.291(3)	4	88
	$(n\text{-}C_4H_9)_4N^+Cl^-\cdot2(NH_2)_2CS$	3.206(7)	3.321(7)	4	41c
	$(n\text{-}C_4H_9)_3N(CH_3)^+Br^-\cdot2(NH_2)_2CS$	3.353(6)	3.459(6)	4	88
	$(n\text{-}C_3H_7)_4N^+I^-\cdot(NH_2)_2CS$	3.564(5)	3.680(5)	2	88
O–H⋯X	**In water-halide host lattices**				
	$4(C_2H_5)_4N^+F^-\cdot11H_2O$	2.59(1)	2.76(1)	4, F(1)	96
		2.63(1)	2.72(1)	4, F(2)	
		2.65(1)	2.85(1)	4, F(3)	
		2.61(1)	2.70(1)	4, F(4)	
	$(C_2H_5)_4N^+Cl^-\cdot4H_2O$	3.177(7)	3.244(7)	4	97
	In urea-halide-water host lattices				
	$2(n\text{-}C_3H_7)_4N^+F^-\cdot7(NH_2)_2CO\cdot3H_2O$	2.620(8)	2.716(8)	2, F(1)	40f
		2.471(8)	2.705(8)	2, F(2)	
	$(C_2H_5)_4N^+Cl^-\cdot(NH_2)_2CO\cdot2H_2O$	3.178(5)	3.223(5)	3	92
	$(C_2H_5)_4N^+Br^-\cdot(NH_2)_2CO\cdot2H_2O$	3.16(2)	3.36(2)	3	92
	$(n\text{-}C_3H_7)_4N^+Br^-\cdot3(NH_2)_2CO\cdot H_2O$	3.361(3)	3.390(3)	2	40f
	$(n\text{-}C_3H_7)_4N^+I^-\cdot3(NH_2)_2CO\cdot H_2O$	3.570(4)	3.672(4)	2	40f

Table 13. Structural Parameters for (H₂O–X⁻)₂ Systems in Crystalline Complexes

Compound	O(W)···X (Å)	O(W)···X' (Å)	X···O···X' (°)	Site Symmetry	Ref.
(H₂O–F⁻)₂ in **1a**	2.586	2.679		none	89
(H₂O–F⁻)₂ in **1.1**	2.620(8)	2.716(8)	100.6(5)	Ī	40f
(H₂O–Cl⁻)₂ in **2a**	3.178(5)	3.223(5)	107.2(4)	Ī	92
(H₂O–Cl⁻)₂ in **2b**	3.204(4)	3.247(4)	103.8(1)	Ī	90
(H₂O–Cl⁻)₂ in **2c**	3.287(8)	3.311(8)	108.9(4)	Ī	91
(H₂O–Br⁻)₂ in **3a**	3.17(2)	3.36(2)	101(1)	Ī	92
(H₂O–Br⁻)₂ in **3b**	3.338(5)	3.386(5)	105.7(4)	Ī	93
(H₂O–Br⁻)₂ in **1.4**	3.361(3)	3.390(3)	113.1(2)	Ī	40f
(H₂O–I⁻)₂ in **1.5**	3.570(8)	3.672(8)	110.0(5)	Ī	40f

donor capability. In the inclusion compound (CH₃)₃N⁺(CH₂)₂OH·NH₂CONHCO₂⁻ ·(NH₂)₂CO (**1.14**) the allophanate O atoms of the resulting hydrogen-bonded hydrophilic host lattice form acceptor hydrogen bonds with the hydroxyl groups of choline guests [Figure 35b]. This may be contrasted with the classical urea or thiourea inclusion compounds formed by self-assembly, as well as inclusion

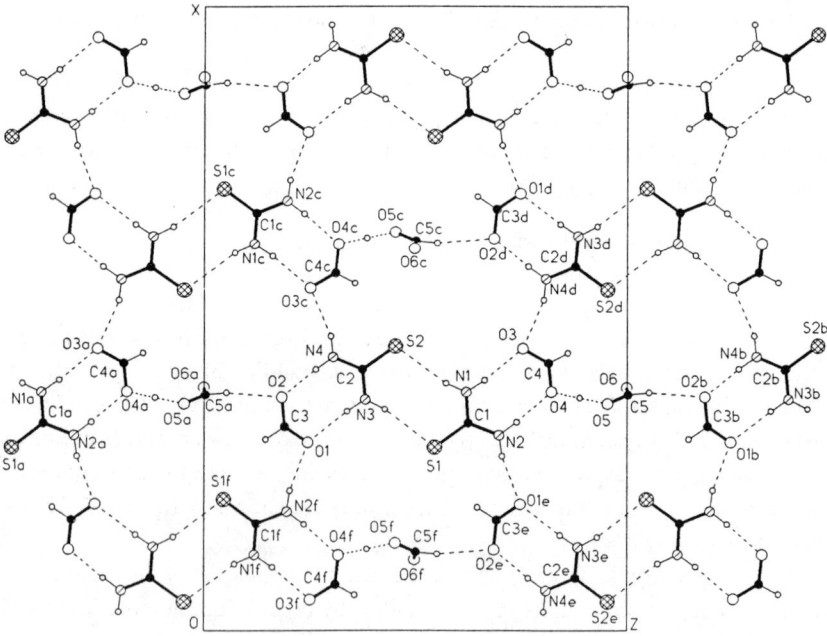

Figure 34. Hydrogen-bonded layer in 2(C₂H₅)₄N⁺HCO₂⁻·2(NH₂)₂CS·HCO₂H (**2.11**) formed by ribbons constructed from thiourea dimers and protonated formate trimers. Broken lines represent hydrogen bonds.

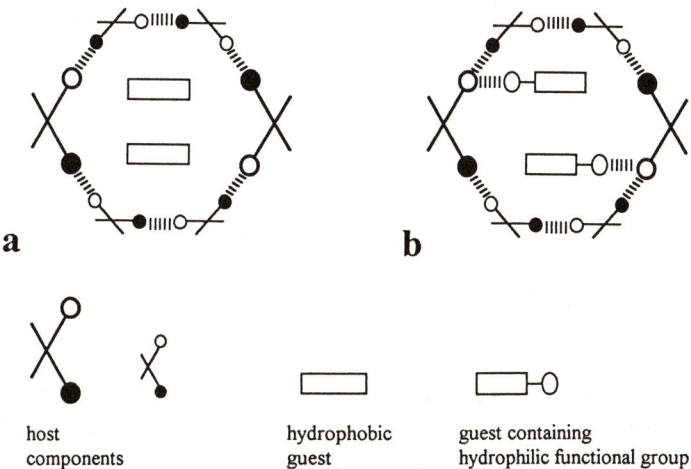

Figure 35. Diagrammatic (projection) representation of two different modes of channel inclusion involving hydrogen bonding interactions (indicated by broken lines): (a) guest molecules retained by steric barriers formed by the hydrogen-bonded host lattice (b) hydrogen bonds also exist between host and guest components.

compounds built of urea or thiourea together with various anions, in which the hydrophobic guest species are retained by steric barriers formed by the hydrogen-bonded host lattice [Figure 35a].

C. Linkage Modes of Urea and Thiourea Molecules

Our studies on the generation of novel urea, thiourea, and selenourea–anion inclusion compounds have shown that novel anionic host lattices can be constructed from urea derivatives and many types of anions as building blocks. The versatility of urea, thiourea, or selenourea as a key component in the construction of novel host lattices is clearly demonstrated by the occurrence of many new types of linkage modes, which include various discrete units, chains or ribbons, and composite ribbons with corresponding anions. These motifs are shown in Tables 14, 15, and 16, respectively.

The hydrogen-bonded layers can be generated by replication of the isostructural and independent discrete units (**UC** in **2.14**) or by other bridging moieties (**UE** in **2.24**). The hydrogen-bonded layers can also be built from a parallel arrangement of ribbons (**RC** in **2.24**) or composite ribbons (**LC** in **2.5**), ribbons bridged by some anions (**RD** in **1.16**, **LF** in **1.11**), or an alternate arrangement of two different types of ribbons (**RC** and **LA** in **1.7**). The three-dimensional channel-type network may

Table 14. Typical Linkage Modes of Discrete Units Built from Urea or Thiourea Molecules, or Together with Some Associated Anions in Their Inclusion Compounds

Type **UA**		shoulder-to-shoulder dimer
Type **UB**		doughnut-like macrocyclic ring formed with molecules of urea, boric acid and $B_3O_3(OH)_2$ fragments of the pentaborate ion
Type **UC**		centrosymmetric hexamer formed with molecules of thiourea and acetate ions
Type **UD**		tetramer formed by thiourea dimer and thiourea-carbonate hetero-dimer
Type **UE**		cyclic, centrosymmetric thiourea hexamer

be formed by cross-linkage of two types of layers (**2.6**) or by two crisscross series of ribbons (**2.21**).

As shown in Table 17, the above types of combining modes of urea or thiourea exist in most of the present inclusion compounds. In an idealized scheme of host design, the well-knit hexagonal ($P6_122$ or $P6_522$) urea (Figure 5) or pseudo-hexagonal ($R\bar{3}$) thiourea channel framework (Figure 6) is first divided into hydrogen-bonded chains or ribbons, and hydrogen bonding between these units with suitable anionic and neutral species would then lead to new varieties of host systems.

Table 15. Typical Linkage Modes of Self-Assembled Hydrogen-Bonded Chains and Ribbons of Urea, Thiourea, or Selenourea Molecules in Their Inclusion Compounds

Type **RA**	singly H-bonded chain (X = S, **2.14** X = Se, **3.1**)

Type **RB**	head-to-tail chain

Type **RC**	shoulder-to-shoulder ribbon (X = O, **1.2** X = S, **2.2** X = Se, **3.4**)

Type **RD**	wide ribbon (X = O, **1.4, 1.16** X = Se, **3.2, 3.3**)

Type **RE**	twisted double ribbon

Type **RF**	mixed shoulder-to-shoulder/head-to-head ribbon

Table 16. Typical Composite Ribbons in Urea/Thiourea-Anion Inclusion Compounds

Type **LA**		*trans*-linked urea dimer/bicarbonate dimer ribbon
Type **LB**		*cis*-linked urea dimer/bicarbonate dimer wide ribbon
Type **LC**		thiourea trimer/nitrate twisted chain
Type **LD**		bridging (thiourea-formate)$_2$ ribbon
Type **LE**		thiourea dimer/hydrogen fumarate-water tetramer wide ribbon
Type **LF**		urea/allophanate double ribbon
Type **LG**		Y-shaped junction built from urea ribbons and dihydrogen borate ribbon

Table 17. Discrete Units, Chains and Ribbons in Urea/Thiourea–Anion Inclusion Compounds

Type	Compound	Structural Unit	Figure No.
UA	**1.8**		11
UB	**1.13**		16
UC	**2.14**		24a
UD	**2.24**		
UE	**2.24**		
RA	**2.14**	one thiourea molecule	24b
	3.1	one selenourea molecule	30
RB	**1.8**	two independent urea molecules	12
	1.14	one urea molecule	
RC	**2.2, 2.7, 2.8, 2.9**	one thiourea molecule	
	2.1, 2.16, 2.17	one thiourea molecule	
	3.4	one selenourea molecule	
	1.2, 1.10	two independent urea molecules	
	2.3, 2.6, 2.12, 2.13, 2.25	two independent thiourea molecules	21, 29
	1.3	three independent urea molecules	
	1.1	five independent urea molecules	
	2.24	a pair of thiourea pentamers	
RD	**1.16, 3.2, 3.3**	two independent urea molecules or selenourea molecules	
RE	**2.4, 2.23**	three independent thiourea molecules	
RF	**1.17**	two independent urea molecules	18
LA	**1.7**	Centrosymmetric cyclic urea dimer and bicarbonate dimer	37
LB	**1.6**	urea dimer and bicarbonate dimer	36
	2.15	urea is replaced by thiourea and bicarbonate is replaced by hydrogen diacetate ion	42
LC	**2.5**	thiourea trimer and a nitrate ion	
	1.12	thiourea is replaced by urea and nitrate ion is replaced by $B_3O_3(OH)_2$ fragment of the pentaborate ion	40
LD	**2.10**	(thiourea-formate)$_2$ tetramer and a water molecule	41
	2.11	formate is replaced by acetate and water is replaced by acetate ion	34
LE	**2.21**	cyclic thiourea dimer, (fumarate-water)$_2$ tetramer and a water molecule	
LF	**1.8**	one independent urea molecule and an allophanate ion	11
	1.11	allophanate is replaced by $B(OH)_3-CO_3^{2-}-B(OH)_3$	39
LG	**1.10**	two independent urea molecules or one independent $BO(OH)_2^-$ ion	14

D. Comolecular Aggregates of Urea/Thiourea and Other Host Components

Comolecular Aggregates of Urea and Other Host Components

As mentioned above, in urea/thiourea–anion inclusion compounds both urea and thiourea molecules adopt various linkage modes not only of self-assembled hydrogen-bonded chains or ribbons but also of composite ribbons with the corresponding anions. The results show that comolecular aggregates of urea or thiourea with other neutral molecules or anionic moieties can be considered as fundamental building blocks for the constructions of various types of novel host lattices.

Comparison of the urea–anion inclusion compounds with the thiourea–anion inclusion compounds indicates that urea molecules more readily combine with other anionic moieties to generate composite dimers or ribbons. Composite dimers or ribbons exist in seven complexes among the nine urea–anion inclusion compounds studied here. For instance, both urea-bicarbonate complexes **1.6** and **1.7** (Figure 36 and Figure 37) contain a hydrogen-bonded $[((NH_2)_2CO)_2(HCO_3^-)_2]_\infty$ ribbon built of alternating urea dimers and cyclic dimeric bicarbonate moieties (**LB**

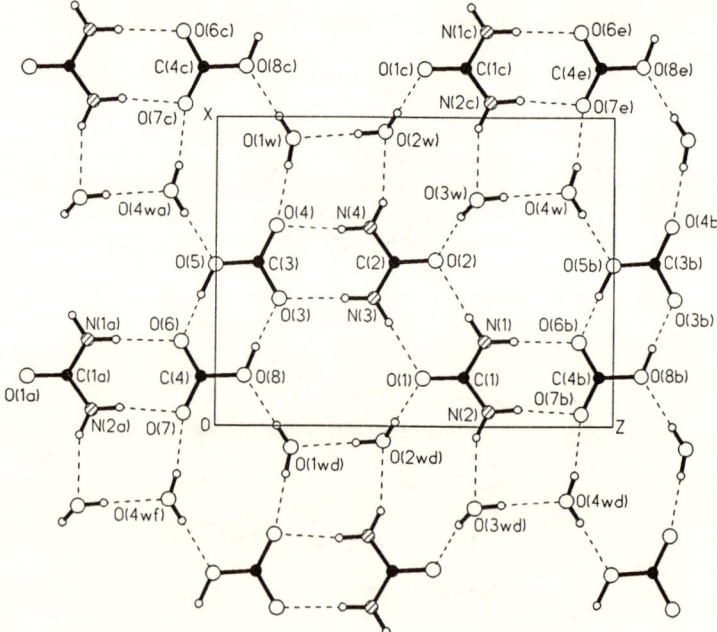

Figure 36. Hydrogen-bonded layer in $(C_2H_5)_4N^+HCO_3^-\cdot(NH_2)_2CO\cdot2H_2O$ (**1.6**) formed by urea dimers and bicarbonate dimers bridged by water molecules. Broken lines represent hydrogen bonds.

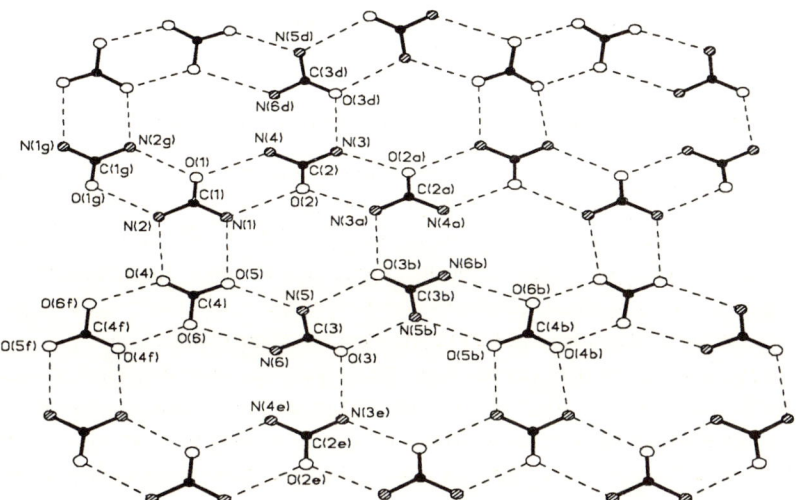

Figure 37. Hydrogen-bonded layer in $(n\text{-}C_4H_9)_4N^+HCO_3^-\cdot(NH_2)_2CO$ (**1.7**) formed by the cross-linkage of urea ribbons and $[(HCO_3^-)_2((NH_2)_2CO)_2]_\infty$ ribbons. Broken lines represent hydrogen bonds.

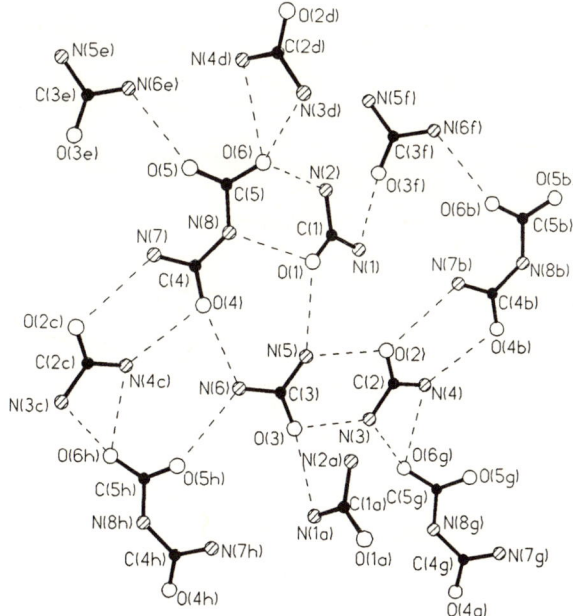

Figure 38. Perspective view of a portion of the wide urea-allophanate ribbon linked by hydrogen bonds in $(n\text{-}C_3H_7)_4N^+NH_2CONHCO_2^-\cdot3(NH_2)_2CO$ (**1.9**).

and **LA** type, respectively), but different additional molecular components, namely water molecules in **1.6** and urea ribbons in **1.7**, are used in the construction of the hydrophilic layer.

In one of the two urea–allophanate compounds (**1.8**, Figure 11), hydrogen-bonded zigzag ribbons are built up from repetition of the allophanate ion and one independent urea molecule joined together by a pair of N–H···O hydrogen bonds (**LF** type), whereas wide puckered urea and allophanate ribbons connected by hydrogen bonds exist in the other (**1.9**, Figure 38).

In the urea–boric acid–carbonate host lattice (**1.11**, Figure 39) the boric acid molecules link the carbonate anions and urea molecules to form a double ribbon, and are further connected to water molecules located between adjacent double ribbons to form a hydrogen-bonded planar layer, and three independent urea molecules and a $B_3O_3(OH)_2$ fragment of the pentaborate ion are linked by hydrogen bonds to form an infinite twisted ribbon in one urea–pentaborate complex (**1.12**, Figure 40). Two-dimensional infinite layers of interconnected urea molecules, pentaborate ions, and neutral $B(OH)_3$ molecules generate the host lattice of another urea–pentaborate complex (**1.13**, Figure 16).

It is of interest to note that the allophanate ion, which is known in the form of its derivatives, and the dihydrogen borate anion $BO(OH)_2^-$, the fugitive conjugate base of boric acid, can be generated in situ and stabilized in the crystalline state through

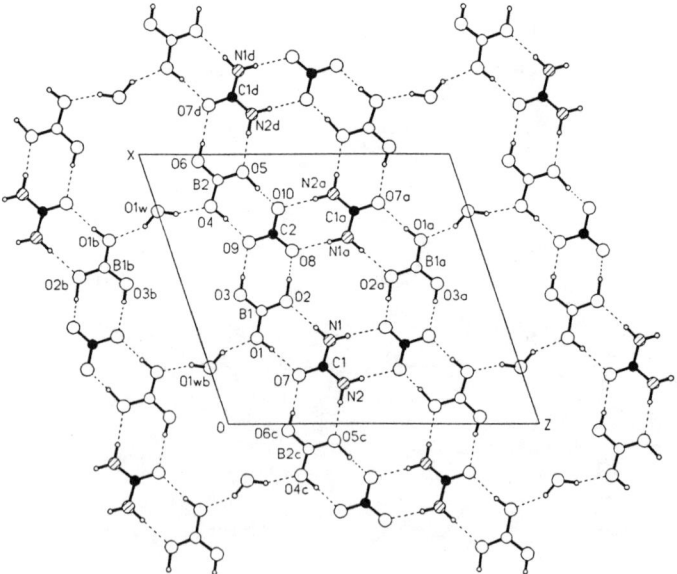

Figure 39. Hydrogen-bonded layer in **1.11** formed by urea–boric acid–carbonate double ribbons cross-linked by water molecules. Broken lines represent hydrogen bonds.

hydrogen-bonding interactions with its nearest neighbors. Notably the bulky and hydrophobic tetraalkylammonium ion plays a dual and complementary role, i.e. it serves as a template for the self-assembly of the anionic host lattice and, abetted by the presence of urea, provides a suitably alkaline aqueous medium that induces boric acid to function as a Brønsted acid.

Comolecular Aggregates of Thiourea and Other Host Components

As compared with urea, the thiourea molecule has a lesser tendency to combine with other host components to form a dimer or composted ribbon. This occurs in only six cases among 23 thiourea–anion inclusion compounds, typical examples of which are the ribbons composed of twisted (thiourea-formate)$_2$ tetramers bridged by water molecules or formate ion in two thiourea–formate inclusion compounds (**2.10**, Figure 41 and **2.11**, Figure 34), the centrosymmetric thiourea–acetate hexamer constituted from two independent thiourea molecules and an acetate ion as a repeated unit of the zigzag ribbons (**2.14**, Figure 24), and the puckered $[((NH_2)_2CS)_2 ((CH_3CO_2)_2H)^-]_\infty$ ribbons in thiourea tetra-n-propylammonium acetate complex (**2.15**, Figure 42).

In most thiourea–anion inclusion compounds the host lattices are built from connections of thiourea ribbons and the anionic dimers or just the ribbons only.

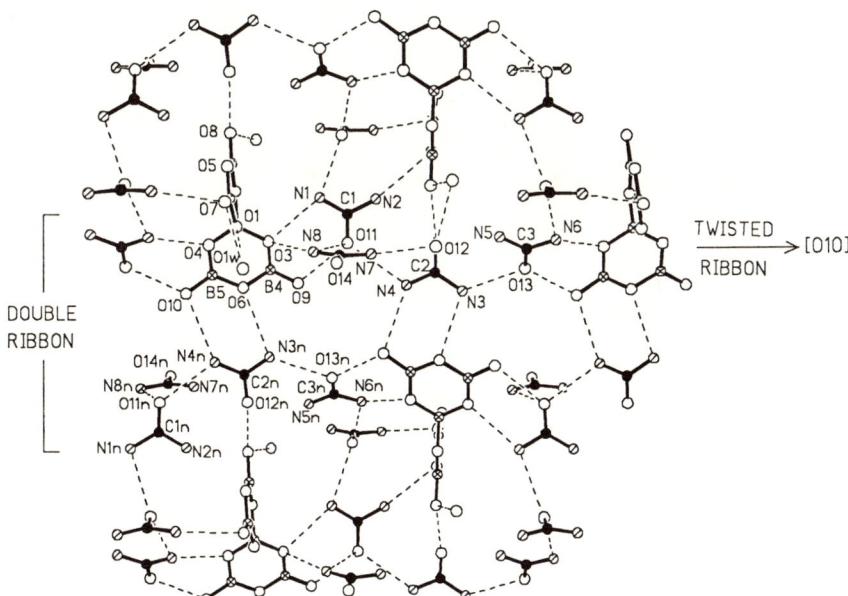

Figure 40. Hydrogen-bonded layer in **1.12** formed by the linkage of urea molecules and pentaborate anions. Note the double ribbon that runs parallel to [010]. Broken lines represent hydrogen bonds.

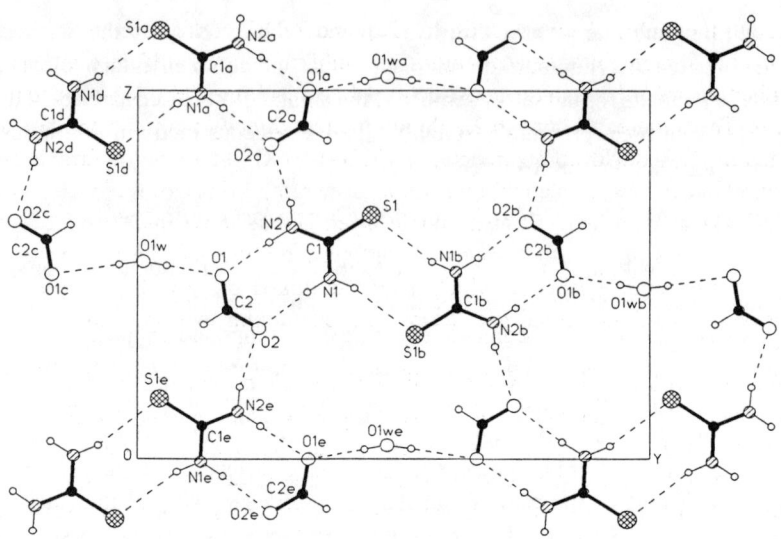

Figure 41. Projection diagram showing a hydrogen-bonded layer constructed from (thiourea–formate)$_2$ tetramers in the crystal structure of $(C_2H_5)_4N^+HCO_2^-\cdot(NH_2)_2$ $CS\cdot H_2O$ (**2.10**). Broken lines represent hydrogen bonds.

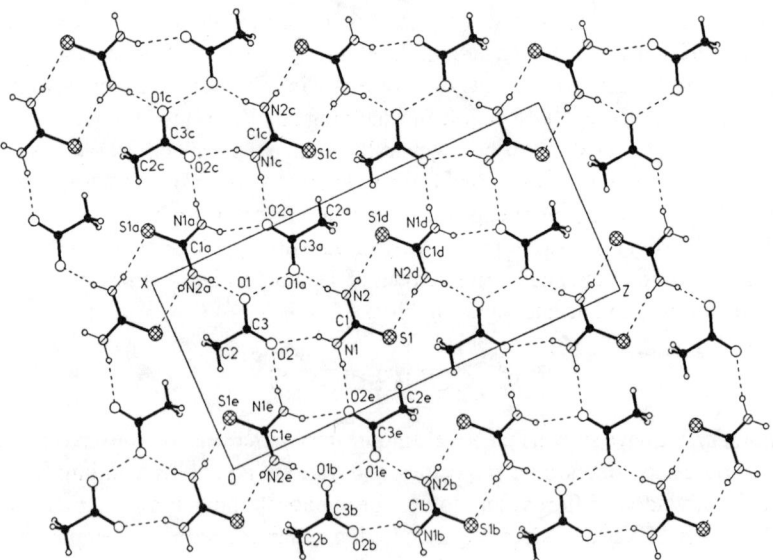

Figure 42. Hydrogen-bonded layer in $(n\text{-}C_3H_7)_4N^+[(CH_3CO_2)_2H]^-\cdot2(NH_2)_2CS$ (**2.15**) formed by ribbons constructed from thiourea dimers and dimeric acetate anions. Broken lines represent hydrogen bonds.

Although the inclusion compounds **2.2**, **2.3**, and **2.4** all contain thiourea ribbons and cyclic dimeric bicarbonate anions as building blocks, different modes of combination are utilized in the construction of the resulting hydrogen-bonded host lattices. The thiourea ribbons in **2.2** do not interact directly with one another, but are laterally linked with complementary $[(HCO_3^-)_2(H_2O)_2]_\infty$ ribbons to form a layer structure. In **2.3** the separated thiourea ribbons are bridged in an orthogonal manner by $(HCO_3^-)_2$ units to give a novel double layer. Unlike those in **2.2** and **2.3**, the thiourea ribbons in **2.4** adopt a twisted configuration; they are further organized into puckered double ribbons which are cross-linked by $(HCO_3^-)_2$ moieties to generate a three-dimensional host framework containing two open-channel systems.

In comparing the thiourea–nitrate inclusion compounds with the thiourea–bicarbonate inclusion compounds, it is noted that the oxygen atoms of the NO_3^- ion, unlike the HCO_3^- ion, can only form acceptor hydrogen bonds with other potential donors. Therefore, the construction of a two- or three-dimensional host framework requires a higher thiourea/nitrate molar ratio, with or without cocrystallized water molecules, as is the case in complexes **2.5** and **2.6**. In the 1:1 complexes **2.7**, **2.8**, and **2.9** only a parallel arrangement of separate ribbons can be constructed from the thiourea and nitrate building blocks.

The $CH_3CO_2^-$ ion, unlike the HCO_3^- ion, can only form acceptor hydrogen bonds with other potential donors and furthermore, unlike both HCO_3^- and NO_3^-, it has only two oxygen atoms functioning as acceptors in the hydrogen-bonded host framework, and accommodation of its hydrophobic methyl group must be made. Therefore, the construction of a two- or three-dimensional host framework requires a higher thiourea/acetate molar ratio, as in the case in compound **2.14**, or the dimeric acetic anion necessarily takes the role of a building block as is the case in complexes **2.15** and **2.16**. The resulting ribbons generally adopt a twisted configuration so that the thiourea molecules can form additional donor hydrogen bonds with atoms in adjacent ribbons, or the more exposed oxygen atoms can form acceptor hydrogen bonds with nitrogen atoms of neighboring thiourea molecules. In the 1:1 complex **2.17** only a "pesudo-channel" arrangement of separate ribbons can be constructed from the thiourea and acetate building blocks.

V. CONCLUSION

In summary, we have shown that novel anionic host lattices can be constructed from urea, thiourea, or selenourea molecules and various anions as building blocks, which readily adopt different topologies for the accommodation of tetraalkylammonium ions of various sizes. By employing organic cations as templates and suitable counter anions as an ancillary host material, with or without neutral molecules such as H_2O as a third component, the "lattice engineering" of new urea, thiourea, and selenourea inclusion compounds by self-assembly may be further explored.

ACKNOWLEDGMENTS

This review is mainly based on work supported by Hong Kong Research Grants Council Earmarked Grant No. CUHK 456/95P. QL wishes to acknowledge the award of a Postdoctoral Fellowship tenable at The Chinese University of Hong Kong.

REFERENCES

1. Mandelcorn, L., Ed. *Non-Stoichiometric Compounds*; Academic Press: New York, 1964.
2. Davies, J. E. D.; Kemula, W.; Powell, H. M.; Smith, N. O. *J. Incl. Phenom.* **1983**, *1*, 3.
3. (a) Weber, E., (Ed.) *Topics in Current Chemistry 140: Molecular Inclusion and Molecular Recognition—Clathrates I*; Springer-Verlag: Berlin, 1987; (b) Weber, E. (Ed.). *Topics in Current Chemistry 149: Molecular Inclusion and Molecular Recognition—Clathrates II*; Springer-Verlag: Berlin, 1988.
4. (a) Atwood, J. L.; Davies, J. E. D.; MacNicol, D. D. (Eds.). *Inclusion Compounds*; Academic Press: London, 1984, Vol. I–III; (b) Atwood, J. L.; Davies, J. E. D.; MacNicol, D. D. (Eds.). *Inclusion Compounds*; Oxford University Press: Oxford, 1991, Vol. IV, V.
5. Iwamoto, T. *J. Incl. Phenom.* **1996**, *24*, 61.
6. (a) MacNicol, D. D.; McKendric, J. J.; Wilson, D. R. *Chem. Soc. Rev.* **1980**, *7*, 65; (b) Frampton, C. S.; McGregor, W. M.; MacNicol, D. D.; Mallinson, P. R.; Plevey, R. G.; Rowan, S. J. *Supramol. Chem.* **1994**, *3*, 223.
7. Saenger, W. In ref. 4a, Vol. II, pp. 231–259.
8. (a) Cram, D. J. *Angew. Chem., Int. Ed. Engl.* **1988**, *27*, 1009; (b) Cram, D. J.; Cram, J. M. *Container Molecules and Their Guests*; RSC: Cambridge, 1994.
9. (a) Lehn, J.-M. *Angew. Chem. Int. Ed. Engl.* **1988**, *27*, 1095; (b) Lehn, J.-M. *Supramolecular Chemistry*; VCH: Weinheim, 1995.
10. Pedersen, C. J. *Angew. Chem. Int. Ed. Engl.* **1988**, *27*, 1021.
11. Weber, E.; Hager, O.; Foces-Foces, C.; Llamas-Saiz, A. L. *J. Phys. Org. Chem.* **1996**, *9*, 50.
12. Meissner, R. S.; Rebek, J. Jr.; de Mendoza, J. *Science (Washington)* **1995**, *270*, 1485.
13. (a) Toda, F. In ref. 3a, p. 43; (b) Toda, F.; Tohi, Y. *J. Chem. Soc., Chem. Commun.* **1993**, 1238; (c) Toda, F. *Supramol. Chem.* **1995**, *6*, 159; (d) Toda, F. *J. Mol. Struct.* **1996**, *374*, 313; (e) Kaupp, G. *Angew. Chem. Int. Ed. Engl.* **1994**, *33*, 728.
14. Lehn, J.-M. (Ser. Ed.). *Comprehensive Supramolecular Chemistry*; Pergamon: Oxford, 1996, Vols. 1–11.
15. Schmidt, G. M. *J. Pure Appl. Chem.* **1971**, *27*, 647.
16. Zimmerman, S. C. *Science (Washington)* **1997**, *276*, 543.
17. (a) Desiraju, G. R. *Crystal Engineering: The Design of Organic Solids*; Elsevier: Amsterdam, 1989; (b) Desiraju, G. R. (Ed.). *The Crystal as a Supramolecular Entity*; Wiley: Chichester, 1995.
18. Zaworotko, M. J. *Chem. Soc. Rev.* **1994**, *23*, 283.
19. (a) Mak, T. C. W.; Bracke, R. F. In ref. 14, Vol. 6, pp. 23–60; (b) Ermer, O. *Helv. Chim. Acta* **1991**, *74*, 1339; (c) Ermer, O.; Röbke, C. *J. Am. Chem. Soc.* **1993**, *115*, 10077; (d) Ermer, O.; Eling, A. *Angew. Chem., Int. Ed. Engl.* **1988**, *27*, 829; (e) Ermer, O. *J. Am. Chem. Soc.* **1988**, *110*, 3747.
20. Wuest, J. D. In *Mesomolecules: From Molecules to Materials*; Mendenhall, G. D.; Greenberg, A.; Liebman, J. F., Eds.; Chapman & Hall: New York, 1995, pp. 107–131.
21. Russell, V. A.; Evans, C. C.; Li, W.; Ward, M. D. *Science (Washington)* **1997**, *276*, 575.
22. (a) Mark, H.; Weissenberg, K. *Z. Phys.* **1923**, *16*, 1; (b) Hendricks, S. B. *J. Am. Chem. Soc.* **1928**, *50*, 2455; (c) Wyckoff, R. *Z. Kristallogr.* **1930**, *75*, 529; (d) Wyckoff, R. *Z. Kristallogr.* **1932**, *81*, 102; (e) Wyckoff, R.; Corey, R. B. *Z. Kristallogr.* **1934**, *89*, 462.
23. (a) Vaughan, P.; Donohue, J. *Acta Crystallogr.* **1952**, *5*, 530; (b) Worsham, J. E.; Levy, H. A.; Peterson, S. W. *Acta Crystallogr.* **1957**, *10*, 319; (c) Sklar, N.; Senko, M.; Post, B. *Acta Crystallogr.*

1961, *14*, 716; (d) Caron, A.; Donohue, J. *Acta Crystallogr. Sect. B* **1964**, *17*, 544; (e) Pryor, A.; Sanger, P. L. *Acta Crystallogr. Sect. A* **1970**, *26*, 543; (f) Mullen, D.; Hellner, E. *Acta Crystallogr. Sect. B* **1978**, *34*, 1624; (g) Mullen, D. *Acta Crystallogr. Sect. B* **1980**, *36*, 1610.

24. Natalie, S.; Michael, E. S.; Ben, P. *Acta Crystallogr.* **1961**, *14*, 716.

25. (a) Swaminathan, S.; Craven, B. M.; Spackman, M. A.; Stewart, R. F. *Acta Crystallogr. Sect. B* **1984**, *40*, 398; (b) Swaminathan, S.; Craven, B. M.; McMullan, R. K. *Acta Crystallogr. Sect. B* **1984**, *40*, 300.

26. Solomon, A. L. *Phys. Rev.* **1956**, *104*, 1191.

27. (a) Takahashi, I.; Onodera, A.; Shiozaki, Y. *Acta Crystallogr. Sect. B* **1990**, *46*, 661; (b) Tanisaki, S.; Mashiyama, H. *Acta Crystallogr. Sect. B* **1988**, *44*, 441; (c) Moudden, A. H.; Denoyer, F.; Lambert, M.; Fitzgerald, W. *Solid State Commun.* **1979**, *32*, 933; (d) Elcombe, M. M.; Taylor, J. C. *Acta Crystallogr. Sect. A* **1968**, *24*, 410; (e) Goldsmith, G. J.; White, J. G. *J. Chem. Phys.* **1959**, *31*, 1175; (f) Tanisaki, S.; Mashiyama, H.; Hasebe, K. *39th Annu. Meet. Phys. Soc. Jpn, Fukuoka, Advanced Abstracts*; 1984, Vol. 2, p. 101.

28. (a) Takemoto, K.; Sonoda N. In ref. 4a, Vol. II, pp. 47–67; (b) Bishop, R.; Dance, I. G.; In ref. 3b, pp. 153–159.

29. (a) Yeo, L.; Harris, K. D. M.; Guillaume, F. *J. Solid State Chem.* **1997**, *128*, 273; (b) Harris, K. D. M.; Gameson, I.; Thomas, J. M. *J. Chem. Soc., Faraday Trans.* **1990**, *86*, 3135.

30. (a) Chatani, Y.; Taki Y.; Tadokoro, H. *Acta Crystallogr. Sect. B* **1977**, *33*, 309; (b) Chatani, Y.; Anraku H.; Taki, Y. *Mol. Cryst. Liq. Cryst.* **1978**, *48*, 219; (c) Forst, R.; Jagodzimki, H.; Frey, F. *Z. Kristallogr.* **1986**, *174*, 56; **1986**, *174*, 58; (d) Forst, R.; Boysen, H.; Frey, F.; Jagodzinski, H.; Zeyen, C. *J. Phys. Chem. Solids* **1986**, *47*, 1089.

31. Harris, K. D. M.; Smart, S. P.; Hollingsworth, M. D. *J. Chem. Soc., Faraday Trans.* **1991**, *87*, 3423.

32. Rennie, A. J. O.; Harris, K. D. M. *J. Chem. Phys.* **1992**, *96*, 7117.

33. Forst, R.; Jagodzinski, H.; Boysen, H.; Frey, F. *Acta Crystallogr. Sect. B* **1990**, *46*, 70.

34. Mahdyarfar, A.; Harris, K. D. M. *J. Chem. Soc., Chem. Commun.* **1993**, 51.

35. Schofield, P. A.; Harris, K.D. M.; Shannon, I. J.; Rennie, A. J. O. *J. Chem. Soc., Chem. Commun.* **1993**, 1293.

36. (a) Harris, K. D. M.; Thomas, J. M. *J. Chem. Soc., Faraday Trans.* **1990**, *86*, 2985; (b) Harris, K. D. M.; Thomas, J. M. *J. Chem. Soc., Faraday Trans.* **1990**, *86*, 1095.

37. Harris, K. D. M. *J. Mol. Struct.* **1996**, *374*, 241.

38. (a) Brown, M. E.; Hollingsworth, M. D. *Nature* **1995**, *376*, 323; (b) Hollingsworth, M. D.; Harris, K. D. M. In ref. 14, Vol. 6, pp. 177–237.

39. (a) Vasanthan, N.; Shin, I. D.; Huang, L.; Nojima, S.; Tonelli, A. E. *Macromolecules* **1997**, *30*, 3014; (b) Shin, I. D.; Vasanthan, N.; Nojima, S.; Tonelli, A. E. *Polym. Prepr.* **1997**, *38*, 882; (c) Huang, L.; Vasanthan, N.; Tonelli, A. E. *Polym. Mater. Sci. Eng.* **1997**, *76*, 449; (d) Huang, L.; Vasanthan, N.; Tonelli, A. E. *J. Appl. Polym. Sci.* **1997**, *64(2)*, 281; (e) Eaton, P.; Vasanthan, N.; Shin, I. D.; Tonelli, A. E. *Macromolecules* **1996**, *29*, 2531.

40. (a) Mak, T. C. W.; Yip, W. H.; Li, Q. *J. Am. Chem. Soc.* **1995**, *117*, 11995; (b) Li, Q.; Yip, W. H.; Mak, T. C. W. *J. Incl. Phenom.* **1995**, *23*, 233; (c) Li, Q.; Mak, T. C. W. *Supramol. Chem.* **1996**, *8*, 73; (d) Li, Q.; Mak, T. C. W. *Supramol. Chem.* **1996**, *8*, 147; (e) Li, Q.; Mak, T. C. W. *J. Incl. Phenom.* **1997**, *28*, 151; (f) Li, Q.; Mak, T. C. W. *Acta Crystallogr. Sect. B* **1997**, *54*, 180; (g) Li, Q.; Xue, F.; Mak, T. C. W. *Inorg. Chem.* **1998**, accepted; (h) Xue, F.; Mak, T. C. W. To be published.

41. (a) Li, Q.; Mak, T. C. W. *J. Incl. Phenom.* **1995**, *20*, 73; (b) Li, Q.; Mak, T. C. W. *Acta Crystallogr. Sect. B* **1996**, *52*, 989; (c) Li, Q.; Mak, T. C. W. *Acta Crystallogr. Sect. C* **1996**, *52*, 2830; (d) Li, Q.; Mak, T. C. W. *Acta Crystallogr. Sect. B* **1997**, *53*, 252; (e) Li, Q.; Mak, T. C. W. *J. Incl. Phenom.* **1997**, *27*, 319; (f) Li, Q.; Mak, T. C. W. *J. Incl. Phenom.* **1997**, *28*, 183; (g) Li, Q.; Mak, T. C. W. *Supramol. Chem.* **1998**. In press; (h) Lam, C. K.; Mak, T. C. W. To be published.

42. Li, Q.; Mak, T. C. W. *Acta Crystallogr. Sect. B* **1997**, *53*, 262.

43. (a) Keller, W. E. *J. Chem. Phys.* **1948**, *16*, 1003; (b) Waldron, R. D.; Badger, R. M. *J. Chem. Phys.* **1950**, *18*, 566.

44. (a) Andrew, M. R.; Hyndman, D. *Proc. Phys. Soc. A.* **1953**, *66*, 1187; (b) Andrew, M. R.; Hyndman, D. *Disc. Faraday Soc.* **1955**, *19*, 195; (c) Kromhout, R. A.; Moulton, W. G. *J. Chem. Phys.* **1955**, *23*, 1673.

45. Lobachev, A. N.; Vainshtein, B. K. *Soviet Phys. Crystallogr.* **1961**, *6*, 313.

46. Mak, T. C. W.; Zhou, G.-D. *Crystallography in Modern Chemistry: A Resource Book of Crystal Structures*; Wiley-Interscience: New York, 1992, p. 175.

47. Wyckoff, R.; Corey, R. B. *Z. Kristallogr.* **1932**, *81*, 386.

48. Kunchur, N. R.; Truter, M. R. *J. Chem. Soc.* **1958**, 2551.

49. Truter, M. R. *Acta Crystallogr.* **1967**, *22*, 556.

50. (a) Tanisaki, S.; Nakamura, N. *J. Phys. Soc. Jpn.* **1970**, *28*, suppl. 293; (b) Shiozaki, Y. *Ferroelectrics* **1971**, *2*, 245.

51. (a) Kondrashev, Yu. D.; Andreeva, N. A. *J. Struct. Chem.* **1963**, *4*, 413; (b) Rodriguez, P.; Cubero, M.; Lopez-Castro, A. *Nature* **1964**, *201*, 180; (c) Hope, H. *Acta Crystallogr.* **1965**, *18*, 259.

52. Rutherford, J. S.; Calvo, C. *Z. Kristallogr.* **1969**, *128*, 229.

53. (a) Dvoryankin, V. F.; Ruchkin, E. D. *J. Struct. Chem.* **1962**, *3*, 325; (b) Dvoryankin, V. F.; Vainshtein, B. K. *Kristallogr.* **1960**, *5*, 589.

54. Bengen, M. F. DP Patent 12 438 18, 1940.

55. Angela, B. *C. R. Acad. Sci.* **1947**, *244*, 402.

56. (a) Shannon, I. J.; Harris, K. D. M.; Rennie, A. J. O.; Webster, M. B. *J. Chem. Soc., Faraday Trans.* **1993**, *89*, 2023; (b) Shannon, I. J.; Jones, M. J.; Harris, K. D. M.; Siddiquie, M. R. H.; Joyner, R. W. *J. Chem. Soc., Faraday Trans.* **1995**, *91*, 1497.

57. (a) Smith, A. E. *Acta Crystallogr.* **1952**, *5*, 224; (b) Smith, A. E. *J. Chem. Phys.* **1950**, *18*, 150.

58. Schlenk, W. *Liebigs Ann. Chem.* **1949**, *565*, 204.

59. Paul, J.; Nicolás, Y.; Guillermo, G. *J. Incl. Phenom.* **1995**, *22*, 203.

60. (a) Hadicke, E.; Schlenk, W. *Liebigs Ann. Chem.* **1972**, *764*, 103; (b) Lenné, H. U. *Z. Kristallogr.* **1963**, *118*, 454; (c) Lenné, H. U.; Mez, H. C.; Schlenk, W. *Chem. Ber.* **1968**, *101*, 2435.

61. Otto, J. *Acta Crystallogr. Sect. B* **1972**, *28*, 543.

62. Yeo, L.; Harris, K. D. M. *Acta Crystallogr. Sect. B* **1997**, *53*, 822.

63. (a) Kinight, M. B.; Witnaüer, L. P.; Coleman, J. F.; Noble, W. R.; Swern, D. *Anal. Chem.* **1952**, *24*, 1331; (b) McAdie, H. G. *Can. J. Chem.* **1962**, *40*, 2195; (c) McAdie, H. G. *Can. J. Chem.* **1963**, *41*, 2144; (d) McAdie, H. G.; Frost, G. B. *Can. J. Chem.* **1958**, *36*, 635.

64. Harris, K. D. M. *J. Phys. Chem. Solids* **1992**, *53*, 529.

65. Lenné, H.-U. *Acta Crystallogr.* **1954**, *7*, 1.

66. Schlenk Jr, W. *Justus Liebigs Ann. Chem.* **1951**, *573*, 142.

67. Hough, E.; Nicholson, D. G. *J. Chem. Soc., Dalton Trans.* **1978**, 15.

68. Fait, J. F.; Fitzgerald, A.; Caughlan, C. N.; McCandless, F. P. *Acta Crystallogr. Sect. C* **1991**, 332.

69. Nicholaides, N.; Laves, F. *Z. Kristallogr.* **1965**, *121*, 283.

70. Chatani, Y.; Nakatani, S. *Z. Kristallogr.* **1976**, *144*, 175.

71. Garneau, I.; Raymond, S.; Brisse, F. *Acta Crystallogr. Sect. C* **1995**, *51*, 538.

72. Clement, R.; Jegoudez, J.; Mazieres, C. *J. Solid State Chem.* **1974**, *10*, 46.

73. George, A. R.; Harris, K. D. M. *J. Mol. Graphics* **1995**, *13*, 138.

74. Schiessler, R. W.; Flitter, D. *J. Am. Chem. Soc.* **1952**, *74*, 1720.

75. Fetterly, L. C. In ref. 1, pp. 491–567.

76. Drew, M. G. B.; Lund, A.; Nicholson, D. G. *Supramol. Chem.* **1997**, *8*, 197.

77. Shannon, I. J.; Jones, M. J.; Harris, K. D. M. *J. Chem. Soc., Faraday Trans.* **1995**, *91*, 1497.

78. Anderson, A. G.; Calabrese, J. C.; Tam, W.; Williams, I. D. *Chem. Phys. Lett.* **1987**, *134*, 392.

79. van Bekkum, H.; Remijnse, J. D.; Wepster, B. M. *J. Chem. Soc., Chem. Commun.* **1969**, 67.

80. Gopal, R.; Robertson, B. E.; Rutherford, J. S. *Acta Crystallogr. Sect. C* **1989**, *45*, 257.

81. Nicolaides, N.; Laves, F. *J. Am. Chem. Soc.* **1958**, *80*, 5752.

82. Barlow, G. B.; Corish, P. S. *J. Chem. Soc.* **1959**, 1706.

83. Gasu, B. *Nature* **1961**, *191*, 802.

84. Ha, T.-K.; Puebla, C. *Chem. Phys.* **1994**, *181*, 47.
85. Frind, M.; Scheibe, O.; Fischer, S.; Ross, A.; Dreyer, R. *ZfI-Mitt.* **1991**, *165*, 40.
86. Schster, P.; Zundel, G.; Sandorfy, C. *The Hydrogen Bond*; North-Holland: Amsterdam, 1976, Vol. II.
87. Wallwork, S. C. *Acta Crystallogr.* **1962**, *15*, 758.
88. Mak, T. C. W. *J. Incl. Phenom.* **1990**, *8*, 199.
89. Emsley, J.; Arif, M.; Bates, P. A.; Hursthouse, M. B. *J. Chem. Soc., Chem. Commun.* **1989**, 738.
90. Loehlin, J. H.; Kvick, Å. *Acta Crystallogr. Sect. B* **1978**, *34*, 3488.
91. Mak, T. C. W.; So, S. P.; Chieh, C.; Jasim, K. S. *J. Mol. Struct.* **1985**, *127*, 375.
92. Mak, T. C. W.; McMullan, R. K. *J. Incl. Phenom.* **1988**, *6*, 473.
93. Mak, T. C. W. *Inorg. Chem.* **1984**, *23*, 620.
94. Jeffrey, G. A.; Saenger, W. *Hydrogen Bonding in Biological Structures*; Springer Verlag: Berlin, 1991.
95. Bernstein, J.; Etter, M. C.; Leiserowitz, L. In *Structure Correlation*; Bürgi, H.-B.; Dunitz, J. D., Eds.; VCH: Weinheim, 1994, Vol. 2.
96. Mak, T. C. W. *J. Incl. Phenom.* **1985**, *3*, 347.
97. Mak, T. C. W.; Bruins Slot, H. J.; Beurskens, P. T. *J. Incl. Phenom.* **1986**, *4*, 295.

ROLES OF ZINC AND MAGNESIUM
IONS IN ENZYMES

Amy Kaufman Katz and Jenny P. Glusker

Advances in Molecular Structure Research
Volume 4, pages 227–279
Copyright © 1998 by JAI Press Inc.
ISBN: 0-7623-0348-4

ABSTRACT

The roles of the magnesium and zinc ions in proteins, particularly enzymes, are described in terms of their structural and catalytic functions. Among magnesium-utilizing enzymes the major role of the metal ion appears to be structural, in view of the precisely octahedral coordination of six oxygen atoms normally found in magnesium salts and complexes. This suggests that the magnesium ion serves to orient the substrate in the required orientation for the catalytic mechanism to take place. Some polarization of the substrate by the magnesium ion may also aid in the catalytic reaction. Zinc ions behave differently from magnesium ions, even though both have approximately the same size and charge. The innermost coordination number of zinc can vary between 4, 5, and 6, and zinc tends to bind sulfur and nitrogen, rather than oxygen atoms when its coordination number is low (4). Some enzymes employ both magnesium and zinc ions and provide different types of binding sites appropriate for each cation. Examples are presented. These findings on metal ion coordination and preferred ligands can be used to design active sites for enzymes that alter or enhance their activities by the introduction of appropriate active-site binding groups.

I. INTRODUCTION

Metal ions can serve a variety of functions in the mechanisms of action of metalloenzymes. They may polarize functional groups both in the substrate and in amino acid side chains in the active site. As a result, the reaction being catalyzed can be facilitated. If the metal ion can undergo a change in oxidation number (such as is found for copper and iron), this may further aid in catalysis. Metal ions may also serve as a means of stiffening the geometry of the active site so that appropriate functional groups in it are lined up with respect to the substrate in a finely tuned manner dictated by the stereochemical requirements of the biochemical reaction to be catalyzed. Catalysis proceeds most efficiently in an enzyme when the transition state of the reaction is stabilized with respect to substrate and product.

Metal ions, which are positively charged, act as electrophiles, that is, they seek the possibility of sharing electron pairs on adjacent atoms so that electrical neutrality can thereby be achieved. In these types of interactions a metal ion acts similarly to a hydrogen ion (proton) which is also positively charged. Most metal ions are, however, different from protons in that they can have positive charges greater than unity and, by virtue of their larger ionic volume, can bind more ligands at the same time. Metal-ion concentrations at neutral pH values can be high, while for hydrogen ions these concentrations are low. As a result, metal-ion catalysis can occur in pH ranges at which hydrogen-ion catalysis would be much less effective [*1*]. The positive charge of a metal ion can also act in an electrostatic manner to shield negative charges on the substrate or cofactor so that they do not cause a substantial repulsion of the electron pairs of an approaching nucleophile.

We have chosen two metal ions, zinc and magnesium, for description here. The properties of these two ions are of significant utility to the structure and function of many proteins, particularly enzymes. The preferred ligands of these ions will be reviewed, together with a description of some of the types of binding motifs in which magnesium and zinc ions are found in proteins. We have concentrated in this chapter on structural information obtained by X-ray diffraction analyses of crystals. The results of such protein structure determinations are used to produce the figures. Several other physical techniques give complementary information. NMR, EPR, and XAFS techniques provide valuable data to be combined with crystallographic information from the enzyme, its complexes with substrate and inhibitors, and various mutant enzymes.

Some metalloenzymes require one specific metal ion for activity, while others are active when one or more of a set of metal ions is bound. Which metal ions does a given enzyme select and how does it prevent other metals ions that might have an inactivating function from binding to the same site? This question can be answered in the light of which metal ions are readily available and which are rare. For example, the metal ion composition of a human adult weighing about 70 kg is as follows for the commoner elements: 1.7 kg calcium, 42 g magnesium, 70 g sodium, and 250 g potassium [*2*]. Transition metal ions are found in significantly lower amounts as follows (for an individual of the same weight): 4 g of iron, 3 g of zinc, and 0.2 g of copper [*3*]. Therefore, while magnesium is an element that is found in reasonably high concentrations in the body, zinc is not.

Life evolved in seawater. Therefore a consideration of the relative compositions of seawater and extracellular and intracellular fluids is relevant to an analysis of metal ion utilization. Seawater contains a high concentration of sodium ions, and, in lesser amounts, potassium, calcium, and magnesium ions. These cations are also found in living cells in varying amounts. It is necessary, however, for the cell to maintain pumps to keep the individual concentration of these ions within it to appropriate limits. For example, sodium ions are found in high concentrations in seawater and in extracellular fluids, but potassium ions are concentrated within living cells. Sodium ions must be pumped out of the cells, and systems are available to do this. Pumps are also available to control the intracellular concentrations of other cations. The transition metal ions, such as zinc, are also found in seawater, but in much, much lower concentrations, and they are, as described above, equally rare within cells.

Sodium ions are so pervasive in the cellular environment that it would be expected to be difficult to control their concentration. Therefore only a few enzymes use sodium ions in their action, and, when they do, the function of the sodium ion is rarely, if ever, catalytic. Potassium ions are used in proteins more often, although their role is seldom catalytic. On the other hand, magnesium and calcium ions are bound by many enzymes and proteins and often play important roles in any catalytic event. Magnesium salts are very soluble, but many calcium salts are not, and calcium is used in the formation of bone. It is difficult for the body to utilize a metal ion if this usage might, as a result, compromise the stability of bone. Calcium, therefore, is found in many structural proteins, but not often in the catalytic regions of proteins. Transition elements, such as zinc, are of great importance in the action of many metalloenzymes. Zinc is catalytically active in many enzymes, participating directly in the biochemical reaction (such as that of carboxypeptidase). Alternatively it can act as a structural focus, holding different parts of the protein in a rigid arrangement (as in zinc-finger proteins). In horse plasma the concentrations of free magnesium and zinc is reported as 0.5 mM and 2×10^{-10} M respectively, total values for each ion being 0.6 mM and 8 μM [4].

In order to carry out most biochemical reactions, metalloenzymes generally utilize the rarer transition metal ions. Elements such as zinc, copper, iron, nickel, and cobalt are found in low concentrations in plasma and seawater and yet the enzyme has to select the appropriate metal ion from them. There is evidence for the existence of proteins that can chaperone specific metal ions to their appropriate sites in apoenzymes, protecting the metal ions from adverse reactions as they are guided to their required location [5]. How does the enzyme attempt to select out the one metal ion it requires? The answer is that the chemistry of the metal ion is used as a basis for selection. Each metal ion has some property that is different from that of most others, but, in fact, there is often considerable overlap in these properties so that a given enzyme may bind one of several different cations in one specific site. Some relevant data are provided in Tables 1 and 2. The metalloenzyme contains within its overall design an arrangement of preferred side-chain functional groups with the correct size hole to bind the required metal ions in an appropriate hydrophobic or hydrophilic environment. Thus the metalloenzyme binds metal ions

Table 1. Amino Acid Side Chains That Bind Mg^{2+} and Zn^{2+} Ions

		Mg^{2+}	Zn^{2+}
Negatively charged oxygen	Asp, Glu	yes	yes
Hydroxyl groups	Ser, Thr, Tyr	yes	yes
Water		yes	yes
Carbonyl (C=O) groups	Asn, Gln	yes	rarely
N	His, Arg, Lys, Asn, Gln, Trp, main chain NH	very rarely	yes
S	Cys	no	yes

Table 2. Cation Radii and Average Coordination Numbers[a]

	Cation Radius (Å)	Inner Coordination Number
Mg(II)	0.65	6.0
Zn(II)	0.71	5.0
Na(I)	0.95	6.7
Ca(II)	0.99	7.3
K(I)	1.33	9.0

Note: [a]From Ref. 16.

selectively because the nature of the ligands and their relative geometry is optimal for that metal ion. Selective delivery of the optimum metal ion will also assist this process.

Ligands are the atoms or groups of atoms that surround the metal and which are close enough to be chemically bonded [6]. They donate a pair of electrons to the metal ion and so create a bond in which both electrons come from one atom (a dative bond in old parlance). Ligands are generally negatively charged (e.g. the carboxylate oxygen atoms of aspartic acid) or neutral with a slight partial negative charge (e.g. an oxygen atom in a water molecule or in a carbonyl group of the protein backbone). The metal ion acts like a Lewis acid, accepting a pair of electrons from the ligand. The number of such ligating atoms surrounding a central metal ion is termed the coordination number of the metal ion. When a metal ion coordinates a ligand in this way, it can affect the electron distribution of the ligand and therefore its reactivity. What is important in the study of metal–ligand interactions are the polarizabilities of both the metal ion and the ligand (that is, how readily the electron cloud is distorted), the number of ligands surrounding each metal ion, and the stereochemistry of the resulting arrangement [7–9].

The differing abilities of metal cations to gain and lose water molecules contains information on their inherent reactivities. The exchange of coordinated water with bulk solvent by various cations has been categorized by Langford and Gray [10] into four groups: (1) those for which the exchange rate is greater than 10^8 per second include alkali and alkaline earth metal ions (except for beryllium and magnesium), together with Cr^{2+}, Cu^{2+}, Cd^{2+}, and Hg^{2+}; (2) intermediate rate constants (from 10^4 to 10^8 per second) are found for divalent first-row transition metal ions (except for V^{2+}, Cr^{2+}, and Cu^{2+}), Ti^{3+}, and Mg^{2+}; (3) those with slow rate constants (from 1 to 10^4 per second) include Be^{2+}, Al^{3+}, V^{2+}, and Ga^{3+} and some trivalent first-row transition metal ions; and (4) the "inert group" with rates from 10^{-6} to 10^{-2} per second are Cr^{3+}, Co^{3+}, Rh^{3+}, Ru^{2+}, Ir^{3+}, and Pt^{2+} [11–13]. Trivalent lanthanide cations are also found in the first two groups. In this list Zn^{2+} and Mg^{2+} are both in the second group; but zinc ions exchange water faster than do magnesium ions. The rates are 10^5 per second for magnesium, and 10^7 to 10^8 per second for zinc. One of the factors involved in rates of exchange is the charge-to-radius; if this ratio is high the exchange rate is generally low, as found for Al^{3+} (for which the rate is 10 per second).

The charge distribution in the active site of an enzyme is designed to stabilize the transition state of the catalyzed reaction relative to that of the substrate. In enzyme-catalyzed reactions it is essential that the reactants be brought together with the correct spatial orientation, otherwise the chance of the reaction taking place is diminished and the reaction rate will be low. The electrostatic environment in the active site is a major factor for ensuring the correct orientation of substrate. Thus an enzyme will bind its substrate in a manner that will result in immobilization and alignment of the substrate, will facilitate the formation of the transition state of the reaction to be catalyzed, and will make possible an easy release of the product will result.

The reactions that a metal ion can undergo, and its preferred partners in ligand binding, are related to the behavior of its outer-sphere electrons (its "electron cloud"). The word "hard" has been introduced to indicate a low polarizability so that the electron cloud is difficult to deform it (like a hard sphere) [14] By contrast, "soft" means high polarizability so that the electron cloud is readily deformed. A hard acid or base holds tightly to its electrons, while soft acids and bases contain electrons that are not held so tightly and therefore are easily distorted or removed. A hard acid prefers to combine with a hard base, and it does so by ionic forces. By contrast, a soft acid prefers to bind with a soft base, and it may do so by partially forming covalent bonds, such as π-bonds. Mildvan [15] wrote, "Cations that indulge in ionic binding prefer ligands that so indulge: cations that indulge in covalent binding prefer ligands that so indulge." These types of binding are related to the energies of the highest occupied molecular orbital (HOMO) of the electron-pair donor (a Lewis base, the ligand) and the lowest unoccupied molecular orbital (LUMO) of the electron-pair acceptor (a Lewis acid, the metal ion). If these have similar energies, then electron transfer will give a covalent (soft–soft) interaction. If the energy difference is large, electron transfer does not readily take place and the interaction is mainly electrostatic (hard–hard). These hard–soft categorizations are a help in understanding the relative binding preferences of various cations.

II. MAGNESIUM THE CONSERVATIVE

The magnesium ion is a simple alkaline earth cation which is found in several enzymes. It has an ionic radius of approximately 0.65 Å [16]. In an examination of the crystal structures of magnesium salts and complexes listed in the Cambridge Crystallographic Database (CSD) [17] two major findings emerged [18]; a magnesium ion prefers to bind six ligands in a regular octahedral arrangement (in about 80% of such complexes), and it prefers to bind to ligands by way of oxygen atoms (again in about 80% of Mg^{2+} complexes). It was also found that magnesium binds readily to water molecules. The hexaaquated ion, $Mg[H_2O]_n^{2+}$, is found in crystal structures containing magnesium ions, even in the presence of a crown ether or a strong anion which might have been expected to bind more firmly to the magnesium ion, but which does not [18].

The strong affinity of the magnesium ion for water may, in part, be a result of the size of this cation, so that, when it is surrounded by six water molecules, the oxygen atoms of the water molecules (at an $Mg^{2+}\cdots O$ distance of about 2.07 Å) are in contact with each other ($O\cdots O$) approximately 2.9 Å apart), as shown in Figure 1. It was found by ab initio molecular orbital calculations that the energetic penalty for changing the coordination number of the inner sphere of a magnesium ion from six to any other number is high [18]; it prefers to keep this coordination number of six. The energy penalty of transferring water molecules from the inner to the next coordination shell of the hexahydrate of divalent magnesium is 4, 9, and 34 kcal per mole, respectively, as shown later in Figure 2 (filled circles). This is why magnesium can be considered a conservative structural binder [19]; it has very specific binding preferences.

These trends also extend to protein crystal structures which show that magnesium overwhelmingly prefers to bind to oxygen atoms. The oxygen atoms in proteins that bind magnesium ions are generally those in the carboxylate groups of aspartate and glutamate side chains, the carbonyl groups in the amides asparagine and

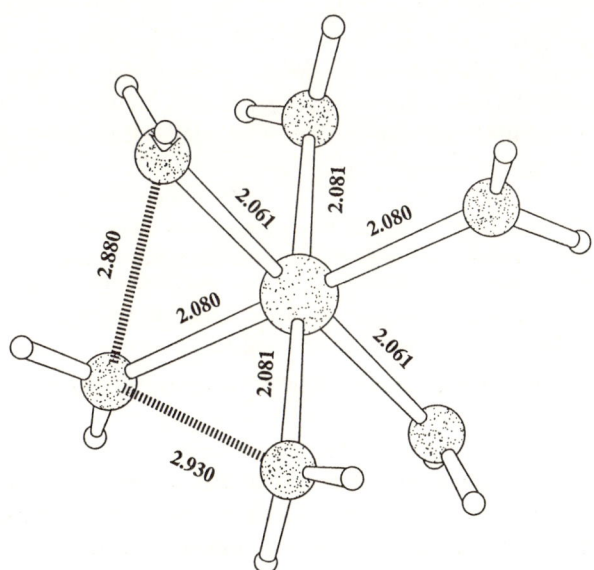

Figure 1. Six water molecules around a magnesium ion, showing $Mg^{2+}\cdots O$ distances (2.06–2.08 Å) and $O\cdots O$ distances (around 2.9 Å) [142].

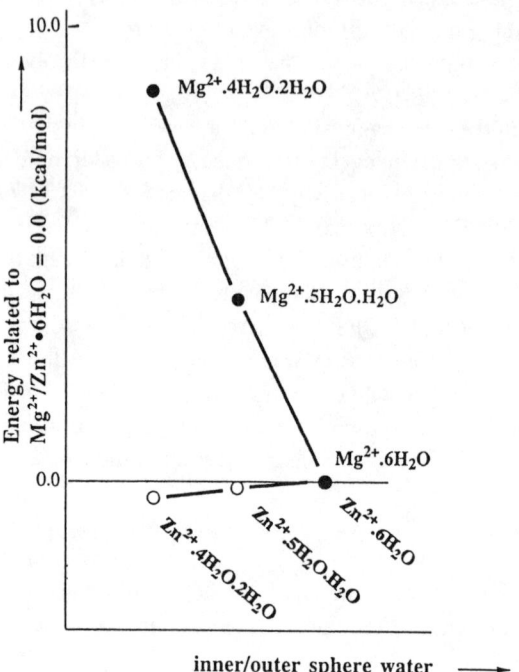

Figure 2. The energies of inner versus second (outer) sphere water molecules around Mg^{2+} (*filled circles*) and Zn^{2+} (*open circles*). $Mg^{2+} \cdot xH_2O \cdot yH_2O$ implies x molecules of water in the innermost coordination sphere and y molecules of water in the next shell. Energies are drawn relative to a value of zero for the ion surrounded by six water molecules in its innermost coordination sphere (see Figure 1) [21].

glutamine, and the hydroxyl groups of serine and threonine side chains. The magnesium ion generally binds near the plane of a carboxylate group or water molecule [20]. Main-chain carbonyl groups and the oxygen and nitrogen atoms of amide groups do not often bind to magnesium ions. In proteins if two water molecules bind to the magnesium ion they can be displaced by substrate; this is found for many magnesium-binding proteins.

The charge on the magnesium ion is +2 and this can serve to neutralize local negative charge (to which it is attracted). This is important in enzyme reactions in that magnesium ions can help neutralize the negative charges of phosphate and carboxylate groups in the active sites of enzymes so that other negatively charged groups can approach. Magnesium ions can also serve to lower the pK_a of metal ion-bound water.

III. ZINC THE ACTIVATOR

Divalent zinc has an ionic radius of 0.71 Å and binds to nitrogen, oxygen, and sulfur atoms in ligands [16]. It lies on the border between soft and hard cations. Its *d* shell is filled and therefore it shows both tetrahedral and octahedral binding geometries, since it is not subject to ligand field effects that stabilize octahedral coordination. Zinc, like magnesium, does not show any redox activity. It was found that 76% of crystal structures containing divalent zinc coordinated to oxygen, nitrogen, sulfur, and/or bromine have a coordination number of 4 or 6 [21]. The energy penalty for changing the innermost coordination number of zinc between 4, 5, and 6 is found to be very small, as shown in Figure 2 (open circles). Thus, magnesium and zinc ions have different binding preferences, as shown in Table 1; magnesium prefers to bind oxygen, while zinc binds sulfur and nitrogen when the coordination number is low (4), and also oxygen at higher coordination numbers (5 and 6). Zinc is also better than magnesium at activating water to a metal-bound hydroxide group [22].

Zinc is an essential component of a large number of enzymes. Zinc-containing enzymes are involved in the synthesis and degradation of major metabolites, and serve to regulate the replication, transcription, and translation of genetic material. In some proteins zinc binds with a coordination number of 4 to sulfur and nitrogen, while in others in which it also binds oxygen the coordination number is 5 or 6.

IV. MAGNESIUM-UTILIZING ENZYMES

Magnesium ions may interact directly with the substrate so that the enzyme binds the magnesium-substrate complex, or it may bind directly to the enzyme, taking on a structural or catalytic role [23, 24]. Most of the enzymes that magnesium is involved with are regulators of the biochemistry of nucleic acids. Magnesium also binds to oligonucleotides and plays a significant role in ribozyme function. A prominent activity of magnesium ions, which are weaker Lewis acids than are zinc ions, is in phosphate ester hydrolysis and phosphoryl transfer. Further details are provided in the text by J.A. Cowan [23]. Protein structures described, and for which ball-and-stick diagrams are available, are part of the Protein Data Bank [25] and atomic coordinates are obtained for the listed file (PDB file).

A. Mandelate Racemase from *Pseudomonas putida*

The enzyme mandelate racemase from the soil bacterium *Pseudomonas putida* catalyzes the racemization of the (*R*)- and (*S*)-enantiomers of mandelate, as shown in Figure 3. In the action of the enzyme a hydrogen atom (α-proton) is extracted from the aliphatic carbon atom of the mandelate ion. Two active-site bases are involved in this: one to extract the α-proton from the (*S*)-isomer and the other to extract the α-proton from the (*R*)-isomer [26]. A proton from a solvent molecule in the active site then adds to the deprotonated mandelate in one or other of the two

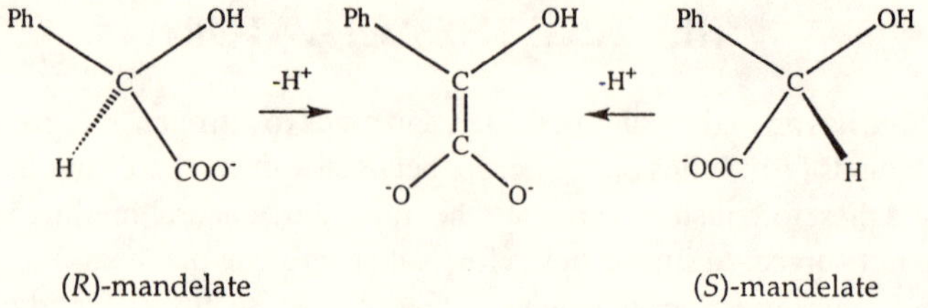

(R)-mandelate   (S)-mandelate

Figure 3. Interconversion of (R)- and (S)-mandelate by the action of the enzyme mandelate racemase. The addition of a proton (H$^+$) to one or other of the two faces of the enolate intermediate gives one of the two enantiomers of mandelate.

possible directions to give the product which is eventually composed of equal amounts of the two enantiomers of mandelate. Therefore, the overall role of the enzyme is to bind mandelate and facilitate the extraction of the carbon-bound α-proton from it. This would be done by an active-site base. The enzyme has an absolute requirement for a divalent metal ion, and magnesium is the best for this, although divalent cobalt, nickel, manganese, and iron can take its place, although less effectively.

The crystal structure of mandelate racemase with bound (S)-atrolactate has been determined (to 2.1 Å resolution), and two views are presented in Figures 4 and 5 (PDB file 1MDR) [27]. Atrolactate, a potent competitive inhibitor of the enzyme is a homologue of (S)-mandelate in which a methyl group has replaced the α-proton so that racemization cannot take place. Mandelate racemase in its crystal structure is found to fold into two structural domains, one of which is a $(\beta\alpha)_8$ barrel of the type that is found in triose phosphate isomerase [28–30]; this is the domain containing the active site. As shown in Figure 4, the negatively charged carboxylate group of atrolactate binds near the positively charged ε-amino group of Lys164, while the hydroxyl group of atrolactate, with its hydrogen-bond donating capacity, binds near the carboxylate group of Glu247. The magnesium ion is coordinated to Asp195, Glu221, and Glu247, in addition to one water molecule and the two ligand sites on atrolactate. Two lysine residues, Lys164 and Lys166, a histidine side chain, His297, and two carboxylate groups, Glu317 and Asp270, lie nearby, as illustrated in Figure 6(a).

The atrolactate is bound in the crystal structure to the magnesium ion in a bidentate manner by way of one of its carboxylate oxygen atoms and its α-hydroxyl group (see Figure 4). The α-methyl group of the atrolactate ligand points directly at Lys166, as shown in Figure 5; if the (R)-isomer of atrolactate were to be bound to the enzyme, the α-methyl group would point directly towards His297 [31] This is diagrammed in Figure 6(a). Lys166 is the (S)-specific acid/base catalyst [extracting the (S)-α-proton], while His297 is the (R)-specific acid/base catalyst [27, 32]. The importance of Lys166 was highlighted by X-ray diffraction studies of the product of stereospecific alkylation of the enzyme at this lysine residue by (R)-α-

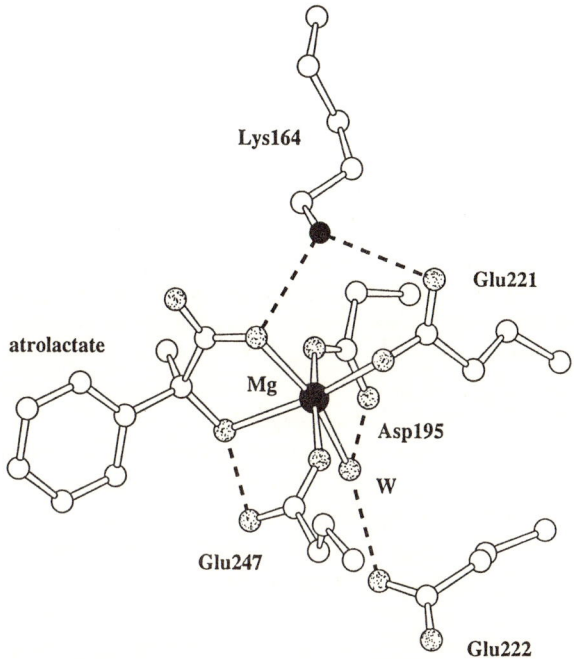

Lys164

Glu221

atrolactate

Mg

Asp195

W

Glu247

Glu222

Figure 4. Immediate surroundings of the magnesium ion in the active site of the enzyme mandelate racemase (PDB file 1MDR). In this and all following ball-and-stick type diagrams, drawn by the computer program ICRVIEW [*143*], zinc ions and oxygen atoms are stippled, magnesium ions and nitrogen atoms are black, carbon atoms are white, and hydrogen bonds are dashed lines. Water molecules are designated W. Atomic coordinates for each ball-and-stick diagram are taken from the Protein Databank files [*25*].

phenylglycidate [*33*]. It was also found that a K166R mutant enzyme (Lys166 replaced by arginine) racemizes mandelate very slowly; therefore its crystal structure in a complex with (*S*)-mandelate could be determined [*32*]. In this (*S*)-mandelate was bound in the same manner as (*S*)-atrolactate.

The reaction to be catalyzed poses problems for the enzyme [*34*]. The abstraction of a proton from an aliphatic carbon atom is generally difficult and slow. The pK_a of the carbon-bound α-proton of mandelic acid is 22.0 [*35*], while the pK_a of the α-proton of the mandelate anion (as for the phenylacetate anion) is approximately 29 [*36, 37*]. In spite of this, mandelate racemase increases the rate of the racemization reaction by a factor of 1.7×10^{15} to approximately 1000 per second at 25 °C at pH 7 [*31, 38*]. Interactions of mandelate with enzyme, analogous to those with inhibitor (*S*)-atrolactate in which one carboxylate oxygen atom is coordinated to the magnesium ion and also hydrogen-bonded to the ε-ammonium group of Lys164,

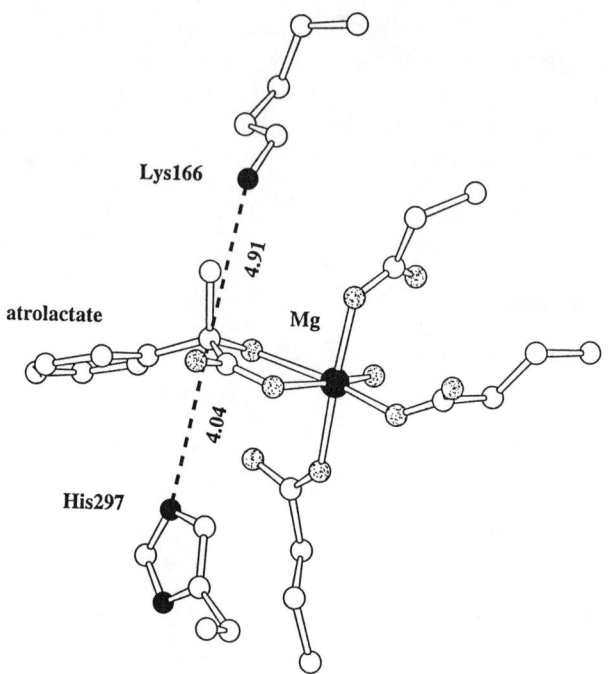

Figure 5. The locations of proton-abstracting groups in the enzyme mandelate racemase (PDB file 1MDR). The binding of (*S*)-atrolactate, which has an additional methyl group that causes it to inhibit the reaction of mandelate racemase, is shown. Presumably when mandelate is bound, one of the bases (Lys166 or His297) approaches the α-carbon atom of the mandelate ion.

allow the bound mandelate anion to resemble bound mandelic acid, thereby reducing the pK_a of its α-proton from 29 to 22 [*35*].

The ease of loss of the α-proton of mandelic acid is further enhanced by the stabilization of the enolic tautomeric intermediate by Glu317 and Lys164. These two hydrogen-bonding groups favor the enolate form (the putative intermediate in the reaction). In the crystal structure the carboxylic acid group of Glu317 and the ε-amino group of Lys164 are each hydrogen-bonded to one of the two oxygen atoms of the atrolactate carboxylate group, as diagrammed in Figure 6(b). This implies that Glu317 participates as a general-acid (electrophilic) catalyst in the formation of an enolic tautomer of mandelic acid as a reaction intermediate [*39*]. The enolic intermediate is stabilized by a short, strong hydrogen bond between its hydroxyl group (pK_a 6.6) and the carboxylic acid group of Glu317 (pK_a near 6) [*40, 41*], and by the proximity of the positively charged Lys164 side chain. This stabilization of the enolic intermediate leads to a lowering of the pK_a of the α-proton so that it is more readily removed. The hydrophilic/hydrophobic environment around Lys166 lowers its pK_a value so that it is similar (about 6) to that of the other α-proton-extracting base, His297 [*37*].

The magnesium ion serves to hold the hydroxyl and carboxylate groups of the mandelate ion in one plane, as would be required by an ene intermediate. This is

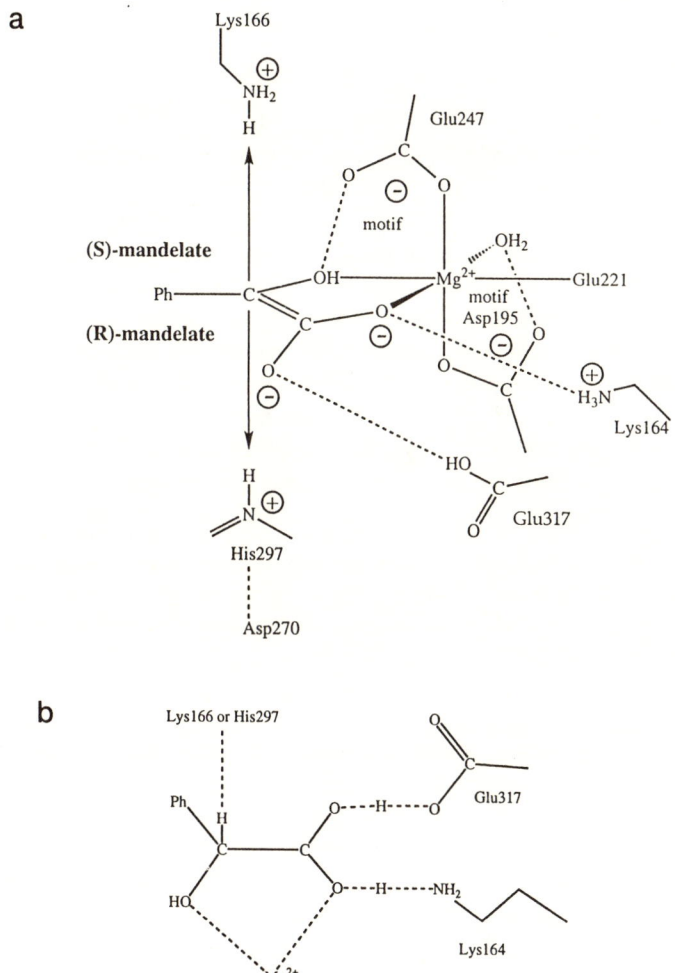

Figure 6. (a) Diagram of the arrangement of magnesium ion-binding groups and the bases Lys166 and His297 that extract a hydrogen atom from the carbon atom of mandelate. The diagram is approximately in the same orientation as that in Figure 5. (b) The hydrogen bonding interactions of Glu317 and Lys164 which help to stabilize the enolate intermediate.

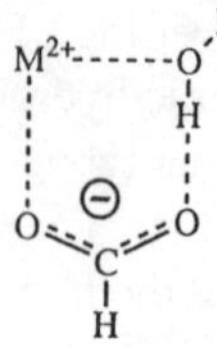

Figure 7. A metal ion–water–carboxylate motif found in many magnesium-utilizing enzymes. This motif probably helps to orient the fairly rigid coordination octahedron of magnesium.

also facilitated by the proximity of the positively charged Lys164 side chain. The coordination octahedron of the magnesium ion contains two copies of a motif in which a carboxylate group spans (by its two oxygen atoms) the magnesium ion and one of its metal ion-bound water molecules, as diagrammed in Figure 7. The interactions in these motifs appear to be mainly electrostatic, but they appear to serve to align the magnesium-ion octahedron so that its six coordination positions are precisely oriented. Therefore, as a result, the orientations of the two remaining sites available for the binding of the substrate or inhibitor are precisely controlled in space. This may help define and stabilize the transition state of the reaction. Bearne and Wolfenden wrote, " . . . general acid-general base catalysis, inefficient in simple model systems, becomes an efficient mode of catalysis when structural complementarity between an enzyme and its substrate is optimized in the transition state" [*34*].

B.   Rat Liver Catechol O-Methyltransferase

Methyltransferases catalyze the transfer of the S-methyl group of S-adenosyl-methionine to the oxygen, nitrogen or carbon atoms of a nucleophile. Catechol O-methyltransferase catalyzes the first of these three possibilities, transferring an S-methyl group to one of the two hydroxyl groups of a catechol to give a methylated catechol and S-adenosyl-L-homocysteine, Figure 8. Catechol O-methyltransferase is a monomeric enzyme that catalyzes the O-methylation of a variety of catecho-lamine neurotransmitters such as dopamine. It requires magnesium ions [*42*], and the rate-determining step appears to be transfer of the methyl group [*43, 44*].

The crystal structure of the enzyme (at 2.0 Å resolution) consists of a single domain of eight α-helices and seven β-strands (PDB file 1VID) [*45*]. The fourth to eighth α-helices and the first five β-strands provide a nucleotide-binding motif (the Rossmann fold) to bind the adenine of the cofactor [*46*]. The reported crystal structure contains the enzyme, the coenzyme S-adenosylmethionine, a magnesium ion, and a competitive inhibitor 3,5-dinitrocatechol [*45*].

In this structure the chelation of the substrate by the magnesium ion appears to bring the S-adenosylmethionine and catechol together in the active site, poised

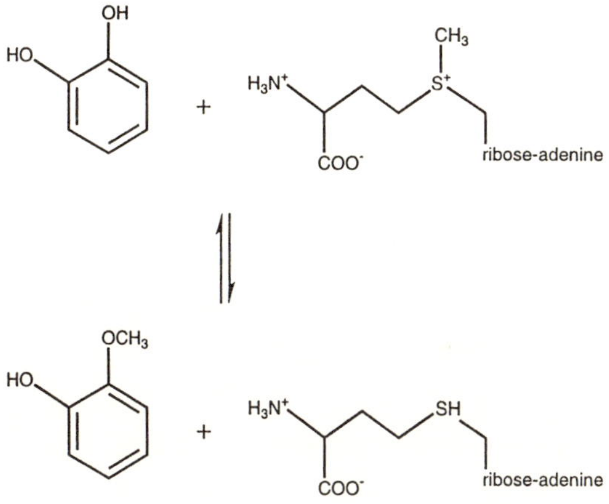

Figure 8. The reaction catalyzed by the enzyme catechol *O*-methyltransferase. The methyl group is transferred from *S*-adenosylmethionine to the catechol molecule.

ready for reaction, as shown (in a simplified version) in Figure 9(a). The magnesium ion is bound to the two hydroxyl groups of the 3,5-dinitrocatechol, one carboxylate oxygen atom of each Asp141, Asp169, the carbonyl group of Asn170, and a water molecule, as shown in Figure 9(b). One hydroxyl group of the inhibitor is near a carboxyl oxygen atom of Glu199, and the other is near the methyl group to be donated by *S*-adenosylmethionine, as diagrammed in Figure 10(a). It appears that the *S*-adenosylmethionine binds first, then the magnesium ion, and finally the catechol substrate, although there are no direct interactions between the *S*-adeno-sylmethionine and the magnesium ion (see Figure 9) [*47*].

The mechanism of action of this enzyme appears to be similar to that of many DNA methyltransferases [*48*]. A direct transfer of the methyl group from the sulfur atom of *S*-adenosylmethionine to an oxygen atom of one of the hydroxyl groups is presumed to occur by an S_N2 mechanism [*44, 49*]. Thus the reaction proceeds with inversion of configuration. Once the catechol has bound to the enzyme, one of its hydroxyl groups is surrounded by two positively charged groups (Mg^{2+} and AdoMet). These encourage the catechol to give up one of the hydroxyl hydrogen atoms (presumably to Lys144) and become, as a result of this deprotonation, a negatively charged catecholate ion. The remaining hydroxyl group of the catecho-late helps to stabilize the oxyanion, and is, itself, hydrogen bonded to the negatively charged carboxylate group of Glu199. The oxyanion (ionized hydroxyl group) then attacks the electron-deficient methyl group of AdoMet in a direct nucleophilic manner. The methyl ether of the catechol substrate then dissociates from the enzyme, its methyl group being repelled by Lys144. It has been suggested that the side chain amino group of Lys144 is the base that abstracts the proton from enzyme-bound catechol, as diagrammed in Figure 10(b) [*50*], and illustrated in

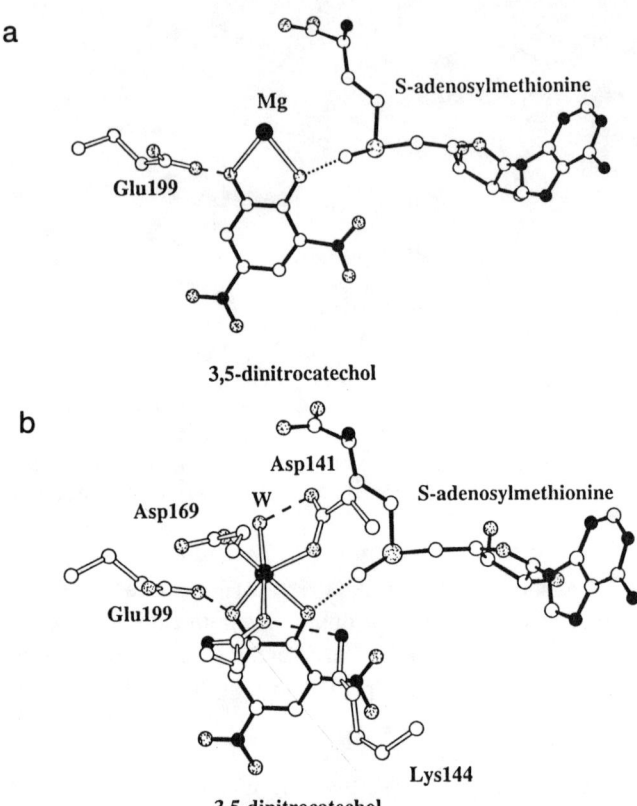

Figure 9. The active site of the enzyme catechol *O*-methyltransferase (PDB file 1VID), showing the binding of 3,4-dinitrocatechol and the relative dispositions of negatively charged Glu199 and positively charged *S*-adenosylmethionine. (**a**) A simplified view. (**b**) A more extensive view of the active site, drawn in approximately the orientation of (**a**).

Figure 11. Methylation of catechol reduces the interaction between magnesium ions and catechol so that the product is more readily released.

C. The Chemotaxis Protein CheY from *Escherichia coli*

CheY is a response regulator protein in the signal transduction system for bacterial chemotaxis. This protein enables the swimming behavior of the bacterium (smooth swimming, when all of the flagellae rotate counterclockwise, to tumbling, when some of the flagellae rotate clockwise) to be altered in response to changes in the chemical composition of its environment. In this way the bacterium can avoid noxious materials and approach advantageous compounds. A signal is transmitted

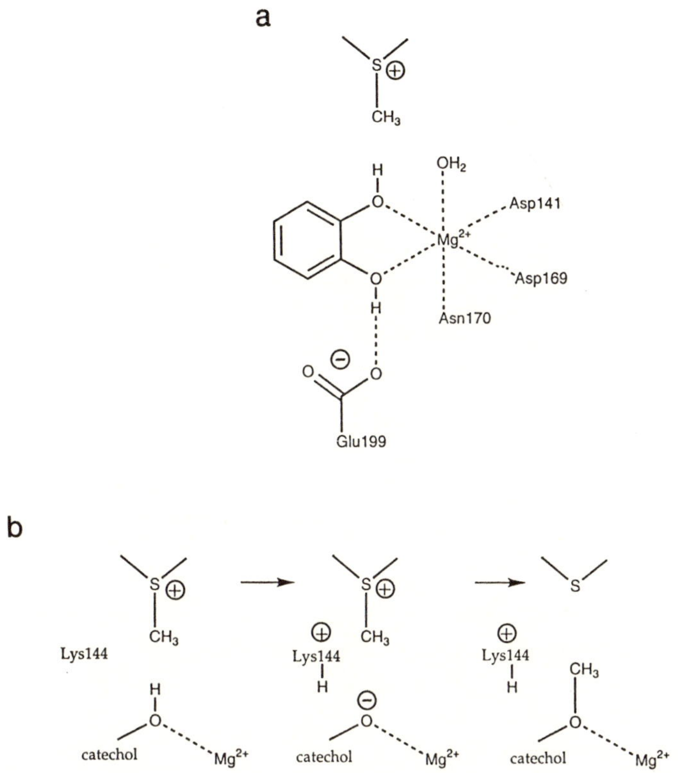

Figure 10. Catechol *O*-methyltransferase. Binding of catechol to catechol *O*-methyltransferase. (a) The overall mode of binding, and (b) the proposed course of the reaction in which the hydroxyl group on the catechol is converted to an oxyanion and then to a methylated derivative.

from the receptor to an effector protein CheA, a histidine kinase, which phosphorylates one of its own histidine residues [51]. In the presence of magnesium ions this phosphoryl group is transferred from CheA to a CheY aspartate residue and the product, phosphorylated Mg^{2+}-CheY, binds to the flagellar switch proteins in such a manner that the chances that the motor will turn clockwise are increased and the type of motion of the bacterium is altered. Dephosphorylation regenerates the initial system.

Crystal structures have been reported for the metal-free protein from *Salmonella typhimurium* [52–54] and *Escherichia coli* [55–58]. We have used the *E. coli* protein as a basis for the following discussion. CheY is a $(\beta/\alpha)_5$ protein, folded in

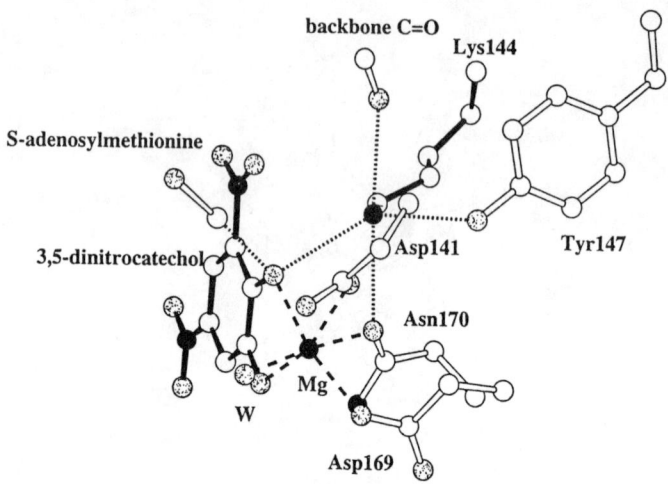

Figure 11. Surroundings of Lys144 (with filled bonds) in catechol *O*-methyltransferase. The 3,4-nitrocatechol molecule is also drawn with filled bonds.

the form of a five-stranded β-sheet core flanked by five α-helices (PDB file 1CHN). The phosphorylation site lies in a cavity on one face of the protein. The Lys109 side chain extends into the active site; this is made possible by the *cis* conformation of the peptide bond preceding Pro110. The ε-amino group of Lys109 interacts with Asp57, the site of phosphorylation upon activation of CheY. Magnesium ions are required for phosphorylation of CheY by the kinase CheA; this phosphorylation occurs at the oxygen atom of the carboxyl group on Asp57 that is not bound to the magnesium ion. Lys109 is displaced by phosphate, and therefore this is presumably the site of the active phosphate.

The active site for phosphorylation consists of Asp13, Asp57, and Lys109. Also in the area are Asp12, Asn59, and Glu93. The magnesium ion is bound to Asp57, Asp13, the carbonyl oxygen of Asn59, and three water molecules in an octahedral arrangement, as shown in Figure 12(a) and (b) [57]. Lys109 is positioned by the *cis* configuration of the peptide bond between it and Pro110. The interaction between Lys109 and Asp57 is similar in orientation to that between Asp57 and the phosphate group of CheA that will be covalently linked to Asp57, as shown in Figures 13(a) and (b). It is proposed, on the basis of other examples of phosphoryl transfer and on model building, that the Asp57 acyl phosphate of CheY is dephosphorylated by a mechanism initiated by nucleophilic attack on the phosphorus atom [59]. It appears that the magnesium ion provides a binding point for the carboxylate group of Asp57 so that its other carboxylate oxygen atom can provide an in-line path for attack of the phosphorus atom by nucleophilic attack, for instance by water [60].

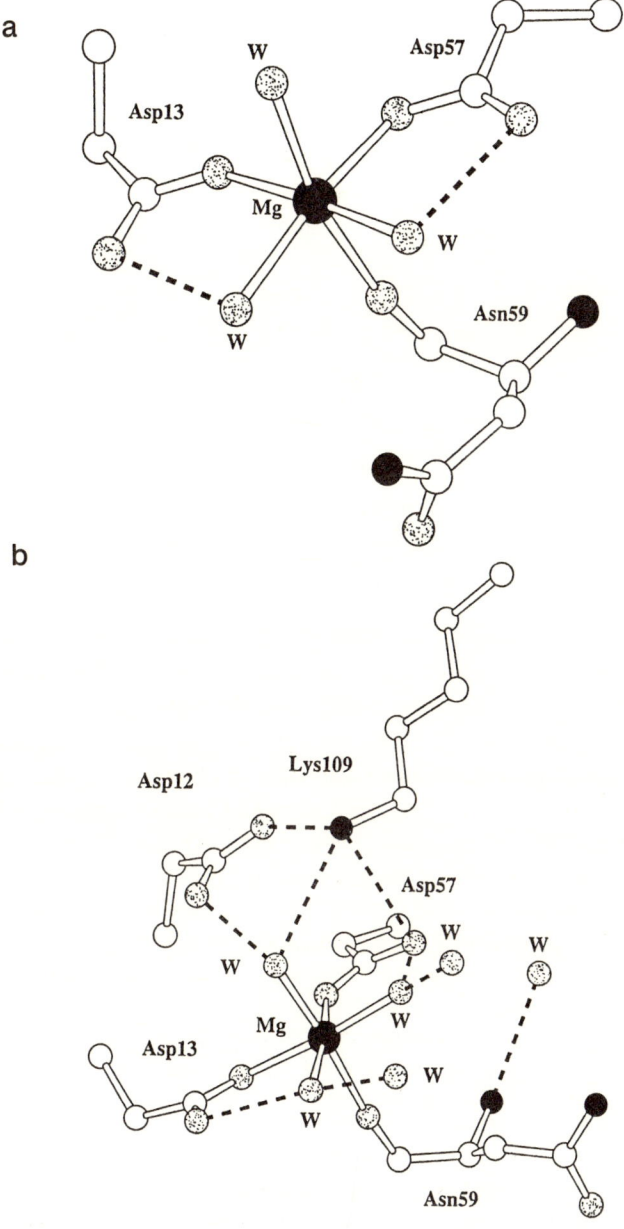

Figure 12. CheY (PDB file 1CHN). (**a**) Immediate surroundings of the magnesium ion in the crystal structure of CheY. Note the magnesium–water–carboxylate motif of Figure 7 occurs twice in this magnesium coordination octahedron. (**b**) Further details of the active site of CheY.

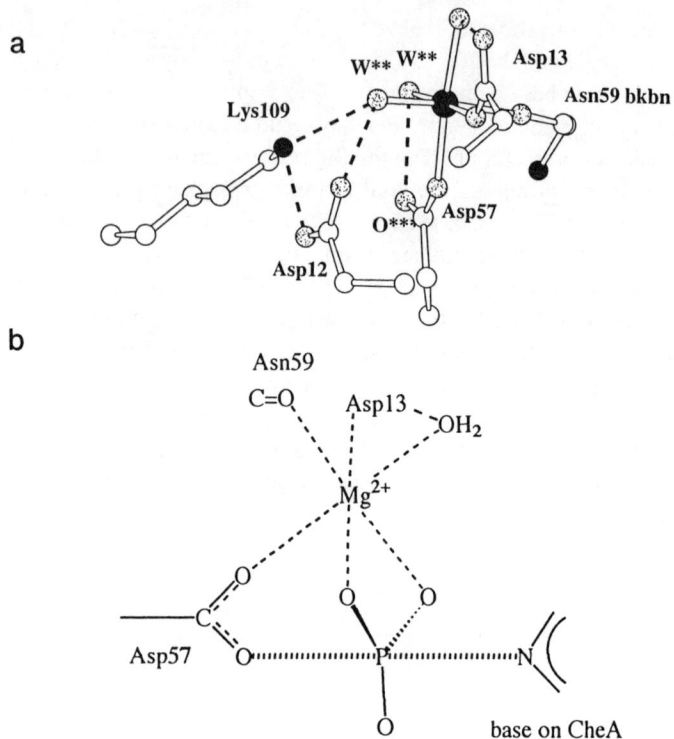

Figure 13. CheY active site. (a) Surroundings of Asp57. Asterisked atoms are probably involved in the putative trigonal bipyramidal transition state in the reaction which is shown in (b).

The phosphoryl group on Asp57 lies (in the model used) between Asp12 and Lys109 (which is replaced by phosphate).

It appears that phosphorylation of CheY causes it to change its conformation. NMR studies of CheY in its active phosphorylated state find significant chemical shifts which involve residues that must be on or near the surface of the molecule that interacts with CheA and with motor switch components [61]. The half-life of the aspartyl phosphate group on the protein is short, less than a minute, making an X-ray diffraction analysis difficult.

D. Avian Sarcoma Virus Integrase

The retroviral integrase (called IN) is a virus-encoded diesterase that catalyzes the incorporation (integration) of viral DNA into host DNA. Two biochemical reactions are catalyzed by this enzyme—the "processing" of the viral DNA ends (an endonucleolytic cleavage reaction) and then the "joining" of these ends to host

DNA (a polynucleotide transfer reaction). During the processing reaction the enzyme introduces nicks in both of the viral DNA strands; these nicks usually are made at a position two base pairs from each 3'-end, immediately 3'- of the highly conserved CA dinucleotide sequence, and they result in newly formed 3'-hydroxyl groups at the ends of both strands. The joining reaction involves a direct attack of these new 3'-hydroxyl groups of the viral DNA on phosphate groups on the host DNA that are separated by six base pairs. As a result of these two reactions the 3'-ends of both viral DNA strands are covalently joined to the host DNA. The missing portion is then repaired in an as-yet undetermined manner. The overall result of this coupled cleavage-ligation reaction is the covalent linking of the 3'-ends of both strands of viral DNA to host DNA, as diagrammed in Figure 14.

Avian sarcoma virus (ASV) integrase contains 286 amino acids in three domains. The catalytic domain (referred as the "catalytic core") is composed of residues 52 to 207 and contains three carboxylate groups, Asp64, Asp121, and Glu157, that bind the necessary metal ion cofactors (magnesium or manganese ions). These three acidic residues are conserved in many retroviral and retrotransposon integrases, as

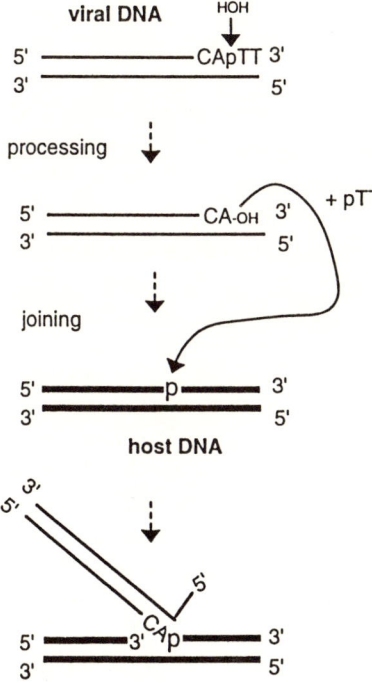

Figure 14. Mode of action of avian sarcoma virus integrase (courtesy R.A. Katz). The viral strands are drawn with thin lines, the host DNA is drawn with heavy lines. This diagram is highly simplified in that there is another insertion site on the other strand 6 base pairs away. The resultant nucleic acid is repaired after the action of integrase.

well as in some bacterial transposases. This suggests that they may play a role in DNA processing and joining activities.

The crystal structure of the catalytic core of the ASV IN has been determined to 1.7 Å resolution, in the presence of magnesium and manganese ions (PDB file 1VSF at 2.05 Å resolution) [62–64]. This catalytic core portion of the integrase enzyme encodes a subset of activities of the total enzyme; it can perform a nonspecific nicking reaction, but cannot catalyze the joining reaction. The catalytic domain can also carry out a type of joining called "disintegration" or "DNA splicing." Each subunit of this dimeric protein folds as a five-stranded mixed β-sheet flanked by five α-helices. The active site is shallow and lies on the surface of the enzyme. In the crystal structure single magnesium or manganese cation interacts with the carboxylate groups of the active site residues Asp64 and Asp121. In addition, four water molecules bind to the metal ion to complete its octahedral coordination sphere, as shown in Figure 15(a). It is presumed that the wild-type enzyme, which contains magnesium, binds in the same manner as the manganese-containing enzyme described here.

ASV integrase acts on the phosphodiester DNA. By analogy to crystallographic studies of the 3′-5′-exonuclease domain of *E. coli* DNA polymerase [65], *E. coli* ribonuclease H [66, 67], and the ribonuclease H domain of HIV-1 reverse transcriptase [68], it is suggested [62] that an enzyme-metal ion complex may possibly stabilize a pentacoordinate phosphate intermediate during two successive nucleophilic attacks. These two reactions are:

1. Attack by a water molecule (a nucleophile) on the viral DNA phosphate (the processing reaction).
2. Attack by the new 3′-hydroxyl group (a nucleophile) on host DNA phosphate (the joining reaction).

The proposed mechanism for the action of the enzyme ribonuclease H, an enzyme that hydrolyzes the RNA strand of DNA/RNA hybrids to yield 5′-phosphates, is shown in Figure 16(a). This enzyme only carries out the first of the reactions listed above.

Figure 15. Avian sarcoma virus (ASV) integrase. (a) Manganese ions in the active site of ASV integrase. Note that there are two examples of the metal ion–water–carboxylate motif in the coordination octahedron. (b) Zinc ions in the active site of ASV integrase. The two zinc ions are separated by water and the carboxylate group of Asp64. (c) Cadmium ions in the active site of ASV integrase. The overall geometry is similar to that for the zinc complex. Neighboring water molecules with long $Cd^{2+}\cdots O$ distances are also indicated, but with single lines.

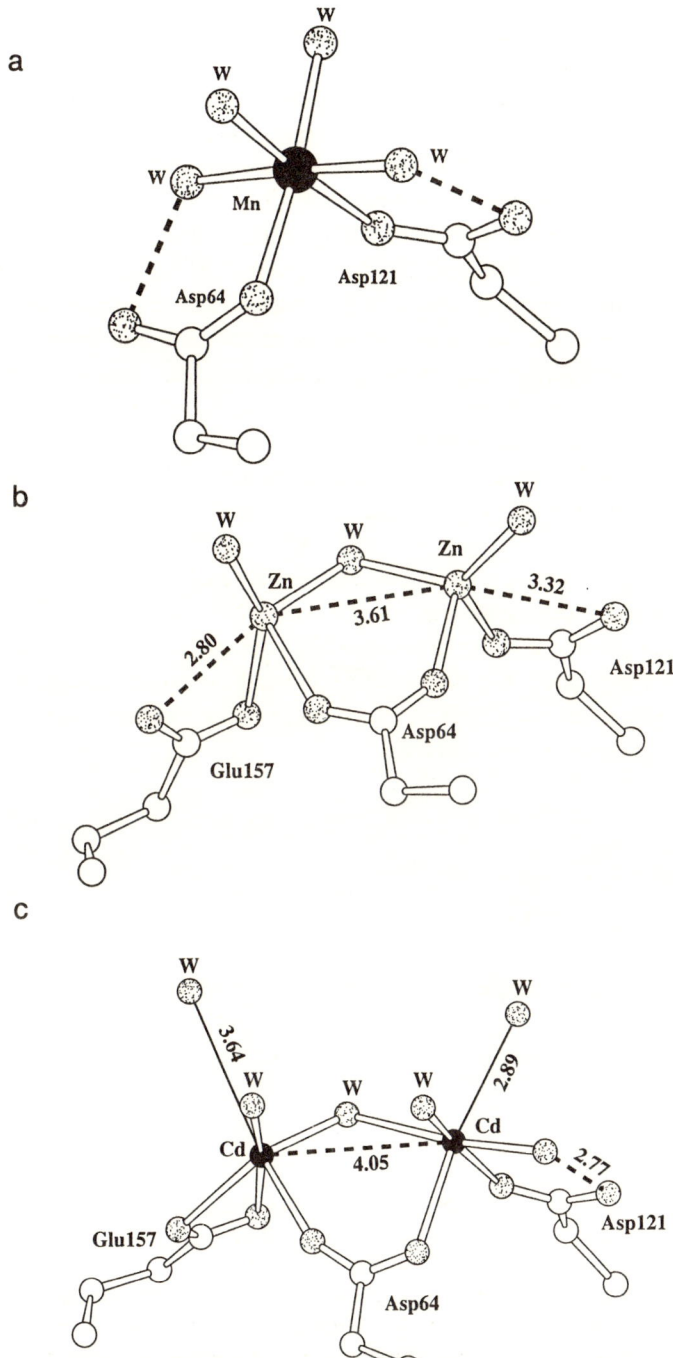

It has been found that other divalent cations such as Ca^{2+}, Zn^{2+}, and Cd^{2+} can also bind to this ASV IN catalytic core, and crystal structures have been determined to resolutions of 1.75 to 2.5 Å [64]. While manganese and magnesium are the normally encountered metal ions in this enzyme, zinc will also work for the processing reaction, while cadmium and calcium do not. The results for Zn^{2+} (PDB file 1VSN at 1.95 Å resolution) and Cd^{2+} (PDB file 1VSJ at 2.1 Å resolution) are different from those of other cations in that two metal binding sites are found, a distance 3.61 Å apart for the Zn^{2+} [Figure 15(b)] and 4.05 Å for the Cd^{2+} complex [Figure

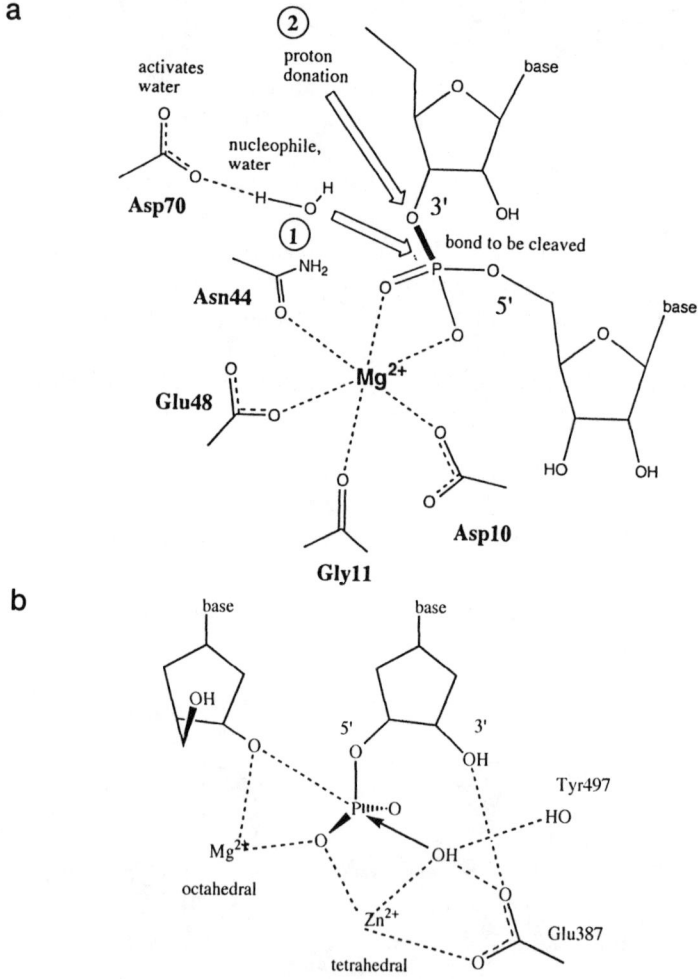

Figure 16. (a) Proposed mechanism of action of ribonuclease H [66]. The reactions are labeled "1" for the attack by water and "2" for the proton donation so that the P–O bond (drawn with a heavy line) is cleaved. In ASV integrase this corresponds to the processing reaction. (b) Proposed mechanism of action of 3′-5′-exonuclease [65].

15(c)]. These Zn^{2+} and Mg^{2+} two-metal sites are almost identical and superimposable. One metal ion is bound by Asp121 and the other by Glu157, while both are shared by Asp64 and a water molecule. One of the sites is octahedral, very similar to that of the Mg^{2+} and Mn^{2+} derivatives. The other site has a deformed octahedral coordination. It is possible that this provides a model of an active site like that found for the 3'-5'-exonuclease of the Klenow fragment [65], illustrated in Figure 16(b), and that this gives clues to the mode of action of integrase.

E. Enolase from *Saccharomyces cerevisae* (Baker's Yeast)

Enolase is one of the enzymes in the glycolytic pathway in which glucose is converted to pyruvate with the formation of two molecules of ATP from ADP per molecule of glucose. The reaction catalyzed by enolase is the reversible dehydration of 2-phospho-D-glycerate to phosphoenolpyruvate (see Figure 17). The phosphate group is then removed by the action of the enzyme pyruvate kinase. The reaction intermediate is believed to be a carbanionic intermediate at C2 of substrate (*aci*-carboxylate), formed by interaction with a base in the active site, so that the hydrogen atom on C2 is removed. The hydroxyl group is eliminated from C3 of a carbanion (enolate) intermediate. The *aci*-carboxylate is stabilized by the metal ions.

X-ray diffraction studies of this enzyme (PDB file 1EBG) show that enolase forms an atypical $(\beta\alpha)_8$ barrel that is better described as $\beta_2\alpha_2(\beta\alpha)_6$ [69–76]. The crystal structure of the complex of yeast enolase with an inhibitor phosphonoacetohydroxamate has been reported at 2.1 Å resolution [77–79]. This inhibitor is a mimic of the *aci*-carboxylate intermediate in the catalyzed reaction. The formulae of these are shown in Figure 18. There are two magnesium ions in the enzyme, and both are necessary for full activity. The hydroxyl group of phosphoglycerate is coordinated to one of the magnesium ions. This makes the hydroxyl group a better leaving group when the C–O bond has been broken. The surroundings of these two

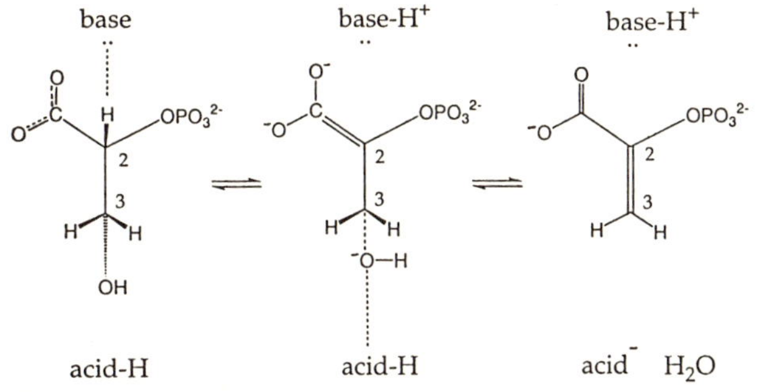

Figure 17. (a) The reaction catalyzed by enolase, and (b) the proposed mechanism of action of the enzyme.

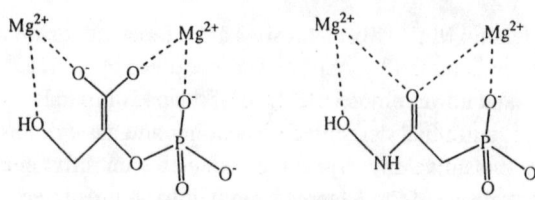

Figure 18. The enolase reaction. (a) The binding of 2-phosphoglycerate and (b) phosphonoacetohydroxamate in crystal structures (see Figure 19).

magnesium ions which are 4.05 Å apart and bridged by a μ-oxyl ligand from the carbonyl moiety of phosphonoacetohydroxamate, are shown in Figure 19. One of the magnesium-binding sites, the high-affinity site, contains Asp246, Glu295, Asp320, water, and the hydroxamate and carbonyl oxygen atom of phosphonoacetohydroxamate, as shown in Figure 19. The second magnesium ion binds the phosphonyl oxygen atom of phosphonoacetohydroxamate, two water molecules, the μ-bridge carbonyl oxygen atom of phosphonoacetohydroxamate, and the carbonyl and γ-oxygen atoms of Ser-39, also shown in Figure 19. 2-Phosphoglycerate binds in an analogous manner [74], diagrammed in Figure 18, although its hydroxyl group is not involved in metal ion binding. Chelation of Ser39 to Mg^{2+} "latches" a

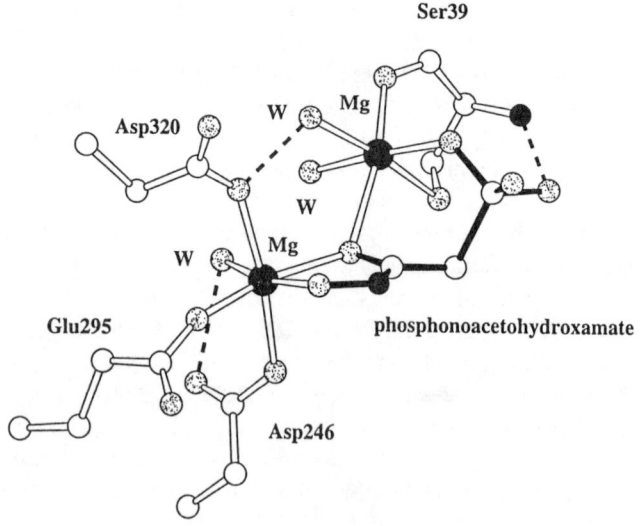

Figure 19. Enolase with bound phosphonoacetohydroxamate (the latter with filled bonds). The phosphorus atom is an open circle. The magnesium ion nearer the bottom of the view is the high-affinity magnesium ion.

flexible loop extending from Gly37 through His43 and closes off the entrance to the active site.

The requirements of this enzyme are similar to those of mandelate racemase, but enolase employs a two-metal ion center, whereas mandelate racemase only requires one metal ion in the active site. Since the pK_a of a carbon-bound proton is high (28–32), the enzyme needs to promote rapid ionization of the carbon acid, to stabilize the *aci*-carboxylate form of the carbanion intermediate, and to aid the hydroxyl group in leaving the active site. The rate-limiting step is the elimination of the C3-hydroxyl group from the carbanion (enolate) intermediate. The turnover of yeast enolase is 80 per second suggesting considerable reduction of the pK_a of the α-proton occurs in the enzyme. Thus the role of enolase is to bind the carboxylate group of 2-phosphoglycerate in the proper conformation for enolate formation.

The nature of the base that abstracts the proton on C2 in the various forms of enolase is not yet settled. For example, in the mechanism proposed on the basis of the crystal structure determination of the yeast enzyme, the base that is invoked in the forward reaction is the ε-amino group of Lys345, while Glu211 interacts with the 3-hydroxyl group of substrate in the next stage of the reaction [74, 76]. In lobster enolase, on the other hand, the water molecule is part of the proton relay system that keeps the substrate in the carboxylic acid form; this makes the pK_a of the C2 proton low enough for proton transfer to the base which, in this enzyme, may be His157 [80].

F.   Rat Guanine Nucleotide-Binding Protein $G_{i\alpha1}$

G proteins (named for their ability to bind guanine nucleotides) hydrolyze GTP in order to transduce extracellular signals. They are molecular on-off switches, enabling messages to be relayed (in the "on" state with bound GTP) and stopped (in the "off" state with bound GDP). When they receive a signal, they exchange GTP for GDP with the assistance of helper proteins (exchange factors). Then, with GTP bound to them, the G proteins send out their own signals within the cell. As a result external signals, such as ligand binding to a receptor, can be interpreted inside the cell leading to the required events. G proteins in the inactive state are found in the membrane as $\alpha\beta\gamma$-heterotrimers with GDP tightly bound to their α-subunit. Upon activation by extracellular signals (such as the binding of hormones), the receptor catalyzes the exchange of bound GDP for GTP. The newly formed GTP-bound form of the heterotrimer is unstable and dissociates into a single α-subunit with bound GTP, leaving the other ($\beta\gamma$) subunits as a dimer. These two species, α and $\beta\gamma$, bind and modulate the activities of a target protein. The G proteins are then released when GTP is hydrolyzed as a result of slow GTPase activity by the α-subunit. The resulting inactive α-subunit can then combine with the $\beta\gamma$-subunit to form the heterotrimer again, and this can now reassociate with its receptor to undergo a new cycle of signal transduction.

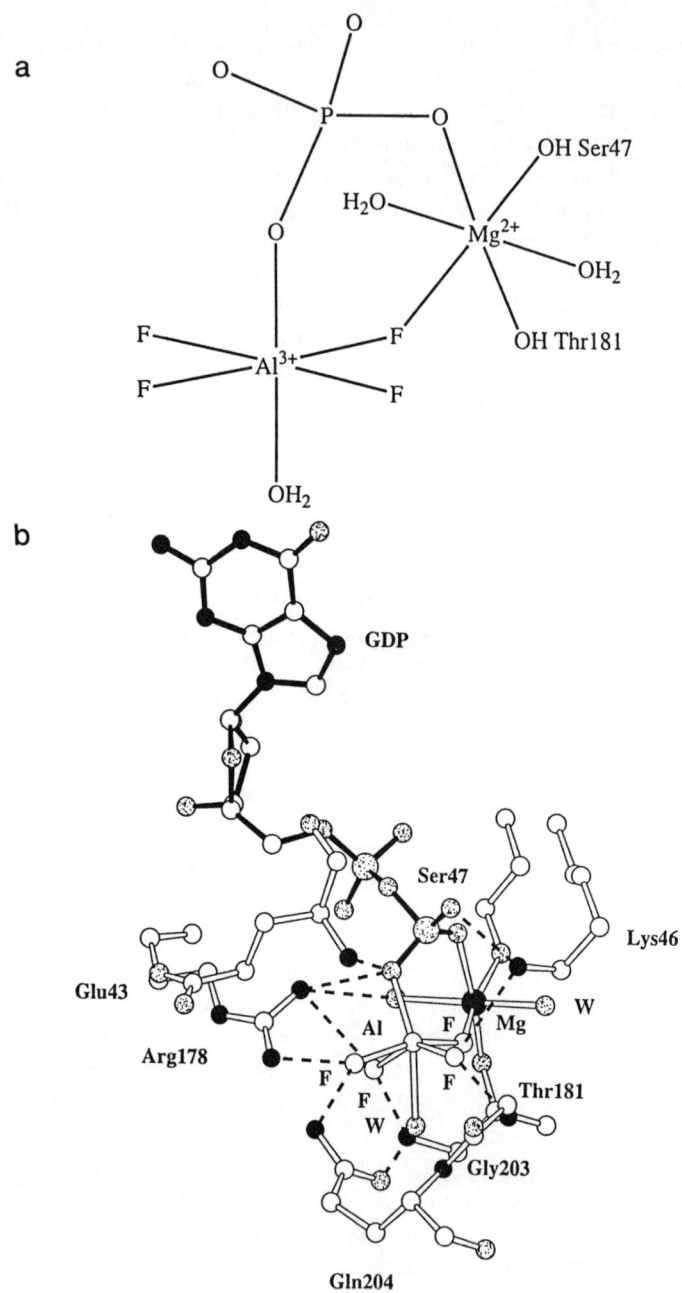

Figure 20. The binding of aluminum fluoride to the Giα1 protein (PDB file 121P) [*82, 83*]. (**a**) Diagram of the surroundings of Al³⁺ and Mg²⁺. (**b**) The metal ion surroundings are shown in (**c**) in a simplified view.

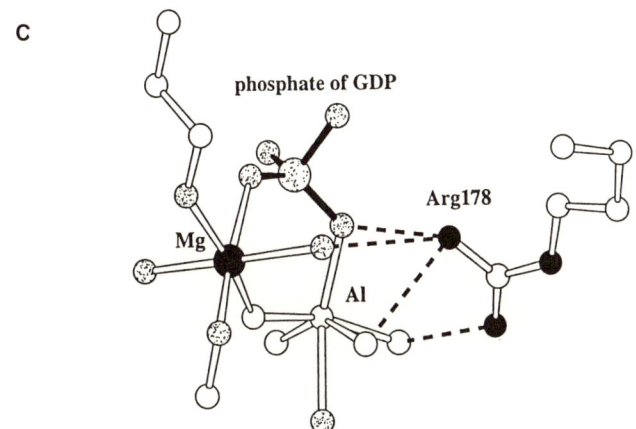

Figure 20. Continued

Structural studies of α-subunits have been focused on $G_{t\alpha}$ (transducin, which takes part in vertebrate vision) and $G_{i\alpha1}$ (for the hormone-regulated inhibition of adenylate cyclase). The X-ray crystal structures of $G_{t\alpha}$ [81] and $G_{i\alpha1}$ [82, 83], activated with the magnesium ion complex of GTPγS, reveals that the α-subunit is comprised of two domains: a GTP-GDP binding domain that is similar in structure to that in the $p21_{ras}$ protein [84], and an α-helical domain. Both of these crystal structures and the crystal structure of $G_{i\alpha1}$ with bound GDP-AlF$_4^-$ suggest that Arg178 is responsible for stabilizing the developing charge in the transition state, while Gln204 orients and polarizes the water molecule that attacks GTP during hydrolysis to GDP [see Figures 20(b) and (c)]. Tetrafluoroaluminate (AlF$_4^-$) is also an activator; it binds to inactive Gα-GDP near the site occupied by the γ-phosphate group when GTP is bound [82, 85]. Thus it confers a GTP-bound geometry to the active site without the possibility of hydrolysis of the bound species. The aluminum is octahedrally coordinated by four fluoride ions in a plane and two axially disposed oxygen atoms, one from the β-phosphate which emulates the β-γ bridging oxygen of the triphosphate, and the other from a water molecule, as shown in Figure 20(c) (PDG file 1GFI, resolution 2.2 Å).

G. D-Xylose Isomerase from *Streptomyces rubiginosus*

The enzyme D-xylose isomerase catalyzes the interconversion of D-xylose to D-xylulose and D-glucose to D-fructose by transferring a hydrogen atom between C1 and C2. Various mechanisms have been suggested including base-catalyzed transfer of a proton with a *cis*-ene diol as an intermediate, a hydride transfer, and a metal-assisted hydride shift [86–89]. The last of these three suggestions is the preferred mechanism at this time, but more studies of the mechanism are needed.

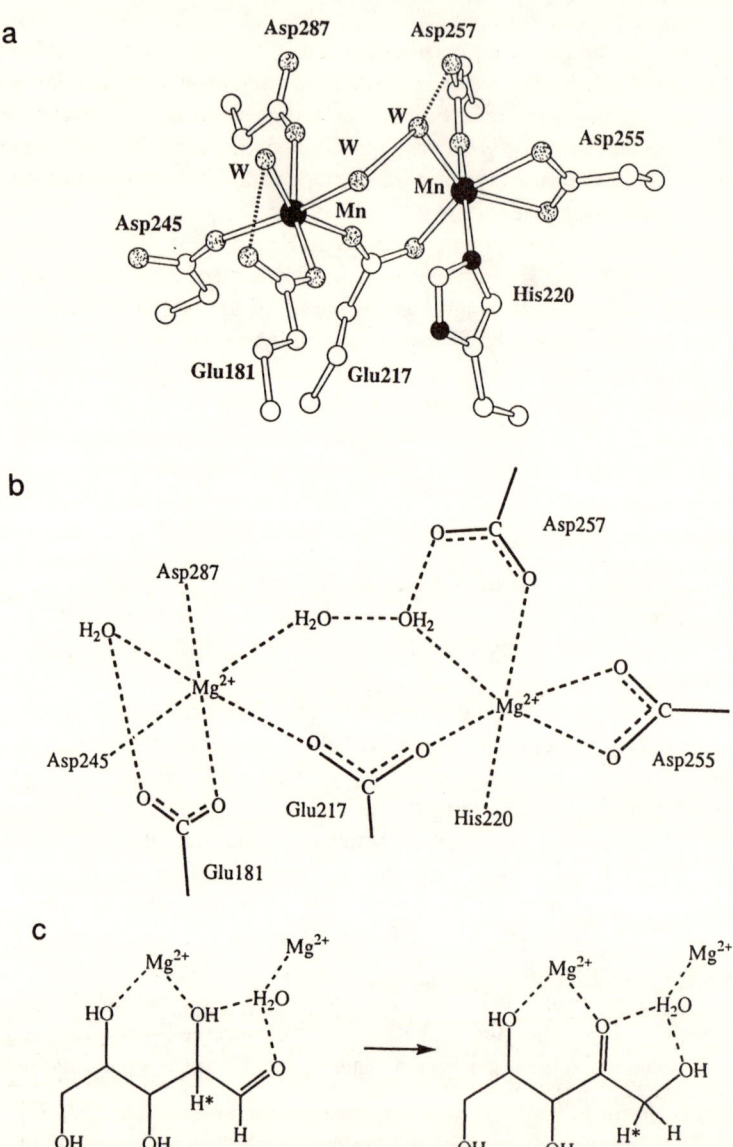

Figure 21. (a) The active site of the enzyme D-xylose isomerase (PDB file 1XIB) [86], see diagram in (b). (b) The metal ion that binds substrate by the displacement of two water molecules from its inner coordination sphere is on the left. This metal ion is generally magnesium or manganese when the enzyme is active. The metal ion that is proposed to activate a water molecule to attack the substrate is on the right [89]. This metal ion may be magnesium, manganese, or divalent cobalt for activity. (c) The proposed catalyzed reaction.

The crystalline enzyme contains a $(\beta\alpha)_8$ or TIM barrel which is constructed of eight β-sheets surrounded by eight α-helices. The β-strands can be considered to form a lining for the barrel, while the α-helices pack around this β-lining and provide a very stable structure with loops extending from it. The average twist between successive β-strands is 26° as found in open β-sheets. The active site of the enzyme is at the C-terminal end of the β-strands and involves residues that are part of the β-sheet or very near it.

This enzyme binds two metal ions, as shown in Figure 21(a) and diagrammed in Figure 21(b) (PDB file 1XIB, resolution 1.6 Å) [86]; both of these metal ions may be replaced by other cations. Magnesium is the metal ion normally found in this enzyme, but manganese also binds to give an active enzyme and is the metal ion present in the reported crystal structure. One site binds carboxylate groups (Glu181, Glu217, Asp245, Asp287) and water molecules, and this is the site to which the substrate binds. The other site, in which the magnesium or manganese ion may also be replaced with divalent cobalt and retain activity, contains carboxylate groups Glu217, bidentate Asp255, Asp257), water, and a histidine residue (His220). The carboxylate group of Glu217 is shared by both metal ions. The mode of action of the enzyme appears to involve attack by the water molecule bound to the second metal ion. The biochemical reaction that results [see Figure 21(c)] is still under investigation.

V. ZINC-UTILIZING ENZYMES

Zinc can play a structural role in proteins and it does this by binding sulfur and/or nitrogen (in cysteine or histidine residues, respectively) [90, 91]. For example, the "zinc finger" motif which is found in several structures of zinc-binding proteins, contains a motif of Cys and His side chains, four in all, that bind the zinc ion. This motif is also used to bind certain other metal ions such as iron (in rubredoxin), copper (in plastocyanin and azurin), and cadmium (in metallothionein) and so is not entirely specific to zinc. On the other hand, it is an entirely different binding site from that preferred by magnesium ions.

Zinc ions are stronger Lewis acids than are magnesium ions [2]. They are useful in enzymes because their coordination geometry is flexible, ligand exchange from zinc ions is reasonably fast, and they lie on the border between hard and soft cations and therefore prefer to bind softer anions than do magnesium ions [91, 92]. The major ligand to zinc ions in proteins is histidine which does not generally bind to magnesium if oxygen-containing ligands are present [93]. Of 27 zinc-containing enzymes with one zinc ion in their active sites that are listed by Lipscomb and Sträter [94], 17 have zinc bound to three histidine residues, some with additional ligands such as water, Tyr, and Asp. The average coordination number is between 4 and 5, and about 64% of all zinc ligands are histidine. When there are two or more metal ions in the active site the percentage of carboxylate side chains (Asp and Glu) is higher, while the amount of zinc-bound water remains fairly constant (zero or

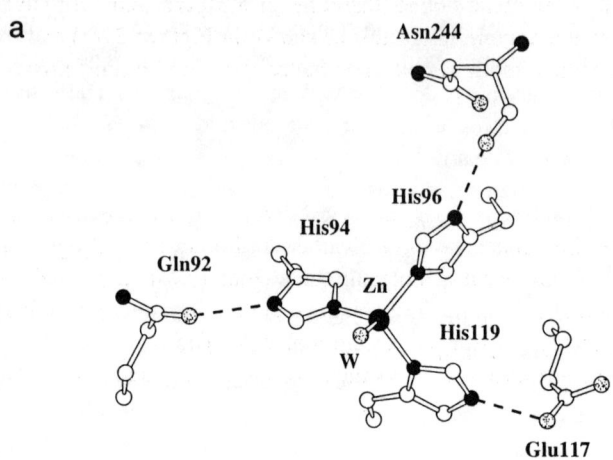

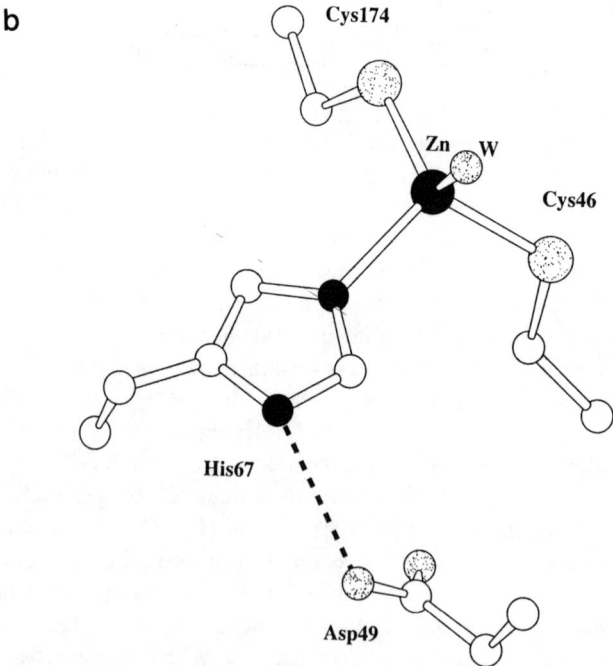

Figure 22. Binding of Asp or Glu to His in (**a**) carbonic anhydrase, and (**b**) alcohol dehydrogenase [96].

one per active site zinc). Metal ions generally bind to histidine residues in the heterocycle plane [95]. Zinc-bound histidine groups are generally hydrogen-bonded to carboxylate-containing groups as shown in Figures 22(a) and (b) [96]. This hydrogen-bonding motif serves to enhance the zinc-binding power of the histidine group.

A. Carbonic Anhydrase from Human Erythrocytes

The action of red cell carbonic anhydrase I is a hydration reaction in which carbon dioxide and water interconvert with a bicarbonate ion and a hydrogen ion. The crystal structure [97, 98] reveals a molecule folded mainly with β-pleated sheet (10 strands, some parallel and some antiparallel) together with seven α-helices. The active site is in a cavity in the center of the molecule. The crystal structure of the enzyme with bound aurocyanide shows the Zn(II) is coordinated to three histidine

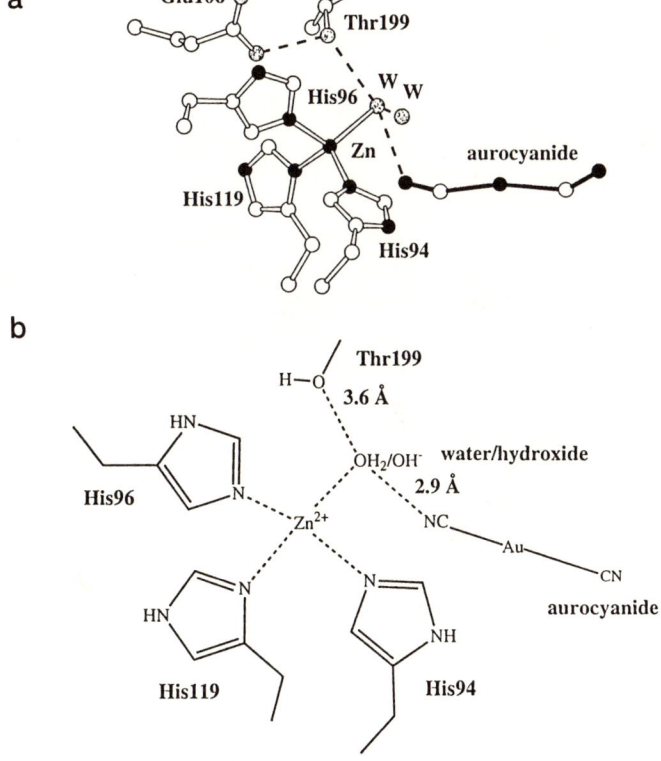

Figure 23. (a) The active site of carbonic anhydrase with bound aurocyanide (indicated by filled bonds). (b) Diagram of the active site of carbonic anhydrase in the same orientation as that in (a) (PDB file 1HUG) [97].

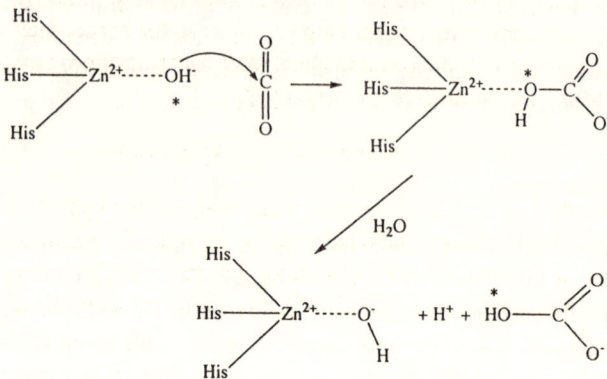

Figure 24. Proposed mechanism of action of carbonic anhydrase.

side chains (His94, His96, and His119) and one water molecule (PDB file 1HUG, 2.0 Å resolution), as shown in Figures 23(a) and (b) [99]. The water molecule is hydrogen bonded to Thr199 which, in turn, is hydrogen bonded to Glu106, as shown in Figure 23(a). This motif makes the oxygen atom of Thr199 more available for hydrogen bonding to the zinc-bound hydroxyl group so that the oxygen atom of this hydroxyl group can attack the carbon atom of carbon dioxide more easily. The water molecule also forms a hydrogen bond to the aurocyanide group. The zinc ion is presumed to polarize the bound water molecule so that it can dissociate into protons (which are hydrated) and hydroxyl groups on the more hydrophilic side of the active site cavity [100, 101]. The resulting metal-ion-coordinated hydroxide group is a good nucleophile and makes a direct attack on a linear incoming carbon dioxide molecule, converting it to bicarbonate, as shown in Figure 24.

The carbon dioxide site is presumed to be about 3.2 Å from the Zn(II) site, which means that there is a ligand between them. Evidence for a zinc-hydroxide mechanism has come from studies of the competition of water and cyanide and the role of cadmium as an inhibitor. At low pH, that is, below the pK_a of HCN, the Zn·OH⁻ model would suggest the reaction:

$$HCN + Zn \cdot OH_2 \rightarrow ZnCN^- + H_2O + H^+$$

This should lead to the release of a proton and this is indeed observed. At high pH the expected reaction is different, that is,

$$CN^- + Zn \cdot OH^- \rightarrow Zn \cdot CN^- + OH^-$$

and the predicted release of hydroxide is observed as a proton uptake [101]. The observed change in the pK_a from 7.5 to near 10 when Cd(II) is substituted for Zn(II) at the active center provides strong evidence that the activity-linked pK_a is that of a coordinated water molecule. Coordination of water to the harder metal ion Zn(II) would be expected to make the aquated metal ion a much stronger Lewis acid than coordination to the softer metal ion, Cd(II) [*102, 103*].

B. Bovine Pancreatic Carboxypeptidase A

Carboxypeptidase A is an exopeptidase that catalyzes the hydrolysis of amino acids, especially those with aromatic side chains, from the C-terminus of a protein or polypeptide. It utilizes zinc in its mode of action [*104*]. The detailed crystal structure of carboxypeptidase (PDB file 5CPA) has been determined to 1.54 Å

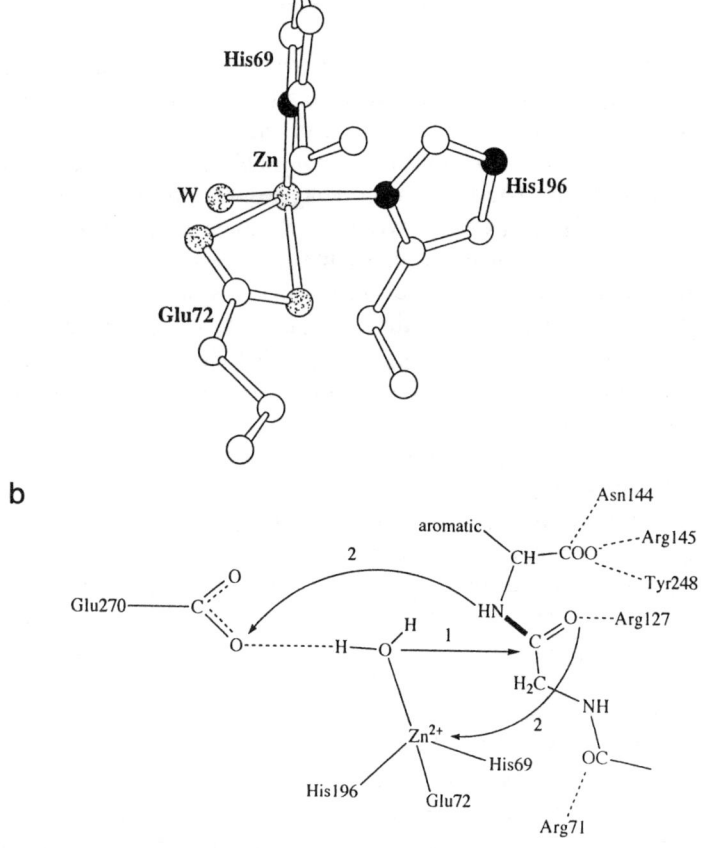

Figure 25. Carboxypeptidase A. (**a**) Surroundings of Zn²⁺, (**b**) proposed mechanism of action of the enzyme.

resolution, and an enzyme–substrate complex was located in a difference map [*105, 106*]. In this structure, as shown in Figure 25(a), the Zn(II) ion is bound to His69, His196, and Glu72, the latter bidentate and a water molecule [*107*]. This water molecule is hydrogen-bonded to Glu270, as shown in Figure 25(b), and this may cause deprotonation of the water molecule, aiding in the formation of what is probably the required nucleophile, a Zn(II)-bound hydroxide group. The hydroxide group attacks the carbonyl carbon atom of the scissile –CO–NH– bond [step 1 in Figure 25(b)] and the carbonyl oxygen atom binds to the Zn(II), which helps to polarize the carbonyl group [step 2 in Figure 25(b)]. The –NH– group of the scissile bond interacts with the carboxyl group of Glu270 and, as a result, peptide bond cleavage occurs. Thus the zinc ion participates directly in the catalytic reaction (polarizing the carbonyl group and aiding in reprotonation of the water nucleophile), and also serves to stabilize the intermediate formed during hydrolysis.

C. Phosphotriesterase from *Pseudomonas diminuta*

The enzyme phosphotriesterase hydrolyzes many different organophosphorus triesters, including several acetylcholinesterase inhibitors. The chemical mechanism involves an activated water molecule that directly attacks the phosphorus center with a resultant inversion of configuration [*108*]. Thus the overall reaction mechanism does not involve the formation of a phosphorylated-enzyme intermediate, but it does involve a Schiff base-type interaction between carbon dioxide and a lysine residue giving a carbamate group in the active site.

The bacterial enzyme requires two divalent metal ions for activity, one playing an important role in the catalytic reaction. Zn^{2+}, Mn^{2+}, Co^{2+}, Ni^{2+}, and Cd^{2+} work in the binuclear metal center of the enzyme [*109, 110*]. On the basis of an analogy with the chemical mechanism for carboxypeptidase it appears that the metal acts as a Lewis acid for the polarization of the phosphorus–phosphoryl oxygen bond and also as a binding site for the hydrolytic water molecule.

The crystal structures of the enzyme (PDB file 1PSC) and its apo- and Cd-substituted forms have been determined at about 2 Å resolution [*111, 112*]. The enzyme is dimeric. One metal ion (the more buried one) is bound to His55, His57, and Asp301, while the more solvent-exposed cadmium is coordinated to His201, His230, and two water molecules. In addition the two metal ions are bridged by an hydroxyl group and a carbamate group, the latter from the reaction of carbon dioxide and the ε-amino group of Lys169; the carbamate group presents two negatively charged oxygen atoms to the metal ions. The active site geometry, with the carbamate as CO_2^-, is shown in Figure 26(a).

The phosphotriesterase-catalyzed reaction involves an S_N2-type mechanism, with inversion of configuration at the phosphorus atom [*108*]; activated water is presumed to attack the electrophilic phosphorus of the substrate, shown as an attack on diethyl 4-nitrophenyl phosphate (a non-biological substrate) in Figure 26(b). Both zinc ions work to activate hydrolytic water and polarize the P–O bond [*113, 114*].

a

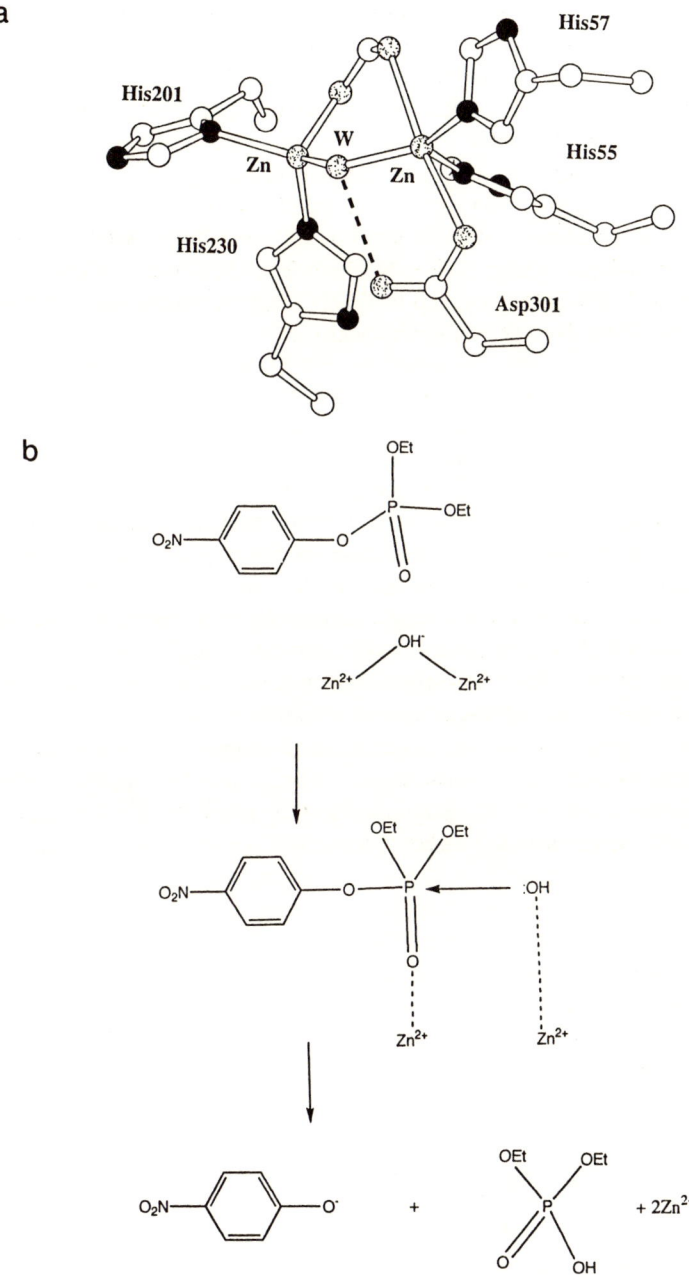

b

Figure 26. (a) The active site of phosphotriesterase. (b) the proposed reaction mechanism of the enzyme.

D.   Zinc-Finger DNA-Binding Proteins Including Zif268

Zinc fingers are DNA-binding motifs that were first identified in a 30-residue sequence in the *Xenopus* transcription factor IIIA [*115–117*]. The motif that was identified contained the sequence:

$$X_3\text{-Cys-}X_{2-4}\text{-Cys-}X_{12}\text{-His-}X_3\text{-His-}X_4$$

where X is any amino acid. This motif binds zinc (and other metal cations) by way of the two invariant cysteine residues and the two invariant histidine residues. Other zinc-binding motifs that have also been identified contain four cysteine residues rather than two cysteines and two histidines [*118*].

The structure of a zinc finger was first reported as a result of NMR studies [*119, 120*]. It was shown to contain an antiparallel β-ribbon and an α-helix. The two residues lie near the turn in the β-ribbon and the two histidine residues lie at the carboxyl terminus of the α-helix. These two cysteine and two histidine are therefore held in place in orientations ideal for binding a zinc ion, as shown in Figure 27. This zinc ion-binding motif was then found in the crystal structure of a protein (called Zif268) which contains three zinc fingers. The crystal structure of this protein (from *Mus musculus*, mouse) was determined in a complex with the portion of DNA to which it binds (an 11-mer and its complementary strand) (PDB file 1AAY, resolution 1.6 Å) [*121*]. Both divalent zinc and cobalt complexes were studied and their structures were found to be approximately the same. The crystal structure of each zinc finger had precisely the conformation found by NMR studies in solution. The α-helix of each zinc finger fits directly into the major groove of the nucleic acid and makes primary contacts with a 3-base-pair region of the nucleic acid. This provides a way to form a protrusion from a protein to give a motif that can form specific and nonspecific interactions (such as by hydrogen bonding) with DNA.

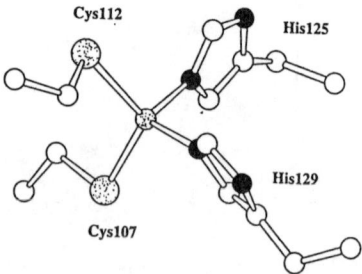

Figure 27. The surroundings of a zinc ion in the Zif268-DNA complex (PDB file 1AAY) [*121*].

E. Horse Liver Alcohol Dehydrogenase

Horse liver alcohol dehydrogenase is an NAD^+-dependent enzyme that catalyzes the oxidation of various primary and secondary alcohols to their corresponding aldehydes. The active enzyme is a dimer composed of two identical subunits. Each

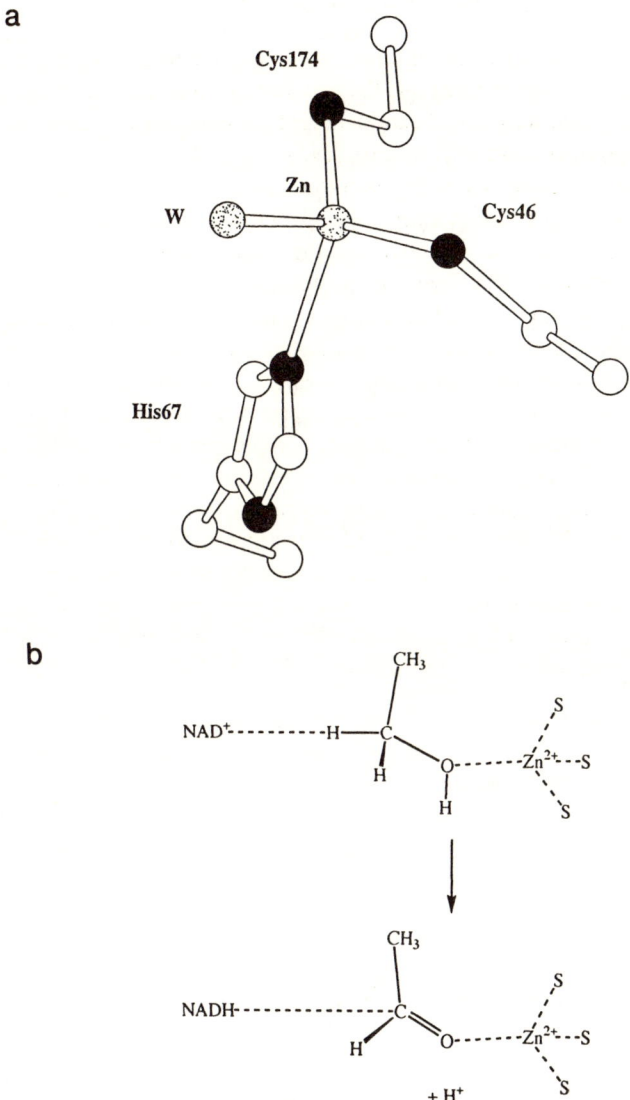

Figure 28. Alcohol dehydrogenase. (a) Surroundings of one zinc ion. (b) the reaction catalyzed by this enzyme.

subunit has one coenzyme binding site and binds two zinc ions, one catalytic and the other structural. The catalytic domain is mainly formed from three distinct antiparallel pleated sheet regions. Residues within this domain provide ligands to the catalytic zinc ion. A second zinc ion is found in the lobe which is formed by residues 95 to 113. This zinc ion is bound to four cysteine residues (Cys97, Cys100, Cys103, and Cys111) in a distorted tetrahedral arrangement.

A zinc-hydroxide mechanism has also been proposed for this enzyme. Crystallographic studies (PDB file 8ADH) [46, 112–124] show that the Zn(II) is bound to the sulfur atoms of Cys46, Cys174, the nitrogen atom of His-67, and one water molecule (or hydroxyl group), as shown in Figure 28(a). While this type of coordination would be expected to lower the pK_a of the bound water molecule, the sulfur ligands should cause less lowering than would the nitrogen and oxygen ligands of the hydrolytic Zn(II) enzymes. The pK_a of the Zn(II)-OH$_2$ complex in the enzyme without bound coenzyme is about 9.2, but this falls to 7.6 when the NAD$^+$ complex is formed [125]. The alcohol substrate binds directly to Zn(II) through the alcohol oxygen atom, displacing a solvent molecule, as shown in Figure 28(b). This should lower the pK_a of the alcohol group as it did for water, suggesting a change in pK_a from 16 in the unbound alcohol to about 6 in the E–NAD$^+$–RCH$_2$OH complex making the alcoholate the major form of the bound substrate, thereby facilitating hydride transfer from substrate to NAD$^+$.

VI. ENZYMES BINDING BOTH MAGNESIUM AND ZINC IONS

Several enzymes bind both magnesium and zinc ions, using them for different purposes. For example, nucleotidyl transferases can use one metal ion, to which is coordinated a phosphate group and a carboxylate group, the latter serving to polarize the water nucleophile. When there are two metal ions present, the carboxylate group also binds the second metal ion which in turn activates the nucleophile. Therefore it appears that the second metal ion aids in the catalysis, replacing the action of an active-site side chain in those enzymes of that activity that only bind one metal ion. Two enzymes will be described here—the alkaline phosphatase and the 3′-5′ exonuclease from E. coli.

A. Alkaline Phosphatase from Escherichia coli

Alkaline phosphatase is a widely distributed nonspecific monoesterase that catalyzes the hydrolysis of a variety of phosphomonoesters. The reaction catalyzed involves a two-step reaction in which a covalent phosphoseryl intermediate (E-P) is formed; in this it is unique among metallophosphatases containing two or more bound metal ions. Its activity is highest near alkaline pH values (8.0), hence its name. The evidence to date is that the transition state of the reaction that is catalyzed is metaphosphate-like. Almost complete bond breaking to the leaving group and

a

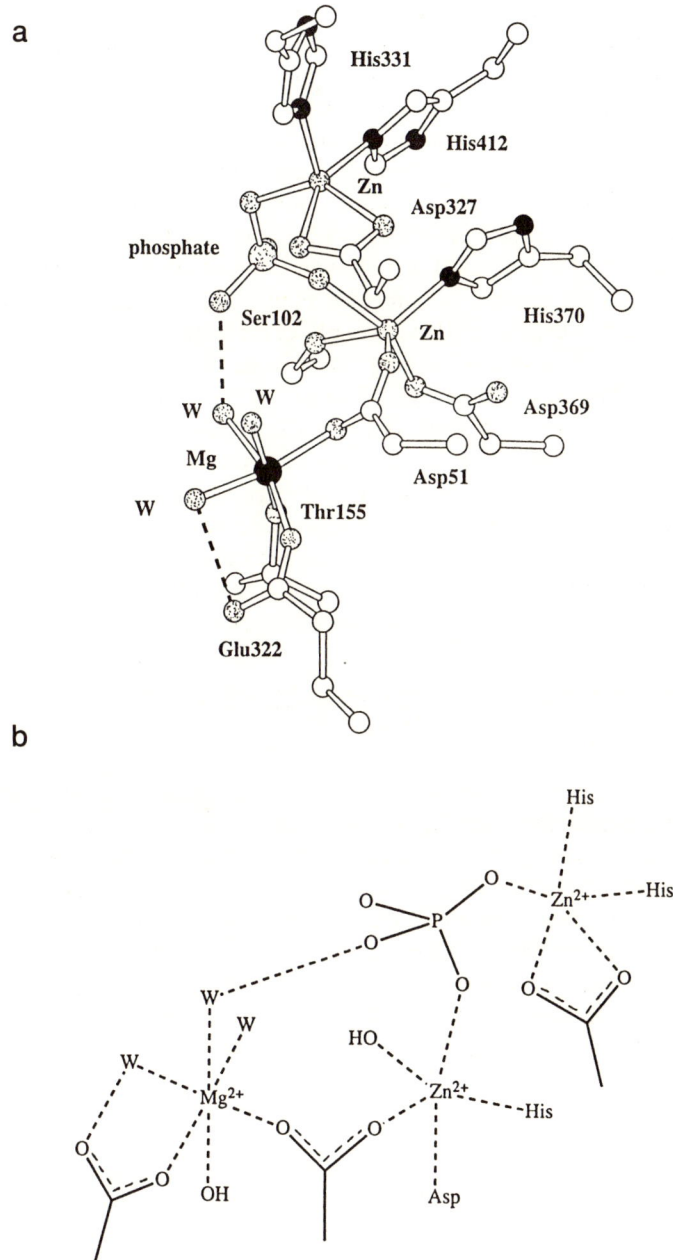

b

Figure 29. Alkaline phosphatase. (**a**) Active site showing the three metal ions. (**b**) the chemical identities of the atoms in (**a**). (**c**) A diagram highlighting the relative arrangements of the three metal ions.

C

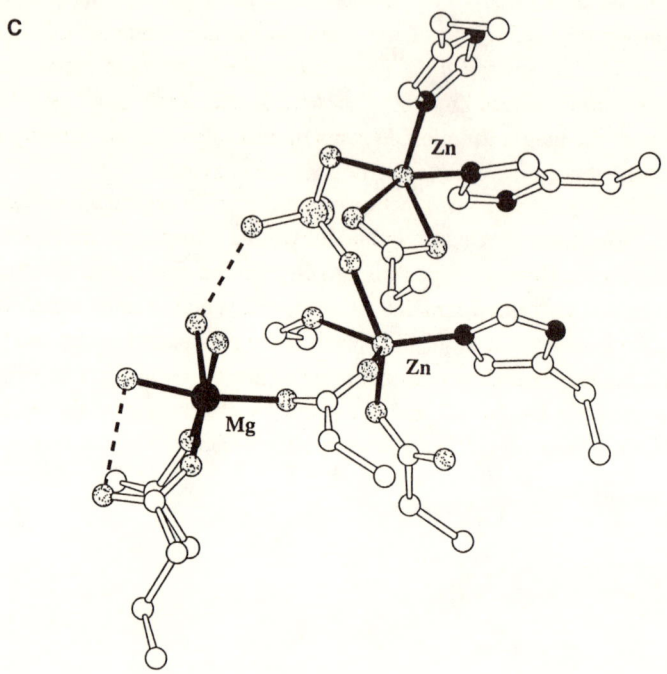

Figure 29. Continued

almost no bond formation to the incoming nucleophile results in metaphosphate (PO_3^-) with long axial distances in the pentacoordinate transition state [*126*].

The crystal structure of the native *E. coli* enzyme complexed with inorganic phosphate has been refined at 2.0 Å resolution (PDB file 1ALK) [*127*]. The enzyme, which is a dimer, has two active sites, located about 30 Å from each other. Each subunit shows α/β topology with a central 10-stranded β-sheet in the middle, flanked by 15 α-helices of various lengths. The enzyme utilizes two zinc and one magnesium ion. The relative arrangement of these cations is shown in Figures 29(a), (b), and (c).

The two zinc ions are bound by histidine and aspartate groups, and one is near Ser102. Coordination to zinc decreases the pK_a of the Ser102 hydroxyl group and facilitates nucleophilic attack at the phosphorus atom. The first zinc ion is penta-coordinated by His331 and His412 and both carboxylate oxygen atoms of Asp327 and one of the phosphate oxygen atoms. The second zinc ion helps to neutralize the developing negative charge on the leaving group. Its coordination geometry includes His370, one of the carboxyl oxygen atoms of Asp51, Asp369, one of the phosphate oxygen atoms, and a disordered Ser102, as shown in Figure 29. The magnesium ion is octahedrally coordinated to the second oxygen atom of the carboxylate group of Asp51, one of the carboxyl oxygen atoms of Glu332, the hydroxyl group of Thr155, and three water molecules. The three metal ions are relatively close with distances $Zn^{2+}\cdots Zn^{2+}$ of 4.12 Å, and $Zn^{2+}\cdots Mg^{2+}$ of 4.68 and 7.00 Å. Although the magnesium ion does not appear to participate directly in the catalytic reaction, it does increase the activity of the enzyme.

The intermediate in the catalytic reaction, diagrammed in Figure 30, is phosphorylated Ser102. The five-coordinate trigonal bipyramid phosphorus transition state is stabilized by the neighboring zinc ions and an arginine side chain. The Ser102-phosphate ester that is formed remains bound to the zinc ion until a zinc-bound hydroxide group from a water molecule displaces the phosphoryl group and leaves the phosphate product bound to both zinc ions. Thus two nucleophilic displacements take place, one by Ser102 and the other by water to give the hydrogen phosphate (HPO_4^{2-}) as product. These two steps, each of which causes an inversion

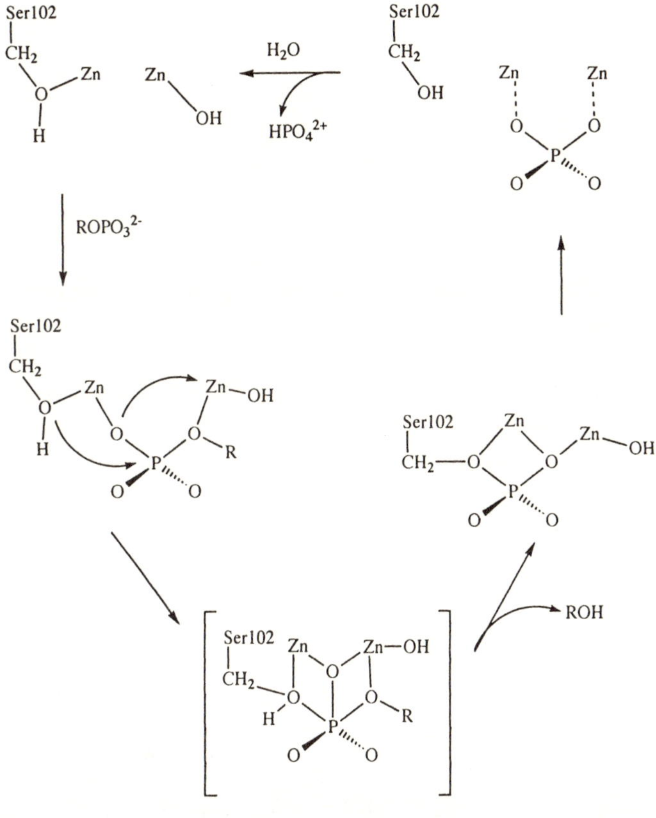

Figure 30. Proposed mechanism of action of alkaline phosphatase.

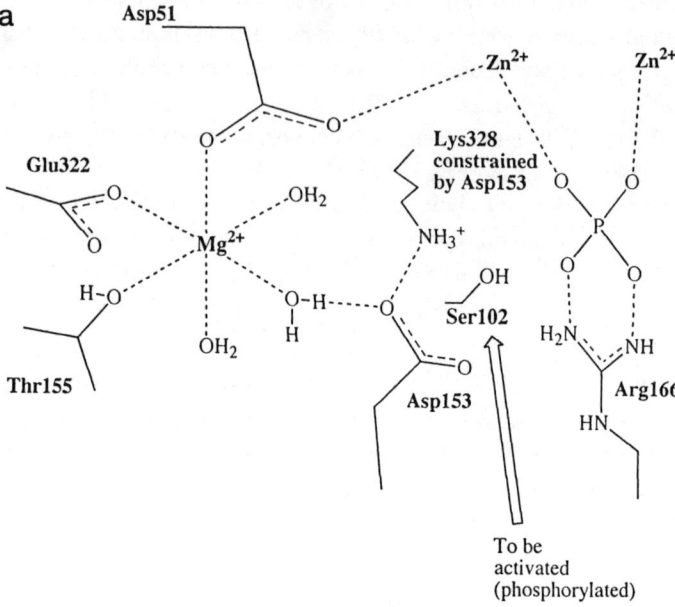

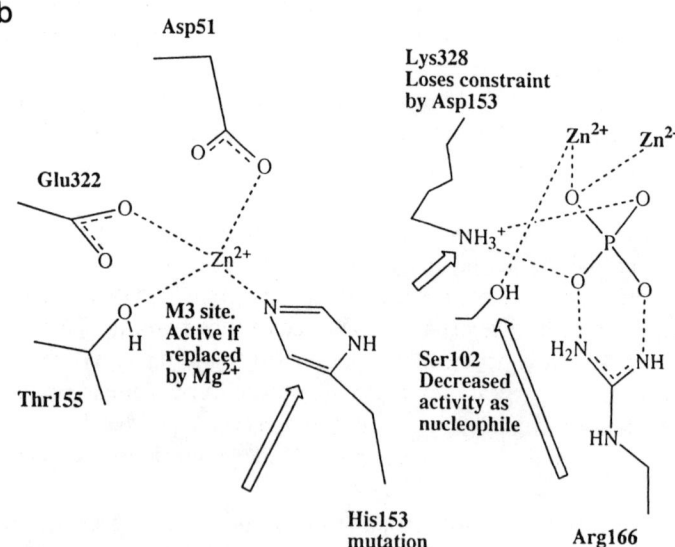

Figure 31. Alkaline phosphatase. (a) The active site of the wild type *E. coli* enzyme showing the two zinc and one magnesium sites [127]. (b) The active site of an inactive mutant enzyme (Asp153His) [128]. The third metal-binding site in this enzyme binds zinc rather than magnesium.

of configuration at the phosphorus atom, cause a net retention of configuration at the phosphorus atom. The increase in pK_a when Cd(II) replaces Zn(II) in the enzyme favors a zinc–hydroxide mechanism, as was described earlier for carbonic anhydrase.

In some mammalian alkaline phosphatases there is an Asp153His mutation which increases the phosphate affinity of the enzyme over that for the *E. coli* enzyme. Asp153 binds a magnesium ion-bound water molecule, as shown in Figure 31(a). In the *E. coli* enzyme Asp153 binds directly to Lys328. The structure of the inactive Asp153His mutant of *E. coli* alkaline phosphatase has been reported [*128*]. The octahedral site that binds magnesium in the wild-type enzyme has now been replaced by a tetrahedral zinc-binding site as a result of this single amino acid change. As a result Asp51 no longer spans two metal ion sites [see Figure 31(b)], but merely binds to the newly added zinc that has replaced magnesium. Lys328 has also lost its constraint and so forms a hydrogen bond to the phosphate group. The metal ion environment of the phosphate group has also been altered in the mutant enzyme, as shown in Figure 31(b), and this is presumed to explain the decreased activity of the mutant enzyme.

B.   3′-5′-Exonuclease from *Escherichia coli*

DNA polymerase I from *E. coli* is a single subunit, molecular weight 103,000. It can be proteolyzed to give an amino terminal portion, molecular weight 35,000, and a "Klenow fragment," molecular weight 69,000 which retains the polymerizing and editing activities of the whole enzyme. The structure of the Klenow fragment has been determined [*129, 130*]. It has two domains. The 200 amino acids in the amino terminus have 3′-5′-exonuclease activity which involves editing out the mismatched terminal nucleotides. The 400 amino acids at the carboxy terminus have polymerase activity.

The crystal structure of the 3′-5′-exonuclease domain shows that it contains two metal ions, probably zinc and magnesium. As shown in Figure 32(a), one metal ion is pentacoordinated by three carboxylate groups (Asp355, Glu357, and Asp501), an oxygen atom of the 5′ phosphate group, and a water molecule. This is a zinc site. The other metal ion is octahedrally coordinated to a carboxylate group (Asp355 shared with the other metal ion), two 5′ phosphate oxygen atoms, and three water molecules, each of which is bound to the protein (via Asp424 and backbone amide groups). This is a magnesium site. A possible mode of activity has been proposed [*65*].

There is inversion of configuration at the 3′- terminal phosphorus, implying a "tight" associative in-line mechanism [*131*] in which the hydroxyl group attacks the phosphorus atom from the opposite side from the 3′-hydroxyl leaving group [Figures 16(b) and 32(b)]. The hydroxyl group (formed from the metal ion-bound water molecule), the hydroxyl group of Tyr497, and the carboxylate group of Glu357 are arranged so that they can position the lone pair on the hydroxyl group towards the phosphorus atom. The zinc ion provides the hydroxyl group to attack the phosphorus atom in a five-coordinate intermediate. The opposite axial ligand in line is a 3′-phosphodiester oxygen atom which is stabilized as an anion by the magnesium ion. Thus two metal ions are required to bind and orient the single-stranded DNA and the attacking water molecule in the correct manner for the reaction to occur.

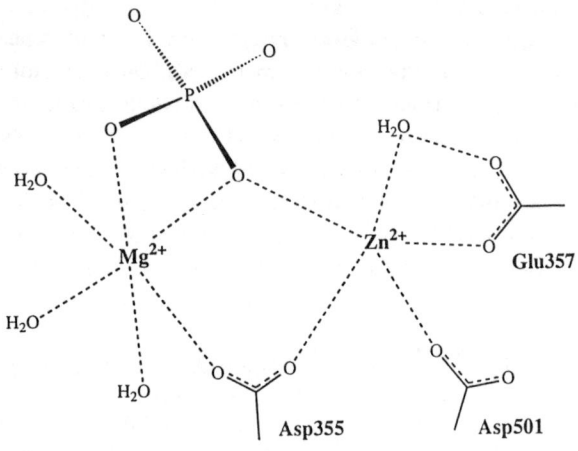

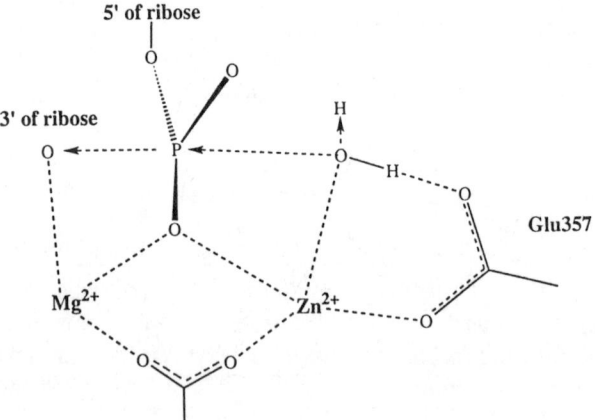

Figure 32. (a) The binding of metal ions to the Klenow fragment portion of DNA polymerase I. The magnesium ion binds the phosphate group on DNA, while the zinc ion polarizes a water molecule which can then attack the phosphorus atom. The overall geometry of the reaction is maintained by the octahedral coordination sphere of the magnesium ion. (b) Details of the proposed mechanism of action of this enzyme (see also Figure 16b) [65].

The structure of the 3′-5′- exonuclease domain of T7 DNA polymerase [*132*] closely resembles that of the Klenow fragment just described. Again there are two metal sites that share a carboxylate ion, and the mechanism of action is probably very similar to that of the 3′-5′-exonuclease.

VII. THE DESIGN OF METAL ION-BINDING SITES IN PROTEINS

When designing a metalloenzyme the first question is what does one want the metal ion to do. This will limit the choice of metal ions. Once a metal ion or group of metal ions has been chosen it is necessary to find out the ligand preferences of that metal ion. It is also necessary to ensure that functional groups are in the correct orientation with respect to the transition state of the reaction to be catalyzed. Some rigidification of the active site environment can aid in this. It is, of course, possible to design a small metallo center that can catalyze the required reaction. But the reality is that the properties of a human cell are such that the protein needs bulk and interactions that enable it to evade (at least initially) degradation, modification, and other changes.

Metal ions mainly form electrostatic interactions with their ligands and have plenty of opportunity to do so since about 65% of the various types of amino acid side chains are potential metal-binding groups. Naturally occurring metal chelation is of significant physiological importance, and the structural and functional analyses of metal-binding proteins has generated a wealth of information regarding the coordination geometry, binding affinity, and chemistry of metal ions. In recent years much research has focused on the use of this knowledge to design novel metal-binding sites in proteins [*133–135*]. Many enzymes are known to utilize metal ions as a necessary component of their catalytic mechanisms at their active sites. If an extrinsic metal binding-site can be introduced into the active site of an enzyme that does not normally bind metal ions, it may be possible to modulate the catalytic activity via metal ion chelation.

The field of metal-binding design includes de novo design of metal ion-containing proteins, the use of metal ions to stabilize proteins and peptide assemblies, the incorporation of metal-binding sites into preexisting proteins, and the use of metal-based protein domains for the construction of proteins with desired functions [*136, 137*]. The experimental results of such metalloprotein design can provide an attractive means of gaining a deeper understanding of the principles that govern the reactivity and structure of metals in proteins.

In order to design a metalloprotein, two sets of factors need to be taken into account: the designed protein must be able to fold in the required manner, and the coordination preferences of the metal must be satisfied [*133*]. In order to achieve these factors:

1. *The stereochemistry of the metal ion-binding functional groups in the protein must satisfy the coordination geometry of the metal ion to be used.* This requires that the appropriate number of protein side chains or solvent molecules be placed around the metal ion, that they be placed at the appropriate bond distances and angles, and that they be in the correct overall coordination geometry.

2. *Second-sphere interactions to the metal ion should not be strained, but should lie within normally encountered limits.* For example, the interactions of secondary shell hydrogen bonds around the metal ion need to be taken into account in addition to the ligands that form direct bonds to the metal ion. This will ensure that there are no unsatisfied bonds in the structure.

3. *Optimal amino acid side-chain geometries must be taken into account in metalloprotein design.* This is a steric requirement because these side chains generally adapt to a conformation that corresponds to an energetic minimum.

4. *Structural and functional specificity must be achieved.* One of the most difficult aspects of metalloprotein design is the formation of a unique metal ion-binding site that attracts the required metal ion and excludes other metal ions. An ideal example was provided by the structure of alkaline phosphatase, described above.

One of the most active areas of research in metalloprotein design involves the engineering of metal-binding sites into existing proteins [138, 139]. In rationally designing metal-binding sites, the availability of a protein with known or presumed structure allows for the use of computer graphics and other modeling methods. The fact that structure or presumed structure is known limits the risk of the designed sequence not being able to fold properly. Several experiments have been completed involving changes in the specificity and properties of naturally occurring Ca(II) and Zn(II) sites. For example, several mutations were made to the zinc-binding site of carbonic anhydrase II to try and change its metal affinity. The natural site consists of a tetrahedrally coordinated zinc that is bound to three histidines and a solvent hydroxyl group. One mutation that was made involved His94 being changed to an aspartate. This change resulted in a four order-of-magnitude decrease in zinc affinity [125]. The X-ray structure of mutant protein revealed subtle differences in the zinc coordination sphere, which may account for the loss of affinity. In an additional experiment, the affinity for zinc was improved slightly (fourfold) by replacing the solvent hydroxyl with a thiol provided from a cysteine. This was achieved, however, at the cost of catalytic activity (reduced by three orders of magnitude). This hydroxyl is hydrogen bonded to Thr199. The Thr199Cys mutant bound zinc more tightly. The X-ray structure of this mutant showed that the zinc-binding site was nearly perfectly tetrahedral. This geometry was achieved by a slight movement of the loop in which Cys199 is located. In keeping with the analyses described here it has proved possible to engineer zinc-binding sites in proteins, particularly if tetrahedral geometries with histidine and/or cysteine ligands are used [*140, 141*].

VIII. SUMMARY

1.   Magnesium will bind six oxygen atoms in a regular tetrahedral arrangement. The edges of the tetrahedron are 2.9 Å. This arrangement is structurally fairly rigid.

2.   Zinc will bind nitrogen and sulfur in a tetrahedral manner. This is also a fairly rigid structural arrangement, but, because it is so different from the binding mode of magnesium, can be used to exclude it from that site.

3.   Zinc bound by at least one oxygen atom plus some histidine residues can provide a catalytic center.

ACKNOWLEDGMENTS

This work was supported by grants CA-10925 (to JPG) and CA-06927 (to ICR) from the National Institutes of Health and by an appropriation from the Commonwealth of Pennsylvania. Its contents are solely the responsibility of the authors and do not necessarily represent the official views of the National Cancer Institute.

REFERENCES

1.   Hughes, M. N. *The Inorganic Chemistry of Biological Processes,* 2nd ed.; Wiley: Chichester, England, 1981.
2.   Cowan, J. A. *Inorganic Biochemistry: An Introduction,* 2nd ed.; Wiley-VCH: New York, 1997.
3.   McCance, R. A.; Widdowson, E. M. *Biochem. J.* **1942**, *36,* 692–696.
4.   Magneson, G. R.; Puvathingal, J. M.; Ray, W. J., Jr. *J. Biol. Chem.* **1987**, *262,* 11140–11148.
5.   Pufahl, R. A.; Singer, C. P.; Peariso, K. L.; Lin, S. -J.; Schmidt, P. J.; Fahrni, C. J.; Culotta, V. C.; Penner-Hahn, J. E.; O'Halloran, T. V. *Science* **1997**, *279,* 853–856.
6.   Kauffman, G. B.; Brock, W. H.; Jensen, K. A.; Jørgensen, C. K. *J. Chem. Educ.* **1983**, *60,* 509–510.
7.   Williams, R. J. P. *Quart. Rev. Chem. Soc.* **1970**, *24,* 331–360.
8.   Silva, J. J. R. F. D.; Williams, R. J. P. *The Biological Chemistry of the Elements*; Clarendon Press: Oxford, 1991.
9.   Glusker, J. P. *Adv. Protein Chem.* **1991**, *42,* 1–73.
10.   Langford, C. H.; Gray, H. B. Benjamin-Cummings: Reading, MA, 1966.
11.   Diebler, H.; Eigen, M.; Ilgenfritz, G.; Maass, G.; Winkler, R. *Pure Appl. Chem.* **1969**, *20,* 93–115.
12.   Ducommum, Y.; Merbach, A. E. *Inorganic High Pressure Chemistry; Kinetics and Mechanisms*; van Eldik, R. Ed.; Elsevier: Amsterdam, Chap. 2, pp. 69–114.
13.   Douglas, B. E.; McDaniel, D. H.; Alexander, J. J. *Concepts and Models of Inorganic Chemistry*; John Wiley: New York, 1994.
14.   Pearson, R. G. *Science* **1966**, *151,* 172–177.
15.   Mildvan, A. S. *Enzymes* **1970**, *2,* 445–536.
16.   Brown, I. D. *Acta Cryst.* **1988**, *B44,* 545–553.

17. Allen, F. H.; Bellard, S.; Brice, M. D.; Cartwright, B. A.; Doubleday, A.; Higgs, H.; Hummelink, T.; Hummelink-Peters, B. G.; Kennard, O.; Motherwell, W. D. S.; Rodgers, J. R.; Watson, D. G. *Acta Cryst.* **1979**, *B35*, 2331–2339.
18. Bock, C. W.; Kaufman, A.; Glusker, J. P. *Inorg. Chem.* **1994**, *33*, 419–427.
19. Jaffe, E. K. *J. Bioenerget. Biomembr.* **1995**, *27*, 169–179.
20. Carrell, C. J.; Carrell, H. L.; Erlebacher, J.; Glusker, J. P. *J. Amer. Chem. Soc.* **1988**, *110*, 8651–8656.
21. Bock, C. W.; Katz, A. K.; Glusker, J. P. *J. Amer. Chem. Soc.* **1995**, *117*, 3754–3765.
22. Bertini, I.; Luchinat, C.; Rosi, M.; Sgamellotti, A.; Tarantelli, F. *Inorg. Chem.* **1990**, *29*, 1460–1463.
23. Cowan, J. A. (Ed.) *The Biological Chemistry of Magnesium;* VCH: New York, Weinheim, Cambridge, 1995.
24. Tari, L. W.; Matte, A.; Goldie, H.; Delbaere, L. T. J. *Nature Struct. Biol.* **1997**, *4*, 990–994.
25. Bernstein, F. C.; Koetzle, T. F.; Williams, G. J. B.; Meyer, E. F. Jr.; Brice, M. D.; Rodgers, J. R.; Kennard, O.; Shimanouchi, T.; Tasumi, M. *J. Mol. Biol.* **1977**, *112*, 535–542.
26. Powers, V. M.; Koo, C. W.; Kenyon, G. L.; Gerlt, J. A.; Kozarich, J. W. *Biochemistry* **1991**, *30*, 9255–9263.
27. Neidhart, D. J.; Howell, P. L.; Petsko, G. A.; Powers, V. M.; Li, R.; Kenyon, G. L.; Gerlt, J. A. *Biochemistry* **1991**, *30*, 9264–9273.
28. Phillips, D. C. *Scient. Amer.* **1966**, *215(5)*, 78–90.
29. Banner, D. W.; Bloomer, A. C.; Petsko, G. A.; Phillips, D. C.; Wilson, I. A. *Biochem. Biophys. Res. Commun.* **1976**, *72*, 146–155.
30. Banner, D. W.; Bloomer, A. C.; Petsko, G. A.; Phillips, D. C.; Pogson, C. I.; Wilson, I. A.; Corran, P. H.; Furth, A. J.; Milman, J. D.; Offord, R. E.; Priddle, J. D.; Waley, S. G. *Nature* **1975**, *255*, 609–614.
31. Kenyon, G. L.; Gerlt, J. A.; Petsko, G. A.; Kozarich, J. W. *Acc. Chem. Res.* **1995**, *28*, 178–186.
32. Kallarakal, A. T.; Mitra, B.; Kozarich, J. W.; Gerlt, J. A.; Clifton, J. G.; Petsko, G. A.; Kenyon, G. L. *Biochemistry* **1995**, *34*, 2788–2797.
33. Landro, J. A.; Gerlt, J. A.; Kozarich, J. W.; Koo, C. W.; Shah, V. J.; Kenyon, G. L.; Neidhart, D. J.; Fujita, S.; Petsko, G. A. *Biochemistry* **1994**, *33*, 635–643.
34. Bearne, S. L.; Wolfenden, R. *Biochemistry* **1997**, *36*, 1646–1656.
35. Chiang, Y.; Kresge, A. J.; Pruszynski, P.; Schepp, N. P.; Wirz, J. *Angew. Chem., Int. Ed. Engl.* **1990**, *29*, 792–794.
36. Renaud, P.; Fox, M. A. *J. Amer. Chem Soc.* **1988**, *110*, 5705–5709.
37. Gerlt, J. A.; Kozarich, J. W.; Kenyon, G. L.; Gassman, P. G. *J. Amer. Chem. Soc.* **1991**, *113*, 9667–9669.
38. Maggio, E. T.; Kenyon, G. L.; Mildvan, A. S.; Hegeman, G. D. *Biochemistry* **1975**, *14*, 1131–1139.
39. Mitra, B.; Kallarakal, A. T.; Kozarich, J. W.; Gerlt, J. A.; Clifton, J. G.; Petsko, G. A.; Kenyon, G. L. *Biochemistry* **1995**, *34*, 2777–2787.
40. Gerlt, J. A.; Gassman, P. G. *J. Amer. Chem. Soc.* **1993a**, *115*, 11552–11568.
41. Gerlt, J. A.; Gassman, P. G. *Biochemistry* **1993b**, *32*, 11943–11952.
42. Borchardt, R. T.; Cheng, C. F. *Biochimica et Biophysica Acta* **1978**, *522*, 49–62.
43. Gray, C. H.; Coward, J. K.; Schowen, K. B.; Schowen, R. L. *J. Amer. Chem. Soc.* **1979**, *101*, 4351–4358.
44. Hegazi, M. F.; Borchardt, R. T.; Schowen, R. L. *J. Amer. Chem. Soc.* **1979**, *101*, 4359–4365.
45. Vidgren, J.; Svensson, L. A.; Liljas, A. *Nature* (London) **1994**, *368*, 354–358.
46. Eklund, H.; Samana, J. -P.; Wallén, L.; Brändén, C. -I.; Åkeson, Å.; Jones, T. A. *J. Mol. Biol.* **1981**, *146*, 561–587.
47. Lotta, T.; Vidgren, J.; Tilgmann, C.; Ulmanen, I.; Melén, Julkunen, I.; Taskinen, J. *Biochemistry* **1995**, *34*, 4202–4210.
48. Schluckebier, G.; O'Gara, M.; Saenger, W.; Cheng, X. *J. Mol. Biol.* **1995**, *247*, 16–20.

49. Woodard, R. W.; Tsai, M. -D.; Floss, H. G.; Crooks, P. A.; Coward, J. K. *J. Biol. Chem.* **1980**, *255*, 9124–9127.
50. Zheng, Y. -J.; Bruice, T. C. *J. Amer. Chem. Soc.* **1997**, *119*, 8137–8145.
51. Sanders, D. A.; Gillece-Castro, B. L.; Stock, A. M.; Burlingame, A. L.; Koshland, D. E. Jr. *J. Biol. Chem.* **1989**, *264*, 21770–21778.
52. Stock, A. M.; Mottonen, J. M.; Stock, J. B.; Schutt, C. E. *Nature* (London) **1989**, *337*, 745–749.
53. Stock, J. B.; Surette, M. G.; McCleary, W. R.; Stock, A. M. *J. Biol. Chem.* **1992**, *267*, 19753–19756.
54. Stock, A. M.; Martinez-Hackert, E.; Rasmussen, B. F.; West, A. H.; Stock, J. B.; Ringe, D.; Petsko, G. A. *Biochemistry* **1993**, *32*, 13375–13380.
55. Volz, K.; Matsumura, P. *J. Biol. Chem.* **1991**, *266*, 15511–15519.
56. Bellsolell, L.; Prieto, J.; Serrano, L.; Coll, M. *J. Mol. Biol.* **1994**, *238*, 489–495.
57. Bellsolell, L.; Cronet, P.; Majolero, M.; Serrano, L.; Coll, M. *J. Mol. Biol.* **1996**, *257*, 116–128.
58. Zhu, X.; Rebello, J.; Matsumura, P.; Volz, K. *J. Biol. Chem.* **1997**, *272*, 5000–5006.
59. Herschlag, D.; Jencks, W. P. *J. Amer. Chem. Soc.* **1990**, *112*, 1942–1950.
60. Volz, K.; Beman, J.; Matsumura, P. *J. Biol. Chem.* **1986**, *261*, 4723–4725.
61. Lowry, D. F.; Roth, A. F.; Rupert, P. B.; Dahlquist, F. W.; Moy, F. J.; Domaille, P. J.; Matsumura, P. *J. Biol. Chem.* **1994**, *269*, 26358–26362.
62. Bujacz, G.; Jaskólski, M.; Alexandratos, J.; Wlodawer, A.; Merkel, G.; Katz, R. A.; Skalka, A.M. *J. Mol. Biol.* **1995**, *253*, 333–346.
63. Bujacz, G.; Jaskólski, M.; Alexandratos, J.; Wlodawer, A.; Merkel, G.; Katz, R. A.; Skalka, A. M. *Structure* **1996**, *4*, 89–96.
64. Bujacz, G.; Alexandratos, J.; Wlodawer, A.; Merkel, G.; Andrake, M.; Katz, R. A.; Skalka, A. M. *J. Biol. Chem.* **1997**, *272*, 18161–18168.
65. Beese, L. S.; Steitz, T. A. *The EMBO Journal* **1991**, *10*, 25–33.
66. Katayanagi, K.; Miyagawa, M.; Matsushima, M.; Ishikawa, M.; Kanaya, S.; Ikehara, M.; Matsuzaki, T.; Morikawa, K. *Nature* **1990**, *347*, 306–309.
67. Kashiwagi, T.; Jeanteur, D.; Haruki, M.; Katayanagi, K.; Kanaya, S.; Morikawa, K. *Protein Eng.* **1996**, *9*, 857–867.
68. Davies, J. F. II; Hostomska, Z.; Hostomsky, Z.; Jordan, S. R.; Matthews, D. A. *Science* **1991**, *252*, 88–95.
69. Stec, B.; Lebioda, L. *J. Mol. Biol.* **1990**, *211*, 235–248.
70. Lebioda, L.; Stec, B. *J. Amer. Chem. Soc.* **1989**, *111*, 8511–8513.
71. Lebioda, L.; Stec, B. *Biochemistry* **1991**, *30*, 2817–2822.
72. Lebioda, L.; Stec, B.; Brewer, J. M.; Tykarska, E. *Biochemistry* **1991**, *30*, 2823–2827.
73. Lebioda, L.; Zhang, E.; Lewinski, K.; Brewer, J. M. *Proteins: Struct. Funct. Genet.* **1993**, *16*, 219–225.
74. Larsen, T. M.; Wedekind, J. E.; Rayment, I.; Reed, G. H. *Biochemistry* **1996**, *35*, 4349–4358.
75. Reed, G. H.; Poyner, R. R.; Larsen, T. M.; Wedekind, J. E.; Rayment, I. *Curr. Opin. Struct. Biol.* **1996**, *6*, 736–743.
76. Poyner, R. R.; Laughlin, L. T.; Sowa, G. A.; Reed, G. H. *Biochemistry* **1996**, *35*, 1692–1699.
77. Wedekind, J. E.; Poyner, R. R.; Reed, G. H.; Rayment, I. *Biochemistry* **1994**, *33*, 9333–9342.
78. Wedekind, J. E.; Reed, G. H.; Rayment, I. *Biochemistry* **1995**, *34*, 4325–4330.
79. Zhang, E.; Hatada, M.; Brewer, J. M.; Lebioda, L. *Biochemistry* **1994**, *33*, 6295–6300.
80. Duquerroy, S.; Camus, C.; Janin, J. *Biochemistry* **1995**, *34*, 12513–12523.
81. Noel, J. P.; Hamm, H. E.; Sigler, P. B. *Nature* **1993**, *366*, 654–663.
82. Coleman, D. E.; Berghuis, A. M.; Lee, E.; Linder, M. E.; Gilman, A. G.; Sprang, S. R. *Science* **1994**, *265*, 1405–1412.
83. Coleman, D. E.; Lee, E.; Mixon, M. B.; Linder, M. E.; Berghuis, A. M.; Gilman, A. G.; Sprang, S. R. *J. Mol. Biol.* **1994**, *238*, 630–634.
84. Pai, E. F.; Krengel, U.; Petsko, G. A.; Goody, R. S.; Kabsch, W.; Wittinghofer, A. *EMBO J.* **1990**, *9*, 2351–2359.

85. Sondek, J.; Lambright, D. G.; Noel, J. P.; Hamm, E.; Sigler, P. B. *Nature* **1994**, *372*, 276–279.
86. Carrell, H. L.; Glusker, J. P.; Burger, V.; Manfre, F.; Tritsch, D.; Biellmann, J. -F. *Proc. Natl. Acad. Sci. USA* **1989**, 86, 4440–4444.
87. Collyer, C. A.; Henrick, K.; Blow, D. M. *J. Mol. Biol.* **1990**, *212*, 211–235.
88. Farber, G. K.; Glasfeld, A.; Tiraby, G.; Ringe, D.; Petsko, G. A. *Biochemistry* **1989**, *28*, 7289–7297.
89. Whitlow, M.; Howard, A. J.; Finzel, B. C.; Poulos, T. L.; Winbourne, E.; Gilliland, G. L. *Proteins: Struct., Funct., Genet.* **1991**, *9*, 153–173.
90. Berg, J. *Science* **1986**, *232*, 485–487.
91. Vallee, B. L.; Auld, D. S. *Biochemistry* **1990**, *29*, 5647–5659.
92. Christianson, D. W. *Adv. Protein Chem.* **1991**, *42*, 281–355.
93. Coleman, J. E. *Annu. Rev. Biochem.* **1992**, *61*, 897–946.
94. Lipscomb, W. N.; Sträter, N. *Chem. Rev.* **1996**, *96*, 2375–2433.
95. Carrell, A. B.; Shimoni, L.; Carrell, C. J.; Bock, C. W.; Murray-Rust, P.; Glusker, J. P. *Receptor* **1993**, *3*, 57–76.
96. Christianson, D. W.; Alexander, R. S. *J. Amer. Chem. Soc.* **1989**, *111*, 6412–6419.
97. Kannan, K. K.; Liljas, A.; Waara, I.; Bergsten, P. -C.; Lövgren, S.; Strandberg, B.; Bengtsson, U.; Carlbom, U.; Fridborg, K.; Järup, L.; Petef, M. *Cold Spring Harbor Symp. Quant. Biol.* **1972**, *36*, 221–231.
98. Kannan, K. K.; Notstrand, B.; Fridborg, K.; Lövgren, S.; Ohlsson, A.; Petef, M. *Proc. Natl. Acad. Sci. USA* **1975**, *72*, 51–55.
99. Kumar, V.; Kannan, K. K.; Sathyamurth, P. *Acta Cryst.* **1994**, *D50*, 731–738.
100. Cook, C. M.; Allen, L. C. *Ann. NY Acad. Sci.* **1984**, *429*, 84–88.
101. Coleman, J. E. *Zinc Enzymes*; Bertini, I.; Luchinat, C.; Maret, W.; Zeppezauer, M., Eds.; Birkhäuser: Boston, Basel, Stuttgart, 1986, Chap. 4, pp. 49–58.
102. Bauer, R.; Limkilde, P.; Johansen, J. T. *Biochemistry* **1976**, *15*, 334–342.
103. Tibell, L.; Lindskog, S. *Biochim. Biophys. Acta* **1984**, *788*, 110–116.
104. Christianson, D. W.; Lipscomb, W. N. *Acc. Chem. Res.* **1989**, *22*, 62–69.
105. Rees, D. C.; Lipscomb, W. N. *Proc. Natl. Acad. Sci. USA* **1981**, *78*, 5455–5459.
106. Rees D. C.; Lewis, M.; Lipscomb, W. N. *J. Mol. Biol.* **1983**, *168*, 367–387.
107. Hardman, K. D.; Lipscomb, W. N. *J. Amer. Chem. Soc.* **1984**, *106*, 463–464.
108. Lewis, V. E.; Donarski, W. J.; Wild, J. R.; Raushel, F. M. *Biochemistry* **1988**, *27*, 1591–1597.
109. Omburo, G. A.; Kuo, J. M.; Mullins, L. S.; Raushel, F. M. *J. Biol. Chem.* **1992**, *267*, 13278–13283.
110. Omburo, G. A.; Mullins, L. S.; Raushel, F. M. *Biochemistry* **1993**, *32*, 9148–9155.
111. Benning, M. M.; Kuo, J. M.; Raushel, F. M.; Holden, H. M. *Biochemistry* **1994**, *33*, 15001–15007.
112. Benning, M. M.; Kuo, J. M.; Raushel, F. M.; Holden, H. M. *Biochemistry* **1995**, *34*, 7973–7978.
113. Vanhooke, J. L.; Benning, M. M.; Raushel, F. M.; Holden, H. M. *Biochemistry* **1996**, *35*, 6020–6025.
114. Hong, S. -B.; Raushel, F. M. *Biochemistry* **1996**, *35*, 10904–10912.
115. Miller, J.; McLachlan, A. D.; Klug, A. *The EMBO J.* **1985**, *4*, 1609–1614.
116. Berg, J. M. *Proc. Natl. Acad. Sci.* **1988**, *USA 85*, 99–102.
117. Klug, A.; Rhodes, D. *Trends Biochem. Sci.* **1987**, *12*, 464–469.
118. Katz, R. A.; Jentoft, J. E. *BioEssays* **1989**, *11*, 176–181.
119. Párraga, G.; Horvath, S. J.; Eisen, A.; Taylor, W. E.; Hood, L.; Young, E. T.; Klevit, R. E. *Science* **1988**, *241*, 1489–1492.
120. Lee, M. S.; Gippert, G. P.; Soman, K. V.; Case, D. A.; Wright, P. E. *Science* **1989**, *245*, 635–637.
121. Pavletich, N. P.; Pabo, C. O. *Science* **1991**, *252*, 809–817.
122. Eklund, H.; Nordström, B.; Zeppezauer, E.; Söderlund, G.; Ohlsson, I.; Boiwe, T.; Söderberg, B.-O.; Tapia, O.; Brändén, C. -I.; Åkeson, Å. *J. Mol. Biol.* **1976**, *102*, 27–59.
123. Eklund, H.; Samama, J. -P.; Jones, T. A. *Biochemistry* **1984**, *23*, 5982–5996.
124. Colonna-Cesari, F.; Perahia, D.; Karplus, M.; Eklund, H.; Brändén, C. I.; Tapia, O. *J. Biol. Chem.* **1986**, *261*, 15273–15280.

125. Pettersson, G. *Zinc Enzymes*; Bertini, I.; Luchinat, C.; Maret, W.; Zeppezauer, M., Eds.; Birkhäuser: Boston, Basel, Stuttgart, 1986, Chap. 32, 451–464.

126. Hollfelder, F.; Herschlag, D. *Biochemistry* **1995**, *34*, 12255–12264.

127. Kim, E. E.; Wyckoff, H. W. *J. Mol. Biol.* **1991**, *218*, 449–464.

128. Murphy, J. E.; Xu, X.; Kantrowitz, E. R. *J. Biol. Chem.* **1993**, *268*, 21497–21500.

129. Ollis, D. L.; Brick, P.; Hamlin, R.; Xuong, N. G.; Steitz, T. A. *Nature* **1985**, *313*, 762–766.

130. Ollis, D. L.; Kline, C.; Steitz, T. A. *Nature* (London) **1985**, *313*, 818–819.

131. Westheimer, F. H. *Science* **1987**, *235*, 1173–1178.

132. Doublié, S.; Tabor, S.; Long, A. M.; Richardson, C. C.; Ellenberger, T. *Nature* **1998**, *391*, 251–258.

133. Hellinga, H. W. *Protein Engineering: Principles and Practice*; Cleland, J. L.; Craik, C. S., Eds.; Wiley-Liss: New York, 1996, Chap. 14, pp. 369–398.

134. Matthews, D. J. *Curr. Opin. Biotechnology* **1995**, *6*, 419–424.

135. Volbeda, A.; Fontecilla-Camps, J. C.; Frey, M. *Curr. Opin. Struct. Biol.* **1996**, *6*, 804–812.

136. Berg, J. M. *Curr. Opin. Struct. Biol.* **1993**, *3*, 585–588.

137. Lu, Y.; Valentine, J. S. *Curr. Opin. Struct. Biol.* **1997**, *7*, 495–500.

138. Hellinga, H. W.; Richards, F. M. *J. Mol. Biol.* **1991**, *222*, 763–785.

139. Hellinga, H. W.; Caradonna, J. P.; Richards, F. M. *J. Mol. Biol.* **1991**, *333*, 787–803.

140. Regan, L.; Clarke, N. D. *Biochemistry* **1990**, *29*, 10878–10883.

141. Regan, L. *Trends in Biochemical Sciences* **1995**, *20*, 280–285.

142. Johnson, C. K. *Acta Cryst.* **1965**, *18*, 1004–1018.

143. Erlebacher, J.; Carrell, H. L. *ICRVIEW*. Program from the Institute for Cancer Research, The Fox Chase Cancer Center, Philadelphia, PA, 1992.

THE ELECTRONIC SPECTRA OF ETHANE AND ETHYLENE

C. Sandorfy

Advances in Molecular Structure Research
Volume 4, pages 281–318
Copyright © 1998 by JAI Press Inc.
All rights of reproduction in any form reserved.
ISBN: 0-7623-0348-4

ABSTRACT

Our knowledge of the electronic spectra and lower excited states of organic molecules has made spectacular progress in the last 30 years. This progress was made possible by the joint efforts of researchers in quantum chemistry, photoelectron-, optical-, and electron-impact spectroscopies, and photochemistry, all extended into the far-ultraviolet. Ethane and ethylene are two typical, thoroughly studied examples that illustrate the many fundamental problems that are encountered. An approximate chronological order is followed to give an idea of the difficulties and conflicting opinions that have arisen in the course of the investigations on the nature of the excited states of these two apparently simple molecules which are still not fully understood. The reader might be surprised to notice that most of the references are to works published in end around the 1970s. Indeed, that was the epoch when most of the substantial contributions to the spectroscopy and quantum chemistry of ethane and ethylene were made. There was a lull between about 1983 and 1993. The reason for this is that with the experimental and computational means available in those years the researchers in this field could have hardly done better. Recently this situation has changed and new hope has arisen. One of the aims of this chapter is to stimulate renewed interest in the Rydberg and valence states of these two key molecules.

I. GENERALITIES

A. Rydberg and Valence States

For a long time the best known electronic absorption spectra of organic molecules were those due to $\pi \rightarrow \pi*$ and $n \rightarrow \pi*$ transitions of unsaturated hydrocarbons and their heteroatomic derivatives, mainly aza-aromatics, aldehydes, ketones, and some other related molecules. The electronic spectra of such molecules, at least their parts located in the visible or in the near-ultraviolet, reflect so-called valence-shell, or intravalency transitions. In the molecular orbital (MO) language these are described in the following way. The total wavefunction of the molecule is taken to be the product of one-electron wavefunctions to each of which one electron is assigned, or two electrons with opposite spins. The one-electron wavefunctions are built as linear combinations of atomic orbitals (AO) which an electron would have if it was restricted to the field of just one given atom (LCAO). An approximate Schrödinger equation is solved minimizing the energy of the ground state for the given wavefunction. This procedure entails a variety of approximations, but modern theoretical chemistry possesses adequate means of obtaining better (or even good) energies and wavefunctions, the most important of these being without doubt configuration interaction. This was and still is one of the main subjects of quantum chemistry but we do not need to go into this here.

The main point is, for the purposes of this chapter, that the simple molecular orbital method yields as many MOs as there were AOs from which they were constructed. We place the electrons into these orbitals in order of increasing energy obtaining first a ground-state "configuration." With four electrons this would be configuration N (Figure 1). The empty orbitals can then be used to build excited configurations V_1 to V_4 considering one-electron excitations only. Configuration interaction then mixes together configurations having the same symmetry, that is, it takes linear combinations of them reminimizing their energy. Then electronic transitions can be considered as transitions from N to any of the configurations V_1 to V_4 or, better, from the ground state to any state of higher energy obtained after applying configuration interaction. (In this computational technique the wavefunction of a *state* is, in general, a linear combination of wavefunctions of *configurations* of the same symmetry.)

Such electronic transitions are called valence-shell (or intravalency) transitions because the wavefunctions of both the ground and excited states are built exclusively from atomically unexcited AOs. The visible and near-ultraviolet spectra of conjugated acyclic and aromatic hydrocarbons and their heteroatomic derivatives have been interpreted in a quite satisfactory manner within this framework.

As is well known, however, many organic molecules do not absorb light in the visible or in the near-ultraviolet. This is the case, in particular, of saturated (paraffin) hydrocarbons, their heteroatomic derivatives, and monoolefins. They do absorb in the far-ultraviolet, that is at wavenumbers between about 50,000 and 100,000 cm^{-1}, or at wavelengths from about 200 to 100 nanometers, or 2000 to 1000 Å, or energies between about 6 and 10 electron volts (1 eV = 8065.73 cm^{-1}).

What electronic transitions can give bands in this far-UV part of the electromagnetic spectrum? Among these are, of course, the higher transitions of π-electrons. Then we have the transitions of the σ-electrons which form the single bonds in organic molecules, the $\sigma \rightarrow \sigma*$ transitions. These are expected to be at relatively high energies; the σ-bonds have generally lower energies (are more stable) than the π-bonds, so the $\sigma*$ are at energies higher than the $\pi*$. Typically, they are expected to be strongly antibonding. This is not all, however.

At this stage we have to ask the question: why do we build our MOs only from atomically unexcited AOs? This is only justified by the fact that they suffice to give a fair description of ground states and the lower valence-shell excited states. The

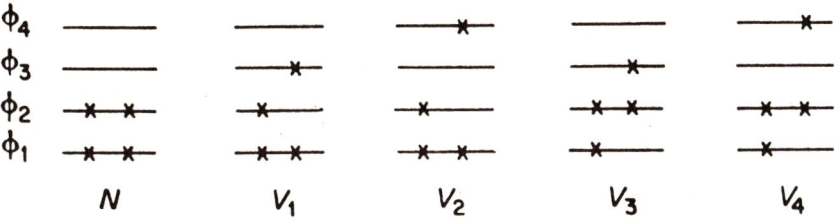

Figure 1. The electron configurations for four electrons using molecular orbitals obtained as linear combinations of atomic orbitals.

example of the hydrogen molecule can illustrate these conditions. Let us call the two hydrogen atoms A and B and the corresponding AOs $1s_A$ and $1s_B$. Omitting normalization factors we obtain the two well-known MOs:

$$\sigma_g 1s = 1s_A + 1s_B \quad \text{and} \quad \sigma_u 1s = 1s_A - 1s_B$$

In the ground state both electrons are assigned to $\sigma_g 1s$. Then an electronic transition can take place whereby one of the electrons is promoted from $\sigma_g 1s$ to $\sigma_u 1s$. If this was the only possible electronic transition only two excited states would result: one singlet and one triplet. The transition would be a valence-shell transition since both $\sigma_g 1s$ and $\sigma_u 1s$ are built from the same atomically unexcited AOs.

Here we can make an important point. At the united atom limit $\sigma_u 1s$ becomes $2p\sigma$ (a dumbbell). Since then the principal quantum number increases to 2, in a sense this would become a Rydberg orbital. Mulliken [1] called this "Rydbergization" and we see that Rydberg or non-Rydberg depends on the distance, that is geometry. At actual molecular distances, $\sigma_u 1s$ can then be reasonably considered as intermediate Rydberg/non-Rydberg.

As the next step we could form MOs from the 2s AOs: $\sigma_g 2s$ or $\sigma_u 2s$. The former, $\sigma_g 2s$ is Rydberg at any internuclear distance. So is $\sigma_u 2s$ and at the united atom limit is becomes $3p\sigma$, even more "Rydberg". At any rate it is seen that we have two kinds of Rydberg orbitals: ones that are truly Rydberg at any molecular geometry, and ones that only become so at the united atom limit and are better described as intermediate. This is an important distinction and it complicates matters considerably.

The transition $\sigma_g 1s \rightarrow \sigma_g 2s$ does take place, although it is, of course, forbidden by the $g \leftrightarrow g$ (Laporte) rule and other selection rules. The interested reader might like to consult Herzberg's Volume 1 on this subject [2] (Page 340, Fig. 160). Electronic transitions to many states of the hydrogen molecule, formed from Rydberg MOs are known. The spectrum of H_2 contains far more bands than the two due to $\sigma_g 1s \rightarrow \sigma_u 1s$.

The transition to $\sigma_u 2s$ also occurs. In other cases, however, the negative sign is expected to create problems. Sinanoglu [3] drew attention to the fact that with the large AOs which are involved near-unity negative overlap integrals are introduced meaning that such MOs would virtually "disappear." A look at some larger molecules can further illustrate the problem.

Let us consider a molecule like cyclohexane or eicosane. Disregarding the hydrogens we could form Rydberg MOs from the 6 or 20 carbon 3s AOs respectively. Linear combinations of these would give 6, respectively 20 MOs. All but the lowest one would contain negative signs. From these we could build a very great number of Rydberg excited states. But do these states really "exist"? Obviously, we are reaching the limits of the usual linear-combination-of-atomic-orbitals (LCAO) technique. For smaller molecules the problem is not so acute but it appears that for large molecules techniques like those applied to Rydberg states of solids [4–8] have to be applied.

Returning for a moment to the hydrogen molecule the question arises as to whether 1s – 1s or 2s + 2s has the lower energy? 1s – 1s is antibonding but it is built from atomically unexcited AOs; on the other hand 2s + 2s is formally bonding but it is built from excited AOs. As is now known this depends on given cases. Furthermore, Rydberg and valence-shell-type MOs having the same symmetry and not widely different energies can mix. This is a very important point; for polyatomic molecules, rather than using united atom considerations or referring to correlation diagrams it is more realistic to treat the excited states on the basis of mixing between Rydberg and valence-shell states and orbitals (ref. 9 and refs. therein). Then we shall encounter three types of excited states: some of them will be purely Rydberg, others will be purely valence-shell (or valence), and many of them will be of intermediate type.

Computational techniques for entering molecular Rydberg orbitals into the LCAO scheme consist in building them from atomic orbitals corresponding to higher principal quantum numbers. Such techniques were elaborated and perfected by Buenker and Peyerimhoff [10–18]. According to these authors the most logical way to classify a Rydberg orbital is to use the procedure relating its principal quantum number to that of the orbital to which it corresponds at the united-atom limit.

Then, in view of their overall character, the Rydberg orbitals are not, in general, attributed to one given atom, but are placed at a chosen point, for example the geometrical midpoint in the molecule. The exponents of these long-range functions can be obtained by usual optimization procedures and are found to be relatively insensitive to the special molecular environment. Then a highly sophisticated ab initio multireference single- and double-excitation configuration interaction procedure is applied. According to cases Rydberg-valence mixing occurs at the orbital or at the configuration interaction stage.

B. Closer to Experiment

Rydberg bands form series converging to an ionization potential related to a given state of the positive ion,

$$\nu = I_p - \frac{R}{(n-\delta)^2} \tag{1}$$

where ν is the wavenumber of the given band in cm^{-1}, R is the Rydberg constant ($R_\infty = 109,737.3 \ cm^{-1}$), n is the principal quantum number of the excited state, and δ is the "quantum defect." δ is zero for the hydrogen atom and ions with only one electron, like He^+, Li^{++}, ... For any other atomic species, however, the energy of a state depends on both the principal and the azimuthal quantum number, even in the absence of an external field. The quantum defect takes this into account; the value of δ for *s, p, d,* . . . type orbitals reflects the degree of "penetration" of the Rydberg orbital into the core. (The core in this case is the atom without the electron which is promoted.) It is to be noted that the nuclear charge Z is absent from Eq. 1; this implies that the Rydberg electron is attracted by only one positive charge; that is the other electrons are screening perfectly all the others. δ is zero for hydrogen but for other atoms the screening is not perfect and it is different for *s, p, d,* . . . type orbitals. The most penetrating ones are the *s* orbitals. This can be understood in the following way. The orbitals (wavefunctions) must be mutually orthogonal; for example 3*s* must be orthogonal to 2*s* and 1*s*. However, since their angular parts are the same, orthogonality can only be achieved by the radial parts of the respective functions. This introduces two local maxima and two radial nodes in the electron density function close to the nucleus. So the Rydberg electron (in this example 3*s*) has some probability of presence near the nucleus, amounting to less efficient screening and stronger attraction of the Rydberg electron by the nucleus. To take this into account δ is subtracted from the principal quantum number; it is about 0.9–1.2 for *s* orbitals. Orbitals *p, d,* . . . are less penetrating, since they are more diffuse and point away from the nucleus. For the *p*, δ is about 0.3–0.5, and for the *d* it is about 0.1 [122 *p*, 341]. In actual fact penetration is not the only factor determining δ but we do not need to go into more detail at this point. Mulliken [19–21] examined these conditions thoroughly.

For simple atoms the Rydberg Eq. 1 usually works well, although large deviations are often observed for the first member of the series which corresponds to the most penetrating Rydberg orbital. By fitting the observed spectral lines to Eq. 1, one can obtain accurate values for the ionization potential.

For molecules the situation is much more difficult. The Rydberg formula is still valid; the principal quantum number *n,* for the first member, can be taken for that of the lowest empty Rydberg orbital of the atom from whose field the electron is thought to be excited—3 for carbon, 4 for chlorine, for example. However, long series of Rydberg bands are seldom found. For polyatomic molecules they are usually found if they contain an atom which dominates the given part of the spectrum, like CH_3I, or for π-electron systems, like the one of benzene, for example. In many cases we find only the first or first and second members of the "series." The bands are often broadened out by unresolved vibrational and rotational fine structure, predissociation, or dissociation in the excited states. The higher members of the series are often very weak and lost in the background. Yet, the Rydberg-valence distinction can still be maintained. The quantum defect, δ, still has characteristic values for *s, p, d,* . . . type series [22, 24]. For organic molecules containing only carbon, hydrogen, nitrogen, and oxygen it is 0.9–1.0 for 3*s*, 0.4–0.6 for 3*p*, and about 0.1 for 3*d*. Robin [23] has shown that because of the difficulties to give a definite value to *n* it is preferable to use the term value that is, $R/(n - \delta)^2$. A large amount of data show that there are characteristic ranges for 3*s*, 3*p*, and 3*d*; about 30,000 to 23,000 cm^{-1} for 3*s*, 20,000 to 16,000 for 3*p*, and 14,000 to 10,000 for 3*d*. It is, of course not possible to determine ionization potentials under the given conditions. To the contrary, we need an independent source of ionization potentials to locate the Rydberg bands at least approximately.

This has been provided by UV photoelectron (PE) spectroscopy. Since Turner's breakthrough [25–27], such spectra can be obtained with sufficient accuracy to be used together with optical absorption or electron impact spectra and results of up to date quantum chemical calculations. PE spectroscopy yields bands due to transitions from the electronic ground state to various states of the positive ion.

The interpretation of PE spectra often implies the validity of Koopmans' theorem [28] or, better, its differential form [29]. This assumes that if $E(A)$ and $E(A^+)$ are the total energies of the neutral molecule A and one of its positive ions then, approximately,

$$E(A) = E(A^+) + \varepsilon \tag{2}$$

where ε is the energy of the molecular orbital from which the electron is taken. This neglects changes in the energies of the other electrons due to the removal of an electron from the orbital corresponding to ε (rearrangement) and the changes it produces in correlation energies. These two are supposed to cancel each other in a fair approximation, so that the negative of ε can be taken for the energy of ionization from the given orbital. This gives us a convenient way of relating PE spectra to the orbital energies obtained by MO calculations. If cautiously handled this can be very useful. It is to be noted, however, that in particular for close-lying MOs or when more than one state is issued from the same configuration the order of the PE bands is not necessarily the same as the order of the MOs obtained by calculations.

When the electron is taken out from an orbital which is bonding in a given bond, then ionization or excitation from that orbital is expected to weaken that bond; if taken from an antibonding orbital it may actually strengthen that bond, and if taken from a nonbonding orbital it may hardly affect its strength. This is likely to be one of the factors determining the photochemical primary step in photolysis (see below).

To conclude this section we might assert that the main techniques for assessing the problem of Rydberg and valence excited states of organic molecules are, in practice, far-UV absorption spectra, photoelectron spectra, and advanced quantum chemical calculations. It should be added that the electronic spectra can also be obtained from electron-impact (energy loss) and circular dichroism spectra. The reader is referred to Robin's Volumes I to III [8, 23]; he systematically considers the three kinds of spectra when they are available.

C. The Lowest Excited States

Several years ago the present writer proposed a rather simple classification for organic molecules [30] for the use of far-UV spectroscopists:

1. Molecules with bonding σ-electrons only (like saturated hydrocarbons).

2. Molecules with bonding σ- and lone-pair-type electrons (like saturated alcohols or amines).

3. Molecules with bonding σ- and π-electrons (like olefins or aromatics).

4. Molecules with bonding σ-, π-, and lone-pair-type electrons (unsaturated heteroatomic molecules).

On the other hand, the spectra could be divided into four categories: σ-type, lone-pair-type, π-type, or mixed. Indeed in the part of the spectrum which extends from the visible to the far-UV (down to about 120 nm or 10 eV, or 85000 cm^{-1}) the lowest bands are due to excitation from bonding σ-orbitals (a), lone-pair orbitals (b), and π-orbitals (c), respectively, while the lowest bands for unsaturated heteroatomic molecules (d) are from either π- or lone-pair orbitals. This concerns the initiating (or originating) orbitals. The excited orbitals can be from either the Rydberg or the valence type, or can be of intermediate character.

In this chapter two small organic molecules, ethane and ethylene, will be treated in some detail. They both played an important role in the evolution of far-UV spectroscopy. However, we first have a look at an even simpler molecule, methane; this may be helpful in clarifying the above discussion.

Mulliken [31] who first treated methane in 1935, took the united atom (neon) as a starting point. Then the first singlet-singlet electronic transition is,

$$[sa_1]^2 \, [pf_2]^6 \, ^1A_1 \rightarrow [sa_1]^2 \, [pf_2]^5 \, [3sa_1]^1 \, F_2$$

where a_1 and f_2 (or t_2) correspond to T_d symmetry. At molecular internuclear distances this would become σ → σ*. So this is a united atom Rydberg (the principal quantum number increased) transition which would become formally non-Rydberg at the actual molecular geometry. $A_1 \rightarrow F_2$ is, of course, the only allowed transition under T_d symmetry (x, y, z). The band is a broad, diffuse band centered at about 128 nm (about 80,000 cm^{-1}). It is fairly strong, with an oscillator strength $f = 0.26$, as it should be for an allowed transition. The band is affected by the Jahn-Teller effect—the excited state being degenerate [32].

However, one may treat methane by the usual MO method as well (see for example refs. 33–35). The following MOs are obtained where f_2 is triply degenerate:

Table 1. Correlations for Transitions Departing from the f_2 MO of Methane[a]

United Atom	T_d		Molecular Orbital	
$1s^2 2s^2 2p^6$	^1S_g	$^1A_1$		$^1A_1$
$1s^2 2s^2 2p^5 3s$	$^{3,1}P_u$	$^{3,1}F_2$	$a_1 - f_2$	$^{3,1}F_2$
$1s^2 2s^2 2p^5 3p$	$^{3,1}D_g$	$^{3,1}E + ^{3,1}F_2$		
	$^{3,1}P_g$	$^{3,1}F_1$	$f_2 - f_2$	$^{3,1}A_1 + ^{3,1}E +$
	$^{3,1}S_g$	$^{3,1}A_1$		$^{3,1}F_1 + ^{3,1}F_2$

Note: [a] Ref. 30. Reproduced by permission from Elsevier Science-NL.

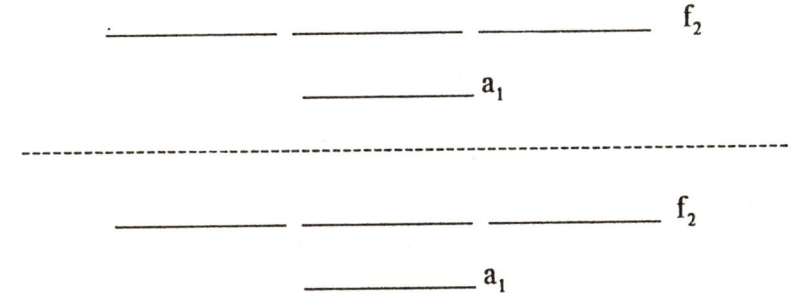

Figure 2. The LCAO molecular orbitals of methane.

Then the first transition is again $f_2 \rightarrow a_1$ in terms of orbitals and $^1A_1 \rightarrow {}^1F_2$ in terms of states. Table 1 shows the excited states obtained by both the united atom and the MO approaches which originate from the f_2 level (Figure 2). They are the same! So the actual excited states must be of the intermediate type. Since, however, the σ^*-orbitals are very antibonding and the Rydberg $3s$ and $3p$ are mainly nonbonding the mainly Rydberg states are expected to lie at lower energies. This is the situation for all saturated hydrocarbons. Thus we may consider their lowest excited states as being of intermediate character, but essentially Rydberg.

II. ETHANE

A. The Initiating Orbitals: The Photoelectron Spectrum

The electronic absorption spectrum of ethane was not known until 1967 when Lombos et al. [*36, 37*] and Raymonda and W. T. Simpson [*38*] finally published it. Prior to that date it was believed that ethane has no discrete bands in the UV or far-UV part of the spectrum which is of main concern to us [*39*] (see ref. 22, pp. 413 and 545). Lassettre et al. [*40, 41*], Lipsky and J. A. Simpson [*42*], and Johnson et al. [*43*] determined the spectrum by electron-impact and Koch and Skibowski the optical spectrum using a synchrotron radiation source [*44*]. The surprising fact is that the band of lowest frequency (and the next one) has vibrational fine structure, so it has one (or more) stable excited states. This is a unique case among saturated *n*-alkanes and is of very rare occurrence in the spectroscopy of organic molecules not having π-electrons. The widely held view had been that excitation or ionization from a bonding σ-orbital necessarily leads to immediate dissociation. So this molecule played an important role in the evolution of the far-UV spectroscopy of organic molecules (as did ethylene which will be treated in the next section). In what follows we are trying to review the evolution of our ideas on the excited states of ethane in the last 30 years.

Mulliken [31] focused his attention on ethane well before the epoch of large scale quantum chemical calculations. The f_2 frontier orbital of methane splits into $a_1 + e$ under the lower symmetry of a methyl group. Then taking the united molecule approach, we have for the ground state of ethane the following configuration:

$$[sa_1]^2 \, [sa_1]^2 \, [\pi e]^4 \, [\pi e]^4 \, [\sigma + \sigma, a_1]^2 \, {}^1A_1$$
$$\text{CH}_3 \quad \text{CH}_3 \quad \text{CH}_3 \quad \text{CH}_3 \qquad \text{C–C}$$

In the ground state, ethane was shown to have D_{3d} (staggered) symmetry [45]. So, if we allow interactions between the two methyl groups we obtain the following orbital structure,

$$KK[2a_{2g}]^2 \, [2a_{2u}]^2 \, [1e_u]^4 \, [3a_{1g}]^2 \, [1e_g]^4 \, {}^1A_1$$

in order of increasing energy. Here $3a_{1g}$ is essentially the C–C bond which methane, of course, does not have; $1e_u$ and $1e_g$ are mainly populated in the C–H bonds and they are split by the interaction between the two methyl groups. Whether or not $3a_{1g}$ is the frontier orbital depends on the extent of this split. The above order agrees with a series of quantum chemical calculations [46–56]. The existence of two close-lying frontier orbitals is the distinctive feature which determines the ethane spectrum and is "inherited" by the higher n-alkanes. It is also in line with the observed photoelectron spectrum [25, 57].

The first ionization potential of ethane was measured by photoionization techniques more than 30 years ago [56–62] and it was found to be between 11.4 and 11.65 eV. Some indication of vibrational fine structure was found by Chupka and Berkowitz [62]. The classic HeI photoelectron spectrum of ethane was recorded by Turner's group (Baker et al. [25, 57]; it is reproduced on Figure 3.

At the low-energy side there are three bands centered at 12.0, 12.7, and 13.4 eV (96,800, 102,425, and 108,070 cm^{-1}, respectively). The first band has vibrational fine structure with an apparent progression of about 1170 cm^{-1}. The adiabatic Ip is at 11.56 eV (93,230 cm^{-1}). These three bands can only belong to ionization from either the $1e_g$ or the $3a_{1g}$ levels; the ionization from the $1e_g$ orbitals yields a positive ion in a degenerate state, so it must be split by the Jahn–Teller effect [32]; the 0.7 eV splitting is of the expected order of magnitude [57]. This, however, does not tell us which one of the bands is due to $3a_{1g}$ and which ones to $1e_g$. Even if we knew the assignments of the UV absorption bands (see below), this would not help. The two orbitals are so close to one other that Koopmans' theorem may not apply; the order of the UV bands may or may not be the same as the order of the PE bands.

The observed vibrational fine structure is evidently only partly resolved. It could correspond to any of the following vibrations of ethane:

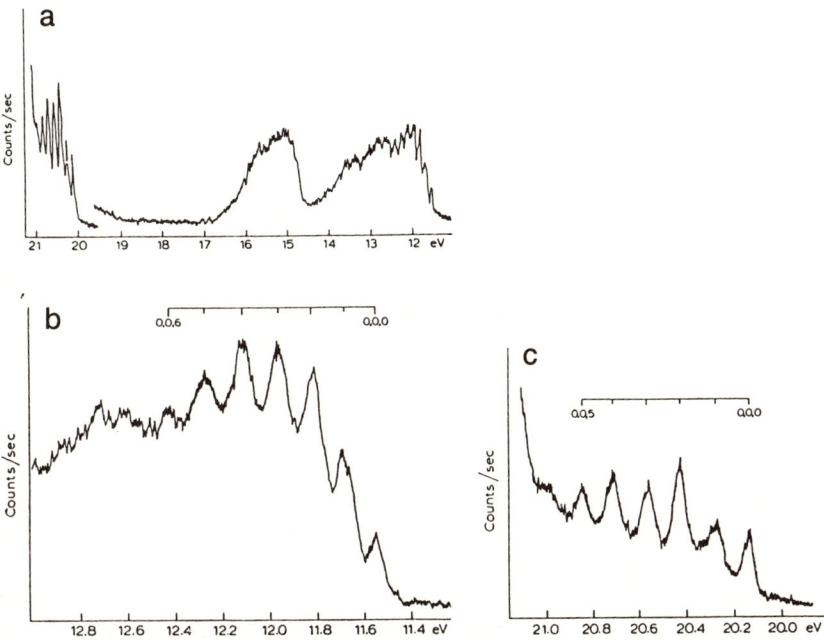

Figure 3. (a) Helium 584 Å photoelectron spectrum of ethane. (b) The first band, (c) the fourth band of the spectrum, with expanded electron energy scale. Baker et al. [*25, 57*]. Reproduced by permission from Elsevier Science-NL.

ν_2	$1375\ \text{cm}^{-1}$	CH sym. bending	a_{1g}
ν_6	$1379\ \text{cm}^{-1}$	CH sym. bending	a_{2u}
ν_8	$1486\ \text{cm}^{-1}$	CH antisym. bending	e_u
ν_{11}	$1460\ \text{cm}^{-1}$	CH antisym. bending	e_g
ν_3	$993\ \text{cm}^{-1}$	C–C stretching	a_{1g}

These are neutral ground-state values. If upon excitation the electron is mainly taken from the C–C bond, the latter is expected to become considerably weaker, so the frequency of the excited state ν_3 would be much lower than 993. If, on the other hand, the excitation is from an orbital which is C–H bonding but C–C antibonding the CH frequencies would be lowered but the C–C frequency would increase. Thus the observed interval of about $1170\ \text{cm}^{-1}$ could be a C–H bending vibration but it could also be the ν_3 with an increased frequency. So this does not settle the issue (but see below).

Some help is provided by deuteration. In the ground state the C_2H_6/C_2D_6 ratio is 1.187 for ν_2 and for ν_3 it is 1.165, not very different. Clearly, the C–C stretching and C–H motions are mixed in the ground state, probably more so in C_2D_6 than in C_2H_6. In our laboratory we measured the PE spectra of a few deuterated ethanes. For C_2D_6 the measured vibrational interval in the PE spectrum was 920 cm^{-1}, giving an isotopic ratio of 1.27. This is closer to what one would expect for a mainly CH bending vibration than for a mainly C–C vibration but it still does not amount to a decisive proof. The CH assignment is made more likely by the observation that similar vibrational intervals are found for CH_3F [ref. 25, p. 221] which has no C–C bond. (In the other methylhalides the first PE band originates with the halogen lone pair orbital.)

A stronger experimental indication against the $^2A_{1g}$ assignment is the absence of a long vibrational sequence in the (decreased) C–C frequency in the observed fine structure [64, 65] (from 11.56 to about 12.4 eV). Indeed, normally the vibrations which show up prominently are those which are in the direction of the geometrical change that occurs upon electronic excitation or ionization.

At higher energies in the PE spectrum a double band is found at about 15.0 and 15.8 eV which is readily assigned to the Jahn–Teller split ^2E_u state of the $C_2H_6^+$ ion, originating with the $1e_u$ orbital. At even higher energies a fine structured band is found at about 20.6 eV which corresponds to $^2A_{2u}$ originating with the $2a_{2u}$ orbital. The spacing in the vibrational fine structure is about 1160 cm^{-1} for $C_2H_6^+$ and 1070 cm^{-1} for $C_2D_6^+$ [57]. This moderate shift occurred upon deuteration fits a C–C stretching vibration. There are no known UV absorption bands originating with $1e_u$ and $2a_{2u}$; they could be at very high frequencies in the far UV, probably a diffuse part of the spectrum.

Later Narayan [63] and Rabalais and Katrib [64] further examined the assignment of the low frequency PE bands; as Turner et al. [57] they all came to the conclusion that the first two bands must be due to the components of the Jahn–Teller split ^2E_g state of $C_2H_6^+$. They pointed out that the vibration causing Jahn–Teller instability must be one of the degenerate CH_3 deformation vibrations, ν_8 (e_u) or ν_{11} (e_g). Their frequencies in the neutral molecule are 1486 and 1460 cm^{-1}, respectively.

More help comes from the theoretical side. Lathan et al. [55] carried out ab initio open-shell SCF calculations on ethane. They obtained the result that $C_2H_6^+$ has $^2A_{1g}$ symmetry in its ground state. This implies ionization from the $3a_{1g}$ orbital leading to a large C–C bond length and nearly planar CH_3 groups. The energy minimum they computed was 10.24 eV above that of neutral C_2H_6 while the experimental 0–0 band is at 11.56 eV in the PE spectrum. Lathan et al., however, did not include configuration interaction (correlation energy) in their calculations. As it turned out this does change the order of the states of the ion in this case.

A few years later Richartz et al. [66] made a highly successful effort to clarify these conditions. They carried out large-scale ab initio SCF + CI calculations with a double zeta AO basis set augmented by polarization functions (see ref. 66 for details). After having made sound use of the Mulliken–Walsh type correlation diagrams they computed potential surfaces for ethane and the three lowest lying states of its positive ion and this for all the vibrations which are likely to bring about geometrical change upon ionization. This was followed by a Franck–Condon analysis.

Now ionization from the $3a_{1g}$ orbital yields a $^2A_{1g}$ ion while ionization from the $1e_g$ orbital produces a ^2E_g ion with a degenerate ground state. The latter is subject to the Jahn–Teller effect; to achieve stability this ionic state must split into two. D_{3d} ethane can deform itself without losing its center of inversion to C_{2h}; under this symmetry ^2E_g would split into ^2A_g and ^2B_g where ^2A_g interacts with the $2A_{1g}$ state issued from $3a_{1g}$ ionization.

The Franck–Condon analysis of the ^2B_g state shows that progressions of both the totally symmetrical ν_3 C–C stretch and the asymmetric HCH bending vibrations can contribute to the observed progression of 1170 cm^{-1} (cf. the discussion of the UV bands below). This accounts for the first, structured, PE band whose maximum is at about 12.0 eV. The two other maxima at 12.7 and 13.4 eV then belong to the two ^2A_g states whose interaction is likely to be one of the causes of the irregular, diffuse character of these bands.

Unfortunately the resolution in the PE spectra is not sufficient to detect all the vibrations which are present in the observed fine structure.

B. The Far-Ultraviolet Spectrum

In 1967–1968 three substantial papers appeared on the absorption and electron-impact spectra of ethane by Raymonda and Simpson [*38*] and Lombos et al. [*36, 37*] who measured the optical absorption spectra, and by Lassettre et al. [*40, 41*] who determined the electron impact-energy loss spectra. There is good agreement among their spectra. Later Koch and Skibowski [*44*] confirmed their results by using a synchrotron radiation source. All three groups presented the surprising result that vibrational fine structure is observed in the bands, indicating stable excited states due to the excitation of a binding σ-electron.

Figure 4 shows the lowest frequency absorption band of ethane and fully deuterated ethane measured by Lombos et al. [*36*] and Figure 5 gives the electron impact spectrum of Lassettre et al. [*40*]. There is a weak band or shoulder at the low-frequency end.

Electron energy-loss spectroscopy has some specific advantages over optical spectroscopy. At high-impact energy and zero scattering angle the energy-loss spectra are like the optical spectra and the same selection rules apply. On the other hand at impact energies close to the excitation energies to be measured (the "threshold" energies) and at scattering angles far from zero, the optical selection rules are no longer valid. Therefore singlet–triplet and other forbidden transitions can be observed in electron energy-loss spectroscopy. Furthermore, for different bands the Franck–Condon factors do not, in general, have the same angle dependence, a fact that may allow to uncover mutually overlapping bands. The interested reader may like to consult reviews by Dillon [*67*], Compton [*68*], and Robin [*23*] on this subject.

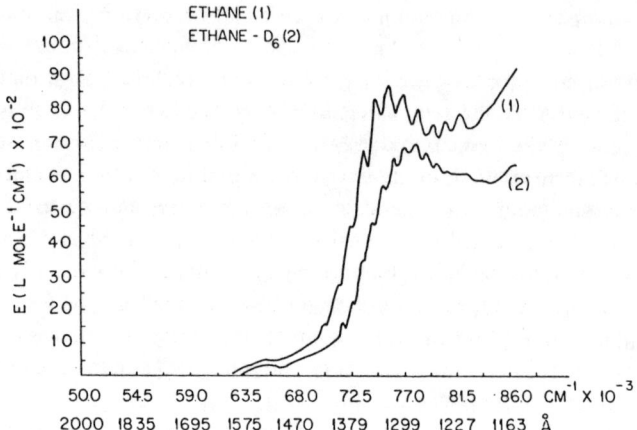

Figure 4. The far-ultraviolet absorption spectrum of ethane and fully deuterated ethane. Lombos et al. [*36*]. Reproduced by permission from Academic Press.

While the spectra published by the above-mentioned three groups were very similar, their interpretation of them entailed important differences. Raymonda and Simpson [*38*] argued that saturated hydrocarbons having no lone pairs and with strongly localized bonds are unlikely to have Rydberg-type transitions and worked

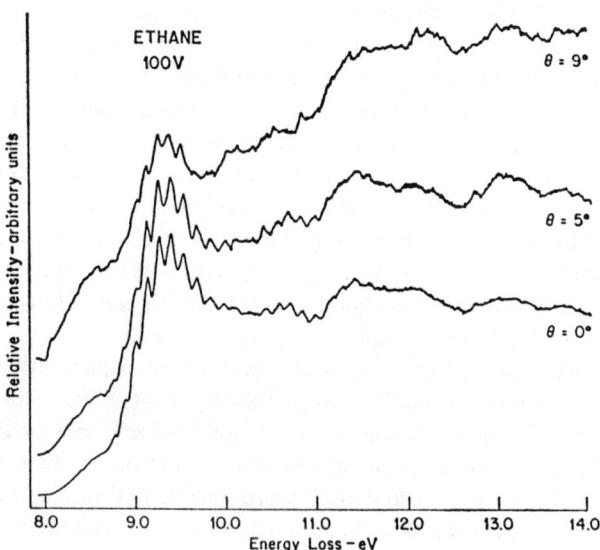

Figure 5. The electron impact energy loss spectrum of ethane. Lassettre et al. [*40*]. Reproduced by permission from the American Institute of Physics.

out a method in order to interpret the observed bands as valence-shell transitions. In this independent systems theory the Hamiltonian of a group of bonds is written as a sum of two parts containing a local Hamiltonian for each bond i and a part representing the electrostatic interaction involving bonds i and j. Then in zeroth order approximation the first excited state is n fold degenerate (for a system of n bonds), since an electron can be promoted from any of the n bonds. The electrostatic interaction potential breaks the degeneracy and the energies and wavefunctions of the excited states are obtained by perturbation techniques (cf. McRae and Kasha [69]). Raymonda and Simpson first took into account the C–C bonds only and then both the C–C and C–H bonds. According to these calculations the C–C bond is mainly involved in the 75,800 cm^{-1} excitation of ethane (the first structured band). The main vibrational interval is taken to be the totally symmetrical CH bending mode (1380 cm^{-1} in the ground state).

It is indeed tempting to consider the paraffin hydrocarbons as being built of a certain number of C–C and C–H bonds; additivity rules work well for a number of ground-state properties. In the opinion of the writer, however, this cannot be hoped for excited-state properties. For example, the pronounced bathochromic shift observed in the series methane, ethane, propane, butane, and pentane is not unlike the shifting observed for conjugated C=C double-bonded systems. The independent systems method as applied to normal paraffins by Raymonda and Simpson [38] is of more than casual interest, however.

Lombos et al. [36, 37] went back to Mulliken's original idea who considered the excited-state orbitals of saturated hydrocarbons as having Rydberg character [31]. Here is what he said in 1935 about methane: "In previous papers, certain LCAO approximate forms were given for CH$_4$ orbitals, namely, C–H bonding forms corresponding to [sa_1] and [pt_2] and certain complementary forms called [$s*a_1$] and [$p*t_2$], similar to [s] and [p], respectively, but of consistently C–H antibonding instead of C–H bonding character. Examination of the nodal surfaces of the LCAO forms [$s*$] and [$p*$] shows that they are qualitatively very similar to those of $3s$ and $3p$ of an atom. It seems altogether probable, therefore, that [$s*$] and [$p*$] should be identified with Rydberg orbitals $3s$ and $3p$; and also probable that in this case the real molecular orbitals can be much better approximated by suitable $3s$ and $3p$ atomic orbitals, i.e. penetrating central field orbitals, than by LCAO forms." The case of ethane is analogous.

Then the lowest energy bands of ethane should be due to the transitions from the originating $3a_{1g}$ and $1e_g$ orbitals to $3s$ and $3p$ Rydberg orbitals centered at the midpoint of the C–C bond. Transitions to non-Rydberg antibonding σ* (the corresponding σ are strongly bonding) are expected to intervene at significantly higher frequencies yielding continuous absorption due to dissociation and/or pre-dissociation. To simplify matters one could put it this way: at the level of the lower excited states of ethane (and other alkanes) Rydberg and valence-excited orbitals, having the same symmetry, can mix. For each Rydberg–non-Rydberg pair this yields two states: one mainly Rydberg and one mainly valence-shell. Of these the

Rydberg, typically non-bonding, has the lower energy and the other, typically antibonding, has the higher energy. In the case of ethane this is strongly supported by the great resemblance of the vibrational fine structures observed in the photo-electron and in the ultraviolet absorption or energy-loss spectra. Indeed, Rydberg states only differ from the ion by the addition of an essentially nonbonding orbital.

Then it is logical to expect that the weak, lowest frequency band observed near 69,760 (8.65 eV) is due to a transition from $3a_{1g}$ and/or $1e_g$ to the $3s$ Rydberg orbital, a transition forbidden by the Laporte rule, being g ← g. The structured, intense band could then be assigned to transitions from the same originating orbitals to $3p$ Rydberg orbitals. Because of degeneracies in both the ground and excited levels this will then produce a great number of states.

Since the ground state is totally symmetrical this gives formally the following assignments for the orbitals and states which are involved supposing D_{3d} symmetry and disregarding for the moment the Jahn–Teller effect. For the $3s$,

$$3s(a_{1g}) \leftarrow 1e_g, \, ^1E_g \leftarrow \, ^1A_{1g}$$

$$3s(a_{1g}) \leftarrow 3a_{1g}, \, ^1E_{1g} \leftarrow \, ^1A_{1g}$$

and for the $3p$,

$$3p\sigma(a_{2u}) \leftarrow 1e_g, \, ^1E_u \leftarrow \, ^1A_{1g}$$

$$3p\pi(e_u) \leftarrow 1e_g, \, ^1A_{1u} \leftarrow \, ^1A_{1g}$$

$$^1A_{2u} \leftarrow \, ^1A_{1g}$$

$$^1E_u \leftarrow \, ^1A_{1g}$$

since $(e_u \times e_g) = A_{1u} + A_{2u} + E_u$; $3p\sigma(a_{2u}) \leftarrow 3a_{1g}$, $^1A_{2u} \leftarrow \, ^1A_{1g}$; and $3p\pi(e_u) \leftarrow 3a_{1g}$, $^1E_u \leftarrow \, ^1A_{1g}$

This is a complicated situation due to the existence of two close-lying originating orbitals for ethane. Lombos et al. [36] also discussed these conditions in terms of a theoretical treatment by Katagiri and Sandorfy [70] which led to similar results. This was the first Pariser–Parr type [71] calculation applied to σ-electron systems.

The matter of the electronic assignments will be pursued below, in connection with the advanced quantum chemical treatments that later became available.

The vibrational interval observed by Lombos et al. [36] had an average value of 1150 cm^{-1}, the same as the one given by Raymonda and Simpson [38]. It actually varies between 1100 and 1250 cm^{-1}. In those times Lombos et al., not taking into account the Jahn–Teller effect, thought that the main progression is likely to be due to one of the totally symmetrical vibrations of ethane. These are the CH stretching motion, ν_1, 2899 cm^{-1}; the symmetrical CH bending motion, ν_2, 1375 cm^{-1}, and, the C–C stretching motion, ν_3, 993 cm^{-1}, all in the ground state. The two latter are eligible. In fully deuterated ethane the respective values are 1158 cm^{-1} for ν_2 and 852 cm^{-1} for ν_3. As we have seen in the case of the photoelectron spectrum, the two modes are mixed. According to the normal coordinate calculations of Schachschneider and Snyder [72] for the deuterated species they are highly mixed. More importantly, it has to be emphasized at this point that the observed vibrational components are obviously not simple; they receive contributions from more than one vibration which are not resolved at the given resolution. This also applies to the photoelectron spectra.

The spectra of best quality of ethane and deuterated ethane are probably those of Lassettre et al. [40]. They used accelerating voltages from 50 to 180 V and scattering angles from 0° to 9°. Their spectra are less diffuse and somewhat better resolved than the absorption spectra. The shoulder at the low-frequency end is distinctly present and the bands at higher energies are also better defined. Lipsky and J. A. Simpson also measured these spectra [42]. In addition to the 75,800 cm^{-1} band both these groups and Lombos et al. [36] found some fine structure at about 10.8 eV, so ethane might have another stable excited state.

Lassettre et al. argued that the progression observed in the 75,800 cm^{-1} band is due to the quanta of the C–C stretching vibration of the excited state but, since the frequency appeared to be higher than in the ground state, they concluded that the C–C bond is shorter in the excited state. Raymonda and Simpson [38] rejected this idea on the ground that vibrational frequencies do not usually increase in excited states. Lassettre et al. [40] were certainly right on this point (because sometimes they do increase) but they did not go as far as suggesting an originating orbital of CH character [$1e_g$]. They suggested $^1A_{2u}$ for the symmetry of the excited state while Lombos et al. suggested ^1E_g.

By this time the reader is probably confused by these different assignments for the originating orbital of the band having the well-developed vibrational fine structure and the vibration to which the main progression is due. Some clarification should come from the following.

In 1969 Pearson and Innes [73] remeasured the structured band at much higher resolution using a 21 foot vacuum spectrograph, 15 μ slits, and long exposure times. They found that the longest wavelength peak of ethane-d_6 is the sharpest one observed and that the difference between the energies of the 0–0 bands of ethane and ethane-d_6 is 731 cm^{-1}. This is unusually large (70,368 for ethane and 71,099 for ethane-d_6). They found no other bands at lower frequencies belonging to this system and argue that these bands *are* 0–0 bands and not false origins since the band is strong ($f = 0.3$) and it is unlikely to be a forbidden band. This means that a large reduction of some vibrational frequency must occur upon excitation. For ethane-d_6 they found distinct peaks at $\nu_{0-0} + nx$ 781 ($n = 0,1,2$) and at $\nu_{0-0} + nx$ 1751 cm^{-1}. They assigned 781 cm^{-1} to the totally symmetric CD$_3$ deformation, ν_2, of the excited state (1154,5 in the ground state), 1751 cm^{-1} to the symmetrical CD$_3$ stretching, ν_1, of the excited state (2083 cm^{-1} in the ground state), and 616 cm^{-1} to C–C stretching (843 in the ground state). They favored an interpretation in terms of the three totally symmetric vibrations only, supposing that the Jahn–Teller effect is slight.

The situation is similar in non-deuterated ethane: $\nu_2 = 1053$ (1388); $\nu_1 = 2215$ (2954); $\nu_3 = 755$ (995 cm^{-1}) where the ground-state values are given in brackets. These values lead to a satisfactory agreement with the large shift of the 0-0 band.

Very importantly, Pearson and Innes found distinctive rotational contour in the 0–0 band of ethane-d_6. The band contour analysis using computer simulation showed that the transition moment for the 0–0 band is perpendicular to the C–C bond. This indicates that the transition is $^1A_{1g} \rightarrow {}^1E_u$. This is believed to be definite. It is however, still compatible with more than one orbital assignment. It will be further discussed below.

The large shift of the CH stretching and bending frequencies is particularly significant, because it shows that the CH distance and the HCH angle increased considerably upon excitation. Another result is that the ν_2(CH) progression is longer than the ν_3(C–C) progression; this seems to show that the more important changes occur in the HCH angles.

We are also interested in the "orbital assignment." How can the ^1E_u state be obtained? If departure takes place from $1e_g$ the excited orbital could be a_{1u}, a_{2u}, or e_u (in the latter case the direct product is $A_{1u} + A_{2u} + E_u$). If the level of departure is $3a_{1g}$ the excited orbitals must be e_u.

Now, a very important point is the Rydberg character of the transition which is strongly indicated by the fargoing resemblance between the fine structures of the PE and UV bands. Under D_{3d} symmetry the *ns, np,* and *nd* Rydberg orbitals have symmetries a_{1g}, $a_{2u} + e_u$, and $a_{1g} + 2e_g$, respectively. If the molecule keeps its center of symmetry in the excited state, *ns* and *nd* type transitions are g ↔ g and forbidden by the Laporte rule from both $3a_{1g}$ and $1e_g$. In view of the high intensity of the band this leaves us with the possibility of an *np* type Rydberg excited state. This is possible from both $3a_{1g}$ and $1e_g$.

From the former we have $3a_{1g} \rightarrow 2e_u$, from the latter $1e_g \rightarrow 2e_u$ or $1e_g \rightarrow 3a_{2u}$, all with perpendicular polarization. The term value computed from the first adiabatic I_p (11.56 ev = 93.243 cm^{-1}) and the 0–0 band of Pearson and Innes (70,368 cm^{-1}) is 22,875 cm^{-1}.

First, let us suppose that the orbital of departure is the same for both the ionization and first electronic transition of lowest energy. The great similarity between the vibrational fine structure of the first PE and first UV bands leads us naturally to this supposition. Then, if the transition was $3a_{1g} \rightarrow 2e_u$ the excited Rydberg state would be related to the $^2A_{1g}$ ion which possesses a very long C–C bond but C–H bonds of near normal length. The $2e_u$ Rydberg orbital has some C–C bonding and C–H antibonding character but being of Rydberg character it is unlikely to be very far from being nonbonding. Cleavage could be expected in the long C–C bond through its highly anharmonic stretching vibration.

If, on the other hand, the excitation takes place from the $1e_g$ orbital the resulting Rydberg state should resemble the ^2E_g ion which has longer C–H bonds and a relatively short C–C bond. The above argument is certainly oversimplified, however.

Then, it is logical to ask: how many electronic transitions can contribute to the absorption in the spectral area of the structured band? If more than one, which one is responsible for the structure (maximum at 75,800 cm^{-1} or about 1400 Å)?

In 1975 Buenker and Peyerimhoff [74] carried out large-scale double-excitation configuration interaction calculations on the ground and excited states of ethane. The atomic basis set was augmented by one-component long-range *s, px, py,* and *pz* functions for the description of the corresponding ($n = 3$) Rydberg orbitals which are located at the inversion center of the molecule. The details of the configuration interaction procedure are described in this and previous papers [75] of the same authors. First, a series of main or reference configurations found to be critical for the representation of a given state are generated and then single- and double-excited configuration interaction is applied to each of the reference configurations. In this way each state is treated in a very nearly equivalent manner and the ground state is not privileged with respect to the other states. SCF calculations alone do not secure results concerning the relative stabilities of the states resulting from $3a_{1g}$ and $1e_g$ excitations.

Table 2 adapted from Table 3 in ref. 74 contains the results in eV's (the Z axis is taken along the C–C bond). The two $3s$ Rydberg singlets only differ by 0.05 eV. (about 400 cm^{-1}). They might well account for the weak band (or shoulder) which is best seen in Lassettre's electron-impact spectrum centered at about 8.65 eV (69,750 cm^{-1}). The most striking result, however, concerns the $3p$ Rydbergs. Between 9.91 and 10.04 eV (about 1000 cm^{-1}) there are five $3p$-type singlet excited states! These are the $1e_g \rightarrow 3p\sigma$ and $3a_{1g} \rightarrow 3p\sigma$ transitions; the former being perpendicularly polarized, is one of the obvious candidates with ^1E_u excited states to be assigned to the structured band. Among the $3p\pi$ there is another yielding ^1E_u excited states: $1e_g \rightarrow 3p\pi$. In addition there are two allowed transitions yielding $^1A_{2u}$ excited states ($3a_{1g} \rightarrow 3p\sigma$ and $1e_g \rightarrow 3p\pi$) which are polarized along the C–C bond. Naturally for each singlet there is a triplet; as expected they are at slightly lower frequencies than the corresponding singlets. Since the excited Rydberg orbitals are large it makes little difference as to whether the excited electron has the same or the opposite spin as the one remaining in the originating orbital. Some of these bands were reported by Brongersma and Oosterhoff [76]. The $3a_{1g} \rightarrow \infty$ and $1e_g \rightarrow \infty$ ionization potentials would only differ by 0.03 eV, but this seems to be unrealistic. We have to remember that these calculations do not take account of the Jahn–Teller effect.

Buenker and Peyerimhoff also calculated the oscillator strengths of these different transitions. They found that the strongest of them is the $^1A_{1g} \rightarrow {}^1A_{2u}$, $3a_{1g} \rightarrow 3p\sigma$ transition with $f = 0,14$ (9.86 eV, parallel polarized). This is a high value for a Rydberg transition making the authors believe that the transition is mixed with a $\sigma \rightarrow \sigma^*$ valence transition of the same symmetry properties. The other parallel transition is considerably weaker ($f = 0.02$): it is $1e_g \rightarrow 3p\pi(e_u)$ at 9.99 eV.

Table 2. Calculated Vertical Excitation Energies (in eV) for Ethane by Buenker and Peyerimhoff[a]

Excitation	State	Polarization	Energy of Excitation (eV)	Oscillator Strength
Ground state	$^1A_{1g}$		0.0	
$3s \leftarrow 1e_g$	$^1E_{1g}$	Forbidden	9.16	
$3s \leftarrow 3a_{1g}$	$^1A_{1g}$	Forbidden	9.21	
$3p\sigma \leftarrow 1e_g$	^1E_u	x,y	9.91	$2 \times (0.02 - 0.03)$
$3p\sigma \leftarrow 3a_{1g}$	$^1A_{2u}$	z	9.86	0.14
$3p\pi \leftarrow 1e_g$	^1E_u	x,y	9.99	0.00
$3p\pi \leftarrow 1e_g$	$^1A_{2u}$	z	9.99	0.02
$3p\pi \leftarrow 1e_g$	$^1A_{1u}$	Forbidden	10.04	
$3p\pi \leftarrow 3a_{1g}$	^1E_u	x,y	10.00	$2 \times (0.02 - 0.03)$
$\infty \leftarrow 3a_{1g}$	$^2A_{1g}$		12.22	
$\infty \leftarrow 1e_g$	^2E_g		12.25	

Note: [a]Ref. 74. Reproduced by permission from Elsevier Science-NL.

$1e_g \rightarrow 3p\sigma$ and $3a_{1g} \rightarrow 3p\pi$ have f numbers of 0.02–0.03 per component, so together they would account for 0.08–0.12. The third perpendicular transition, $1e_g \rightarrow 3p\pi(e_u)$ seems to be very much weaker.

These results strongly suggest that more than one transition is active in the (8 to 10 eV) part of the spectrum including and underlying the structured band. This is also indicated by Custer and Simpson's band analysis [77] and also by one of Robin's suggestions [23].

Custer and Simpson [77] carried out an extensive study on the effect of internal rotation on electronic spectra. They presented the spectra of C_2H_6 and C_2D_6 recorded at temperatures ranging from −78 to 260 °C. From these they deduced an excited-state barrier height of 1.7 kcal/mol which compares favorably with the calculated value of 1.3 kcal/mol obtained by Salahub and Sandorfy [78]. Band contour analysis led them to a new assessment of the electronic assignments; while accepting that the structured band is perpendicularly polarized (^1E_u excited state), they suggested that the "transition lies over an intense parallel transition in the same region." This is entirely in line with the theoretical results of Buenker and Peyerimhoff [74]. The three following transitions are the most eligible: (1) $3a_{1g} \rightarrow 3p\sigma$ (to $^1A_{2u}$) which is predicted to be intense and should be diffuse since it originates from the C–C bond—it should contribute continuous intensity underneath the fine structured band; (2) the $3a_{1g} \rightarrow 3p\pi$ (to ^1E_u) transition should also yield a diffuse band (calculated to 9.99 eV) but it is perpendicularly polarized; (3) then $1e_g \rightarrow 3p\sigma$ (to ^1E_u, 9.91 eV) should be responsible for the fine structure.

It would be tempting to venture that both (2) and (3) are contributing to be observed fine structure; while this is possible, there is no evidence for this in the high resolution spectrum of Pearson and Innes [73]; the observed C–C stretching and CH_3 deformation quanta can well belong to the same electronic transition. This latter point needs further clarification. On the other hand the total oscillator strength of these transitions is about 0.28 in very satisfactory agreement with the observed 0.30 [36].

The calculated excitation energy for the strongest band ($^1A_{2u}$) is 9.86 eV while the experimental value is 9.4 eV. According to Buenker and Peyerimhoff this discrepancy can be explained by vibrational effects and reasons linked to the procedure of calculation.

We shall have to return to the assignments of the broad bands found in all the spectra beyond 10 eV.

The 3s Rydberg states are calculated to be about 0.65 eV lower than the 3p, in good agreement with the measured spectra.

All in all while certain afterthoughts remain to be dissipated, these calculations by Buenker and Peyerimhoff were highly informative. They terminate their paper by stating that "... the most fruitful direction for further calculations on this general subject lies in determining the nature of the potential surfaces of both ground and excited states of this system, and hence no longer restricting such studies to the nuclear conformation of the ground state at equilibrium." The writer would like to add that there would also be a need to measure the spectra at even higher resolution in order to know more about the vibrational and rotational structures of these rather elusive band systems.

As we have seen the structured band at 75,800 cm^{-1} (9.4 eV) has been the object of most of the research on the ethane spectrum. This is doubtless due to the intriguing vibrational fine structure of this band system. Several bands can be seen, however, at higher energies. While the optical spectra tend to be diffuse in that part of the spectrum, these bands are seen to advantage in the electron-impact energy-loss spectra. In the 1968 paper of Lassettre et al. [40] they were given at several accelerating voltages and scattering angles (ref. 40, figures 3 to 6). It clearly appears that there is another fine-structured band system at about 10.8, and possibly another between the 9.4 and 10.8 eV systems. Then structureless bands follow at about 11.5, 12.2, 13.2, and 13.8 eV. In this part of the spectrum we might expect transitions from $1e_g$ and $3a_{1g}$ to 3d, 4s, and 4p Rydberg orbitals, and also transitions originating from $1e_u$ and $\sigma \rightarrow \sigma^*$ valence-shell transitions.

Koch and Skibowski [44] measured the optical absorption spectrum of ethane from 8 to 35 eV using synchrotron radiation, as well as reflectance spectra of the solid phase. For the 9.4 eV band their results are in good agreement with those of Lombos et al. [36] and also with previous results by Schoen [79]. In the 10.8 eV band system they could identify seven vibrational peaks. These are followed by much stronger absorption which reaches its maximum at about 15 eV and then goes diminishingly up to about 30 eV. This spectral region is likely to contain all the transitions originating from the $1e_g$, $3a_{1g}$, $1e_u$, and $2a_{2u}$ and $2a_{2g}$ orbitals. Those from the K shell of carbon, $1a_{1g}$ and $1a_{1u}$, are expected to appear at much higher energies, around 290 eV. The spectrum of the solid is slightly shifted to higher frequencies and the vibrational structure is washed out. This can be taken as a confirmation of the mainly Rydberg character of this band. A large part of this absorption is beyond the first ionization potential (12.25 eV) and the limit of photodissociation [80, 81].

Lipsky and coworkers [41, 43] also did extensive work on the electron-impact spectra of ethane and other hydrocarbons. They noted the extreme sensitivity of these spectra to the scattering angle. They confirmed the appearance of the "first clearly discernible feature" at 8.56 eV (about 69,000 cm^{-1}) (the $3s$) and three structured absorption systems with envelope maxima at 9.40, 10.7, and 11.4 eV which we discussed before. The broad features lying at higher energies, 13.0, 13.7, and also 11.4 and 12.1 eV are separated by 0.7 eV. Interestingly this splitting is also found in the photoelectron spectrum and it is characteristic of Jahn–Teller distortion (two of 12.0, 12.7, and 13.4 for $1e_g$ and 15.0 and 15.7 for $1e_u$). Johnson et al. [43], "granting that the 0.7 eV intervals of both impact and photoelectron spectra have a common origin", suggested assignments which are in line with the previous discussion.

Before concluding this section we have to acknowledge the fact that in his monumental work on "The Higher Excited States of Polyatomic Molecules" Robin [8, 23] made thorough assessments on the state of our knowledge on the electronic spectra of ethane, first up to 1974 (Vol. I) and then up to 1985 (Vol. III).

C. Further Theoretical Work

Exploratory attempts to include higher (Rydberg) atomic orbitals into semiempirical quantum mechanical calculations were done by several groups. Salahub and Sandorfy [78] introduced the RCNDO (Rydberg-complete neglect of differential overlap) method with first-order configuration interaction and computed energy levels and charge distribution for a series of normal and branched alkanes. Saatzer et al. [82] studied the low-lying excited states of n-alkanes using a semi-empirical INDO method with configuration interaction. These methods interpreted well some of the spectral properties of those compounds and were probably helpful for those who later applied much more sophisticated methods to these problems. More recently Lindholm et al. [83] worked out a semiempirical, CNDO type method parametrized to yield the energies of σ*-orbitals. These authors interpreted the electronic spectra of saturated hydrocarbons in terms of σ → σ* valence transitions instead of transitions to Rydberg type orbitals. While for these molecules the lower excited states are admittedly of mixed Rydberg-valence character (but predominantly Rydberg), the present writer cannot go along with this interpretation of the alkane spectra. (For one thing, let us remember the great similarity of the vibrational fine structures of the photoelectron and ultraviolet absorption bands of ethane.)

Interesting contributions were made by Caldwell and Gordon [*84–86*]. Their basis sets consisted of STO 4G plus a set of one s and three p Rydberg functions centered at the midpoint of the C–C bond and simply excited configuration interaction was applied. In addition to excitation energies, they computed vibrational frequencies for the ground and excited states, potential energy surfaces for the low-lying states, Franck–Condon factors, as well as electronic charge densities in the Rydberg orbitals. Their results show that all vertical states below the first ionization potential (12.25 eV) are predominantly Rydberg in character, as are all minima detected on the potential energy surfaces. Table 3 taken from Caldwell and Gordon [*85*] compares the vertical energies of the states obtained by STO-4G$^+$, to those obtained by Buenker and Peyerimhoff [*74*], and also gives the net Mulliken populations in Rydberg basis functions. It is seen that the energies are consistently higher than those of [*74*]; the difference is 0.45 eV for $1E_g$, 0.90 eV for $^2A_{1g}$, 0.25 for $1E_u$, and 0.54 eV for $1A_{2u}$, in particular. If this corresponds to reality, it is an important result.

The ordering of the vertical states is the same as that determined by previous authors. However, upon geometry optimization within D_{3d} symmetry the $2A_g(3a_{1g} \rightarrow 3s)$ state undergoes a large degree of stabilization and becomes the lowest excited state at about 8.45 eV. The C–C band lengthens to 1.90 Å (!) and the methyl group becomes nearly planar (as in $C_2H_6^+$ [*66*]). The state is stable with

Table 3. Vertical Singlet States of Ethane by Caldwell and Gordon[a]

State	STOR-4G+	BP[b]	P_R^c	Dominant Configuration
		ΔE, eV		
$1A_{1g}$	0.0	0.0	0.017	c
$1E_g$	9.61	9.16	0.970	$1e_g \rightarrow 3s$
$2A_{1g}$	10.11	9.21	0.967	$3a_{1g} \rightarrow 3s$
$1e_u$	10.16	9.91	0.990	$1e_g \rightarrow 3p\sigma$
$1A_{2u}$	10.40	9.86	0.984	$1e_g \rightarrow 3p\pi$
$2E_u$	10.48	9.99	0.994	$1e_g \rightarrow 3p\sigma$
$1A_{1u}$	10.50	10.04	0.991	$1e_g \rightarrow 3p\pi$
$2A_{2u}$	10.69	9.99	0.991	$1e_g \rightarrow 3p\sigma$
$3E_u$	10.69	10.00	0.981	$3a_{1g} \rightarrow 3p\sigma$ $3a_{1g} \rightarrow 3p\pi$
IP(^2E_g)	12.86	12.25		$1e_g \rightarrow \infty$

Notes: [a] Ref. 85. Reproduced by permission from the American Chemical Society.

   [b] BP: Buenker and Peyerimhoft [*74*].

   [c] P_R: Net Mulliken population in Rydberg basis functions.

respect to the D_{3d} minimum for all internal coordinates; ^1E_g being doubly degenerate is distorted by the Jahn–Teller effect. In C_{2h} symmetry ^1E_g splits into $2A_g$ and $1B_g$. The $1B_g$ state is still unstable to further distortion to C_s where it becomes $1A$. It has one pair of long C–H bonds while the other CH_3 group has normal geometry. This would predispose the molecule to photodissociate into H_2 and ethylidene ($HCCH_3$). As the two hydrogen atoms pull away the energy reaches a maximum at 1.52 Å and becomes sharply repulsive for larger distances. At the same time the highest orbital changes gradually from Rydberg to valence antibonding ("derydbergisation"). All this concerns the $3s$ type states which are assigned to the weak band at about 8.65 eV; these are expected to be the photochemically active states.

As to the vertical states $1E_u$ and $1A_{2u}$ the calculations suggest that they are unstable and are bound to distort into C_{2h} symmetry, "wherein they are characterized by significantly lengthened C–C bonds. The vibrational structure does appear to be due to the overlapping progressions from two vertical ($1E_u$ and $1A_{2u}$) states. The progressions are in the C–C stretching mode of the excited molecule, the molecular structure has the very long C–C bond and nearly planar methyl groups characteristic of ethane ($C_2H_6^+$) ionic structures. This is not surprising since all minima are strongly Rydberg in character" [86].

This may appear to be revolutionary. It has been hitherto supposed that upon excitation from the e_g orbital which is C–C antibonding this bond must *shorten* not lengthen. Furthermore, the observed fine structure has always been interpreted as belonging to only one excited state, not two. To alleviate these objections let us say that our intuitive ideas are based on the early low-resolution spectra which seemed to indicate that the observed progression is in a CH bending (decreased with respect to the ground state) or a C–C stretching (increased) frequency which may be mutually overlapped. Pearson and Innes [73] were able to distinguish them, but ν(C–C) *did not* increase. They found, for C_2H_6 $\nu_2' = 1053$ cm^{-1} (1388.4 for the ground state) and $\nu_3' = 755$ cm^{-1} (994.8 for the ground state); for C_2D_6 they found $\nu_2' = 781$ cm^{-1} (1154.5 for the ground state) and $\nu_3' = 616$ cm^{-1} (843.0 in the ground state), where ν_2 is a CH deformation frequency and ν_3 is the C–C stretching frequency. They gave 2215 (2953.7) and 1751 (2083.0) for the CH stretching frequency.

If these vibrational frequencies belong to the same excited electronic states, then clearly the C–C stretching frequencies did not *increase* as would be expected upon excitation from the $1e_g$ orbital and the C–C bond did not become *shorter*. However, if the symmetry of the excited state is only C_{2h} (or less), it is conceivable that the C–C bond which would become shorter when the electron leaves the $1e_g$ orbital would still become longer, even much more longer when the lowering of symmetry is achieved (cf. Caldwell and Gordon [84–86]). Then how about the Jahn–Teller effect? Pearson and Innes did not find any sure manifestation of it and so suppose that it must be very slight and so the observed vibrational fine structure can be interpreted in terms of totally symmetric vibrations only.

If, on the other hand, the vibrational fine structure is due to two different, intermingling electronic states, the two vibrations could both be C–C stretching frequencies, the one increased (1053 cm^{-1}) and the other decreased (755 cm^{-1}). The first one could correspond to excitation from $1e_g$, the second one from $3a_{1g}$. This may appear to be satisfactory to the mind, but then why do we not see the CH deformations (ν_2) (that there be no changes in HCH angles is hard to believe) and why do we see only one CH stretching frequency (ν_1)?

Clearly, the extent and influence of the Jahn–Teller effect should be examined in more detail, both theoretically and experimentally.

The ultimate theoretical effort which is presently available is due to Buenker's group (Chantranupong et al. [87]). These authors carried out ab initio multirefer-ence single- and double-excitation configuration interaction calculations (MRD-CI) for the ground and 32 electronic excited states of ethane, as well as its two lowest ionic states, ^2E_g and $^2A_{1g}$. A double-zeta-plus-polarization atomic orbital basis was employed in a large configuration interaction treatment. The basis was augmented by diffuse functions of s and p type. These highly advanced calculations confirmed the Rydberg character of the lower excited states of ethane and also the near-degeneracy of the $1e_g$ and $3a_{1g}$ frontier orbitals. The D_{3d} (staggered) geometry was adopted but for the excited states the calculations were carried out in the C_{2h} geometry what amounts to a partial account of the Jahn–Teller effect. The calcula-tions extended to 32 states including triplets and the $4s$ and $4p$ Rydberg states, the latter for the first time. The $3s$ excitation energy is 0.57 eV lower for $1e_g$ than for $3a_{1g}$ but the difference is only 0.27 for the $3p\sigma$ and even smaller for the $3p\pi$. The (vertical) $3a_{1g}$ and $1e_g$ ionization potentials are 12.90 and 12.67 eV, respectively, higher than in previous calculations. This, the authors attribute mainly to the polarization functions included in the basis set.

An important result is that the energies of the excited states are higher in this treatment than in Buenker and Peyerimhoff's previous treatment [74] and higher even than those of Caldwell and Gordon [85]. The vertical center of the structured band system is experimentally 9.4 eV ($75,800 \text{ cm}^{-1}$). Table 4 taken from [87] compares the new results to those obtained earlier by Buenker and Peyerimhoff and by Caldwell and Gordon. For the degenerate states E_g and E_u two C_{2h} components are given, A_g/B_g and A_u/B_u, respectively. GSMO stands for ground-state molecular orbital basis set and TSMO for triplet state molecular orbital basis set. The strongest transition, $3a_{1g} \to 3p\sigma$, $^1A_{1g} \to {}^1A_{2g}$ excitation is at 10.58 eV with an oscillator strength $f = 0.115$. This is about 1.2 or 1,4 eV higher than the experimental 9.4 eV and although the previous results were also higher (but less), Buenker et al. [87] argue that 1.2 is well above the possible error inherent in the calculations and must largely reflect reality. If this is so the strongest transition cannot be assigned to the center of the structured band (9.4 eV) but must belong to the next strong band at 10.7 eV. The same applies to all the $3a_{1g} \to 3p\sigma$, $3a_{1g} \to 3p\pi$, $1e_g \to 3p\sigma$, and $1e_g \to 3p\sigma$ transitions which still fall close together . One of these could account for the second structured band, probably $1e_g \to 3p\sigma$. If so, then the first structured

Table 4. Computed Vertical Excitation Energies to Various Singlet States of Ethane by Chantranupong et al.[a]

Excitation	State	Earlier Results		This Work GSMO
		[5]	[6]	
$1e_g - 4a_{1g}(3s)$	^1E_g	9.16	9.61	9.373/9.388
$3a_{1g} - 4a_{1g}(3s)$	$^1A_{1g}$	9.21	10.11	9.933
$1e_g - 3a_{2u}(3p\sigma)$	^1E_u	9.91	10.16	10.310/10.310
$3a_{1g} - 3a_{2u}(3p\sigma)$	$^1A_{2u}$	9.86	10.40	10.578
$1e_g - 2e_u(3p\pi)$	$^1A_{2u}$	9.99	10.69	10.637
	^1E_u	9.94	10.69	10.641/10.654
	$^1A_{1u}$	10.04	10.50	10.698
$3a_{1g} - 2e_u(3p\pi)$	^1E_u	10.00	10.48	10.742/10.748
$1e_g - 5a_{1g}(4s)$	^1E_g			11.239/11.242
$1e_g - 4a_{2u}(4p\sigma)$	^1E_u			11.573/11.582
$3a_{1g} - 5a_{1g}(4s)$	$^1A_{1g}$			11.591
$1e_g - 3e_u(4p\pi)$	$^1A_{2u}$			11.827
	^1E_u			11.810/11.827
	$^1A_{1u}$			11.838
$3a_{1g} - 4a_{2u}(4p\sigma)$	$^1A_{2u}$			11.866
$3a_{1g} - 3e_u(4p\pi)$	^1E_u			11.920/11.921

Note: [a] Ref. 87. Reproduced by permission from Elsevier Science-NL.

band (9.4 eV) should be assigned to the transitions to $3s$ orbitals. These are, of course, parity forbidden but Buenker et al. are of the opinion that in the excited state the molecule might be deformed so that it looses its center of symmetry, making the transitions allowed. (This supposes a large deformation, however. It also means that the symmetry should become less than C_{2h}.) Then what is the weak band or shoulder at the low frequency end of the spectrum which we thought to be the $3s$? Well, they could be the adiabatic (relaxed) part of $3s$ bands, now centered at 9.4 eV. While this is quite possible, it is not evident. We are not at the end or our worries.

According to Buenker et al. these changes seem to be caused primarily by the presence of d functions in the atomic orbital basis chosen for their calculations.

The $3a_{1g}$ orbital turned out to be 0.3–0.6 eV more stable than $1e_g$ which is probably exaggerated. As expected, the triplets differ only slightly from the related singlets. The $4s$ and $4p$ levels were computed for the first time; their oscillator strengths are about 40% of the $3s$ and $3p$ and could therefore make a significant contribution to the observed intensity above 11 eV. It should be remembered that there are intense and diffuse bands in that part of the spectrum.

The $3a_{1g} \rightarrow 3p\sigma$ transition to an $^1A_{2u}$ upper state remains the strongest ($f = 0.115$) while the transitions to the ^1E_u upper states, $3a_{1g} \rightarrow 2p\pi$ and $1e_g \rightarrow 3p\sigma$ have f values between 0.05 and 0.06. The $3d$ levels should be close to the $4s$. Since these too are parity forbidden they could make a significant contribution to the intensity observed above 11 eV if the center of symmetry is done away with.

The new results by Buenker et al. [87] constitute a challenge to the interpretation of the ethane spectrum what became "traditional" in the last 30 years. It clearly indicates the need for more theoretical work with an even more extended basis set and a computation of the potential surfaces for all the relevant vibrational motions, implying a closer look at the consequences of the Jahn–Teller effect in the degenerate states as well as possible vibronic couplings between close-lying excited states. Equally needed are experimental works carried out at even higher resolution than has hitherto been possible; more information on the vibrational and rotational structures are badly needed. This concerns not only the absorption spectra but also the electron-impact and photoelectron spectra.

D. The Link to Photochemistry

The connecting link between molecular spectroscopy and photochemistry is the photolytical primary step, the changes that the molecule undergoes during the very short interval in which the energy of the photon becomes a part of the energy of the molecule.

Photolysis, in most cases, takes place at the lowest excited state of a given molecule. Since this is so it is important to realize that the lowest excited state might be of the valence-shell, intermediate, or Rydberg type. Examples for molecules whose lowest excited state is of the valence-shell type are the aromatic molecules (π^* or V states), coordination complexes of transition metals, and alkyl-halides. The lowest excited states of saturated hydrocarbons are best described as Rydberg, those of alcohols and ethers as intermediate. The lowest excited state of ketones and aldehydes is valence-shell (n, π^*) but it is followed by two or three Rydberg states before the next valence-shell (π, π^*) state is reached. It has been suggested that these conditions must have a bearing on the photochemical primary process that follows photon absorption. While every case must be treated according to its particular characteristics it can be expected that the primary process will be different for photolysis occurring in Rydberg states and in valence-shell states [88–90].

Typically, in cases of photolysis, in a valence-shell state an electron is excited from a bonding (or nonbonding) to an antibonding molecular orbital. This usually leads to the cleavage of a bond with the production of free atoms or radicals. With Rydberg states the excitation is generally from a bonding (or nonbonding) orbital of the ground state to an essentially nonbonding Rydberg orbital. With organic molecules, the lowest Rydberg orbital is $3s$ so we have to examine the role that it can play.

In the case of ethane it is observed that the $3s$ orbital encompasses all protons. Furthermore, since s-type orbitals are totally symmetrical the sign of the wavefunction is the same all over and around the molecule. Thus a Rydberg orbital can be considered as a one-electron bond creating weak bonds between nonbonded atoms. These remote interactions might become important in a dissociating molecule, the other interactions in the molecule being weakened. They would help in bringing together nonbonded atoms and lead to primary processes consisting in molecular elimination rather than simple bond cleavages. Actually, the photolysis of ethane and of all normal paraffins yields predominantly H_2 molecules and not H atoms as primary products.

From the many important papers we mention only those of Hampson et al. [91, 92], Hampson and McNesby [93], and Lias et al. [94] who examined the photolysis of ethane using the xenon (8.4 eV), krypton (10.0 eV), and argon (11.6–11.8 eV) resonance lines, respectively. The following discussion will be based on the conclusions of the last paper and of a review by Lias and Ausloos [95]:

(1) $C_2H_6 \rightarrow CH_4 + CH_2$
(2) $C_2H_6 \rightarrow C_2H_5 + H$
(3) $C_2H_6 \rightarrow C_2H_4 + H_2$
(4) $C_2H_6 \rightarrow CH_3 + CH_3$

At 8.4 eV 85% of the decomposition is due to (3); an astonishing result. Is it credible that this reaction occurs in a Rydberg state resembling the $2A_{1g}$ ion which has a long C–C bond length but near normal C–H bond lengths? To the contrary, the great importance of reaction (3) is readily understood if the excited state is the ^1E_u state obtained through excitation from the $1e_g$ orbital. Looking only at this orbital the excitation should result in a shortening of the C–C bond and even if subsequent rearrangement or the addition of the Rydberg orbital lengthens it somewhat it will still be a bond well prepared for the formation of excited ethylene with departure of the hydrogens from the weakened C–H bonds.

The relative importance of the primary steps depends on the energy of the exciting photons but (2) and (3) are always preponderant. That C–H bonds and not the C–C bond is broken is well in line with the assumption that photolysis takes place in a state obtained through excitation from the mainly C–H bonding and C–C anti-bonding $1e_g$ orbital and not from the C–C bonding $3a_{1g}$. Furthermore, at 8.4 eV (67,750 cm^{-1}, the xenon resonance line) 85% of the photolysis yields H_2 in the primary step, not H. Since at 8.4 eV only the $3s$ Rydbergs can be excited this shows that the relatively large $3s$ orbital favors the extraction of *molecular* hydrogen. All six hydrogen atoms being connected by symmetry both 1,1 and 1,2 H_2 extraction could be expected with the shorter H-H distance favoring 1,1. Actually the latter process is preponderant.

We still have to face the difficulty which we have already encountered. Thinking in terms of excitation from an orbital that is, tacitly applying Koopmans' theorem, one would expect that upon excitation from $1e_g$ the C–C bond will shorten, not lengthen. Then how come the C–C stretching frequency (ν_3) is actually lower in the excited state as was found by high resolution work [73]? Actually, the work of Pearson and Innes concerned the structured band of ethane and not the weak band or tail at 8.65 eV, which evidently still has enough intensity at 8.4 eV to be photochemically active. At the center of the structured band, 9.4 eV, we have either $3p$ or the vertical part of $3s$ but the initiating orbital must be taken to be $1e_g$ to understand the overwhelming production of hydrogen. At 8.4 eV we can only have $3s$ or the adiabatic part of it.

However it may be, this shows the limits of the Koopmans' way of thinking. Saatzer et al. [82] insisted on the fact, that "a more quantitative approach is to carry out complete geometry optimizations for each state of interest followed by an examination of the bonding in bound states." In addition to consideration of the bonding—nonbonding—antibonding character of orbitals an additional principle is needed. We have to take some account of electronic rearrangement which occurs upon ionization or excitation. This rearrangement generally tends to compensate for too great changes in bond strength, as has been postulated by Lathan et al. [55] in the case of the positive ion of ethane. The present writer elaborated on this on various occasions [88–90].

The disconcerting conclusion is that we are still not at the end of the ethane saga; neither theoretically, nor experimentally.

III. ETHYLENE

Ethylene has been chosen for our second case study in far-ultraviolet Rydberg-valence spectroscopy. While more details are known about ethylene than about ethane, the main points can be treated more succinctly. The complications in the case of ethane were due to the existence of close-lying initiating σ-orbitals. Yet, for ethylene too, problems remain.

The main point is that the lowest lying valence (π^*) and Rydberg ($3s$) orbitals do not have the same symmetry and therefore (at least under D_{2h} symmetry) they do not mix. The $3s$ state is then seen to be truly Rydberg and does not become Rydberg at the united atom limit only. As in the case of ethane the classic photoelectron spectrum came from Turner's group [25, 57] (Figure 6). They found five bands corresponding to five occupied orbitals in the neutral ground state: at 10.51, 12.38, 14.4(7), 15.6(8), and 18.8(7) eV, respectively. Among these the 10.51 eV band corresponds to the π-orbital and 12.38 eV to the highest occupied σ-orbital. All these bands have vibrational fine structure and all the above values are adiabatic although only for the first two were the 0–0 bands clearly resolved.

In its ground state, ethylene is coplanar with D_{2h} symmetry. The sequence of the orbitals is, taking the Z axis along the C–C bond and the X axis out-of-plane:

$$(\sigma 1a_g)^2 \, (\sigma 1b_{1u})^2 \, (\sigma 2a_g)^2 \, (\sigma 2b_{1u})^2 \, (\sigma 1b_{2u})^2 \, (\sigma 3a_{1g})^2 \, (\sigma 1b_{3g})^2 \, (\pi 1b_{3u})^2 \, {}^1A_{1g}$$

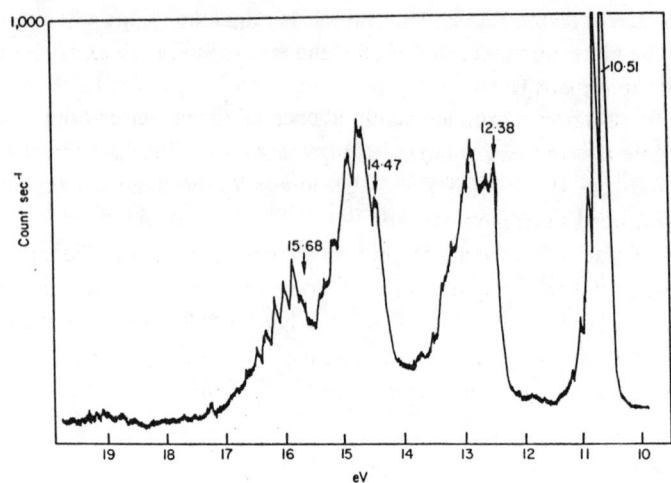

Figure 6. The UV photoelectron spectrum of ethylene. Baker et al. [*25, 57*]. Reproduced by permission from Elsevier Science-NL.

This was called the N (normal, or ground-) state by Mulliken. The vibrations which are relevant for this discussion are:

> ν_1 totally symmetric C–H stretch, 3026 cm^{-1} (a_g)
> ν_2 totally symmetric C–C, 1623 cm^{-1} (a_g)
> ν_3 C–H deformation, 1342 cm^{-1} (a_g)
> ν_4 twisting, 1023 cm^{-1} (a_u)

For the ground state of the positive ion ($^2B_{3u}$), 10.51 eV band of the photoelectron spectrum, only two frequencies appear clearly: ν_2 1230 ± 50 cm^{-1} and ν_4 430 ± 50 cm^{-1}.

The low frequency of ν_2 (C–C) and the prominence and very low frequency of the twisting vibration indicate that the positive ion is twisted in its ground state, so that the two CH$_2$ groups are no longer in the same plane (see below).

Moreover, it is remarkable that at least four bands in the photoelectron spectrum exhibit vibrational fine structure. Thus, the ion possesses as many stable excited states. Only one of them, the first one is due to ionization from a π-orbital. (This makes the fine structure observed in both the photoelectron and electronic spectra of *ethane* less surprising.) ν_1, ν_2, and ν_3 are Raman active, but ν_4 is both Raman and infrared inactive and its frequency had to be determined by indirect methods (ref. 96).

The bonding and symmetry properties of the ground state occupied and the three lowest empty orbitals are shown in Figure 7 taken from Merer and Mulliken [*96*].

(The K shell has been omitted.) These authors summed up all the existing knowledge about the electronic spectrum of ethylene up to 1969; their extensive review is still highly informative.

The (π, π^*) transition is from the bonding π-orbital to the antibonding π^*-orbital and it is dipole allowed and polarized along the Z axis. Since the direct product $(b_{3u} \times b_{2g}) = B_{1u}$ and $[b_{3u} \times 3s(a_{1g})] = B_{3u}$ the lowest Rydberg and valence states are seen to have different symmetries.

In its ground state, ethylene has D_{2h} symmetry, but when it is twisted this reduces to D_2 and when the twist angle is 90° it becomes D_{2d}. For the latter the $1b_2$ and $1b_3$ molecular orbitals become equal and together form a degenerate e orbital. (It is often said that for having degeneracy at least a threefold axis is needed; well, the D_{2d} group is an exception to this.) It is subject to the Jahn–Teller effect.

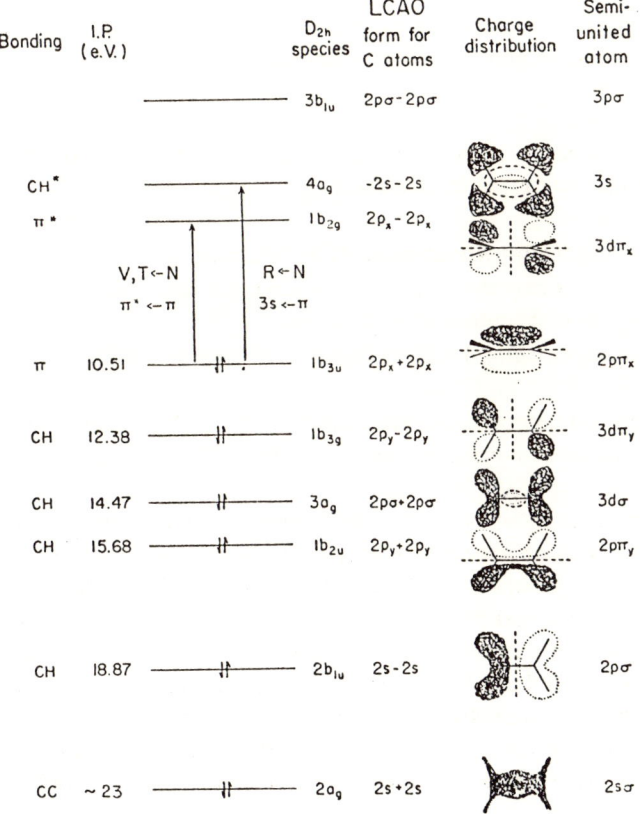

Figure 7. Molecular orbitals for the ethylene molecule (after Merer and Mulliken [96]). Reproduced by permission from the American Chemical Society.

The states resulting from the (π, π^*) transition are called V (the singlet), T (the triplet), and Z (when both π-electrons are in the π^*-orbital). For 90° twisted ethylene the π- and π^*-orbitals ($1b_{3u}$ and $1b_{2g}$) form the degenerate $2e$ and become nonbonding, instead of being bonding and antibonding, respectively. As Mulliken [98] pointed out, however, this does not mean that the C–C bond becomes extremely weak; the π-bond will be partly compensated for by strong hyperconjugation between the CH_2 group on the one side and the π-AO on the other. The N state has the minimum of its potential energy at 0° torsional angle, while for the V and T states it is probably at 90° and 270°. Actually, perpendicular ethylene would have a triplet ground state (like O_2).

The geometry of ethylene in its N ground state (D_{2h}) has been accurately determined by electron diffraction and high-resolution infrared and Raman measurements. For the zero-point level r_0 (C–C) = 1.338 ± 0.003 Å, r_0 (C–H) = 1.086 ± 0.002 Å, and < CCH = 121.3 ± 0.50 Å [99–103].

The optical absorption spectra of ethylene were measured with good resolution by Zelikoff and Watanabe [104] and the energy-loss spectra by Geiger and Wittmaack [105] (Figure 8). These were followed by the works of Wilkinson and Mulliken [105, 106]. (References to earlier works are given in these publications.) Merer and Mulliken [96, 97] and Merer and Schoonveld [108, 109] analyzed them thoroughly and their conclusions are still essentially correct. The $N \rightarrow V$ band begins at about 48,000 cm^{-1} as an extremely weak progression in the v_2 (C–C) vibration of the V state, 800 cm^{-1}; it increases gradually in intensity and eventually becomes a broad, diffuse continuum with maximum at about 61,700 cm^{-1} (1620 Å, 7.65 eV). Each member of the progression is combined with two quanta of the v_4 twisting vibration whose frequency is about 400 cm^{-1} very much lower than in the ground state showing that the suppression of the π part of the C=C bond makes twisting much easier. Previously it was thought that the progression begins at even lower frequencies (about 38,000 cm^{-1}) but McDiarmid [110–112] showed that this long tail was due to traces of oxygen.

The low-frequency end of the spectrum then corresponds to the adiabatic (90° twisted) geometry of the V state, whereas the broad maximum belongs to the vertical transition for which the molecule is still coplanar [97]. The 0–0 band is extremely weak because of the large change of the twisting angle between the N and V states. "Every twisting band in the spectrum is the origin of a progression in the upper state C–C stretching vibration, whose intensity distribution will depend on the average value of the C–C distance for the twisting level in question." Thus, in Merer and Mulliken's interpretation, both v_2 and v_4 are needed to explain the observed vibrational structure [97].

Foo and Innes offered a quite different interpretation [113]. They determined the spectra of C_2H_4, C_2D_4, and several partly deuterated species at high resolution, using the first order of a 21 ft vacuum spectrograph (cf. also the previous work by Schmitt and Brehm [114]).

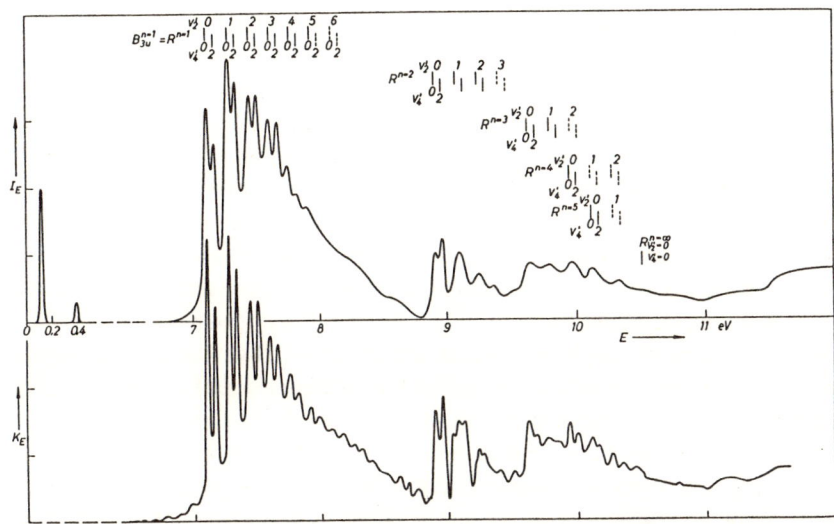

Figure 8. The energy-loss (*upper curve*) and optical absorption spectra (*lower curve*) of ethylene (after Geiger and Wittmaack [*105*] and Zelikoff and Watanabe [*104*]). Reproduced by permission from Verlag von Zeitschrift für Naturforschung.

For the $N \to V$, π,π^*-system they state that: "Most striking is the relative inactivity of the C–C stretching vibration; except in C_2D_4, only the torsional mode is active." They concluded that the progression forming frequency is ν_4' (of the V state), as did also McDiarmid and Charney [*110*]. (Only double jumps in ν_4' appear since ν_4 is not a totally symmetric vibration.) They locate the 0–0 band at 5.5 eV (2285 Å, 44,360 cm^{-1}) and obtain only 1.41 Å for the C–C distance in the V state. While the latter value agrees with the results of theoretical calculations [*115*] and are close to Mulliken's estimate of 1.44 Å [*98*], McDiarmid's value for the 0–0 band is higher, 48,400 cm^{-1} and Petrongolo et al.'s calculation also favor the higher value [*115*]. Foo and Innes [*113*] state that ". . . there is no direct experimental evidence about the electronic assignment of what we have called the π^*-π state;" nor could they confirm that in the V state ethylene is 90° twisted. (But the writer believes in it and in the $^1A_g \to {}^1B_{1u}$ assignment. There is a great deal of theoretical evidence for it.)

As to the transition to the $3s$ Rydberg state Foo and Innes [*113*] carried out band contour analysis—both rotational and vibrational—and confirmed the $^1A_g \to {}^1B_{3u}$ assignment. They found $\nu_{CC} = 1.41$ Å, $\nu_{CH} = 1.08$ Å, HCH = 124.4°, and an azimuthal angle of 37°. The 0–0 band is at 57,338 cm^{-1} (1747 Å, 7.1 eV) atop the $N \to V$ band. Both ν_2' and ν_4' are present in the vibrational fine structure; for ethylene C_2H_4 ν_2 is 1393 cm^{-1}.

The transitions to the triplet states are known and are seen to advantage in energy-loss spectra. The vertical value for the (π, π^*) triplet is 35,200 cm^{-1} (or 4.36 eV) and for the $(\pi, 3s)$ triplet it is at 56,300 cm^{-1} (6.98 eV) [*117*].

In addition to these bands a number of other Rydberg bands have been located. They are of s, p, and d types and all converge to the π-ionization potential. Robin tabulated these in a most helpful manner [ref. 8, p. 214]. A few Rydberg transitions belonging to higher ionization potentials are also known. The p and d series exhibit core splitting and give px and py series, as well as $d\sigma$, $d\delta$, and dxz. The py are forbidden being $u \leftrightarrow u$ for D_{2h} symmetry but they become increasingly allowed with the twisting of the molecule.

While we cannot go into more detail two final points must be made.

As stated above the π^* and $3s$ orbitals do not mix. However, π^* is mixing with one component of the $3d$ manifold which has the same symmetry, namely $3d_{xz}$ [118–119]. There is an important admixture of the $^1(\pi, 3dxz)$ Rydberg configuration into $^1(\pi, \pi^*)$, 45–50% according to Buenker and Peyerimhoff. Yet, for most purposes, the V state can still be considered as a valence state. The mixing of Rydberg and valence excited states is a fact of life in far-ultraviolet spectroscopy.

In certain highly substituted derivatives of ethylene, like tetramethyl [117–120], or di-, tri-, and tetrafluoroethylenes [121] there is a well pronounced band at frequencies lower than that of the singlet $N \rightarrow V$ band. For some time the origin of these bands was unknown and in a semihumoristic manner they were called the "mystery band". At present we know that these are just the $(\pi, 3s)$ Rydberg bands which for such molecules pass to the low frequency side of the (π, π^*) band.

The vibronic structure of the triplet $^3(\pi, \pi*)$ T state is significantly different from that of the singlet V state. Wilden and Comer [122] using high-resolution energy-loss spectroscopy found a long progression in ν_2' (C–C) which dominates the vibrational fine structure, whereas in the singlet ν_4' does. Single jumps in ν_4' which become allowed as the molecule twists are also present.

Wilden and Comer [123] also studied the triplet $^3(\pi, 3s)$ Rydberg state. The origin of the band system is at $56,300 \text{ cm}^{-1}$ (6.98 eV). They found $\nu_2' = 1370 \text{ cm}^{-1}$ for the singlet and 1450 cm^{-1} for the triplet, and $2\nu_2' = 468 \text{ cm}^{-1}$ for the singlet and 400 cm^{-1} for the triplet.

Robin [8, 124] reviewed the problems connected with the spectra of ethylene very thoroughly at two occasions: first in his Volume II in 1975 and then in his Volume III in 1985. The interested reader will find there a great deal more information than what could be taken into the present chapter.

Parallel to experimental developments the theoretical struggle also continued for the conquest of ethylene and played an important role in the evolution of our ideas on the ethylene problem [17, 115, 116, 118, 119, 125–128].

Despite all the good work that has been done, the spectrum of ethylene is still not fully understood. In particular there is no consensus relating to the details of the vibrational fine structure. Perhaps more important than these details is the fact, however, that both ν_2' and ν_4' are involved, so upon (π, π^*) and $(\pi, 3s)$ excitation the C–C bond lengthens (moderately) and the molecule undergoes twisting. Excited ethylenes are not coplanar.

The photochemical primary steps were considered by the writer in his review written in 1979 [*90*]. They all lead to hydrogen production, not to cleavage of the C–C bond. This is understandable since while both C–H and C–C bonds are weaker in the excited states, the sigma part of the double bond remains of course intact. More H_2 is produced than H atoms. Borrell et al. [*129*] found an output of 68% for H_2 production at 185 nm. This is compatible with photolyzis at the perpendicular *V* state. It is possible, however, that a part of the molecules photolyzes at the 3*s* Rydberg state. The totally symmetrical 3*s* orbital would help with the formation of H_2 [*88–90*].

In the triplet states the main primary process is still H_2 formation, both 1,1 and 1,2 [*130*].

IV. CONCLUDING REMARKS

Ethane and ethylene were chosen as the targets of this chapter since they are starting points for the discussion of the spectra and excited states of most organic molecules. Ethane is a prototype for spectra due to the excitation from a bonding σ-orbital, ethylene for those of π-electrons. The third type of important initiating orbitals are those of lone pairs of electrons in both saturated and nonsaturated molecules. Rydberg and valence type excited states intermingle in the part of the spectrum (about 6 to 10 eV) which are of main concern to organic chemical spectroscopists. This is photochemically very important and is yet to be taken fully into account.

Excited molecules change their structures with respect to their structures in their respective ground states to a variable extent, in probably all cases. Except for a few small molecules, these changes have been only qualitatively assessed. Much remains to be done.

Yet we might say, in perspective, that a great deal of progress has been made in recent decades. This is due to the concerted efforts of theoretical chemists, electronic and photoelectron spectroscopists and photochemists, and also to the spectacular improvements in computational and experimental techniques. It is hoped that in the times to come this type of knowledge will penetrate the public opinion of scientists.

Most of the progress reported in this chapter was made in the late 1960s and in the 1970s. Modern high-resolution photoelectron techniques (like ZEKE) and greatly improved computational facilities would warrant a new generation of investigations on the properties of ethane and ethylene in their different states of excitation. This could be of fundamental importance for the understanding of the spectra and photochemical reactions of more complicated molecules.

ABBREVIATIONS

MO molecular orbital
AO atomic orbital

LCAO   linear combination of atomic orbitals
PE   photoelectron

REFERENCES

1. Mulliken, R. S. *Acc. Chem. Res.* **1976**, *7*, 1057.
2. Herzberg, G. *Spectra of Diatomic Molecules,* 2nd ed; Van Nostrand: New York, 1950, Vol. 1.
3. Sinanoglu, O. In *Chemical Spectroscopy and Photochemistry in the Vacuum Ultraviolet;* Sandorfy, C.; Ausloos, P.; Robin, M. B., Eds.; Reidel: Dordrecht, Holland, 1974, p. 337.
4. Jortner, J.; Gaathon, A. *Can. J. Chem.* **1977**, *55*, 1801.
5. Resca, L.; Resta, R. *Phys. Rev.* **1979**, *19B*, 1683.
6. Goodman, J.; Brus, L. E. *J. Chem. Phys.* **1977**, *67*, 933.
7. Goodman, J.; Brus, L. E. *J. Chem. Phys.* **1978**, *69*, 4083.
8. Robin, M. B. *Higher Excited States of Polyatomic Molecules;* Academic Press: New York, Vol III, 1985, p. 72.
9. Peyerimhoff, S. D. *Gazz. Chim. Ital.* **1978**, *108*, 411.
10. Buenker, R. J.; Peyerimhoff, S. D. *Theor. Chim. Acta (Berl.)* **1968**, *12*, 183.
11. Buenker, R. J.; Peyerimhoff, S. D. *Theor. Chim. Acta (Berl.)* **1974**, *35*, 33.
12. Buenker, R. J.; Peyerimhoff, S. D. *Theor. Chim. Acta (Berl.)* **1975**, *39*, 217.
13. Buenker, R. J.; Peyerimhoff, S. D.; Whitten, J. L. *J. Chem. Phys.* **1967**, *46*, 2029.
14. Peyerimhoff, S. D.; Buenker, R. J. In *Chemical Spectroscopy and Photochemistry in the Vacuum Ultraviolet;* Sandorfy, C.; Ausloos, P; Robin, M. B., Eds.; Reidel: Dordrecht, Holland, 1974.
15. Peyerimhoff, S. D.; Buenker, R. J.; Kramer, W. E.; Hsu, H. L. *Chem. Phys. Lett.* **1971**, *8*, 129.
16. Buenker, R. J.; Peyerimhoff, S. D. *Chem. Phys. Letter* **1974**, *29*, 253.
17. Fischbach, U.; Buenker, R. J.; Peyerimhoff, S. D. *Chem. Phys.* **1974**, *5*, 265.
18. Shih, H.; Buenker, R. J.; Peyerimhoff, S. D.; Wirsam, B. *Theoret. Chim. Acta (Berl.)* **1970**, *18*, 277.
19. Mulliken, R. S. *J. Am. Chem. Soc.* **1964**, *86*, 3183; **1966**, *88*, 1849; **1969**, *91*, 4615.
20. Mulliken, R. S. *Acc. Chem. Res.* **1976**, *9*, 7.
21. Mulliken, R. S. *J. Chem. Phys.* **1977**, *66*, 2448.
22. Herzberg, G. *Electronic Spectra and Electronic Structure of Polyatomic Molecules*; Van Nostrand: Princeton, 1966, Vol. I.
23. Robin, M. B. *Higher Excited States of Polyatomic Molecules* . Vol. I. Academic Press, New York, 1974, Vol. I, Chap. 1.
24. Walsh, A. D. *J. Phys. Radium* **1954**, *15*, 501.
25. Turner, D. W.; Baker, A. D.; Baker, C.; Brundle, C. R. *High Resolution Molecular Photoelectron Spectroscopy*; Wiley: New York, 1970.
26. Al-Joboury, M. I.; Turner, D. W. *J. Chem. Soc.* **1963**, 5141.
27. Turner, D.W.; May, D. P. *J. Chem. Phys.* **1966**, *45*, 471.
28. Koopmans, T. *Physica* **1934**, *1*, 104.
29. Buenker, R. J.; Peyerimhoff, S. D. *Chem. Revs.* **1974**, *74*, 127.
30. Sandorfy, C. *J. Mol. Struct.* **1973**, *19*, 183.
31. Mulliken, R. S. *J. Chem. Phys.* **1935**, *3*, 517.
32. Jahn, H. A.; Teller, E. *Proc. Roy. Soc. London* **1937**, *161A*, 220.
33. Palke, W. E.; Lipscomb, W. N. *J. Am. Chem. Soc.* **1966**, *88*, 2384.
34. Moccia, R. *J. Chem. Phys.* **1964**, *40*, 2164.
35. Hoffmann, R. *J. Chem. Phys.* **1963**, *40*, 2047.
36. Lombos, B. A.; Sauvageau, P.; Sandorfy, C. *J. Mol. Spectrosc.* **1967**, *24*, 253.
37. Lombos, B. A.; Sauvageau, P.; Sandorfy, C. *Chem. Phys. Lett.* **1967**, *1*, 42.
38. Raymonda, J. W.; Simpson, W. T. *J. Chem. Phys.* **1967**, *47*, 430.

39. Price, W. C. *Phys. Rev.*, **1935**, *47*, 444.
40. Lassettre, E. N.; Skerbele, A.; Dillon, M. A. *J. Chem. Phys.* **1968**, *49*, 2382.
41. Lassettre, E. N.; Skerbele, A.; Dillon, M. A. *J. Chem. Phys.* **1967**, *46*, 4536; **1968**, *48*, 539.
42. Lipsky, S.; Simpson, J. A. *Fifth Int. Conf. Phys. Electron. Atom. Collisions*; Nauka: Leningrad 1967, p. 575.
43. Johnson, K. E.; Kim, K.; Johnston, D. B.; Lipsky, S. *J. Chem. Phys.* **1979**, *70*, 2189.
44. Koch, E. E.; Skibowski, M. *Chem. Phys. Lett.* **1971**, *9*, 429.
45. Sutton, L. E. (Ed.). *Tables of Interatomic Distances and Configurations in Molecules and Ions*; Supplement Special Publications No. 18; The Chemical Society: London, 1965.
46. Palke, W. E.; Lipscomb, W. N. *J. Am. Chem. Soc.* **1966**, *88*, 2384.
47. Clementi, E.; Davis, D. R. *J. Chem. Phys.* **1966**, *45*, 2593.
48. Clementi, E.; Popkie, H. *J. Chem. Phys.*, **1972**, *57*, 4870.
49. Buenker, R. J.; Peyerimhoff, S. D.; Whitten, J. L. *J. Chem. Phys.* **1967**, *46*, 2029.
50. Fink, W.; Allen, L. C. *J. Chem. Phys.* **1967**, *46*, 2261.
51. Pitzer, R. M. *J. Chem. Phys.* **1967**, *47*, 965.
52. Veillard, A. *Chem. Phys. Lett.* **1969**, *3*, 128, 565.
53. Lathan, W. A.; Hehre, W. J.; Pople, J. A. *J. Am. Chem. Soc.* **1971**, *93*, 808.
54. Buenker, R. J.; Peyerimhoff, S. D.; Allen, L. C.; Whitten, J. L. *J. Chem. Phys.* **1966**, *45*, 2835.
55. Lathan, W. A.; Curtiss, L. A.; Pople, J. A. *Mol. Phys.* **1971**. 22, 1081.
56. Murrell, J. N.; Schmidt, W. *J. Chem. Soc., Faraday Trans. II* **1972**, 1709.
57. Baker, A. D.; Baker, C.; Brundle, C. R.; Turner, D.W. *Int. J. Mass Spectr. Ion Phys.* **1968**, *1*, 285.
58. Nicholson, A. J. C. *J. Chem. Phys.* **1965**, 43, 1171.
59. Watanabe, K. *J. Chem. Phys.* **1957**, *26*, 542.
60. Steiner, B.; Giese, C. F.; Inghram, M. G. *J. Chem. Phys.* **1961**, *34*, 189.
61. Schoen, R. I. *J. Chem. Phys.* **1962**, *37*, 2032.
62. Chupka, W. A.; Berkowitz, J. *J. Chem. Phys.* **1967**, *47*, 2921.
63. Narayan, B. *Mol. Phys.* **1972**, *23*, 281.
64. Rabalais, J. W.; Katrib, A. *Mol. Phys.* **1974**, *27*, 923.
65. Sandorfy, C. In *Chemical Spectroscopy and Photochemistry in the Vacuum Ultraviolet*; Sandorfy, C.; Ausloos, P.; Robin, M. B., Eds.; Reidel: Dordrecht, Holland, 1974, p. 177.
66. Richartz, A.; Buenker, R. J.; Peyerimhoff, S. D. *Chem. Phys.* **1978**, *28* , 305.
67. Dillon, M. A. In *Creation and Detection of the Excited State*; Lamola, A.A., Ed.; Dekker: New York, 1971, p. 375.
68. Compton, R. N.; Huebner, R. H. In *Advances in Radiation Chemistry*; Burton, M.; Magee, J. L., Eds.; Wiley-Interscience: New York, 1970. Vol. II, p. 281.
69. McRae, E. G.; Kasha, M. In *Physical Processes in Radiation Biology*; Academic Press: New York, 1964, pp. 23–42.
70. Katagiri, S.; Sandorfy, C. *Theoret. Chim. Acta (Berl.)* **1966**, *4*, 203.
71. Pariser, R.; Parr, R. G. *J. Chem. Phys.* **1953**, *21*, 466.
72. Schachschneider, J. H.; Snyder, R. G. *Spectrochim. Acta* **1963**, *19*, 117.
73. Pearson, E. F.; Innes, K. K. *J. Mol. Spectrosc.* **1969**, *30*, 232.
74. Buenker, R. J.; Peyerimhoff, S. D. *Chem. Phys.* **1975**, *8*, 56.
75. Buenker, R. J.; Peyerimhoff, S. D. *Theoret. Chim. Acta (Berl.)* **1974**, *35*, 33.
76. Brongersma, H. H.; Oosterhoff, L. J. *Chem. Phys. Lett.* **1969**, *3*, 437.
77. Custer, E. M.; Simpson, W. T. *J. Chem. Phys.* **1974**, *60*, 2012.
78. Salahub, D. R.; Sandorfy, C. *Theoret. Chim. Acta (Berl.)* **1970**, *20*, 227.
79. Schoen, R. J. *J. Chem. Phys.* **1962**, *37*, 2032.
80. Chupka, W. A.; Berkowitz, J. *J. Chem. Phys.* **1967**, *47*, 2921.
81. Chupka, W. A. *J. Chem. Phys.* **1968**, *48*, 2337.
82. Saatzer, P. M.; Koob, R. D.; Gordon, M. S. *J. Chem. Soc., Faraday Trans. II* **1977**, *73*, 829.
83. Lindholm, E.; Asbrink, L.; Ljunggren, S. *J. Phys. Chem.* **1991**, *95*, 3923.

84. Caldwell, J. W.; Gordon, M. S. *Chem. Phys. Lett.* **1978**, *59*, 403.

85. Caldwell, J. W.; Gordon, M. S. *J. Phys. Chem.* **1982**, *86*, 4307.

86. Caldwell, J. W.; Gordon, M. S. *J. Mol. Spectrosc.* **1982**, *96*, 383.

87. Chantranupong, L.; Hirsch, G.; Buenker, R. J.; Dillon, M. A. *J. Mol. Struct.* **1993**, *297*, 373.

88. Sandorfy, C. Z. *Phys. Chem. Neue Folge* **1976**, *101*, 307.

89. Sandorfy, C. *Int. Quant. Chem.* **1981**, *19*, 1147.

90. Sandorfy, C. *Topics Curr. Chem.* Springer-Verlag, Berlin, **1979**, *86*, 92.

91. Hampson, R. F.; McNesby, J. R.; Akimoto, H.; Tanaka, I. *J. Chem. Phys.* **1964**, *40*, 1099.

92. Akimoto, H.; Obi, K.; Tanaka, I. *J. Chem. Phys.* **1965**, *42*, 3864.

93. Hampson, R. F.; McNesby, J. R. *J. Chem. Phys.* **1965**, *42*, 2200.

94. Lias, S. G.; Collin, G. J.; Rebbert, R. E.; Ausloos, P. *J. Chem. Phys.* **1972**, *52*, 1841.

95. Ausloos, P. J.; Lias, S. G. *Ann. Revs. Phys. Chem.* **1971**, 85.

96. Merer, A. J.; Mulliken, R. S. *Chem. Rev.* **1969**, *69*, 639.

97. Merer, A. J.; Mulliken, R. S. *J. Chem. Phys.* **1969**, *50*, 1026.

98. Mulliken, R. S. *Phys. Rev.* **1933**, *43*, 279.

99. Gallaway, W. S.; Barker, E. F. *J. Chem. Phys.* **1942**, *10*, 88.

100. Allen, H. C.; Plyler, E. K. *J. Am. Chem. Soc.* **1958**, *80*, 2673.

101. Smith, W. L.; Mills, I. M. *J. Chem. Phys.* **1964**, *40*, 2095.

102. Dowling, J. M.; Stoicheff, B. P. *Can. J. Phys.* **1959**, *37*, 703.

103. Bartell, L. S.; Roth, E. A.; Hollowell, C. D.; Kuchitsu, K.; Young, J. E. *J. Chem. Phys.* **1965**, *42*, 2683.

104. Zelikoff, M.; Watanabe, K. *J. Opt. Soc. Amer.* **1953**, *43*, 756.

105. Geiger, J.; Wittmaack, K. Z. *Naturforsch.* **1965**, *20A*, 628.

106. Wilkinson, P. G.; Mulliken, R. S. *J. Chem. Phys.* **1955**, *23*, 1895.

107. Wilkinson, P. G. *Can. J. Phys.* **1956**, *34*, 643.

108. Merer, A. J.; Schoonveld, L. *J. Chem. Phys.* **1968**, *48*, 522.

109. Merer, A. J.; Schoonveld, L. *Can. J. Phys.* **1969**, *47*, 1731.

110. McDiarmid, R.; Charney, E. *J. Chem. Phys.* **1967**, *47*, 1517.

111. McDiarmid, R. *J. Chem. Phys.* **1969**, *50*, 1794.

112. McDiarmid, R. *J. Chem. Phys.* **1971**, *55*, 4669.

113. Foo, P. D.; Innes, K. K. *J. Chem. Phys.* **1974**, *60*, 4582.

114. Schmitt, R. G.; Brehm, R. K. *Appl. Opt.* **1966**, *5*, 1111.

115. Buenker, R. J.; Peyerimhoff, S. D. *Chem. Phys.* **1976**, *9*, 75.

116. Petrongolo, C.; Buenker, R. J.; Peyerimhoff, S. D. *J. Chem. Phys.* **1982**, *76*, 3655.

117. Johnson, K. E.; Johnston, D. B.; Lipsky, S. *J. Chem. Phys.* **1979**, *70*, 3844.

118. Buenker, R. J.; Peyerimhoff, S. D. *Chem. Phys. Lett.* **1975**, *36*, 415.

119. Buenker, R. J.; Peyerimhoff, S. D. *Chem. Phys.* **1976**, *9*, 75.

120. Robin, M. B.; Hart, R. R.; Kuebler, N. A. *J. Chem. Phys.* **1966**, *44*, 2664.

121. Bélanger, G; Sandorfy, C. *J. Chem. Phys.* **1971**, *55*, 2055.

122. Wilden, D. G.; Comer, J. *J. Phys. B. Atom. Molec. Phys.* **1979**, *12*, L371.

123. Wilden, D. G.; Comer, J. *J. Phys. B. Atom. Molec. Phys.* **1980**, *13*, 1009.

124. Robin, M. B. In *Higher Excited States of Polyatomic Molecules*; Academic Press: New York, 1975, Vol. II.

125. Clementi, E.; Popkie, H. *J. Chem. Phys.* **1972**, *57*, 4870.

126. Peyerimhoff, S. D.; Buenker, R. J. *Theoret. Chim. Acta (Berl.)* **1972**, 27, 243.

127. Mulliken, R. S. *J. Chem. Phys.* **1977**, *66*, 2448.

128. Mulliken, R. S. *J. Chem. Phys.* **1979**, *71*, 556.

129. Borrell, P.; Cervenka, A.; Turner, J. W. *J. Chem. Soc. B* **1971**, 2293.

130. Cunning, H. E.; Strausz, O. P. *Adv. Photochem.* **1963**, *1*, 209.

FORMATION OF (*E,E*)- AND (*Z,Z*)-MUCONIC ACID IN METABOLISM OF BENZENE:
POSSIBLE ROLES OF PUTATIVE 2,3-EPOXYOXEPINS AND PROBES FOR THEIR DETECTION

Arthur Greenberg

Advances in Molecular Structure Research
Volume 4, pages 319–341
Copyright © 1998 by JAI Press Inc.
All rights of reproduction in any form reserved.
ISBN: 0-7623-0348-4

ABSTRACT

2,3-Epoxyoxepin is a postulated intermediate in the ring-opening metabolism of benzene by eukaryotes to form muconaldehyde en route to (E,E)-muconic acid. The activation energy for ring opening to muconaldehyde is calculated to be 16.5 kcal/mol. This chapter explores two issues related to this putative intermediate. First, the fact that prokaryotes produce (Z,Z)-muconic acid in contrast to the (E,E)-isomer for the eukaryotes is explicable by the 2,3-epoxyoxepin "paradigm." The "paradigm" also explains why eukaryotes do not appear to metabolically ring open naphthalene and higher polycyclic aromatic hydrocarbons. Similarly, the degenerate homo-Cope rearrangement of 2,3-epoxyoxepin is described using energetics arguments particularly relevant to the structure and energy of the transition state.

I. INTRODUCTION

Benzene is an important industrial compound that is widely distributed in the environment. It has long been known to be a human carcinogen and mechanisms of metabolism and toxicology are of great interest [1–4]. At the same time, bacterial metabolism of benzene is an important pathway for its disappearance from the environment. Studies of mechanisms of benzene metabolism by strains of *Pseudomonas* and *Arthrobacter* bacteria are thus also of great interest to scientists interested in environmental decontamination. *Eukaryotes* (humans, mammals, etc.) employ monooxygenases, specifically the cytochrome P450 (CYP) family of enzymes, mostly present in liver to oxidize benzene in order to eventually generate water-soluble derivatives such as sulfate and glucuronide conjugates for elimination. The CYP family of isozymes is quite complex and these may be induced selectively by chemical agents. Thus, CYP2E1 is considered to be the most important isozyme in liver metabolism of benzene in mouse, rat, and human [5]. The fact that ethanol also induces CYP2E1 supports the synergetic effect of ethanol on benzene toxicity [5]. *Prokaryotes* (bacteria) have a different natural strategy. They "burn" benzene to yield carbon dioxide, water, and energy and, in turn, rely on dioxygenases [6–14] although monoxygenases have been observed in some bacteria [15]. Figure 1 depicts the general aspects of *prokaryotic* and *eukaryotic* metabolism of benzene [16].

Eukaryotes metabolize benzene overwhelmingly to phenol (**10**), its conjugates (**11,12**), and various other ring-intact species such as dihydroquinone (**13**) (Figure 1). There remains considerable debate concerning which metabolites and reactive oxygen species are involved in the toxicity of benzene. Toxic effects include myelotoxicity (bone marrow toxicity) and carcinogencity [4]. In addition, oxidative metabolites of benzene, notably benzoquinone, cause destruction of the CYP enzymes that created them [5]. However, a small (<10%) fraction of benzene is metabolized to (E,E)-muconic acid (**4c**) which is found as a urinary metabolite [1–4]. Experimental evidence [17–21] strongly suggests that muconaldehyde

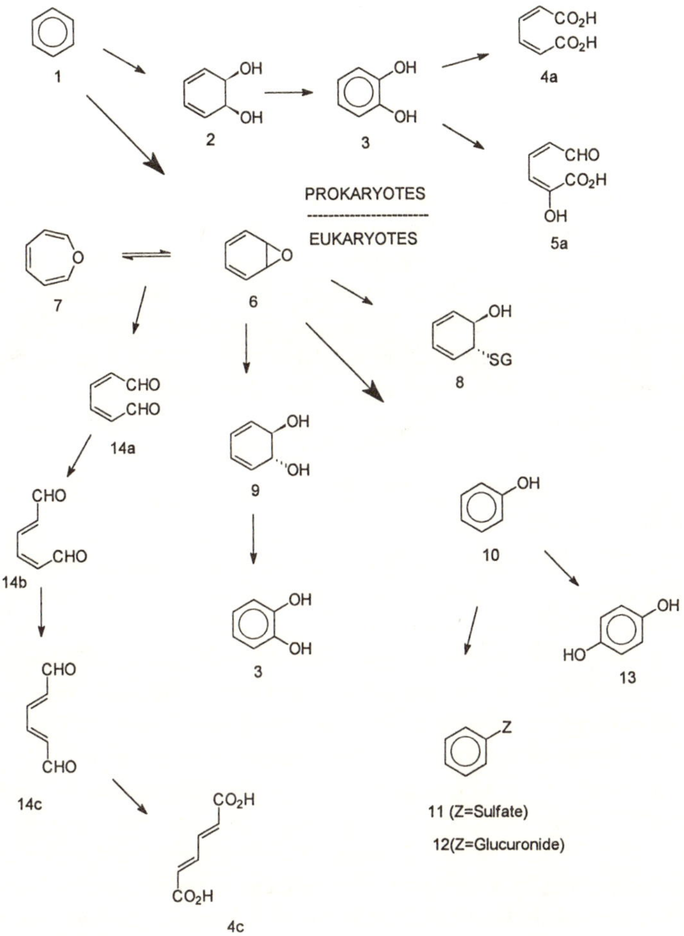

Figure 1. Comparison of the overall pathways for benzene metabolism in prokaryotes and eukaryotes (see ref. 16).

(presumably starting with the (Z,Z)-isomer **14a** and continuing through **14b** and **14c**) is the precursor to (E,E)-muconic acid (**14c**). Muconaldehyde exhibits myelotoxicity as does benzene and is cytotoxic and genotoxic to mammalian cells [19–23].

Various mechanisms have been advanced for generation of muconaldehyde [19–21]. For example, one postulate is that singlet oxygen reacts with benzene to form the dioxetane **15** [4,24]. This intermediate is unlikely, however, on a number of grounds [16]. First, there is simply no precedent for reaction of benzene itself in a symmetry-forbidden $_\pi2_s + _\pi2_s$ thermal reaction with singlet oxygen [16]. Second,

15   16   17

while more highly substituted electron-rich benzene derivatives (e.g. hexamethylbenzene) do react with singlet oxygen, the result is thermally allowed $_\pi2_s + _\pi4_s$
Diels–Alder reaction to form the endoperoxide (e.g. **16**) [25]. Third, the activation
barriers for thermally forbidden retro ($_\pi2_s + _\pi2_s$) of 1,2-dioxetanes to carbonyl
compounds are typically 25 kcal/mol [16]. When added to the calculated endothermicity of 7.6 kcal/mol in the reaction of benzene and singlet oxygen to produce **15**,
one can estimate an activation barrier over 30 kcal/mol for the conversion of
benzene and singlet oxygen to muconaldehyde [16]. The possibility that hydroxyl
radical attacks benzene [21] and that subsequent reaction with (triplet) oxygen leads
to an intermediate such as **17** which thermalizes to muconaldehyde remains an
interesting hypothesis.

We have been intrigued by the work of Davies and Whitham [26] which involved
early investigations of the intermediate "2,3-epoxyoxepin" (**18**, 2,8-dioxabicy-
clo[5.1.0]octa-3,5-diene). Davies and Whitham were, in all likelihood, successful
in generating 2,3-epoxyoxepin although they never isolated or even spectroscopically observed it [26].

18   14a   (1)

Our ab initio (MP2/6-31G*) study of **18** indicated its ring opening isomerization
to (Z,Z)-muconaldehyde is exothermic by 17.0 kcal/mol and has an energy of
activation of only 17.7 kcal/mol (Eq. 1) from the more stable *cisoid* conformer and
proceeding through the 1.2 kcal/mol less stable *transoid* conformer [27]. In Figure
2 [27] we depict the relative energetics for ring opening of 2,3-epoxyoxepin (**18**)
to the initial conformer of (Z,Z)-muconaldehyde. Although the more stable ring
conformation for **18** is the cisoid, the transoid conformation is the one calculated
to ring open [27]. The surprisingly low barrier for uncatalyzed, thermal ring opening
of **18** is likely due to the fact that the reaction is concerted and proceeds through a
Mobius–8π–electron transition state [27]. The structure calculated for the transition
state is depicted in Figure 3. Moreover, traces of acid greatly accelerate this
isomerization and we have obtained spectroscopic evidence for **18** in solutions of
methyl(trifluoromethyl)dioxirane at temperatures between −70 and 0 °C only in
the presence of bases such as Na_2HPO_4 and 2,6-di-*t*-butylpyridine [28].

The present chapter focuses on two very specific aspects of benzene metabolism.
First, there is the intriguing point that eukaryotic metabolism produces (*E,E*)-

muconic acid (**4c**) while prokaryotic metabolism produces the (Z,Z)-isomer (**4a**). A mechanism based upon initial formation of (Z,Z)-muconaldehyde (**14a**) will nicely explain the formation of **4c** by eukaryotes. Although the intermediacy of 2,3-epoxyoxepin (**18**) is consistent with this mechanism, it is not a requirement. We will employ simplified Benson-style approximations [29,30] of $\Delta H_f^o(g)$ values to help make these arguments [16]. The exclusive formation of (Z,Z)-muconic acid (**4a**) by prokaryotes en route to CO_2 and H_2O will also be rationalized. Admittedly, these explanations will be somewhat speculative.

The second focus of the chapter, albeit a brief one, will be on a potential stereochemical test for the actual existence of 2,3-epoxyoxepin (**18**) in the metabolism of benzene. While **18** is observable at ca. 0 °C and below in the absence of

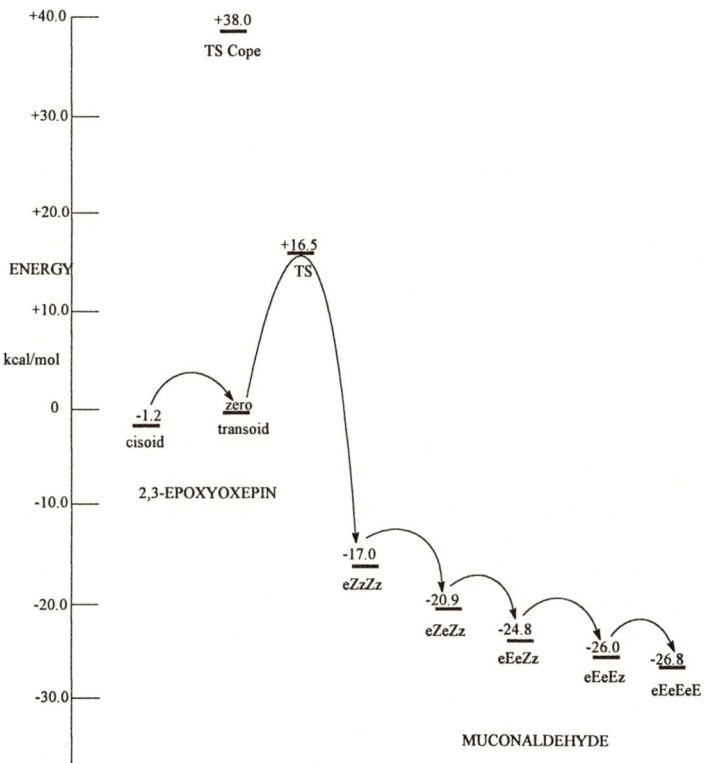

Figure 2. Calculated relative energies of the cisoid and transoid—conformers of 2,3-epoxyoxepin (**18**), the transition state for ring opening (TS) and the relative energetics of geometric isomers and conformers (see ref. 27). "*E*" and "*Z*" represent geometric isomer geometries while "e" and "z" refer to conformations described in ref. 27. The potential Cope rearrangement transition state (TS Cope), is also discussed in this chapter.

a)

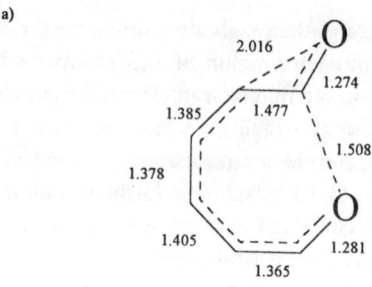

b)

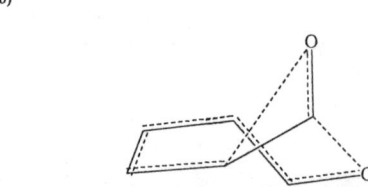

Figure 3. Calculated transition state for ring opening of the transoid conformer of 2,3-epoxyoxepin (**18**) to (*Z,Z*)-muconaldehyde. (See ref. 27).

strong acid [*28*], conclusions about its occurrence as a metabolic intermediate remain speculative. Again, our arguments will employ Benson-style approximations.

II. MUCONALDEHYDE ISOMERIZATION: WHY DO BACTERIA PRODUCE (*Z,Z*)-MUCONIC ACID WHILE MAMMALS PRODUCE THE (*E,E*)-ISOMER?

If muconaldehyde is the initial product of metabolic ring opening of benzene in eukaryotic organisms, then one would expect the first isomer to be (*Z,Z*) (**14a**). How does this compound isomerize to the (*E,E*)-isomer **14c**? Is the isomerization of muconaldehyde the reason why eukaryotes produce (*E,E*)-muconic acid? Prokaryotes produce (*Z,Z*)-muconic acid. How do the mechanistic details relate to this dichotomy? Before exploring simplified mechanistic alternatives, it is important to note that isomerization of muconaldehyde might be catalyzed by proteins and nucleophiles present in solution and may not originate in a unimolecular reaction whose profile we will attempt to calculate. Specifically, we recall the kinetic evidence for the reversible formation of an adduct between (*E,E*)-muconaldehyde and glutathione [*31*]. If this was a Michael-addition adduct, then a bimolecular *Z,E* isomerization pathway might be envisioned starting with the initial (Z,Z)-isomer. Nonetheless, it is worthwhile to investigate the insights gained through exploration of even simpler mechanistic alternatives.

The first relevant studies on the relative rates of muconaldehyde isomerization and oxidation were published almost forty years ago [17]. It was reported that (Z,Z)-muconaldehyde (14a), formed via lead tetraacetate oxidation of cis-3,5-cyclohexane-1,2-diol (2), isomerized to the (Z,E)-isomer 14b overnight in glacial acetic acid (see Scheme 1). Evidently, the (Z,E)- to (E,E)-muconaldehyde isomerization was much slower. Oxidation of 14b overnight in cold perbenzoic acid/chloroform yielded (Z,E)-muconic acid (4b). If the same experiment is repeated overnight at room temperature the product is (Z,Z)-muconic acid (4a). This last result is unusual and will be explored later in this section. These observations are, in part, explicable in terms of: (a) greater stability of (Z,E)-muconaldehyde over the (Z,Z)-isomer; (b) a significant, although not very high, activation barrier toward geometric isomerization of 14a; (c) high barrier for isomerization of Z,E-muconaldehyde (14b) to the (E,E)-isomer 14c; (d) high barriers to isomerization of (Z,Z)-muconic acid (4a) to the Z,E (4b) and then to the (E,E)-isomer (4c); and (e) the possibility of either much enhanced reactivity of (Z,Z)-muconaldehyde with perbenzoic acid compared with the (Z,E)-isomer, special stability of 4a, or a mechanistic alternative combining regioselective isomerizations and oxidation to be described later. Among these interesting points is the idea that, if (Z,Z)-muconic acid is initially formed, it should not isomerize to (E,E)-muconic acid [17].

In their 1977 paper Davies and Whitham [26] studied muconaldehyde and reported that the Z,Z to Z,E isomerization was slower than the Z,E to E,E isomerization. This contradicted the initial study by Nakajima et al. [17] and all subsequent studies. Soon afterwards, Adam and Balci [32] noted that (Z,Z)-6-keto-2,4-heptadienal isomerized in 3 h at 135 °C to the (Z,E)-isomer. This supports the view that Z,E to E,E isomerization is much slower. These authors also noted that Z,Z to E,E isomerization is catalyzed by thiourea, presumably via reversible Michael addition

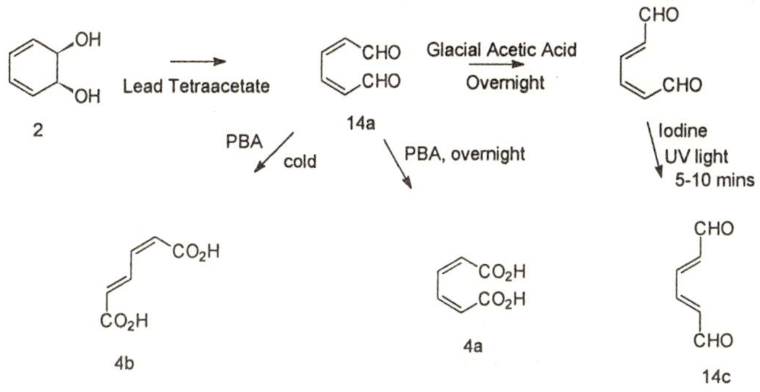

Scheme 1.

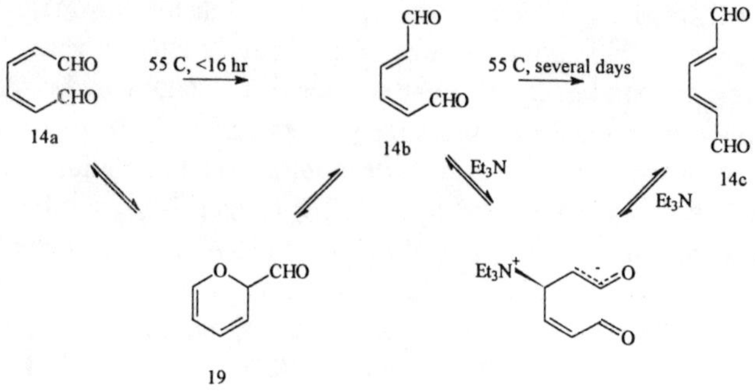

Scheme 2.

[*32*]. A more recent study of this issue by Golding et al. [*33*] found that, in the absence of added nucleophiles, (*Z,Z*)-muconaldehyde isomerized to the(*Z,E*)-isomer in under 16 h at 55 °C, while further isomerization to the (*E,E*)-isomer did not occur under these conditions. They postulated a thermally allowed electrocyclization to form a minute concentration of 2*H*-pyran-2-carboxaldehyde (**19**) to explain this interesting stereospecificity (see Scheme 2) [*33*]. Rotations of C=C bonds are expected to have barriers too high for this isomerization mode to occur [*34*]. If such rotation were to occur, then it would be expected to allow *Z,E* to *E,E* isomerization. The reversible formation of **19** would explain why *Z,E* to *E,E* thermal isomerization is too slow to observe. Golding et al. [*33*] also found that triethylamine catalyzes isomerization of (*Z,E*)-muconaldehyde to the (*E,E*)-isomer, presumably by reversible Michael addition and they corroborated the earlier-cited results for 6-keto-2,4-hexadienal.

The Cope-like ring closure to **19** is not the only feasible thermal rearrangement pathway for muconaldehyde. Vogel noted the utility of entering the muconate series via ring opening of cyclobutene-3,4-dicarboxylates [*35*]. Indeed, the isomerization of **20c** to **20a** (Eq. 2) is known to proceed through cyclobutene **21b** [*36*]. The

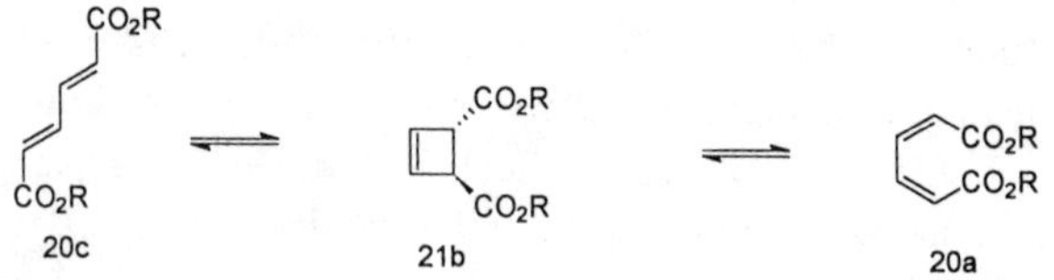

$$(2)$$

attempted synthesis of *cis*-cyclobutene-3,4-dicarboxaldehyde (**19a**) gave instead (*E,Z*)-muconaldehyde (**14b**) through concerted thermal conrotatory opening (Eq. 3) [*37*]. This is an interesting isomerization since it could, in principle, convert (*Z,Z*)-muconaldehyde (**14a**) directly into (*E,E*)-muconaldehyde (**14c**) via the *trans*-isomer **22b** (Eq. 4).

$$(3)$$

22a   14b

$$(4)$$

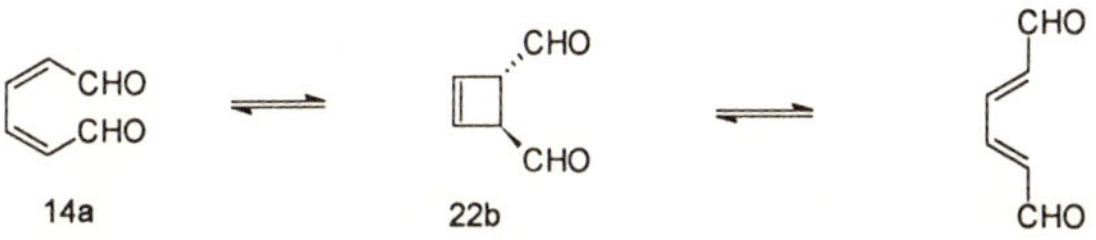

14a   22b   14c

Relevant estimated values for $\Delta H_f^{\circ}(g)$ are found in Table 1. The first point to note is that (Z,Z)-muconaldehyde (**14a**) is estimated to be about 6 kcal/mol more stable than $2H$-pyran-2-carboxaldehyde (**19**) (Eq. 5). Although the two are nearly isoen-

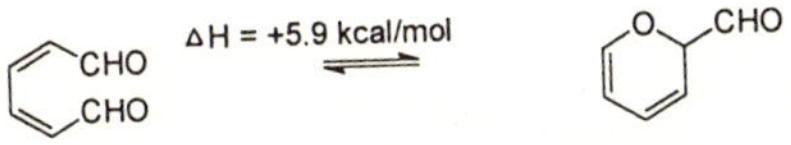

$$\Delta H = +5.9 \text{ kcal/mol} \qquad (5)$$

14a   19

ergetic, the question of the activation barrier for this (presumably) concerted reaction arises. In order to provide an indication of the accessibility of this rearrangement, we note that the enthalpy of activation for the isomerization **23** to **24** (Eq. 6) is 20.6 kcal/mol ($\Delta S = -4.8$ eu) [38]. Thus, this mechanism, advanced

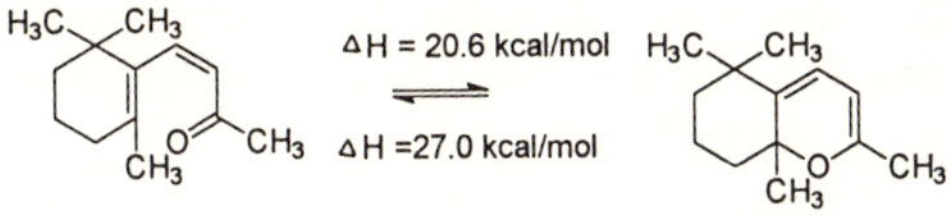

$$\Delta H = 20.6 \text{ kcal/mol}$$
$$\Delta H = 27.0 \text{ kcal/mol} \qquad (6)$$

23   24

by Golding et al. [33] to explain the **14a** to **14b** isomerization, appears to be quite plausible.

If we compare the estimated enthalpies of formation (Table 1) for (Z,Z)-muconaldehyde (**14a**) and *trans*-cyclobutene-1,2-dicarboxaldehyde (see Eq. 4), we find that the isomerization to **22b** is endothermic by 16.9 kcal/mol, about 11 kcal/mol more endothermic than isomerization to the $2H$-pyran **19**. This plausibly explains the rapid Z,Z to Z,E isomerization coupled with very slow Z,E to E,E isomerization in the absence of nucleophiles. Thus, the first step could be unimolecular thermal rearrangement, while the second step under metabolic conditions could be the reversible Michael addition-isomerization. Of course, both steps could result from attack by nucleophiles either in the medium or on biological macromolecules.

It is interesting to compare the isomerization of muconaldehyde with that of muconic acid. First, we may compare the energetics of the Cope-like rearrangement

Table 1. Estimated Gas-Phase Standard Enthalpies of Formation $[\Delta H_f(g)]$[a]

Compound	$\Delta H_f(g)$
(Z,Z)-2,4-Hexadienedioic Acid [(Z,Z)-Muconic Acid] (4a)[b]	-154.6[c]
(Z,E)-Muconic Acid (4b)	-155.6[c]
(E,E)-Muconic Acid (4c)	-156.6[c]
(Z,Z)-2,4-Hexadienedial [(Z,Z)-Muconaldehyde] (14a)[b]	-32.6[c]
(Z,E)-Muconaldehyde (14b)	-33.6[c]
(E,E)-Muconaldehyde (14c)	-34.6[c]
2,8-Dioxabicyclo[5.1.0]octa-3,5-diene (18)	-12.2[c]
2H-pyran-2-carboxaldehyde (19)	-26.7[d]
trans-3,4-Cyclobutenedicarboxaldehyde (22b)	-15.7[e]
6-Hydroxy-2H-pyran-2-carboxylic Acid (25)	-137.0[d]
trans-3,4-Cyclobutenedicarboxylic acid (26b)	-142.5[f]
2H-pyran-2-carboxylic acid (27)	-90.1[d]
6-Hydroxy-2H-pyran-2-carboxaldehyde (28)	-73.6[d]
Methyl-2H-pyran-2-carboxylate (29)	-85.1[d]
6-Methoxy-2H-pyran-2-carboxaldehyde (30)	-67.1[d]
Ethyl (Z,Z)-6-oxo-2,4-hexadienoate (31a)[b]	-96.9[c]
Ethyl (E,Z)-6-oxo-2,4-hexadienoate (31b)	-97.9[c]
Ethyl (Z,E)-6-oxo-2,4-hexadienoate (31c)	-97.9[c]
Ethyl (E,E)-6-oxo-2,4-hexadienoate (31d)	-98.9[c]
Ethyl 2H-pyran-2-carboxylate (32)	-90.1[g]
6-Ethoxy-2H-pyran-2-carboxaldehyde (33)	-72.1[h]
(Z,Z)-6-Oxo-2,4-hexadienoic acid (34a)[b]	-93.6[c]
(E,Z)-6-Oxo-2,4-hexadienoic acid (34b)	-94.6[c]
(Z,E)-6-Oxo-2,4-hexadienoic acid (34c)	-94.6[c]
(E,E)-6-Oxo-2,4-hexadienoic acid (34d)	-95.6[c]
4-Carboxymethylbut-2-en-1,4-olide (muconolactone) (37)	-163[i]
Bicyclo[5.1.0]octa-2,4-diene (41)	$+46.7$[j]
(Z,Z)-Octa-1,3,5,7-tetraene (42)[b]	$+55.1$[k]
4,8-Dioxabicyclo[5.1.0]octa-2,5-diene (60)	$+3.8$[l]

Notes: [a]In kcakl/mol (1 kcal/mol = 4.184 kJ/mol). The methodologies for these estimates are discussed in ref. 16, related data are found in Table 1 of ref. 16 and $\Delta H_f^o(g)$ values for **19, 25** and **27–30** are explicitly derived in the text of ref. 16.

[b]The various Z,Z-dienes in this Chapter will have extra steric repulsions which are relieved by significant twisting about the central C–C bond. In particular, the *e,Z,z,Z,z*-isomers initially formed from the various bicyclo[5.1.0]octane frameworks are ca. 9–10 kcal/mol higher in energy than the most stable stereoisomers (*e,E,e,E,e*) and, therefore, about 7–8 kcal/mol less stable than calculated here for Z,Z-isomers (See ref. 27).

[c]See text of ref. 16.

[d]Use of Benson Group Increments (ref. 29).

[e]-15.7 kcal/mo l = cyclobutene ($+37.5$) + 2[isobutyraldehyde(-51.6)] $-$ 2[propane (-25.0)].

[f]-142.5 kcal/mol = cyclobutene ($+37.5$) + 2[isobutyric acid (-115)²] $-$ 2[propane (-25.0)].

[g]Add a $C(H)_2(C)(O)$ increment (-5.0 kcal/mol) (ref. 29) to the $\Delta H_f(g)$ for **29**.

[h]Add a $C(H)_2(C)(O)$ increment (-5.0 kcal/mol) (ref. 29) to the $\Delta H_f(g)$ for **30**.

[i]Estimated using Benson group increments[29] and estimated 4.0 kcal/mol strain.

[j]$+46.7$ kcal/mol = bicyclo[5.1.0]octane (-4.0) + 1,3-cycloheptadiene (22.5) $-$ cycloheptane (-28.2).

[k]$+55.1$ kcal/mol = (Z,Z)-2,4-hexadiene ($+12.5$) + 2[1,3-butadiene ($+26.3$)] $-$ 2[propene (4.8)]. The $\Delta H_f(g)$ for (Z,Z)-2,4-hexadiene is obtained from: Fang, W.; Rogers, D. W. (1992) Enthalpy of Hydrogenation of the Hexadienes and *cis*- and *trans*-1,3,5-Hexatriene, *J. Org. Chem.*, **57**, 2294.

[l]$+3.8$ kcal/mol = bicyclo[5.1.0]octane (-4.0) + oxirane (-12.6) $-$ cyclopropane (12.7) + hexahydropyran (-53.4) $-$ cyclohexane (-29.5) + divinyl ether (-3.3) $-$ diethyl ether (-60.3).

$$\text{4a} \quad \xrightarrow{\Delta H = +17.6 \text{ kcal/mol}} \quad \text{25} \tag{7}$$

leading to *2H*-pyran derivative **25**. From the estimated $\Delta H_f(g)$ data in Table 1, the isomerization of **4a** to **25** is calculated to be endothermic by 17.6 kcal/mol (Eq. 7). This is 11–12 kcal/mol more endothermic than for the corresponding rearrangement of muconaldehyde. The reason is, of course, the loss of the resonance inherent in one COOH group, wherein the corresponding recovered enol-type resonance is much smaller. When we compare ring closure via cyclobutene **26b** (Eq. 8) this

$$\text{4a} \quad \xrightarrow{\Delta H = +12.1 \text{ kcal/mol}} \quad \text{26b} \quad \longrightarrow \quad \text{4c} \tag{8}$$

reaction is calculated to be endothermic by only 12.1 kcal/mol and thus possibly the preferred thermal pathway. With these points in mind, it is noteworthy that recrystallization of (*Z,Z*)-muconic acid in boiling water produces the (*E,Z*)- but not the (*E,E*)-isomer (Eq. 9) [*39*]. Thus, it appears that **25** (actually the ionized form or

$$\text{4a} \quad \xrightarrow{\text{Water, 100 C}} \quad \text{4b} \tag{9}$$

partly ionized form of this and muconic acid as well) is probably produced, but not **26b** and, as noted, the conditions are much more extreme than needed to yield **19**. From Table 1, **27** is estimated to be 16.5 kcal/mol more stable than the isomer **28** which would yield, upon isomerization, the mixed aldehyde–acid in which the aldehyde-bearing olefin linkage is *E* and the carboxyl-bearing linkage is *Z*. Another indication of this tendency is the observed regioselectivity depicted in Scheme 3 [*40*]. If one uses **29** and **30** as models, then **32** is 18 kcal/mol more stable than **33**.

It is interesting to note that these considerations suggest that if the intermediate **34a** were to occur, it would isomerize regioselectively in the manner shown in Eq. 10. One may thus rationalize the aforementioned observation by Nakajima et al. [*17*] by postulating initial oxidation of (*Z,E*)-muconaldehyde (**14b**) to **34b** followed by thermal isomerization through **27** to form **34a** (Eq. 10) which is then oxidized

27   28   29   30

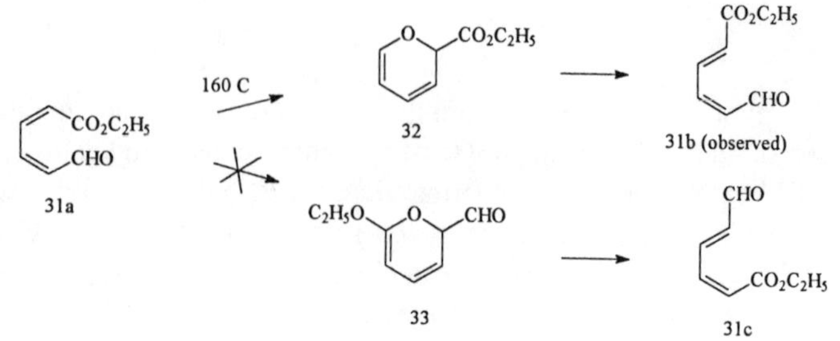

Scheme 3.

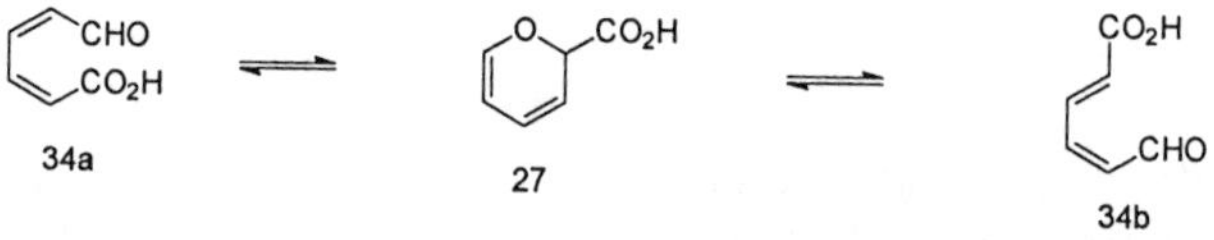

$$(10)$$

to (Z,Z)-muconic acid (**4a**), in its nonionized form in $CHCl_3$, which could be stabilized by intramolecular H-bonding. Michael additions to unsaturated carboxylic acids are known to be very sluggish at best [41]. Thus, the nucleophile-catalyzed isomerization of muconic acids appears to be unlikely.

In contrast to eukaryotes, bacteria such as *Pseudomonas putida* employ non-heme-containing dioxygenases in benzene metabolism [6–14]. Harpel and Lipscomb [42] postulated that intradiol dioxygenases chelate the substrate (see **35**) and the ring-opened product (see **36**) releasing (Z,Z)-muconic acid (**4a**) (see Scheme 4). In contrast to muconaldehyde, Z-E isomerization of muconic acid is very slow as noted earlier.

Thus, bacterial synthesis of (Z,Z)-muconic acid is a consequence of its generation as the first free acyclic benzene metabolite coupled with its sluggishness toward E-Z isomerization. In contrast, the observation of (E,E)-muconic acid as a mammalian urinary metabolite strongly supports the intermediacy of a precursor capable of rapid Z-E isomerization. Initial formation of (Z,Z)-muconaldehyde (**14a**), via 2,3-epoxyoxepin (**18**), should be followed by rapid isomerization [33] and oxidation pathways [22, 23, 31], to yield (E,E)-muconic acid.

The initial metabolic fate of (Z,Z)-muconic acid in bacteria is enzyme-catalyzed ring closure to muconolactone (**37**) [43] (see Scheme 4) en route to succinate and acetyl CoA. The data in Table 1 indicate that this reaction is exothermic by ca. 12 kcal/mol. This simple ring closure is, of course, not available to the (E,E)-isomer some of which is found in mammalian urine, although metabolism of muconic acid yields some CO_2 [1].

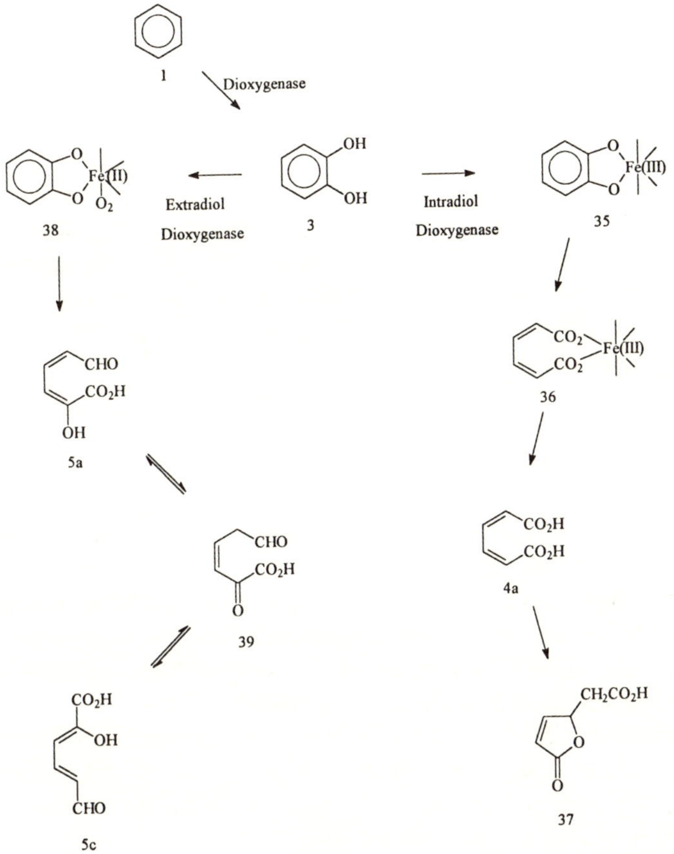

Scheme 4.

Pseudomonas putida also ring opens catechol via an extradiol dioxygenase, possibly through **38** [*42*], to yield 2-hydroxy-6-oxo-2,4-hexadienoic acid (Scheme 4) [*6–14*]. The stereochemistry of this conversion has been a matter for debate, further complicated by the possibility of artifacts resulting from laboratory workup. Initially, the stereochemistry was reported to be *E,Z* (**5a**, Scheme 4) [*44*]. Subsequently, the (*Z,Z*)-isomer was claimed [*45*]. More recent work has established that cleavage of 2,3-dihydroxybenzoate yields the (*Z,E*)-structure **5c** [*46*] (see Scheme 4). Although the thermal and catalytic mechanisms previously described might be invoked, the most likely scenario involves the intermediacy of keto isomer **39** in the conversion of the (presumably initially generated) (*E,Z*)-isomer **5a**. This pathway, involving tautomerization of a "slow-reacting dienol" (lifetime ca. 1 min) [*47*], should be competitive with the other pathways previously described.

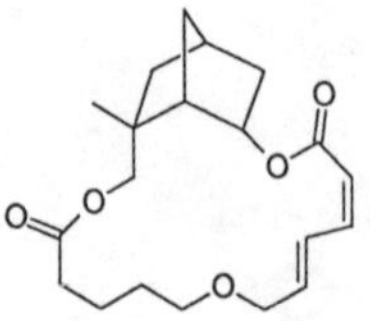

40

Before leaving this section, we note that Trost and McDougal [*48*] have used *trans*-3,4-cyclobutene dicarboxylates to generate the (*Z*,*E*)-muconate segment in macrocyclic trichothecenes such as **40** which are antitumor agents.

III. DEGENERATE REARRANGEMENT OF 18 TO PROBE ITS INTERMEDIACY

A. Comparison of Homo-Cope Rearrangements of **18** and **41**

Once synthesized and observed spectroscopically by NMR at low temperature, 2,3-epoxyoxepin (**18**) may manifest a particularly useful property. It is known that the hydrocarbon analogue (bicyclo[5.1.0]octa-2,4-diene, **41**) undergoes the series of sigmatropic Cope-like 1,3-butadienyl-cyclopropane rearrangements and 1,5-hydrogen shifts depicted in Scheme 5 [*49*]. The fact that such rearrangements can occur in **41** is a consequence of its thermodynamic, and therefore thermal, stability toward ring opening to 1,3,5,7-octatetraene. One can imagine Cope rearrangement of **18**, which is not likely to suffer 1,5-hydrogen shift since the methylenecyclopropane formed would be much more strained than **18**. The consequences of rearrangement starting with doubly deuterated oxepin are shown in Scheme 6 and suggest that, if Cope rearrangement is much faster than ring opening, the label will be scrambled in the manner shown for muconaldehyde. A cautionary note must be

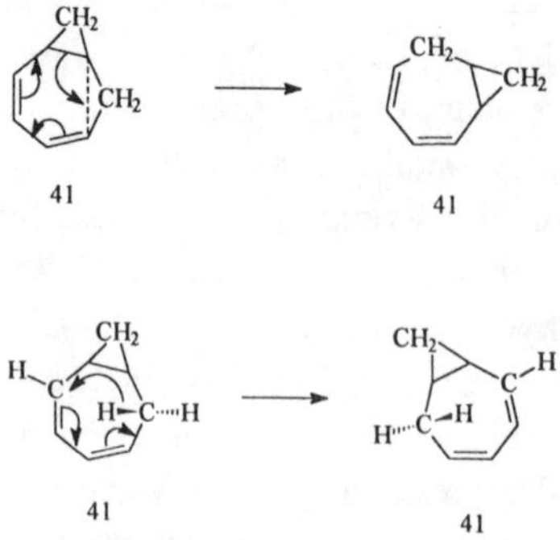

Scheme 5.

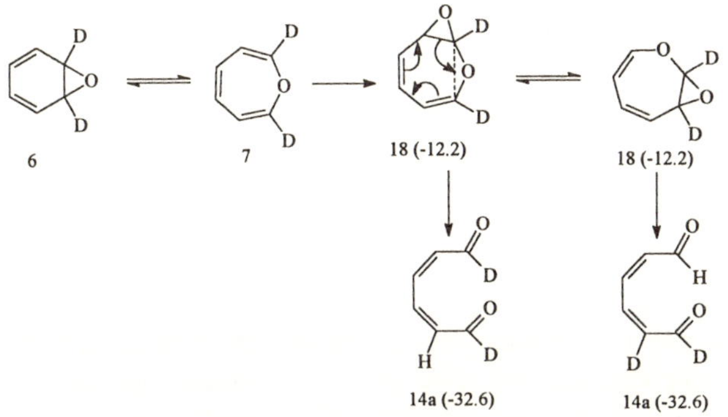

Scheme 6. (calculated enthalpies, kcal/mol, in parentheses)

interjected at this point. According to the data in Table 1, ring opening of **41** to yield (Z,Z)-1,3,5,7-octatetraene (**42**) is endothermic by at least 8.4 kcal/mol. Thus, the Cope rearrangement readily competes with ring opening.

B. Comparison of Transition States for Homo-Cope Rearrangements

Calculation (MP2/6-31G*//6-31G* with thermal corrections) of the activation energy for the homo-Cope degenerate rearrangement of hydrocarbon **41** indicates a barrier of 28.6 kcal/mol (starting from the cisoid conformer) [*34*] in excellent agreement with the experimental value for ΔG^* 31.8 kcal/mol [*49*]. Scheme 7 depicts the structure of the transition state **43** for this rearrangement [*34*]. A particularly striking aspect of the C_2-symmetric **43** is the two exceedingly long predicted (2.19 Å) C---C distances [*34*]. The structural features suggest formulation of **43** as a singlet diradical having almost noninteracting isopropyl radical and pentadienyl radical systems as depicted in Scheme 7.

If one takes the noninteracting diradical model literally, then it is possible to attempt to estimate $\Delta H_f(g)$ for **43** since as noted earlier (a) this transition state has been assigned a structure amenable to parameterization (e.g. isopropyl radical plus pentadienyl radical), and (b) suitable parameters are available. The goal here is not only to rationalize the behavior of **41** but also to attempt to understand the relative rates of rearrangement of the dioxa analogue 2,3-epoxyoxepin.

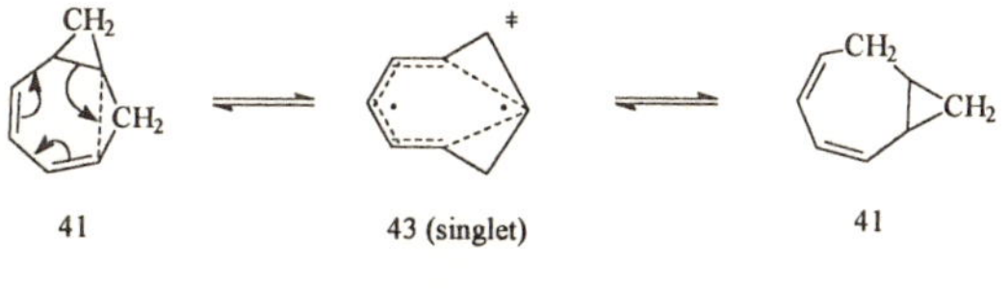

Scheme 7.

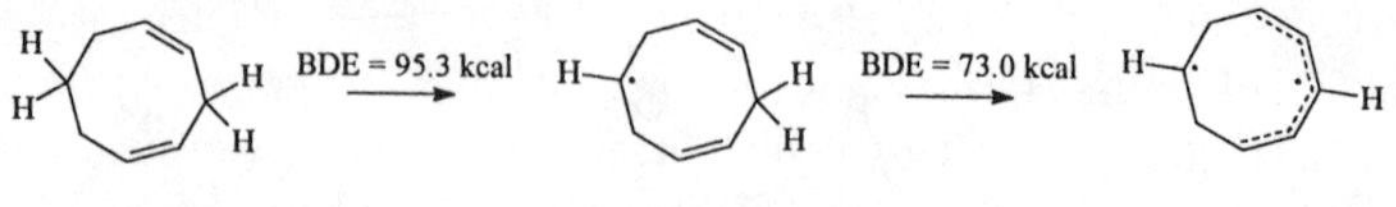

44 (+22.6)   45 (+65.8)   43 (+86.7)

Scheme 8.

We proceed here in a manner reminiscent of the treatment of 1,4-benzyne (or 1,4-benzenediyl) by Jones and Bergman [50]. One can imagine generating the noninteracting diradical which we dub "43" by breaking two C–H bonds in 1,4-cyclooctadiene (44, see Scheme 8). Although the $\Delta H_f(g)$ value for this compound is not known, we could estimate it using Benson group increments. However, we will take the published $\Delta H_f(g)$ for 1,5-cyclooctadiene (+24.2 kcal/mol) [51] and subtract the difference in the heats of hydrogenation[52] of 1,5-cyclooctadiene (53.7 kcal/mol) and 1,4-cyclooctadiene (52.1 kcal/mol) to yield a value of +22.6 kcal/mol for 44. The bond dissociation energy (BDE) for $(CH_3)_2CH$-H is 95.1 ± 1 kcal/mol and that for cyclohexane is 95.5 ± 1 kcal/mol[53]. Using the average (95.3 kcal/mol) and $\Delta H_f(g)(H\cdot)$ = 52.1 kcal/mol [53], the $\Delta H_f(g)$ for radical 45 is calculated according to Eq. 11. The bond dissociation energy for the doubly allylic

$$\Delta H_f(g)(45) = BDE(\text{isopropyl-H}) - \Delta H_f(H\cdot) + \Delta H_f(44)$$

$$+ 65.8 \text{ kcal/mol} = (95.3) \qquad\qquad - (+52.1) + (+24.2) \qquad (11)$$

C–H bond in 1,4-CHD (1,4-cyclohexadiene) is 73.0 ± 2.0 kcal/mol [54]. If one assumes that this value is applicable to the doubly allylic C–H in 45 (Scheme 8), then the energy of "43" is calculated to be 86.7 kcal/mol (Eq. 12). Comparison of

$$\Delta H_f(g)("43") = BDE(\text{1,4-CHD}) - \Delta H_f(g)(H\cdot) + \Delta H_f(g)(45)$$

$$+86.7 \text{ kcal/mol} = (73.0) \qquad\qquad - (+52.1) \quad + (+65.8) \qquad (12)$$

this value with $\Delta H_f(g)(41)$(+46.7 kcal/mol) allows one to predict an energy of activation of about 40 kcal/mol for this rearrangement. This is only 11–12 kcal/mol higher than the barrier calculated [34] at the MP2/6-31G*//6-31G* level which is, as noted previously, in harmony with experiment. This 11–12 kcal/mol discrepancy may represent the extent of interaction between the two formally separated radical systems in 43. Coincidentally, this discrepancy is equal within experimental error to that between the actual $\Delta H_f(g)$ [55] for 1,4-benzyne and that calculated by Jones and Bergman [50] for "1,4-benzenediyl" lacking interaction between the radical centers. In any case, it appears that formulation of the transition state for homo-Cope rearrangement of 41 in terms of singlet diradical "43" is sensible both in terms of the energetics and the calculated structure (see Scheme 7). This also suggests a rational model for understanding the rearrangement of 2,3-epoxyoxepin.

Ab initio molecular orbital calculation of the transition state for homo-Cope rearrangement of 2,3-epoxyoxepin yields the transition state **46** (see Scheme 9) which is similar to the hydrocarbon transition state **43** [27]. The calculated transition state structures **43** and **46** are compared in Figure 4. Unfortunately, the calculated energy of activation, 38 kcal/mol [27], is significantly higher than that for the hydrocarbon. Thus, it would at first appear that the labeling experiment (see Scheme 6) is not practical for implicating the intermediacy of 2,3-epoxyoxepin since ring opening to muconaldehyde has an activation barrier some 20 kcal/mol lower.

What is the origin of the higher activation barrier for homo-Cope rearrangement of 2,3-epoxyoxepin? Now that we have developed a workable model for the transition state, the challenge is to see if one can apply molecular energy estimation schemes toward understanding this behavior. In the present case, the major issue appears to simplify understanding the relative stability of a dialkoxy-substituted free radical. Thus, while α-alkoxy radicals such as $\cdot CH_2OCH_3$

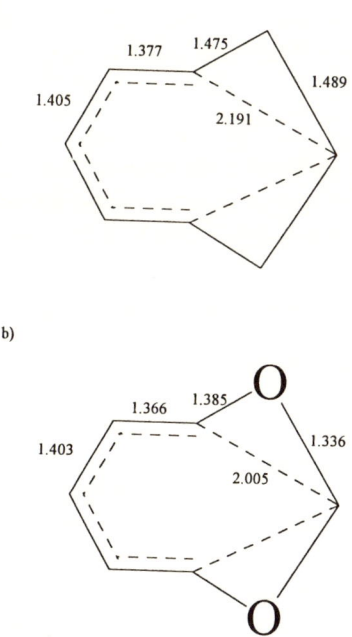

Figure 4. Calculated transition state structures for degenerate Cope rearrangements of hydrocarbon **41** via transition state **43** and 2,3-epoxyoxepin (**18**) via transition state **46** (see ref. 34).

and 2-tetrahydrofuranyl have significant stabilization relative to the analogous hydrocarbon radicals [53, 56, 57], the 1,3-dioxan-2-yl radical is not stabilized and very possibly destabilized [57]. At present, experimental comparison between 1,3-dioxan-2-yl and cyclohexyl radicals is not available.

Although the calculated 38 kcal/mol barrier for homo-Cope rearrangement of 2,3-epoxyoxepin is cause for pessimism concerning the potential success of the labeling experiment in Scheme 6, there remains some hope for its potential for success. It happens that the transition state can be represented both by singlet diradical **46** as well as by zwitterionic **46′** [58]. This latter structure should be

considerably stabilized in polar media. Indeed, the rearrangement depicted in Eq. 13 is thought to proceed via zwitterionic structure **48** and the observed activation

(13)

barriers are only 22–26 kcal/mol [59]. Since **46′** has a pentadienyl anion segment, it is arguably more stable than **48** which has an allylic anion segment. Thus, while **46** may best represent the transition state for rearrangement in the gas phase, **46′** may best represent the transition state in aqueous media. Of additional interest is the epoxyoxepin that can be derived from 2,7-dimethyloxepin (**49**). This oxepin is readily synthesized [60] and shows no detectable benzene oxide using NMR [61]. The transition state corresponding to **46′** would have a stabilizing methyl group at the carbocationic position, thus stabilizing the oxy-Cope rearrangement transition state. In addition, the methyl group should sterically hinder ring opening, thus raising the activation barrier of the competing reaction.

IV. CONCLUDING REMARKS

The strategy for synthesizing the "2-7-dimethyl-2,3-epoxyoxepin" (**49**, 1,3-di-methyl-2,8-dioxabicyclo[5.1.0]octa-3,5-diene) is based upon the use of dioxirane reagents such as dimethyldioxirane (**50a**) and methyl(trifluoromethyl)dioxirane (**50b**) [*62, 63*] (see Scheme 10). These dioxiranes are extremely reactive epoxidiz-ing agents and can (in principle) epoxidize at low temperatures under neutral conditions. In practice, we found that measurable epoxidation of **49** to yield **52a** was initiated by dimethyldioxirane (**50a**) in acetone at ca. −30 °C [*28*]. A small amount of epoxyoxepin **51** is seen in the NMR spectrum at temperatures close to 0 °C but it disappears at ca. 20 °C [*28*]. In contrast, methyl(trifluoromethyl)dioxi-rane (**50b**) epoxidizes **49** at temperatures around −70 °C. However, unless base is added (2,6-di-*t*-butylpyridine or suspended crystalline Na_2HPO_4), one cannot spec-troscopically observe **51**. The reason is that dioxirane **50b** rearranges to form methyltrifluoroacetate which hydrolyzes in the water present in the hygroscopic 1,1,1-trifluoroacetone [*28*]. However, in the presence of these bases, **51** is observ-able up to ca. 0 °C and is gone at ca. 20 °C [*28*]. The ring-opened (Z,Z)-3,5-octadi-ene-2,7-dione (**52a**) isomerizes relatively rapidly (overnight at room temperature in neutral solution) to the (E,Z)-isomer **52b** and more slowly to the (E,E)-isomer (**52c**). It epoxidizes overnight in the presence of dimethyldioxirane to form the

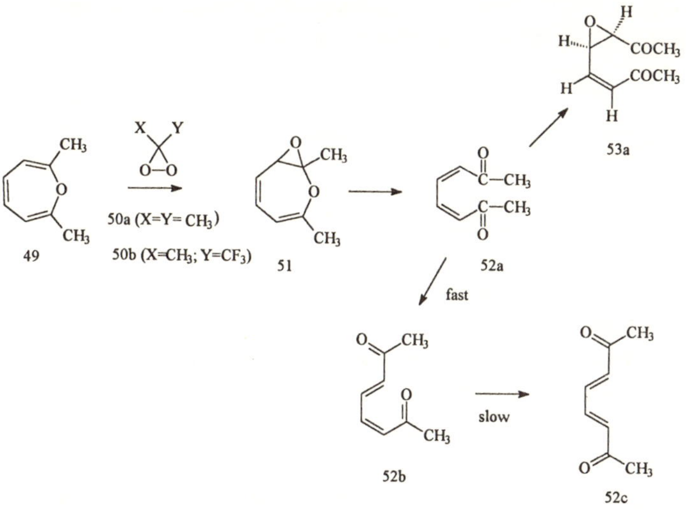

Scheme 10.

syn-2-epoxy derivative **53a**. Thus, epoxidation under these circumstances is more rapid than conversion of **52a** to **52b** [*28*]. Similar studies of muconaldehyde would be interesting. Presumably, the formation of (*E*,*E*)-muconic acid implies that isomerization of muconaldehyde is faster than epoxidation by the monooxygenase system. Of course, one suspects that the isomerization of **52a** to **52b** to **52c** is catalyzed. Nevertheless, the *Z*,*Z*/*E*,*Z*/*E*,*E* isomerization "clock" may provide insight into the epoxidizing power of the monooxygenase system.

It is also interesting to note a dichotomy in the behavior of substituted oxepins toward epoxidizing agents. Specifically, Murray et al. observed that when hexamethylbenzene (**54**) was reacted with 4 equivalents of dimethyl dioxirane (**50a**), the products were diepoxy and triepoxyoxepins **56** and **57** (Scheme 11) [*64*]. Apparently, monoepoxidation of oxepin **55** did not lead to ring opening. One can speculate that initial epoxidation of **55** may have given the 4,5-epoxyoxepin **58**. Rastetter [*65*] synthesized and characterized the unsubstituted compound **60** and found that it was stable to thermal decomposition to 116 °C. Thus, even though 4,5-epoxyoxepin is thermodynamically less stable than 2,3-epoxyoxepin [*16, 27*], it lacks the latter's facile rearrangement pathway and is thus much more kinetically stable. Thus, initial formation of **58**, which has reduced steric repulsions compared to **59**, could be the reason why ring opening does not occur. However, **56** should not arise from **58** (at least directly). Its formation would appear to be best explained via the intermediacy of **59**. If so, then perhaps nonbonded repulsions between methyls significantly raises the ring-opening isomerization barrier to allow a second epoxidation to compete. This is not unrealistic since the rate of monoepoxidation of 2,7-dimethyloxepin with dimethyldioxirane is very comparable (slightly faster) than the rate of ring opening [*28*]. Obviously, the reaction of hexamethylbenzene with 2 mole equivalents of dimethyldioxirane will be of interest.

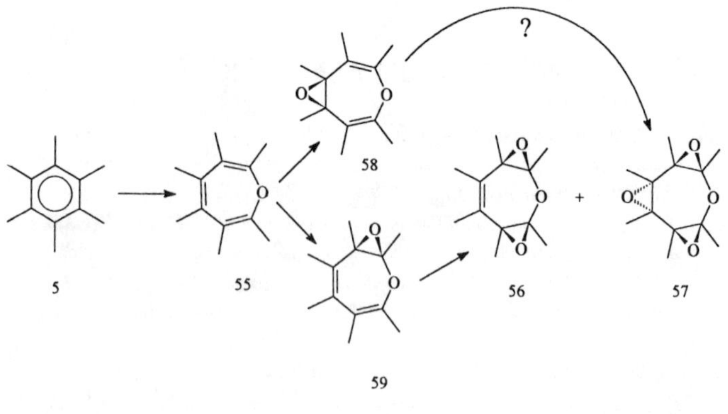

Scheme 11.

60

Thus, in addition to the interesting stereochemical questions raised here for muconaldehyde, muconic acid, and dienediones such as **52**, we have the additional question of how substitution on the benzene ring controls ring-opening or ring-closed products. However, we need to conclude by noting that although the postulation of 2,3-epoxyoxepins seems to be useful in explaining metabolism of benzene as well as polycyclic aromatic hydrocarbons [*16, 66*], there is no proof yet that 2,3-epoxyoxepins are true metabolic intermediates. However, 2,3-epoxyoxepin (**18**) now has more support than the hypothesized dioxetane **15** [*24*]. The latter is unprecedented, whereas we have spectroscopic evidence for dimethyl derivative **51**, and its formation via 2+2 attack by singlet oxygen is illogical. The 2,3-epoxyoxepin pathway is "economical" in that it would logically occur by successive monoxygenations by cytochrome P450, thus employing catalysis rather than stoichiometric reactions with hydroxyl or singlet oxygen. However, hydroxyl (i.e. Fenton) chemistry does pose a potential competing pathway although Zhang et al. suggest that it may be less important than previously assumed [*24*]. They note the possibility of monoxygenation by ferryl ion (FeO^{2+}) [*67*]—a pathway consistent with formation of muconaldehyde via successive monoxygenations of benzene through 2,3-epoxyoxepin.

ACKNOWLEDGMENTS

We are pleased to acknowledge discussions with Drs. Gisela Witz, Stan Kline, Joel F. Liebman, Charles W. Bock, Philip George, Jenny P. Glusker, Charles Shevlin, Robert Snyder, C. S. Yang, and Bernard D. Goldstein. This work was supported in part by Grant No. 1R15ESOD0880901 from the National Institutes of Health.

NOTES AND REFERENCES

1. Williams, R. T. *Detoxification Mechanisms: The Metabolism of Detoxification of Drugs, Toxic Substances and Other Organic Compounds*, 2nd ed.; J. Wiley: New York, 1959, pp. 188–194.
2. NIEHS *Environ. Health Perspect.* **1989**, *82*.
3. Goldstein, B. D.; Witz, G. In *Critical Reviews of Environmental Toxicants—Human Exposure and Their Health Effects*; Lippmann, M., Ed.; Van Nostrand: New York, 1991, Chap. 3.
4. Snyder, R.; Witz, G.; Goldstein, B. D. *Environ. Health Perspect.* **1993**, *100*, 293.
5. Gut, I.; Nedelcheva, V.; Soucek, P.; Stopka, P.; Tichavská, B. *Environ. Health Perspect.* **1996**, *104 (Suppl. 6)*, 1211.
6. Gibson, D. T. *Science* **1968**, *161*, 1093.
7. Gibson, D. T.; Chapman, P. J. In *CRC Critical Reviews in Microbiology*; CRC Press: Boca Raton, 1971, pp. 199–222.

8. Gibson, D. T.; Yeh, W. K.; Liu, T. N.; Subramanian, V. In *Oxygenases and Oxygen Metabolism*; Nozaki, M.; Yamamoto, S.; Ishimura, Y.; Coon, M. J.; Ernster, L.; Estabrook, R. W., Eds.; Academic Press: New York, 1982.

9. Cerniglia, C. E. In *Petroleum Microbiology*; Atlas, R. M., Ed.; MacMillan, New York, 1984.

10. Dagley, S. In *The Bacteria: A Treatise on Structure and Function, Vol. X, The Biology of Pseudomonas*; Sokatch, J. R., Ed.; Academic Press: Orlando, 1986, pp. 527–555.

11. Young, L.; In Lave, L. B.; Upton, A. C. *Toxic Chemicals, Health and the Environment*; Johns Hopkins University Press: Baltimore, 1987, p. 222.

12. Gibson, D. T.; Subramanian, V. In *Microbial Degradation of Organic Compounds*; Gibson, D. T., Ed.; Dekker: New York, 1984, pp. 181–252.

13. Harayama, S.; Timmis, K. N. In *Genetics of Bacteria Diversity*; Hopwood, D. A.; Chater, K. F., Eds.; Academic Press: London, 1989, pp. 151–174.

14. Haggblom, M. *FEMS Microbiol. Rev.* **1992**, *103*, 29.

15. Whited, G. M.; Gibson, D. T. *J. Bacteriol.* **1991**, *173*, 3010.

16. Greenberg, A. In *Active Oxygen in Biochemistry*; Valentine, J. S.; Foote, C. S.; Greenberg, A.; Liebman, J. F., Eds.; Chapman and Hall: London, 1995, pp. 419–422.

17. Nakajima, M.; Tomida, I.; Takei, S. *Chem. Ber.* **1959**, *92*, 163.

18. Tomida, I.; Nakajima, M. *Z. Physiol. Chem.* **1960**, *318*, 171.

19. Goldstein, B. D.; Witz, G.; Javid, J.; Amoruso, M.; Rossman, T.; Wolder, B. In *Biological Reactive Intermediates*; Snyder, R.; Parke, V. V.; Kocsis, J. J.; Jallow, D.; Gibson, G. G.; Witmer, C. M., Eds.; Plenum: New York, 1982, Vol. 2, Part A, pp. 331–339.

20. Latriano, L.; Goldstein, B. D.; Witz, G. *Proc. Natl. Acad. Sci. USA* **1986**, *83*, 8356.

21. Latriano, L.; Zaccaria, A.; Goldstein, B. D.; Witz, G. *J. Free Radicals Biol. Med.* **1986**, *1*, 363.

22. Kirley, T. A.; Goldstein, B. D.; Maniara, W. M.; Witz, G. *Toxicol. Appl. Pharmacol.* **1989**, *100*, 360.

23. Goon, D.; Cheng, X.; Ruth, J. A.; Petersen, D. R.; Ross, D. *Toxicol. Appl. Pharmacol.* **1992**, *114*, 147.

24. Zhang, Z.; Goldstein, B. D.; Witz, G. *Biochem. Pharmacol.* **1995**, *50*, 1607.

25. Foote, C. S.; Clennan, E. L. In *Active Oxygen in Chemistry*; Foote, C. S.; Valentine, J. S.; Greenberg, A. ; Liebman, J. F., Eds.; Chapman and Hall: London, 1995, pp. 105–140.

26. Davies, S. G.; Whitham, G. H. *J. Chem. Soc., Perkin Trans.* **1977**, 1346.

27. Greenberg, A.; Bock, C.; George, P.; Glusker, J. P. *Chem. Res. Toxicol.* **1993**, *6*, 701.

28. Greenberg, A.; Ozari, A.; Carlin, C. M. *Struct. Chem.* **1998**, *9*, 223.

29. Benson, S. W. *Thermochemical Kinetics*, 2nd ed.; John Wiley & Sons: New York, 1976.

30. Liebman, J. F.; Van Vechten, D. V. In *Molecular Structure and Energetics*; Liebman, J. F.; Greenberg, A., Eds.; VCH: New York, 1987, Vol. 2, pp. 315–374.

31. Kline, S. A.; Xiang, Q.; Goldstein, B. D.; Witz, G. *Chem. Res. Toxicol.* **1993**, *6*, 578.

32. Adam, W.; Balci, M. *J. Amer. Chem. Soc.* **1979**, *101*, 7542.

33. Golding, B. T.; Kennedy, G.; Watson, W. P. *Tetrahedron Lett.* **1988**, *29*, 5991.

34. Bock, C. W.; Greenberg, A.; George, P.; Glusker, J. P.; Gallagher, J. D. *J. Org. Chem.* **1995**, *60*, 4358.

35. Vogel, E. *Angew. Chem.* **1954**, *66*, 640.

36. Dalrymple, D. L.; Russo, W. B. *J. Org. Chem.* **1975**, *40*, 492.

37. Hinshaw, J. C. *J. Org. Chem.* **1974**, *39*, 3951.

38. Marvell, E. N.; Caple, G.; Gosink, T. A.; Zimmer, G. *J. Amer. Chem. Soc.* **1966**, *88*, 619.

39. Elvidge, J. A.; Linstead, R. P.; Sims, P.; Orkin, B. A. *J. Chem. Soc.* **1950**, 2235.

40. Wenkert, E.; Guo, M.; Lavilla, R.; Porter, B.; Ramachandran, K.; Sheu, J. H. *J. Org. Chem.* **1990**, *55*, 6203.

41. Patai, S.; Rappoport, Z. In *The Chemistry of Alkenes*; Patai, S., Ed.; Interscience: London, 1964, p. 477.

42. Harpel, M. R.; Lipscomb, J. D. *J. Biol. Chem.* **1990**, *265*, 22187.

43. Cain, R. B.; Freer, A. A.; Kirby, G. W.; Rao, G. V *J. Chem. Soc., Perkin Trans.* **1989**, 202.

44. Dagley, S.; Evans, W. C.; Ribbons, D. W. *Nature* **1960**, *188*, 560.

45. Morgan, L. R., Jr. *J. Org. Chem.* **1962**, *27*, 343.

46. Andreoni, V.; Canonica, L.; Galli, E.; Gennari, C.; Treccani, V. *Biochem. J.* **1981**, *194*, 607.

47. Whitman, C. P.; Aird, B. A.; Gillespie, W. R.; Stolowich, N. J. *J. Amer. Chem. Soc.* **1991**, *113*, 3154.

48. Trost, B. M.; McDougal *J. Org. Chem.* **1984**, *49*, 458.

49. Grimme, W.; Doering, W. v. E. *Chem. Ber.* **1973**, *106*, 1765.

50. Jones, R. R.; Bergman, R. G. *J. Amer. Chem. Soc.* **1972**, *94*, 660.

51. Pedley, J. B.; Naylor, R. D.; Kirby, S. P. *Thermochemical Data of Organic Compounds*, 2nd ed.; Chapman and Hall: London, 1986.

52. Turner, R. B.; Mallon, B. J.; Tichy, M.; Doering, W. v. E.; Roth, W. R.; Schroder, G. *J. Amer. Chem. Soc.* **1973**, *95*, 8605.

53. McMillan, D. F.; Golden, D. M. *Ann. Rev. Phys. Chem.* **1982**, *33*, 493.

54. Burkey, T. J.; Majewski, M.; Griller, D. *J. Amer. Chem. Soc.* **1986**, *108*, 2218.

55. Wenthold, P. G.; Squires, R. R. *J. Am. Chem. Soc.* **1994**, *116*, 6401.

56. Kondo, O.; Benson, S. W. *Int. J. Chem. Kinet.* **1984**, *16*, 949.

57. Huie, R. E.; Clifton, C. L.; Kafafi, S. A. *J. Phys. Chem.* **1991**, *95*, 9336.

58. The author is grateful to Professor Joel F. Liebman for suggesting zwitterionic structure **46′** and noting that both **46** and **46′** are resonance contributors while, in contrast, the singlet diradical and zwitterionic structures for oxyallyl, the planar ring-opened isomer of cyclopropanone, are isomers since they possess different overall state symmetries.

59. Nakamura, E.; Yamago, S.; Ejiri, S.; Dorigo, A. E.; Morokuma, K. *J. Amer. Chem. Soc.* **1991**, *113*, 3183.

60. *Organic Syntheses, Collective Volume V*; Baumgarten, H. E., Ed.; Wiley: New York, 1973, p. 467.

61. Vogel, E.; Gunther, H. *Angew. Chem., Int. Ed. Engl.* **1967**, *6*, 385.

62. Murray, R. W. *Chem. Rev.* **1989**, *89*, 1187.

63. Adam, W.; Curci, R.; Edwards, J. O. *Acc. Chem. Res.* **1989**, *22*, 205.

64. Murray, R. W.; Singh, M.; Rath, N. P. *J. Org. Chem.* **1996**, *61*, 7660.

65. Rastetter, W. H. *J. Am. Chem. Soc.* **1976**, *98*, 6350.

66. Greenberg, A.; Bock, C. W.; George, P.; Glusker, J. P. *Polycyclic Arom. Cmpds.* **1994**, *7*, 123.

67. Sugimoto, H.; Sawyer, D. T. *J. Am. Chem. Soc.* **1984**, *106*, 4283.

SOME RELATIONSHIPS BETWEEN MOLECULAR STRUCTURE AND THERMOCHEMISTRY

Joel F. Liebman and Suzanne W. Slayden

Advances in Molecular Structure Research
Volume 4, pages 343–371
Copyright © 1998 by JAI Press Inc.
All rights of reproduction in any form reserved.
ISBN: 0-7623-0348-4

ABSTRACT

Molecular structure and thermochemistry are interrelated here for species chosen from contributions to the earlier Volume 3 of this book series. Discussion includes halogenated species; gaseous nonmetal dioxides; X–Y bond-containing species (X,Y = C, N, O); small carbon molecules; arenols and substituted arenes; steroids; aromatic carbocycles; difluoramines and nitro compounds; selenium- and tellurium-nitrogen compounds.

I. INTRODUCTION

It is well established that molecular structure and thermochemistry are insepa-rably related on both the research and pedagogical levels of our discipline. In self-referential documentation, for Volume 3 in this series [1] we wrote a chapter [2] in which we presented short thermochemical vignettes based on the contribu-tions by other authors in the preceding book in the series, Volume 2 [3]. This chapter in Volume 4 continues this retrospective and recursive approach by providing related studies on the chapters of Volume 3. Even though we remain interested in many of the same species that we had earlier seen, nevertheless it would seem redundant for us to review our own contribution to that volume. We continue our earlier practice of providing a thermochemical perspective without being exhaus-tive and usually limit our attention to enthalpies of formation [4, 5] and of phase change [6, 7].

II. SIMPLE HALOGENATED SPECIES

We commence with Domaison, Wlodarczak and Rudolph's work [8] dealing with the use of rotational constants for the determination of reliable molecular structures. "Reliable" may be "in the eye of the beholder"—uncertainties in some reported internuclear separations are comparable to the size of atomic nuclei, i.e. ca. 10^{-5} Å. Neither in this chapter, nor anywhere else, will we attempt discussion of the energies to that resolution; instead we will content ourselves here with uncertainties in energies on the order of a "few" kJ mol^{-1}.

Let us start with three chlorinated species: $HC\equiv CCl$, $COCl_2$, and CH_3Cl, listed here in order of increasing C–Cl bond length [9], 1.6353, 1.7381, and 1.7760 Å. We have grown accustomed to assuming short bonds are strong ones and so we expect decreasing C–Cl bond strengths. Homolytic bond energies, per se, are not particularly informative here because it is not obvious what form of CO from $COCl_2$ is to be considered that corresponds to the well-known radicals HCC· and CH_3· for $HC\equiv CCl$ and CH_3Cl, respectively. Immediately we confront a dichotomy between the study of molecular structure and molecular energetics. Many compounds well understood by practitioners of one field are all but unknown to practitioners of the other. The tetratomic species $HC\equiv CCl$ (chloroacetylene) is a superb example of this. In counterpoint to the exquisite accuracy of its determined structure, the enthalpy of formation of this formally simple four-atom molecule is not to be found in the cited works collected in refs. 4 and 5. Proceeding to other less general sources for the desired thermochemical data, the JANAF tables [10] give an estimated value of 214 ± 42 kJ mol^{-1} derived by assuming the following alkyne chlorination reactions have the same enthalpy [11]:

$$HC{\equiv}CH + Cl_2 \rightarrow ClCH{=}CHCl \tag{1}$$

$$HC{\equiv}Cl + Cl_2 \rightarrow ClCH{=}CCl_2 \tag{2}$$

By contrast, the GIANT tables [12] recommend a value of 255 kJ mol^{-1} based on appearance energies of the reactions:

$$HC{\equiv}CCl \rightarrow [HCCCl^{+\cdot} + e^-] \rightarrow HCCH + CCl^+ + e^- \tag{3}$$

$$HC{\equiv}CCl \rightarrow [HCCCl^{+\cdot} + e^-] \rightarrow CH^+ + CCl + e^- \tag{4}$$

We will not attempt to quantitatively disentangle the source of the discrepancy: instead we note that the latter value is just narrowly included in the error bars of the former. However, it is interesting to note that the methyl counterparts [13] of reactions 1 and 2, reactions 5 and 6,

$$HC{\equiv}CH + CH_3CH_3 \rightarrow CH_3CH{=}CHCH_3 \tag{5}$$

$$HC{\equiv}CCH_3 + CH_3CH_3 \rightarrow CH_3CH{=}C(CH_3)_2 \tag{6}$$

have exothermicities of 155.8 and 142.9 kJ mol^{-1} belying the simple assumption of equality of reaction enthalpies for reactions 1 and 2. Although enthalpies of formation from appearance energies are often not trusted, we do not share that general pessimism [14]. In the current case, however, the simplest ("least motion") decomposition of the HC$\equiv$CCl$^{+\cdot}$ radical cation is disallowed and because there is an energy barrier, simple threshold measurements (with even simpler analyses) are problematic. We thus conclude neither value for the enthalpy of formation of chloroacetylene is particularly to be trusted.

The use of the methyl counterparts suggests a simple way of comparing the various chlorides. Consider reaction 7,

$$RCl + CH_3H \rightarrow RH + CH_3Cl \tag{7}$$

where R = CH$_3$, COCl, and HC$\equiv$C. For R = CH$_3$, this is an identity reaction and so thermoneutral by definition. There are no reliable determinations of the enthalpy of formation of HCOCl. In its stead we use CO [or more properly 1/2(CO) because there are two chlorines], and find that this reaction is endothermic by 47.7 kJ mol^{-1}. How much to ascribe this stabilization of COCl$_2$ to the stabilization of general acyl derivatives [15] and how much to additional Y-aromaticity [16] is moot. Nonetheless, this ca. 48 kJ mol^{-1} endothermicity is an impressive quantity.

By contrast, consider the above reaction with R = HC≡C. Depending on which enthalpy of formation value is chosen for chloroacetylene, we find an endothermicity of ca. 7 kJ mol^{-1} or an exothermicity of ca. 35 kJ mol^{-1}. Either value clearly documents that the stabilization of chloroacetylene is considerably less than that of carbonyl chloride despite the considerably shorter C–Cl bond in the former. Bond lengths and bond strengths seemingly do not correlate for the above chlorinated species. We hesitate to suggest that it is inherently a problem for chlorinated species as opposed to a too indiscriminate comparison involving compounds with such disparately hybridized carbons.

We conclude the current vignette with a brief discussion of difluoroacetylene, FC≡F, and compare it with the parent and monofluorinated species. The C≡C and C–F bond lengths seem to be almost identical for these compounds. From the above chlorinated compound analysis and earlier experience with multiply fluorinated compounds [17], we know enough not to conclude that the reaction,

$$FC≡CF + HC≡CH → 2HC≡CF \qquad (8)$$

is nearly thermoneutral. Using the JANAF [10] and GIANT [12] tables as data sources, we find this reaction to be endothermic by 3 ± 92 and exothermic by 18 ± 21 kJ mol^{-1}, respectively. We note that these results are consistent with thermoneutrality but then acknowledge that the error bars are so large as to make this conclusion quite meaningless in fact. Indeed, while there are many new reactions that we may write involving difluoroacetylene, we hesitate to do so because of the excessively large uncertainties in reports of its enthalpy of formation [18].

III. GASEOUS NONMETAL DIOXIDES

In this chapter, Spiridonov [19] discusses various models used in structural determination. One such model is the quasidiatomic approximation which is applied to CO_2, SO_2, and ClO_2. This suggests a discussion of each XO and mean XO bond dissociation enthalpy of XO_2 for X = C, S, and Cl, as well as the bond dissociation enthalpy of XO. The first quantity, denoted D(OX-O), is defined as the value of the enthalpy of the following reaction,

$$XO_2 → XO + O \qquad (9)$$

taken here for all species in their ground state without any concern for spin conservation or any other quantum mechanical constraint [20]. Naively, this corresponds to cleavage of a double bond for X = C, and somewhere between a single and a double bond for X = S and Cl. The second quantity, denoted <D(XO_2)> equals one-half of the enthalpy of the reaction:

Table 1. Dissociation Energies for Some Nonmetal Oxides (kJ mol^{-1})

	$D(OX–O)$	$< D(XO_2) >$	$D(X–O)$
X = C	532.2	804.3	1076.4
X = S	552.3	537	521.7
X = Cl	196	233	269

$$XO_2 \rightarrow X + 2O \qquad (10)$$

The third quantity, denoted $D(X–O)$, is the value of the enthalpy of the reaction:

$$XO \rightarrow X + O \qquad (11)$$

These quantities are neither mathematically nor conceptually independent. It is easy to show,

$$D(X–O) + D(OX–O) = 2 < D(XO_2) > \qquad (12)$$

for all X. Table 1 lists all three dissociation quantities for the three choices of X.

There are several interesting features associated with the table values. The first is that $D(OC–O)$ is less than $D(OS–O)$. This would seem to contradict the folkrule that multiple bonds involving third row elements are generically quite weak: the lower anticipated bond order in SO_2 than in CO_2 only adds to the surprise. Admittedly, this disparity may be corrected by considering only spin-allowed dissociation products, e.g. the $^1\Sigma$ CO + ^{1}D O for CO_2 but this may appear *ad hoc* and complicates the definition of the averaged quantity $< D(XO_2) >$ [21]. The second is that the $D(OC–O)/< D(CO_2) >$ ratio is 0.662, quite close to the ratio found for many other AB_2 species where A and B are group 14 and 16 elements [22], wherein all triatomic species are in their electronic ground state. Given the profound differences between CO_2 and SiO_2 as normally found, such Group 14 constancy is surprising. The third feature notes how different are the X = S and Cl cases, both the relative order of the dissociation enthalpies and how much stronger sulfur–oxygen bonds are than chlorine–oxygen bonds [23]. The final comparison among $D(OX–O)$, $<D(XO_2)>$, and $D(X–O)$ recalls that S$^-$ is isoelectronic to Cl. The dissociation enthalpies for the sulfur–oxygen anions [24] are 552, 489, and 426 kJ mol^{-1}, values which are still much larger and in a different relative order than for chlorine-oxygen species.

IV. COMPOUNDS CONTAINING X–Y BONDS (X, Y = C, N, O)

Mack and Oberhammer [25] discussed experimental and theoretical structures and conformations of selected compounds. Likewise, we will interweave these motifs but from a thermochemical vantage point.

The first compound we will discuss is malonyl chloride, $CH_2(COCl)_2$. We note an over 100-year-old reaction calorimetry study [26] from which an enthalpy of formation value of -389 kJ mol^{-1} was derived [17] for the gaseous phase species. An obvious question is how much do the two COCl groups interact (or interfere) with each other? One documentation of substituent interaction is the ca. 23 kJ mol^{-1} endothermicity, instead of thermoneutrality, for the reaction:

$$2CH_3COCl \rightarrow CH_2(COCl)_2 + CH_4 \qquad (13)$$

We would like to include in our discussion the energetics of the related acyl fluorides, $CH_2(COF)_2$ and, $CF_2(COF)_2$. However, the enthalpies of formation of both species remain unreported, and excepting the archetype acetyl fluoride itself, there are no reported enthalpies of formation of any acyl fluoride known to the current authors. As discussed in ref. 27, there is a rather constant (gas phase) enthalpy of formation difference of about 25 kJ mol^{-1}-F when C–F is replaced by C–OH, C($-$F)$_2$ by C=O + H_2O and C($-$F)$_3$ by COOH + H_2O. For example, the differences between the enthalpies of formation per F are all ca. 24 kJ mol^{-1} for CH_3CF_3 and [CH_3COOH + H_2O]; $CH_2(CF_3)_2$ [28] and [$CH_2(COOH)_2$ [29] + $2H_2O$]. Now consider CH_3COF and thus $CH_2(COF)_2$. The difference between the enthalpy of formation of CH_3CF_3 and CH_3COF is the comparable 31 kJ mol^{-1}. Accordingly, we expect the enthalpy of formation of [$CH_2(COF)_2$ + $2H_2O$] to be $4 \cdot 31$ kJ mol^{-1} higher than that of $CH_2(CF_3)_2$. The derived value of $\Delta H_f(CH_2(COF)_2)$ is -798 kJ mol^{-1} and so reaction 14,

$$2CH_3COF \rightarrow CH_2(COF)_2 + CH_4 \qquad (14)$$

is deduced to be ca. 12 kJ mol^{-1} endothermic. For comparison, we find the related reactions for $CH_2(NO_2)_2$ and $CH_2(COCH_3)_2$ [30] to be endothermic by 16 and 26 kJ mol^{-1}. Quite surprisingly, it appears that the more electron withdrawing groups G (COF and NO$_2$) have less endothermic conproportionation reactions (generically reaction 15) than the less electron-withdrawing groups (COCl and COOH):

$$2CH_3G \rightarrow CH_2G_2 + CH_4 \qquad (15)$$

The next class of compounds discussed in our earlier chapter [25], the carbonylisocyanates, even more strongly document our lack of experimental thermochemical data. There are seemingly no members of this class of compounds for which enthalpies of formation are available, and disappointingly few for any isocyanate at all. The energetics of alkyl, aryl, and acyl halides are distinct enough [17] that we hesitate to extrapolate from one type of isocyanate to another. Perhaps the simplest would appear to be the enthalpy of decomposition via reaction 16, but it is not even obvious whether this reaction is exothermic or endothermic for any R we choose.

$$RCONCO \rightarrow RCN + CO_2 \qquad (16)$$

Hypofluorites also represent a largely thermochemically uncharacterized class of organic compounds and so there is little that can be said about either compound of this type discussed in this chapter, $CF_2(OF)_2$ and FCOOF. The enthalpies of formation of both $CF_2(OF)_2$ and CF_3OF are known: −564 and −770 kJ mol^{-1} respectively [31]. In principle, one can consider the formal fluorocarbon reaction:

$$2CF_3OF \rightarrow CF_4 + CF_2(OF)_2 \qquad (17)$$

This reaction is 55 ± 22 kJ mol^{-1} endothermic. Is this value "plausible"? Consider the related hydrocarbon reaction:

$$2CH_3OCH_3 \rightarrow CH_4 + CH_2(OCH_3)_2 \qquad (18)$$

It is 90.6 kJ mol^{-1} exothermic. Now consider the relatively similar, formal fluorocarbon and related hydrocarbon reactions:

$$2CF_3NF_2 \rightarrow CF_4 + CF_2(NF_2)_2 \qquad (19)$$

$$2CF_3CF_3 \rightarrow CF_4 + CF_2(CF_3)_2 \qquad (20)$$

$$2CH_3N(CH_3)_2 \rightarrow CH_4 + CH_2(N(CH_3)_2)_2 \qquad (21)$$

$$2CH_3C(CH_3)_3 \rightarrow CH_2(C(CH_3)_3)_2 + CH_4 \qquad (22)$$

Reactions 19 and 20 are ca. 25 kJ mol^{-1} endothermic and exothermic, respectively [32]. Reactions 21 and 22 are ca. 45 exothermic and 20 endothermic, respectively. Hydrocarbon analogies do not aid our understanding of the fluorocarbon species. Said differently, the hypofluorites stand apart from other compounds for their understanding.

We now turn to the *O*-nitrosohydroxylamine $(CF_3)_2NONO$. No enthalpy of formation is known for this species or of any other *O*-nitrosohydroxylamine. We do not know the formal dimerization enthalpy,

$$2CF_3NO \rightarrow (CF_3)_2NONO \qquad (23)$$

and so the recently measured enthalpy of formation of the above trifluoronitrosomethane [33] is without use in our current study.

We close this vignette with a brief discussion of alkyl and acyl peroxides. Recall the approximate constant exothermic enthalpies of reaction for strainless alkyl peroxides [34],

$$ROOR + H_2 \rightarrow 2ROH \qquad (24)$$

$$ROOR \rightarrow ROR + 1/2O_2 \qquad (25)$$

of ca. -277 and -59 kJ mol^{-1}, respectively. What about acyl peroxides? For the case of the "aryl acyl" of R = C_6H_5CO, these reactions are exothermic by -317 and -47 kJ mol^{-1}. We would have liked to have compared alkyl peroxides with "alkyl acyl" peroxides (i.e. peresters) but we lack enthalpy of vaporization data for these latter species. We likewise would have wanted to compare aryl peroxides with "aryl" acyl peroxides but we lack experimental thermochemical data of any species with an Ar–O–O structural unit.

V. ENERGETICS OF SMALL CARBON MOLECULES

We now turn to the chapter by Cernak, Monninger and Krätschmer [35] on small carbon molecules, i.e. species with the formula C_n for small values of n without any accompanying affixed groups or additional atoms. Acknowledging that there are a variety of possible structures for most n, nonetheless we will assume that the chemistry at 298 K is dominated by the linear structures, :C=C:, :C=C=C:, and :C=C=C=C:. If these compounds are assumed to be normal (poly)olefinic hydrocarbons (except for lacking hydrogen, of course), we should think that the following transformations to form the even more normal (poly)olefinic hydrocarbons, C_nH_4, would have very nearly the same reaction enthalpies:

$$:C=C: + 2H_2 \rightarrow H_2C=CH_2 \tag{26}$$

$$:C=C=C: + 2H_2 \rightarrow H_2C=C=CH_2 \tag{27}$$

$$:C=C=C=C: + 2H_2 \rightarrow H_2C=C=C=CH_2 \tag{28}$$

After all, in each case, two carbon lone pairs and two H–H bonds are converted into four C–H bonds while leaving the number of C–C σ- and π-bonds unchanged. Is this expectation fulfilled?

Some care needs to be taken in the analysis. Whereas an elementary electronic structure description of C_2 gives the singlet ground state and a conventional σ- and π-carbon–carbon double bond to this diatomic molecule, the real C_2 molecule has two π-bonds and no σ-bond connecting the two carbons. Accepting the enthalpy of formation of ground-state C_2 from the JANAF [10] tables, we thus find reaction 26 is exothermic by at least $829 - 52 = 777$ kJ mol^{-1}. Using the JANAF enthalpy of formation for the C_3 species, we find an exothermicity for reaction 27 of $820 - 190 = 630$ kJ mol^{-1}. Thus, reaction 26 is seemingly 140 kJ mol^{-1} more exothermic than reaction 27. The most meaningful comparison, though, is between species with the same bonding description and so we should use the excited C_2 $^1\Delta$ state and its accompanying $\sigma^2\pi^2$ description which results in a greater energy difference between reactions 26 and 27 equal to the sum of the earlier 140 kJ mol^{-1} and the $^1\Delta$-$^1\Sigma$ gap.

What about the reaction enthalpy in the C_4 case? While there are no measured enthalpies of formation for the product butatriene, this quantity may be simply estimated. The enthalpy of formation of its 1,4-dimethyl derivative (for both the *cis*- and *trans*-isomers) has been determined by hydrogenation calorimetry [36] to be ca. 265 kJ mol^{-1}. Demethylation of propene, 1,2-butadiene, (E)-2-butene, (E)-1,3-pentadiene, and 2,3-pentadiene are accompanied by increases in enthalpies of formation of ca. 30 kJ mol^{-1} [37] and so the enthalpy of formation of butatriene is 265 + (2·30) = 325 kJ mol^{-1}. Combined with the enthalpy of formation of C_4 from the JANAF tables we thus deduce the enthalpy of reaction 28 is 971–325 = 646 kJ mol^{-1}. This value is relatively close to that found for reaction 27. Why should C_2 be such a significant outlier? [38]

VI. THE ENERGETICS OF SOLID ARENOLS AND SOME OTHER SUBSTITUTED ARENES

In their chapter [39], Zorky and Zorkaya discuss specific crystal interactions of hydrogen-bonded aromatic species (mostly phenols). What calorimetric evidence is there for such hydrogen-bonding interactions? First, for arbitrary organic hydrocarbon groups, R and R', the enthalpies of vaporization of ethers, ROR', and their corresponding carbon-containing analogues RCH_2R' are generally quite close [40]; in contrast, the enthalpies of vaporization of ROH and RCH_3 are profoundly different. For example, the enthalpy of vaporization of methoxybenzene is 46.9 kJ mol^{-1} and that of ethylbenzene is 42.2 kJ mol^{-1}. The enthalpy of vaporization of *m*-cresol is 61.7 kJ mol^{-1} (its *o*- and *p*-isomers are both solids) while that of its carbon-containing analogue, *m*-xylene, is 42.7 kJ mol^{-1}. This large difference in the enthalpy of vaporization of alcohols compared to hydrocarbons or ethers is attributable to their hydrogen-bonding interactions in the liquid phase.

Since the enthalpy of vaporization depends linearly on the number of carbons to a high degree of accuracy [6], it is fair to suggest benzene, toluene, ethylbenzene (and the isomeric xylenes), form a homologous series, at least with respect to this property. The set of compounds: benzene, phenol, and the various benzene-diols wherein an increasing number of hydroxy substituents putatively defines a homologous series might also define a linear relationship.

The first series composed of liquid hydrocarbons easily qualifies as homologous in this sense. The enthalpy of vaporization of benzene is 33.6 kJ mol^{-1} and of toluene is 38.0 kJ mol^{-1}. The difference is 4.4 kJ mol^{-1}, corresponding quite well to that predicted for an increase of one carbon [41] for which a difference of 4.7 kJ mol^{-1} is predicted. The three xylenes are expected to have comparable enthalpies of vaporization—by linear extrapolation we would expect 42.4 kJ mol^{-1} while by simple carbon count, a value of 40.0 kJ mol^{-1} is expected. The values of the *o*-, *m*-, and *p*-xylenes are 43.5, 42.7, and 42.4 kJ mol^{-1}, respectively, in satisfactory agreement with both approaches.

With respect to the second series, many (most) arenols are solids and so we recall the identity interrelating enthalpies of sublimation, vaporization, and fusion, Eq. 29,

$$\Delta H_{sub}(T) \equiv \Delta H_v(T) + \Delta H_{fus}(T) \tag{29}$$

and the approximate, but more customarily used, "approximate identity" [42],

$$\Delta H_{sub}(298) \approx \Delta H_v(298) + \Delta H_{fus}(T_m) \tag{30}$$

(where "298" is actually 298.15 K or 25 °C, and T_m is the melting point). The enthalpies of sublimation and fusion of phenol are 59.4 and 11.5 kJ mol^{-1}, respectively, from which we derive an enthalpy of vaporization of 47.9 kJ mol^{-1}. Likewise, for the m- and p-diols, the enthalpies of sublimation of 100.8 and 106.7 and enthalpies of fusion of 21 and 27 kJ mol^{-1} yield enthalpies of vaporization of 79 and 80 kJ mol^{-1}. While it is interesting that these last two values are nearly identical, it is disappointing how nonlinear are the enthalpies of vaporization of benzene, phenol, and the two benzenediols.

What can be said about either series when studied as solids? From Eqs. 29 and 30 we see that our question can be answered in terms of the enthalpies of fusion. Consider the first series. For benzene, the enthalpy of fusion is 9.9 kJ mol^{-1} while it is but 6.6 kJ mol^{-1} for toluene. Extrapolating to 3.3 for the xylenes seems wrong if for no other reason than we would further extrapolate to impossible zero or negative values for a species with four or more methyl groups. Indeed, the fusion enthalpies for the three xylenes are 13.6, 11.6, and 17.1 kJ mol^{-1} for the o-, m- and p-isomers, respectively. Homology is lost, at least suspect, or said more optimistically, one must be tolerant of greater disparities.

For the second series and again starting with benzene and its enthalpy of fusion of 9.9 kJ mol^{-1}, the value rises to 11.5 kJ mol^{-1} for phenol and then to 20.8 and 26.5 kJ mol^{-1} for the m- and p-dihydroxybenzenes, respectively. Again, lack of homology is found. Perhaps even more surprisingly, the enthalpies of fusion of benzene and phenol are nearly identical.

What now about polymethylated phenols, the "hybrid" case of the previous two series? We start with phenol and its enthalpy of fusion of 11.5 kJ mol^{-1}, and changing to 15.3, 9.7, and 12.2 kJ mol^{-1} for the o-, m- and p-monomethylated species. There are seven dimethylated phenols for which simple linearity and homology would suggest a value of ca. 14 kJ mol^{-1}. In fact, the following values are found (kJ mol^{-1}): 2,3-, 23.0; 2,6-, 18.9; 3,4-, 18.1; 3.5-, 17.6. No regularities are found for this series either. Perhaps the irregularities can be explained by recalling the specific interactions in ref. 39 but we lack knowledge of direct crystal structure/crystal energy relations.

What about derivatives of the various polynuclear hydrocarbons? Except for phenol and the isomeric naphthols, arenol thermochemical data are essentially absent in any phase and any substituent pattern. So, let us compare the solid-phase enthalpies of formation of some relatively simple polynuclear hydrocarbons and their methylated derivatives. The solid-phase enthalpies of formation of benzene, toluene, the xylenes, 1,2,4,5-tetramethylbenzene, and hexamethylbenzene are roughly linear ($r^2 = 0.98$) with the number of carbons (or alternatively, with the number of methyl groups). From the slope of the line, we conclude that methylation decreases enthalpies of formation in the solid by ca. 34.5 ± 2 kJ mol^{-1}. How general is this? A difference of 40 and 44 kJ mol^{-1} is found for biphenyl upon sequential methylation in the 4- and 4'-positions while differences of 28.5 and 33.0 kJ mol^{-1} are found for 1- and 2-methylation of naphthalene. The various dimethylated naphthalenes are educational: the difference per methyl group for the unhindered 2,3-; 2,6- and 2,7-species are the nearly identical 40, 42, and 43 kJ mol^{-1}, a value decreased to 26.0 kJ mol^{-1} for the strained peri or 1,8-isomer.

There are seemingly no thermochemical data for any methylated anthracenes. For solid methylated phenanthrenes, there are enthalpies of formation of the unhindered 2,7- and 9,10-dimethylated species for which the methylation has incurred an ca. 40 and 35 kJ mol^{-1} per methyl decrease in enthalpies of formation, while for the hindered 4,5-species the decrease was only ca. 14 kJ mol^{-1} per methyl. Consider now the isomeric 3,4,5,6- and 2,4,5,7-tetramethylphenanthrenes. In that it is methylation in both the 4- and 5-positions that results in the smaller decrease of enthalpies of formation, we would expect a decrease of ca. $(2 \cdot 35)+(2 \cdot 14)$ kJ mol^{-1} or ca. 100 kJ mol^{-1} for both of these species from the parent hydrocarbons. Using measured values, the decreases are 93 and 104 kJ mol^{-1}, respectively, in encouraging agreement. For benz[a]phenanthrene derivatives, the strained 1,12-dimethyl derivative exhibits a 30 kJ mol^{-1} total decrease from the parent hydrocarbon while the unstrained 1,12-isomer shows the much more normal decrease for two methyl groups of ca. 76 kJ mol^{-1}. All of this is quite sensible, and so we are not prepared for the nearly identical 41 and 37 kJ mol^{-1} decreases on going from the 1,12- and 7,12-dimethylbenz[a]anthracene to the parent hydrocarbon. We are even more surprised by the only ca. 13 kJ mol^{-1} decrease of the enthalpies of formation upon 5,6-dimethylation of chrysene since we would have thought that isomer is quite unstrained.

What about derivatives that are formed by ring cleavage, such as stilbene which may be so derived from phenanthrene. For *trans*-stilbene, 4,4'- and 2,2'-dimethylation is accompanied by a decrease of ca. 40 and 30 kJ mol^{-1}. These results are quite plausible either in terms of simple changes upon methylation or even an amended steric argument. However, quite enigmatically, solid 4,4'-dimethyl-*cis*-stilbene has a more positive enthalpy of formation than its 2,2'-isomer. Dare we conclude that at least one of these measurements was in error?

VII. INTERCONVERSIONS OF STEROIDS

Many thermochemical structure/energy analyses proceed from recognition of a common molecular entity upon which small structural modifications are imposed.

In their chapter [43], Kálmán and Párkányi discuss main-part isostructurality of cardiotonic steroids. Here, we discuss experimentally measured solid-phase enthalpies of formation of steroids possessing a common (5α)-androstane skeleton (*anti-anti-anti*-perhydrocyclopenta[a]phenanthrene). The enthalpies of formation are listed in Table 2. The steroids shown later in Schemes 1 and 2 are related by simple formal reactions which interconvert the various substructure functional groups: alcohol ⇔ ketone, alkene ⇔ alkane, alkane ⇔ alcohol, and alkane ⇔ ketone. The difference between the enthalpies of formation of two steroids is the enthalpy of reaction(s) which interconverts them (Eq. 31):

$$\Delta H_{(rxn)} = \Delta H_{f\,(product)} - \Delta H_{f\,(reactant)} \tag{31}$$

Ideally, the enthalpy of reaction between two small molecules possessing a pair of the functional groups mentioned above should be approximately equal to the enthalpy of reaction of the same formal reaction taking place on a steroid. For steroid interconversions requiring more than one formal reaction on the steroid skeleton, the enthalpies of reaction of the relevant small molecule conversions should be additive and approximately equal to the steroid enthalpy of reaction. In making these comparisons, we may assess the accuracy of the measured steroid enthalpies of formation.

Table 2. Enthalpies of Formation of Steroids[a]

Steroid	$\Delta H_f(s)\ kJ\ mol^{-1}$	
progesterone	−551 ± 21	
deoxycorticosterone	−518 ± 4	
hydrocortisone (cortisol)	−1070 ± 33	
cortisone	−1069 ± 17	
cholesterol	−675 ± 4[b]	−605 ± 67
3-methoxycholestene	−652 ± 5[c]	
androstane	−314 ± 38	
androsterone	−677 ± 29	
epiandrosterone	−635 ± 50	
androstane-3,17-dione	−546 ± 46	
stanolone	−1212 ± 25	
testosterone	−395 ± 29	
androst-4-ene-3,17-dione	−431 ± 29	

Notes: [a]All enthalpies of formation are from ref. 44 unless otherwise stated.

[b]The enthalpy of formation is from ref. 45.

[c]The enthalpy of formation is from ref. 4.

One impediment to straightforward comparison is the lack of solid-phase enthalpies of formation for most low molecular weight molecules. Examination of the few examples for which there are both solid- and liquid-phase values shows, however, that the liquid- and solid-phase enthalpies of reaction do not vary too much, at least within the uncertainties of the steroid combustion measurements. We are additionally hampered by not having available enthalpies of formation for those small molecules which most resemble the steroid substructure. The prototype small molecules required for the analysis are primarily those of substituted five- or six-membered carbocyclic rings. Although the functional groups on the available prototypes and the steroids are identical, stereochemical aspects (such as axial vs. equatorial substituent positions) and their consequent interactions are different due to the differences in conformational flexibility of the monocyclic and polycyclic systems. The prototype molecules and their liquid-phase enthalpies of formation are listed in Table 3.

Scheme 1 shows three examples of steroid functional group conversions. The first is the hydroxylation of progesterone to deoxycorticosterone with an enthalpy of reaction of $+30 \pm 21$ kJ mol^{-1}. One of these enthalpies of formation must certainly be incorrect. Although we lack a prototype reaction for hydroxylation alpha to a keto group, no example exists where replacement of hydrogen by hydroxyl is endothermic [34]. The enthalpy of oxidation of hydrocortisone to cortisone is seemingly thermoneutral, albeit with very large uncertainty. The prototypical oxidation of *cis*-2-methylcyclohexanol to 2-methylcyclohexanone is 102.1 kJ mol^{-1} endothermic. Even calculating the enthalpy of oxidation using the steroids' enthalpies of formation at the extremes of their uncertainty intervals does not produce a reasonable enthalpy of reaction. Finally, the enthalpy of methylation of cholesterol to the derivative 3-methoxycholestene is either endo- or exothermic, depending on the enthalpy of formation of cholesterol which is used. The enthalpy

Table 3. Enthalpies of Formation of Selected Functionalized Cycloalkanes

5-Membered Ring	$\Delta H_f(l)$ kJ mol^{-1}	6-Membered Ring	$\Delta H_f(l)$ kJ mol^{-1}
cyclopentane	-105.1 ± 0.8	cyclohexane	-156.7 ± 0.8
methylcyclopentane	-137.9 ± 0.8	methylcyclohexane	-190.1 ± 1.0
cyclopentanol	-300.1 ± 1.6	cyclohexanol	-348.2 ± 2.1
cis-2-methylcyclopentanol	$[-353]^a$	*cis*-2-methylcyclohexanol	-390.2 ± 5.0
cyclopentanone	-235.7 ± 1.8	cyclohexanone	-271.2 ± 2.1
2-cyclopentenone	-141.3^b	2-cyclohexenone	-167.7^b
2-methylcyclopentanone	-265.2 ± 5.4	2-methylcyclohexanone	-288.1 ± 3.3

Notes: [a]The enthalpy of formation is estimated. See ref. 46.

[b]The enthalpy of formation is derived from experimental measurement. See ref. 47.

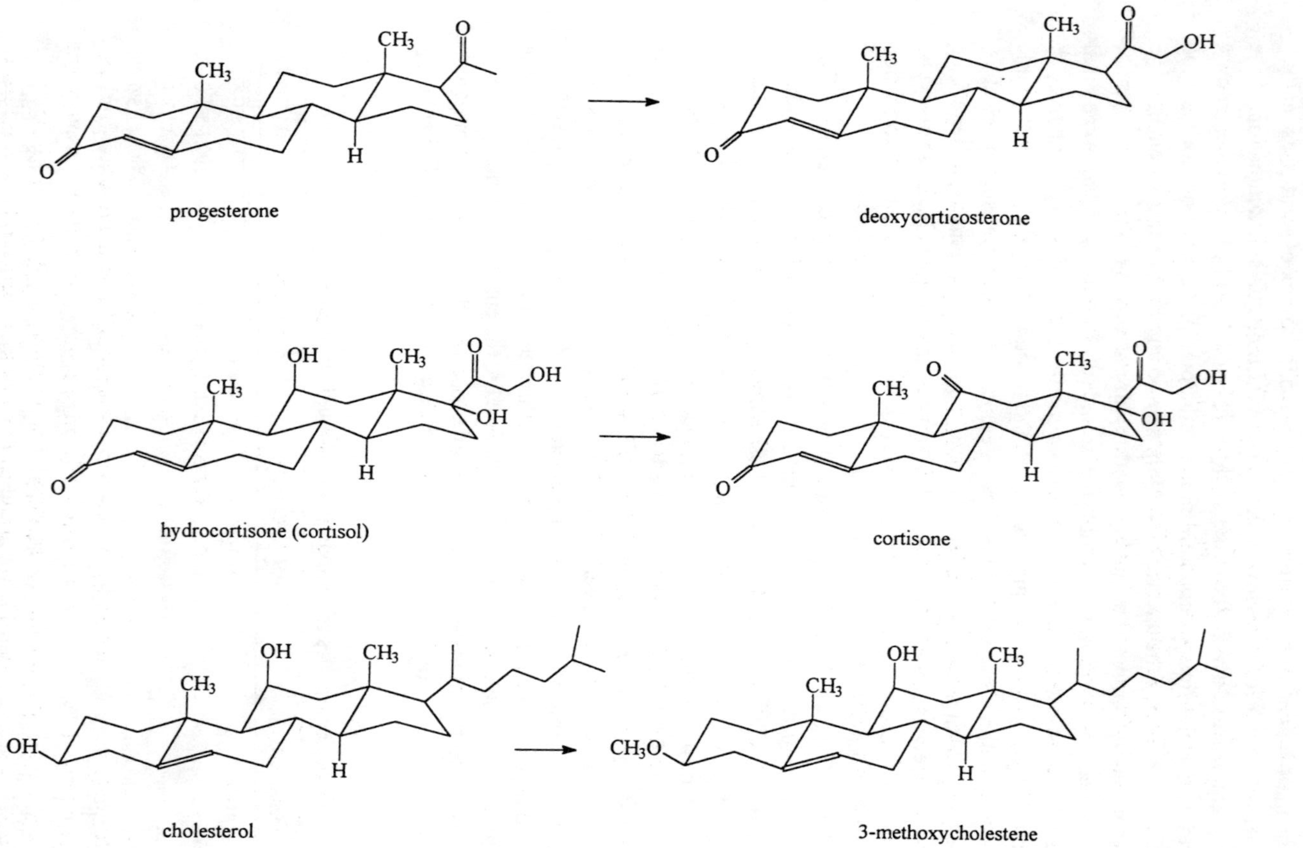

deoxycorticosterone

cortisone

3-methoxycholestene

progesterone

hydrocortisone (cortisol)

cholesterol

Scheme 1.

356

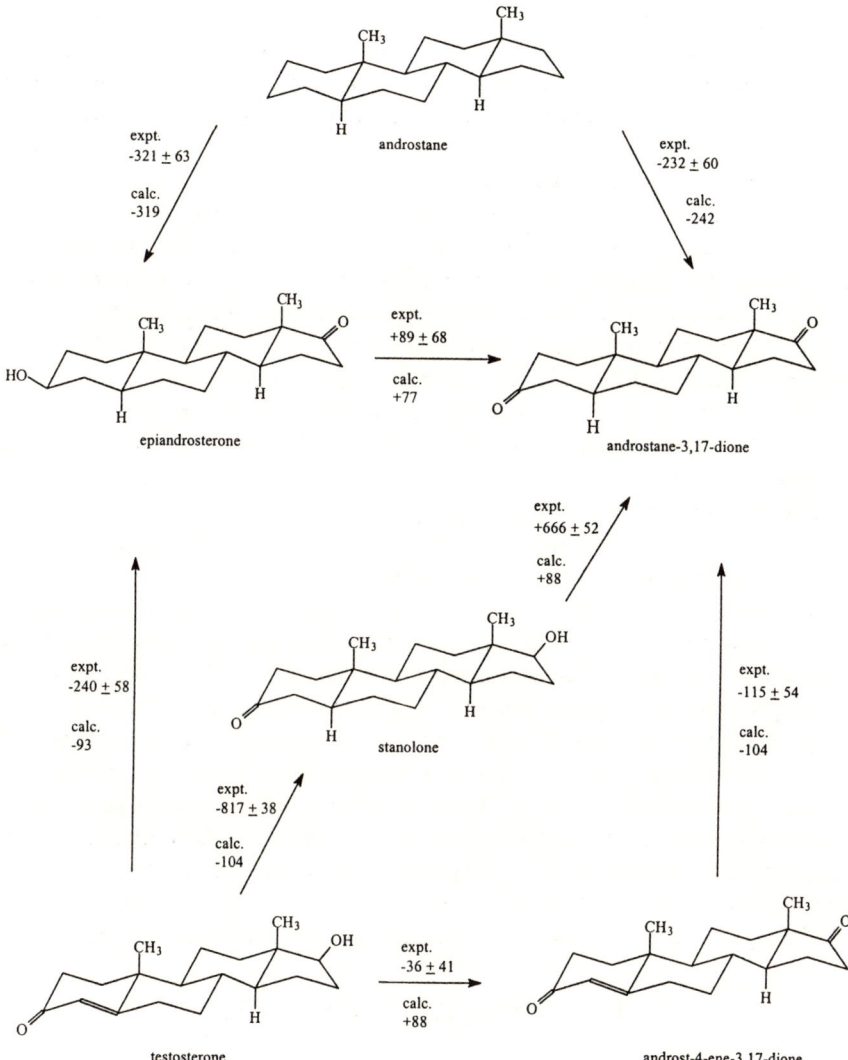

Scheme 2.

of formation for cholesterol [45] of -675 ± 4 kJ mol^{-1} and the resulting enthalpy of reaction of ca. $+23 \pm 6.4$ kJ mol^{-1} is more compatible with the prototypical methylation of isopropanol to isopropyl methyl ether of $+39.4 \pm 1.1$ kJ mol^{-1}.

In Scheme 2, six steroids are arranged to show their structural relationships via functional group interconversion. Next to each arrow connecting two steroids are two numbers: the one labeled "expt." is the enthalpy of reaction derived from Eq.

31 and the solid-phase enthalpies of formation in Table 2; the number labeled "calc." is the sum of the enthalpies of reaction for the prototype small molecules undergoing the same formal reactions as the steroids. Assuming the enthalpies of formation, and thus of reaction, for the prototypes are reasonably accurate, the enthalpies of formation of androstane, epiandrosterone, androstane-3,17-dione, and androst-4-ene-3,17-dione are accurate in that the experimental and calculated enthalpies of reaction are within 2–12 kJ mol^{-1} of each other. The enthalpy of formation of stanolone is clearly incorrect because the calculated and model enthalpies of reaction are wildly discrepant [48]. The enthalpies of reaction for the testosterone conversions diverge from the model enthalpies of reaction by ca. 124 and 147 kJ mol^{-1}. Not shown in Scheme 2 is the conversion of androsterone (3-α-OH) to epiandrosterone (3-β-OH) with an enthalpy of epimerization of $+42 \pm 58$ kJ mol^{-1} in the solid phase. The closest available prototype comparison is the ca. 4 kJ/mol enthalpic preference for an equatorial over an axial hydroxyl group in 4-t-butylcyclohexanol measured in a range of solvents [49]. Within the uncertainty of the steroid measurements, these values are consistent.

Again assuming that the prototype enthalpies of formation are accurate and their enthalpies of reaction are additive and reasonable models for the steroid enthalpies of formation and of reaction, we can derive a self-consistent set of enthalpies of formation for those steroids which seem now to be inaccurately measured. If the enthalpies of formation of epiandrosterone and androst-4-ene-3,17-dione are accurate, then using the prototype reactions for conversion of either of them to testosterone leads to an enthalpy of formation for that compound of ca. −542 or −519 kJ mol^{-1}. The average enthalpy of formation is −531 kJ mol^{-1} which deviates by 136 kJ mol^{-1} from the measured value. From this enthalpy of formation of testosterone an enthalpy of formation of −635 kJ mol^{-1} for stanolone is derived. Likewise, from the probably accurate enthalpy of formation of androstane-3,17-dione, an identical enthalpy of formation of −634 for stanolone is derived. Interestingly, these are slightly greater than one-half the measured enthalpy of formation of stanolone.

VIII. AROMATIC CARBOCYCLES

Of the numerous aromatic polycyclic benzenoid hydrocarbons discussed by Krygowski and Cyranski in their chapter [50], there are enthalpy-of-formation data for fewer than half of them. Despite the paucity of compounds for which there are data, there is no scarcity of data itself: for naphthalene alone there are at least 10 determinations of the enthalpy of formation of the solid and 15 of its enthalpy of sublimation! All but a few of the solid substances have more than one reference for an enthalpy of formation and usually several references for the enthalpy of sublimation. Benzene, a liquid, seems the sole compound for which there are relatively few measurements and consistent results. Our favorite compendium of data [4] does not include some of the more recent values so we begin by recommending gaseous enthalpies of formation by combining enthalpy of formation and enthalpy of sublimation data for a solid compound according to Eq. 32:

$$\Delta H_f(g) = \Delta H_f(c) + \Delta H_{sub} \tag{32}$$

For the compounds we will discuss here (Figure 1), the resulting gaseous enthalpies of formation are (kJ mol^{-1}): benzene [51] (82.6 ± 0.7), naphthalene [52] (151.2 ± 1.0), anthracene [53] (228.5 ± 4.4), phenanthrene [54] (205.1 ± 2.0), naphthacene [55] (291.4 ± 9.4), chrysene [56] (269.6 ± 6.6), and benz[a]anthracene [57] (293.0 ± 4.3).

Naphthalene (**1**), anthracene (**2**), and naphthacene (**4**) constitute an homologous series in which each successive member is formed by benzo-fusion to the [b] side of the preceding member. As with other homologous series [58], a plot of the gaseous enthalpies of formation of these species vs. the number of carbon atoms [59], n_c, results in a reasonably straight line. An unweighted least-squares regression analysis [60] yields a linear equation with a correlation coefficient of 0.997:

$$\Delta H_f(g) \pm 5.9 = [n_c \cdot (17.5 \pm 1.0)] + (-21.7 \pm 14.9) \tag{33}$$

The correlation and the relatively low regression constant errors [61] assure us that our recommended enthalpies of formation for these species are not unreasonable. For naphthalene (**1**), phenanthrene (**3**), and chrysene (**6**), considered as a benz[a]-fused homologous series, the correlation constant for the enthalpy of formation with

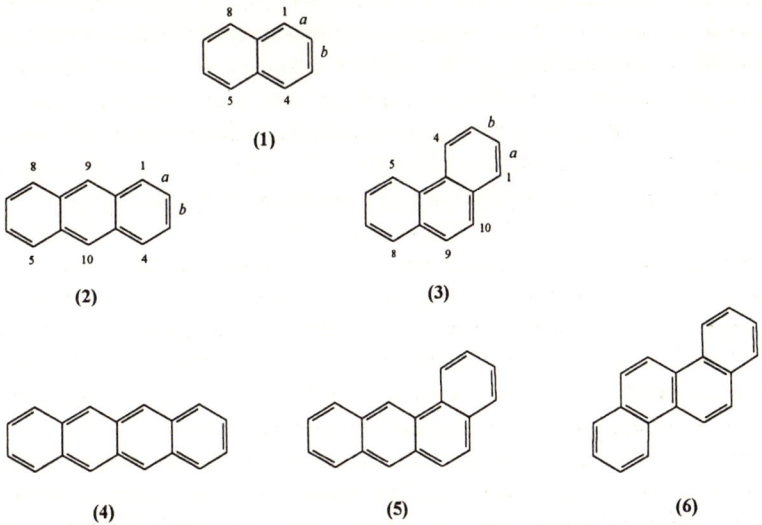

Figure 1. Several polycylic benzenoid hydrocarbons.

the number of carbons is identical to that for the series above. Here too, the errors in the regression constants are not large:

$$\Delta H_f(g) \pm 4.3 = [n_c \cdot (14.8 \pm 0.8)] + (1.4 \pm 11.0) \tag{34}$$

The anthracene/phenanthrene and naphthacene/chrysene isomeric pairs have similar enthalpies of isomerization, a reflection of the similar slopes of Eqs. 33 and 34. However, while the enthalpy of isomerization of anthracene (**2**; "benz[b]naphthalene") to the more stable phenanthrene (**3**; "benz[a]naphthalene") is −23.4 ± 4.8 kJ mol^{-1}, the same structural rearrangement converts naphthacene (**4**; "benz[b]anthracene") into benz[a]anthracene (**5**) with $\Delta H_{isom} = 1.6 \pm 10.3$ kJ mol^{-1}. We cannot dismiss this apparent incongruity of enthalpies of isomerization as due only to the large uncertainties of the enthalpies of formation. Compounds (**3**) and (**5**) are two members of an homologous series whose correlation with the number of carbons [59] could produce regression constants quite different from those in Eq. 33 for the homologous series (**1**, **2**, and **4**). If the slopes of the two series are different the isomerization enthalpies of any two pairs of isomers will be different. Although it is admittedly a circular argument, the slope per carbon atom from Eq. 33 is 17.5 kJ mol^{-1} and the corresponding slope for the only two points in the (**3**, **5**) series is ca. 22 kJ mol^{-1}, a difference which increases quite rapidly for a C$_4$ increase per homologue.

One of the first quantitative measures of aromaticity was a thermochemical one—the determination of the enthalpy of hydrogenation of benzene [62]. The experimentally evolved heat was much less than the exothermicity calculated as three times the enthalpy of hydrogenation of the double bond in cyclohexene. The difference (149.2 kJ mol^{-1} or 49.7 per π-bond) has been called the resonance, or stabilization energy of benzene. Enthalpies of hydrogenation by themselves cannot be used as a reliable quantitative measure of aromaticity for polycyclic aromatic molecules because the aromatic reactants and their hydrogenated products differ in their relative amounts of intramolecular steric interaction and strain. For example, calculated from their respective enthalpies of formation, the exothermicity of formal hydrogenation of *ortho*-xylene to *trans*-1,2-dimethylcyclohexane (−199.0 ± 2.2 kJ mol^{-1}) is less than the hydrogenation of benzene (−206.0 ± 1.1 kJ mol^{-1}). The introduction of the two methyl groups on the benzene ring results in a 63.5 kJ mol^{-1} less positive enthalpy of formation while the introduction of two methyl groups on the cyclohexane ring results in a lowering of the enthalpy of formation by only 56.5 kJ mol^{-1}. It is tempting to state that the dimethylbenzene is thus stabilized relative to benzene by 7 kJ mol^{-1}. Ignoring for the moment the question of steric effects between the two *ortho* groups on benzene, there is unquestionably a *gauche* steric interaction between the two methyl groups in *trans*-1,2-dimethylcyclohexane [32]. The *gauche* substituents' interaction energy is 4.6 ± 2.6 kJ mol^{-1}, calculated from the enthalpy-of-formation difference between *trans*-1,2- and *trans*-1,4-dimethylcyclohexane [2]. The enthalpy-of-formation difference between *ortho-* and *para*-xylene is only 1.1 ± 1.5, essentially thermoneutral, indicating a nearly nonexistent steric effect. The supposed 7 kJ mol^{-1} stabilization of *o*-xylene compared to benzene is now reduced to ca. 2.4 kJ mol^{-1}, nearly vanishing within the uncertainty intervals.

There is another substituent effect which must be accounted for in the calculation of stabilization energy. The enthalpy of hydrogenation of alkenes becomes less exothermic as more alkyl substituent groups are bonded to the double bond (unless steric effects substantially destabilize the alkene). The calculated hydrogenation model for *p*-xylene should not be cyclohexene, a good model for benzene, but should include both 1-methylcyclohexene (-111.4 ± 1.3 kJ mol^{-1}) and cyclohexene (-118.4 ± 1.2 kJ mol^{-1}). For one Kekulé structure of *p*-xylene (1,4-dimethylcyclohexa-1,3,5-triene), the calculated enthalpy of hydrogenation is [$1 \cdot \Delta H_{H2}$ (cyclohexene) $+ 2 \cdot \Delta H_{H2}$ (1-methylcyclohexene) $= -341.2 \pm 2.2$ kJ mol^{-1}], somewhat less exothermic than that calculated for benzene with only cyclohexene-type double bonds ($-355.2 + 1.2$ kJ mol^{-1}). The stabilization energy for *p*-xylene is thus 138.7 ± 3.0 kJ mol^{-1} compared to benzene's 149.2 ± 2.1 kJ mol^{-1}. Accordingly, benzene is reckoned more stable than *o*- and *p*-xylene [64].

How do these conclusions regarding the xylenes relate to hydrogenation of naphthalene which may be considered an *ortho*-disubstituted benzene? The ΔH_{H2} of naphthalene to tetralin is -125.2 ± 2.2 kJ mol^{-1}. Further reduction of tetralin to *trans*-decalin is -208.1 ± 3.0 kJ mol^{-1}. The tetralin-to-decalin hydrogenation is analogous to hydrogenation of *o*-xylene in that account must be taken of steric interactions that may be of different magnitudes in reactant and product. The enthalpy-of-formation difference between benzene and tetralin is 56.6 kJ mol^{-1} and between cyclohexane and *trans*-decalin is 58.7 kJ mol^{-1}. Strain in tetralin is manifested in its 6.9 kJ mol^{-1} *more positive* enthalpy of formation compared to *o*-xylene despite its greater molecular weight. The enthalpy of formation of *trans*-decalin is only 2.2 kJ mol^{-1} more negative than *trans*-1,2-dimethlycyclohexane, showing that it too must be relatively strained. Unfortunately, there are no unstrained analogues for which there are enthalpies of formation to compare to *trans*-decalin. A slight correction can be made in the calculation of the stabilization energy by incorporating the enthalpy of hydrogenation of two 1-methylcyclohexenes and only one cyclohexene as was done for the xylenes. The result is 133.1 ± 3.7 kJ mol^{-1} stabilization, less than for the xylenes. Extending this to the complete reduction of naphthalene to *trans*-decalin while acknowledging the inability to correct for strain, the reference Kekulé structure has three cyclohexene-type double bonds and two methylcyclohexene-type double bonds. The calculated enthalpy of hydrogenation is thus -578.0 ± 2.8 and the stabilization energy is -244.7 ± 3.8 kJ mol^{-1} or 48.9 kJ mol^{-1} per π-bond. The same calculation procedure for anthracene with an enthalpy of hydrogenation to *trans-syn-trans*-perhydroanthacene of -471.7 ± 5.8 kJ mol^{-1} yields a stabilization energy of 47.0 kJ mol^{-1} per π-bond. Unfortunately, there is no enthalpy of formation data for perhydronaphthacene or any other fully reduced benzenoid hydrocarbon. However, the most stable hydrogenation products from the homologous series (**2**, **3**, **5**) also form an homologous series and so their gaseous enthalpies of formation vs. number of carbons would be linearly correlated. From the slope (−15.3) and intercept (−29.4) derived from the enthalpies of formation of decalin and perhydroanthracene, an extrapolated enthalpy of formation of *trans-syn-trans-syn-trans*-perhydronaphthacene is −304.8 kJ mol⁻¹. The stabilization energy is calculated as 47.5 kJ mol⁻¹ per π-bond.

It would be tempting to observe that because the hydrogenation enthalpy is the difference between the enthalpies of formation of the saturated and unsaturated compounds,

$$\Delta H_{H2} = \Delta H_f(\text{sat'd, g}) - \Delta H_f(\text{unsat'd, g}) \tag{35}$$

that we can write a linear equation which includes a constant for the enthalpy of hydrogenation:

$$\Delta H_f(\text{sat'd, g}) = [m \cdot \Delta H_f(\text{unsat'd, g})] + \text{``}\Delta H_{H2}\text{''} \tag{36}$$

However, only in the case where the absolute value of the slope for the saturated compounds is identical to the absolute value of the slope for the unsaturated compounds is the enthalpy of hydrogenation per π-bond constant and equal to the y-intercept ("ΔH_{H2}") of Eq. 36 [65]. If the absolute values of the two slopes are significantly different, the enthalpy of hydrogenation per π-bond will not be constant for the members of the unsaturated series but instead will exhibit a regular increase or decrease. In the present case, the relevant slopes are 17.5 and −15.3 and thus the hydrogenation enthalpies per π-bond, decrease slightly throughout the series.

IX. DIFLUORAMINES: A COMPARISON WITH NITRO COMPOUNDS

In the chapter by Politzer and Lane [66], the authors present the results of their ab initio calculations for the structures and enthalpies of formation of difluoramino and comparable nitro compounds. It is not unreasonable to assume that species with the generic formulas RNF_2 and RNO_2 will have a rather constant enthalpy of formation difference. Both NF_2 and NO_2 are electron-withdrawing substituents, and while the latter has a formal +1 charge on the nitrogen, fluorine is more electronegative than oxygen. The nitro group can conjugatively interact and thus enjoy resonance stabilization with a π-system to which it is affixed, but negative fluorine hyperconjugation can accomplish much of the same for difluoramines. Neither group is particularly bulky. Now what does experiment and *noncalculational* theory tell us about the enthalpies of formation of these two classes of species? Or more precisely, what trends or even constancy is there in the difference quantity δ_{36} where *p* is the number of NF_2 and NO_2 groups:

$$[\Delta H_f(RNF_2) - \Delta H_f(RNO_2)]/p \equiv \delta_{36} \tag{36}$$

Let us commence with simple inorganic compounds containing NF_2 and NO_2 groups. The simplest pair of compounds to compare is composed of the triatomics themselves, NF_2 and NO_2. Their respective enthalpies of formation are 43.1 and 33.2 kJ mol^{-1}, corresponding to a difference of 10.1 kJ mol^{-1}. The difference quantity for their N-bridged dimers, N_2O_4 and N_2F_4, with enthalpies of formation of −7.1 and 9.2, is ca. −8 kJ mol^{-1}. What about R = H? We realize that HNO_2 is in fact nitrous acid or H–O–N=O, and so experiment gives us but a lower bound for the enthalpy of formation of its still unknown and by all accounts less stable isomer H–NO_2 [67]. We thus have only an upper bound of δ_{36}(H). Accepting the enthalpies of formation of HNF_2 from ref. 68 and of nitrous acid from ref. 69 as −65 ± 4 and −79.5 kJ mol^{-1}, respectively, we derive the desired upper bound of 14.5 kJ mol^{-1}. We now turn to the last pair of inorganic compounds, NF_3 and NO_2F, for which the enthalpies of formation (again taking data from the JANAF tables [10]) are −132.089 ± 1.13 and −108.784 ± 20.9 kJ mol^{-1} with a difference of −24 ± 21 kJ mol^{-1}. This last pair is particularly evocative. Consider the first value, that for NF_3. This species is relatively "clean", i.e. it is quite readily purifiable and relatively unreactive. However, the NBS tables [5] recommend an enthalpy of formation of −124.7 kJ mol^{-1} for the enthalpy of formation of NF_3, considerably outside the earlier enunciated error bars. Documentation of the enthalpy of formation of NO_2F by the JANAF editors [10] includes considerable irreproducibility in this measurement.

We thus find a wide range of values for δ_{36}(R). However, the R groups and associated nitro and difluoramino compounds presented above—".", NF_2 and NO_2 themselves, H, and F—are really quite disparate. What about organic derivatives? Our organic thermochemical archive also presents us with a set of rather disparate difluoromine derivatives: we cite results from a generally ignored source as well [70]. Along with their condensed- and gas-phase enthalpies of formation, these compounds are given in Table 4.

What is known about the thermochemistry of the corresponding nitro compounds? The answer is surprisingly little. Consider the first six compounds in Table 4 with only one carbon. Paralleling the first species, tetrafluoroformamidine, is α-nitro-N-fluoroformaldimine. We lack thermochemical data on any compound of this type, and for that matter, thermochemical data on any N-fluoroimine is absent. *A fortiori*, the situation on the third compound, pentafluoroguanidine, is worse since there are additional intersubstituent interactions that we do not know how to simulate. The next four compounds form a nominally homologous series $CF_{4-p}(NF_2)_p$. The first entry, pentafluoromethylamine, corresponds to trifluoronitromethylamine. No experimental measurement of its enthalpy of formation is known to us—the calculated [67] value is −681 kJ mol^{-1}. The enthalpies of formation of the nitro counterparts of the last two members of the series are both known from experiment: as gases, the values are −186.2 ± 8.7 and 82.0 ± 2.8 kJ

Table 4. Enthalpies of Formation of Some Difluoramine Derivatives (kJ mol^{-1})

Formula	Name	$\Delta H_f(l)$	$\Delta H_f(g)$
CF_4N_2	tetrafluoroformamidine		-141 ± 9
CF_5N	pentafluoromethylamine		-707.5 ± 2.2
CF_5N_3	pentafluoroguanidine		95.7 ± 3.6
CF_6N_2	hexafluoromethanediamine		-455.3 ± 4.0
CF_7N_3	heptafluoromethanetriamine		-200.1 ± 2.8
CF_8N_4	octafluoromethanetetramine		1.5 ± 5.6
$C_2F_{11}N_5$	4,4-bis(difluoroamino)hepta-fluorodimethylenetriamine		-362.6 ± 7.2
$C_2H_4F_4N_2$	N,N,N',N'-tetrafluoro-ethane-1,2-diamine[a]	-197 ± 17	-173
$C_3H_6F_4N_2$	N,N,N',N'-tetrafluoro-propane-1,2-diamine[a]	-213 ± 17	-180 ± 17
$C_4H_8N_2F_4$	N,N,N',N'-tetrafluoro-butane-2,2-diamine[a]	-213 ± 5	-179
$C_4H_8N_2F_4$	N,N,N',N'-tetrafluoro-butane-2,3-diamine[a,c]	-236.2 ± 4.2 -233.3 ± 5.0	-192.5 -198.1
$C_5H_7F_4N_5O_6$	N,N,N',N'-tetrafluoro-5,5,5-trinitro-2,2-pentanediamine[a]	-265.5 ± 3.2	
$C_6H_{10}F_4N_2$	N,N,N',N'-tetrafluoro-1,1-cyclohexanediamine	-218.4 ± 3.5	-174.1 ± 3.6
$C_6H_{11}F_3N_2$	N,N,N'-trifluorohexanamidine[b]	-224.1 ± 0.9	-177.6 ± 1.0
$C_6H_{12}F_4N_2$	N,N,N',N'-tetrafluoro-4-methyl-2,2-pentanediamine[a]	-274.5 ± 1.8	-229.5 ± 1.8
$C_6H_{12}F_4N_2$	N,N,N',N'-tetrafluoro-4-methyl-2,3-pentanediamine[a,c]	-285.6 ± 3.2 -307.1 ± 2.6	-242.9 ± 3.2 -260.6 ± 2.6
$C_6H_{12}F_4N_2$	N,N,N',N'-tetrafluoro-4-methyl-1,2-pentanediamine	-264.9 ± 4.4	-220.0 ± 4.4
$C_7H_7F_2N$	N,N-difluorobenzylamine	-37.4 ± 2.7	7.0 ± 3.0
$C_7H_{14}F_4N_2$	N,N,N',N'-tetrafluoro-1,1-heptanediamine	-282.0 ± 3.1	-231.5 ± 3.2

Notes: [a]These data were obtained from ref. 70.

[b]The name given in ref. 4 is incorrectly given as N,N,N'-trifluorohexanediamine, confirmed by checking the original source, ref. 71.

[c]The two sets of values are for different, but otherwise not identified, stereoisomers.

mol^{-1}. The corresponding δ_{36} (remember, it is normalized per NF_2 and NO_2 groups) is -26, -5, and -20 kJ mol^{-1}. Why the middle compound with $p = 3$ should be so discrepant is not apparent: we note that the enthalpy of formation difference for benzyldifluoramine and α-nitrotoluene is -23.7 kJ mol^{-1}. Tentatively, we suggest a normalized gas phase δ_{36} of ca. -23 kJ mol^{-1}.

We now turn to some difluoramino compounds with more carbons and consider these species now as liquids. We do not expect the same δ_{36} value as for gases—to be equal would assume that the contribution to the enthalpy of vaporization of

NO_2 and NF_2 groups would be the same as well. Proceeding through the admittedly rare examples where comparison can be made by use of plausible corrections and estimates, we start with N,N,N',N'-tetrafluoroethane-1,2-diamine and 1,2-dinitroethane for which the direct arithmetic enthalpy of formation difference is -32 ± 17 kJ mol^{-1} and so δ_{36} is -16 ± 13 kJ mol^{-1}. For N,N,N',N'-tetrafluorocyclohexane-1,1-diamine, it is necessary to estimate the enthalpy of formation of 1,1-dinitrocyclohexane. If we assume that the following formal reaction is thermoneutral,

$$\text{cyclo-}(CH_2)_5CO + (CH_3)_2C(NO_2)_2 \rightarrow \text{cyclo-}(CH_2)_5C(NO_2)_2 + (CH_3)_2CO \quad (37)$$

we find the desired estimate to be -204 kJ mol^{-1} from which the normalized value of δ_{36} of -7 kJ mol^{-1} is found.

For liquid benzyldifluoramine and α-nitrotoluene, the corresponding value (from solely experimental measurements) of δ_{36} is -14.6 ± 3.7 kJ mol^{-1}. We close the comparison with a discussion of the thermochemistry of N,N,N',N'-tetrafluoro-1,1-heptanediamine. The enthalpy of formation of 1,1-dinitroheptane is seemingly absent but we know the value for 1,1-dinitropentane, -216.9 ± 1.3 kJ mol^{-1}. To correct for the two liquid-phase CH_2 groups is easy; prior experience (e.g. ref. 34) shows that each contributes ca. -25 kJ mol^{-1} resulting in a value of ca. -267 kJ mol^{-1}. Alternatively, we can use our earlier derived equation [7],

$$\Delta H_f°(1,1,1-(NO_2)_3CR) = -22.9 \cdot n_c - 100.5 \quad (38)$$

where n_c is the total carbon count and derive a value of -261 kJ mol^{-1}. Using an average value of -264, and then normalizing per group, we derive the desired δ_{36} value of -18 ± 3 kJ mol^{-1}.

Summarizing, we suggest liquid- and gas-phase differences of the enthalpies of formation of difluoramino and nitro compounds of ca. -16 and -23 kJ mol^{-1}, respectively.

X. SE–N AND TE–N COMPOUNDS AND SOME THERMOCHEMICAL ANALOGIES

In their chapter, Tornieporth-Oetting and Klapötke [73] discuss complementary aspects of experimental structure determinations and ab initio calculational theory as applied to species containing selenium or tellurium and nitrogen. It is to be acknowledged that experimentally derived thermochemical data is almost nonexistent for such compounds. Admitting now that we are generalizing their scope of compounds for the current vignette, we also acknowledge that few of the original authors' species are to be found in our analysis: the data simply aren't there.

The first and only species in common is written as solid NSe in the NBS tables cited in ref. 5. We assume that it is elsewhere characterized as N_4Se_4 and so arithmetically derive its enthalpy of formation to be a simple multiple of the archival value, $177.0 \cdot 4 = 708$ kJ mol^{-1}. By contrast, we find in the NBS tables [5] an enthalpy of formation of N_4S_4 as 535.6 kJ mol^{-1}, corresponding to a difference of ca. 43 kJ mol^{-1} per group 16 atom. Is that difference plausible? Consider now some gaseous species—H_2S and H_2Se, SCl_2 and $SeCl_2$—to represent singly bonded chalcogen atoms, and SO and SeO, $SOCl_2$ and $SeOCl_2$, to represent doubly bonded ones. The respective S/Se enthalpy of formation differences are 50.3 and -12.1 kJ mol^{-1} for the singly bonded paradigms and -47 and -187 kJ mol^{-1} for the doubly bonded ones. The simplest conclusion is that sulfur and selenium thermochemistry do not correspond. It is also entirely reasonable to assume that selenium compounds still are very poorly characterized with regard to their energetics.

How well do selenium and tellurium compounds correspond thermochemically? Acknowledging the occasional difference in coordination number, e.g. 4 for Se(VI) as found in solid H_2SeO_4 but 6 for Te(VI) in the corresponding solid oxyacid, H_6TeO_6, let us maintain our preference and prejudice of solely considering gaseous species. The differences in the enthalpies of formation of Se_2 and SeTe; SeTe and Te_2; SeO and TeO are nearly identical: 13, 9 and 12 kJ mol^{-1}. However, the differences for the dihydrides and hexafluorides are now ca. 70 and -200 kJ mol^{-1}. Perhaps many of these numbers are wrong—selenium and tellurium thermochemistry is still embryonic and no doubt problematic [74]. It is also acknowledged that there are profound differences between corresponding elements in the 4th and 5th rows in the periodic table: contrast the rich chemistry of xenon with that of its lighter congener, krypton. The current chapter authors hope that new information is sought, found, and eventually integrated into our understanding of the energetics of general nonmetal inorganic compounds.

REFERENCES AND NOTES

1. Hargittai, M.; Hargittai, I. (Eds.). *Advances in Molecular Structure Research*; JAI Press: Greenwich, CT, 1997, Vol. 3.
2. Liebman, J. F.; Slayden, S. W. In: Ref. 1, *Some Relationships Between Molecular Structure and Thermochemistry*, pp. 313–336.
3. Hargittai, M.; Hargittai, I. (Eds.). *Advances in Molecular Structure Research;* JAI Press: Greenwich, CT, 1996, Vol. 2.
4. Our major (and implicit) reference for enthalpies of formation of organic compounds is Pedley, J. B.; Naylor, R. D.; Kirby, S. P. *Thermochemical Data of Organic Compounds*, 2nd ed.; Chapman & Hall: New York, 1986.
5. The primary thermochemical source for inorganic and some small organic compounds is Wagman, D. D.; Evans, W. H.; Parker, V. B.; Schumm, R. H.; Halow, I.; Bailey, S. M.; Churney, K. L.; Nuttall, R. L. *J. Phys. Chem. Ref. Data* 1982, *11*, Supplement 2.
6. Enthalpies of vaporization may be estimated according to the analysis of: Chickos, J. S.; Hyman, A. S.; Ladon, L. H.; Liebman, J. F. *J. Org. Chem.* 1981, *46*, 4294 for hydrocarbons; and Chickos, J. S.; Hesse, D. G.; Liebman, J. F.; Panshin, S. Y. *J. Org. Chem.* 1989, *54*, 3424 for their simple derivatives.
7. Enthalpy of fusion data are generally taken from Domalski, E. S. Hearing, E. D. *J. Phys. Chem. Ref. Data* 1996, *25*, 1 without any attempt to correct them from the value measured at the melting

point by means of measured heat capacity values. Except for the choice of their "best" or "A" values, no further evaluation of Domalski and Hearing's data was made by the current authors.

8. Demaison, J.; Wlodarczak, G.; Rudolph, H. D. In: Ref. 1, *Determination of Reliable Structures from Rotational Constants*, pp. 1–51.

9. Demaison, J.; Wlodarczak, G.; Rudolph H. D. (ref. 8) present numerous "types" of bond lengths, and accordingly numerous values for each compound. The values we have chosen are their recommended values of r_e, the quantity that is most often cited when bond lengths are casually referred to in both the research and textbook literature.

10. Chase, Jr., M. W.; Davies, C. A.; Downey, Jr., J. R.; Frurip, D. J.; McDonald, R. A.; Syverud, A. N. *J. Phys. Chem. Ref. Data*, **1985**, *14*, supplement 1.

11. The enthalpies of formation of (*Z*)- and (*E*)-1,2-dichloroethylene are very nearly the same and so the seeming ambiguity of formal *syn* vs. *anti* addition of Cl_2 to acetylene has no thermochemical consequence.

12. Lias, S. G., Bartmess, J. E., Liebman, J. F., Holmes, J. L., Levin, R. D.; Mallard, W. G. *J. Phys. Chem. Ref. Data* **1988**, *17*, supplement 1.

13. For this analysis we chose the more stable (*E*)-isomer of $CH_3CH=CHCH_3$ for which there is a $4.3 \pm 1.4 \, kJ \, mol^{-1}$ difference between its enthalpy of formation and that of the less stable (*Z*)-isomer. In light of other uncertainties, nothing is qualitatively changed by choice of the 2-butene isomer.

14. As individuals who have used enthalpies of formation derived from the analysis of appearance energy measurements, we do not particularly appreciate this prejudice nor can we give one key reference that either legitimizes nor invalidates the use of thermochemical data from this source.

15. George, P.; Bock, C. W.; Trachtman, M. In *Molecular Structure and Energetics (Biophysical Aspects)*; Liebman, J. F.; Greenberg, A., Eds.; VCH: New York, 1987, Vol. 4.

16. Gund, P. *J. Chem. Educ.* **1972**, *49*, 100.

17. Slayden, S. W.; Liebman, J. F.; Mallard, W. G. In *The Chemistry of Functional Groups Supplement D2: The Chemistry of Organic Halides, Pseudohalides and Azides*: Patai. S.; Rappoport, Z., Eds.; Wiley: Chichester, 1995.

18. Related problems are seemingly inherent in the thermochemistry of all of the mono and dihaloacetylenes and so involving dichloroacetylene in the current discussion provides comparatively little additional insight.

19. Spiridonov, V. P. In: Ref. 1, *Equilibrium Structure and Potential Function: A Goal to Structure Determination*, pp. 53–81.

20. An example of these constraints are the Wigner–Wittmer rules which disallow actual ground-state dissociation of CO_2 into the ground states of CO and O, but instead forces the use of the ^1D state as the atomic oxygen product if the diatomic CO is to be in its electronic ground state. This allowed and, in fact, the lowest lying excited state of atomic oxygen is $15867.862 \, cm^{-1}$ or $189.8 \, kJ \, mol^{-1}$ above the ground state. (See Herzberg, G. *Molecular Spectra and Molecular Structure I. Spectra of Diatomic Molecules, 2nd ed.*; D. Van Nostrand, Princeton, 1966, p. 315 ff. and *Molecular Spectra and Molecular Structure III. Electronic Spectra and Electronic Structure of Polyatomic Molecules*. D. Van Nostrand: Princeton, 1966, p. 281 ff. for a discussion of these rules and Moore, C. E., *Ionization Potentials and Ionization Limits Derived from the Analyses of Optical Spectra, Nat. Stand. Ref. Dat. Ser., Nat. Bur. Stand. (U.S.)*, (NSRDS-NBS 34), 1970.

21. This latter dissociation energy is $722.0 \, kJ \, mol^{-1}$ and so normalcy seemingly returns. However, the mean dissociation energy may naturally be defined by (a) averaging the sequential dissociation energies of CO_2 and CO, and (b) halving the total dissociation energy of CO_2 into (all ground atomic state) $C + 2O$. These two numbers differ by half of the $^3P - {}^1D$ excitation energy of atomic oxygen or numerically ca. $90 \, kJ \, mol^{-1}$.

22. See O'Hare, P. A. G.; Curtiss, L. A. *J. Chem. Thermodyn.* **1995**, *27*, 643 for a discussion of the experimentally measured thermochemistry of other $14(16)_2$ species such as $GeSe_2$. Not enough is known about other classes of AB_2 species to determine whether such near thermochemical

constancy arises in this case with individual (and sensible) values of the ratio for each class, e.g. $17(16)_2$ species such as ClO_2.

23. Oxidizing power generally increases for nonmetal oxygen species as the central element goes to the right in the Periodic Table. For example, one may contrast carbonates and nitrates, or phosphates, sulfates and perchlorates, as salts and *a fortiori*, their corresponding conjugate acids. Nonetheless, this generality should not be taken for granted. Admittedly, there are numerous sulfoxides and sulfones with their >SO and >SO$_2$ substructure, while corresponding organic oxochlorine species are limited to $C_6H_5ClO_3$ and a few other perchloryl derivatives. However, oxychlorine anions ClO_n^- with $n = 1, 2, 3$, and 4 (and, for completeness 0) are all long-known while their dianionic isoelectronic sulfur analogues, SO_n^{-2}, are seemingly and enigmatically limited to $n = 3, 4$ (and 0).

24. The requisite thermochemical data for the sulfur-containing monoanions (SO_n^-) were taken from ref. 12.

25. Mack, H.-G,; Oberhammer, H. In: Ref. 1, *Structures and Conformations of Some Compounds Containing C–C, C–N, C–O, N–O and O–O Single Bonds: Critical Comparison of Experiment and Theory*, pp. 83–115.

26. Berthelot, M. P. E. *Compt. Rendu.* **1891**, *112*, 829.

27. Following an earlier observation by Benson, S. W. *Chem. Rev.* **1978**, *78*, 23, arose the generaliza- tions by Woolf, A. A., *Adv. Inorg. Chem. Radiochem.* **1981**, *24*, 1; *J. Fluor. Chem.* **1978**, *11*, 307; *ibid.* **1982**, *20*, 627; **1986**, *32*, 453, and both extensions and qualifications, by Liebman, J. F. in *Fluorine-Containing Molecules: Structure, Reactivity, Synthesis and Applications*, Liebman, J. F.; Greenberg, A.; Dolbier, Jr., W. R., Eds., VCH: New York, 1988, and Kunkel, D. L.; Fant, A. D.; Liebman, J. F., *J. Mol. Struc.*, **1993**, *300*, 509.

28. The gaseous enthalpy of formation of -1406.1 ± 8.1 kJ mol^{-1} is from Kolesov, V. P.; Kozina, M. P. *Russ. Chem. Rev.* **1986**, *55*, 1603.

29. The gaseous enthalpy of formation of -785.9 ± 1.0 kJ mol^{-1} was obtained by summing the archival enthalpy of formation of solid malonic acid, -891.0 ± 0.4 kJ mol^{-1}, and its sublimation enthalpy, -105.1 ± 0.8 kJ mol^{-1} from Al-Takhin, G.; Pilcher, G.; Bickerton, J.; Zaki, A. A. *J. Chem. Soc., Dalton Trans.*, **1983**, 2657.

30. While the most stable form of 2,4-pentanedione is the keto-enol (β-hydroxyenone), we also note the study by Hacking, J. M.; Pilcher, G. *J. Chem. Thermodyn.* **1979**, *11*, 1015 that presents the enthalpy of formation of both the dione and keto-enol tautomers. The desired diketo enthalpy of formation gas phase value is -374.4 ± 1.3 kJ mol^{-1}.

31. We adapt the value from Foss, G. D.; Pitt, D. A. *J. Phys. Chem.* **1968**, *72*, 3512 for both hypofluorite species even though they did not perform the measurement on CF_3OF. Rather, the thermochemistry of the dihypofluorite is inseparable from the monohypofluorite because of the reaction calorimetry used to determine the values of interest. We consider self-consistency of values to be more important than precise numbers in the current context.

32. Consistent with the hypofluorite study, we have chosen the enthalpy of formation of CF_4 used by Foss and Pitt, *op. cit.*, -922.6 ± 10.5 kJ mol^{-1}, instead of the archival -933.6 ± 1.4 kJ mol^{-1}. Nothing is qualitatively changed by our choice of values.

33. Boyd, A. A.; Nozière, B.; Desclaux, R. *J. Phys. Chem.* **1995**, *99*, 10815.

34. These reactions are adapted from those discussed in Slayden, S. W.; Liebman, J. F. In: *The Chemistry of Functional Groups Supplement E2: The Chemistry of Hydroxyl, Ether and Peroxide Groups*, Patai, S., Ed.; Wiley: Chichester, 1993.

35. Cermak, I.; Monningen, G., Krätschmer, W. In: Ref. 1, *Absorption Spectra of Matrix-Isolated Small Carbon Molecules*, pp. 117–146.

36. Roth, W. R.; Adamczak, O.; Breuckmann, R.; Lennartz, H.-W.; Boese, R. *Chem. Ber.* **1991**, *124*, 2499.

37. Liebman, J. F. In: *The Chemistry of Functional Groups Supplement A2: The Chemistry of Dienes and Polyenes*; Rappoport, Z., Ed.; Wiley, Chichester, 1997, Vol. 1. For reference, the actual numbers are 32.5, 28.2, 31.4, 33.9 and 29.2 kJ mol^{-1} respectively.

38. It is often found that the first member of many thermochemical series is an outlier, e.g. Slayden and Liebman, *op. cit.*, ref. 34. However, since most of the preceding demethylated species so qualify as first members of their respective homologous series, this does not seem to be an adequate reason.

39. Zorky, P.; Zorkaya, O. N. In: Ref. 1, *Specific Intermolecular Interactions in Organic Crystals: Conjugated Hydrogen Bonds and Contacts of Benzene Rings*, pp. 147–188.

40. See, for example, Slayden and Liebman, *op. cit.*, ref. 34.

41. This is a corollary of the approach used by Chickos, Liebman, and their coworkers in ref. 6.

42. Chickos, J. S.; Annunziata, R.; Ladon, L. H.; Hyman, A. S.; Liebman, J. F. *J. Org. Chem.* **1986**, *52*, 4311.

43. Kálmán, A.; Párkányi, L. In: Ref. 1, *Isostructurality of Organic Crystals: A Tool to Estimate the Complementarity of Homo and Heteromolecular Associates*, pp. 189–226.

44. Unless otherwise noted, solid phase enthalpies of formation were derived from the solid phase enthalpies of combustion in Paoli, D.; Garrigues, J.-C.; Patin, H. *C. R. Acad. Sci., Paris, Ser. C* **1969**, *268*, 780.

45. Johnson, W. H. *J. Res. Natl. Bur. Stand. Sect. A* **1975**, *79*, 493.

46. The enthalpy of formation cited for *cis*-2-methylcyclopentanol in ref. 4 is actually the value for 1-methylcyclopentanol. The liquid phase enthalpies of isomerization of 1,1-dimethylcyclopentane and -cyclohexane to their *cis*-1,2-dimethylcycloalkane isomers are +6.7 and +6.9 kJ mol^{-1}, respectively. In contrast to these endothermic enthalpies of isomerization of the hydrocarbons is the exothermic enthalpy of isomerization (–9.2 kJ mol^{-1}) of 1-methylcyclohexanol to the more stable *cis*-2-methylcyclohexanol. Assuming the same energy effects on the isomerization of 1-methylcyclopentanol to *cis*-2-methylcyclopentanol, the liquid enthalpy of formation of the latter is about –353 kJ mol^{-1}. The identical value is generated from a different cycle involving the same compounds: The enthalpy difference between 1,1-dimethylcyclopentane and 1,1-dimethylcyclohexane is –46.5 and that between *cis*-1,2-dimethylcyclopentane and *cis*-1,2-dimethylcyclohexane is –46.7. Assuming the enthalpy difference between *cis*-2-methylcyclohexanol and *cis*-2-methylcyclopentanol is the same as the difference between 1-methylcyclopentanol and 1-methylcyclohexanol (–37.5), the enthalpy of formation of *cis*-2-methylcyclopentanol is –353 kJ mol^{-1}. The derived value is consistent with the observation that replacing a methyl with the similar-size hydroxyl group on cyclopentane is slightly more exothermic than the same replacement on cyclohexane for both the mono- and 1,1-disubstituted cycloalkanes.

47. Rogers, D. W. Unpublished results. Microcalorimetric determination of the enthalpies of hydrogenation of alkenes in dilute alkane solution are equivalent to gas-phase hydrogenation values (See for example Rogers, D. W.; Dagdagan, O. A.; Allinger, N. L. *J. Am. Chem. Soc.* **1979**, *101*, 671; Rogers, D. W.; Dejroongruang, K.; Samuel, S. D.; Fang, W.; Zhao, Y. *J. Chem. Thermodynamics* **1992**, *24*, 561; and references cited therein). Unpublished enthalpies of hydrogenation of 2-cyclopentenone and 2-cyclohexenone are –98.1 ± 1.7 and –107.2 ± 1.7 kJ mol^{-1}, respectively. From these values and the enthalpies of formation of cyclopentanone and cyclohexanone, the gas phase enthalpies of formation of 2-cyclopentenone and 2-cyclohexenone are –94.0 ± 2.5 and –118.9 ± 2.7 kJ mol^{-1} respectively. Conjugation in the enone may affect its polarity and increase the enthalpy of vaporization relative to the ketone. The only comparison conjugated carbonyl is crotonaldehyde whose enthalpy of vaporization is 3.7 kJ mol^{-1} more positive than that for butyraldehyde. From this difference and the enthalpies of vaporization of cyclopentanone and cyclohexanone, the liquid-phase enthalpies formation of 2-cyclopentenone and 2-cyclohexenone are –141.3 and –167.7 kJ mol^{-1}, respectively. We thank D. W. Rogers for these results in advance of publication.

48. The authors of the original publication (ref. 44) noted the inconsistency of the measured enthalpy of combustion of stanolone.

49. Eliel, L. E.; Gilbert, E. C., *J. Amer. Chem. Soc.* **1969**, *91*, 5487.

50. Krygowski, T. M.; Cyranski, M. In: Ref. 1, *Aromatic Character of Carbocyclic π-Electron Systems Deduced from Molecular Geometry*, pp. 227–268.

51. Additional phase-change references not explicitly cited are taken from H. Y. Afeefy, J. F. Liebman, and S. E. Stein, "Neutral Thermochemical Data" in *NIST Standard Reference Database Number 69*; Mallard, W. G. Linstrom, P. J., Eds.; March 1998, National Institute of Standards and Technology: Gaithersburg MD (http://webbook.nist.gov).

52. Of three solid-phase enthalpies of formation for naphthalene determined between 1960 and 1966, ref. 4 recommended two of them. Since then, two other determinations have been made (Ammar, M. M.; El Sayed, N.; Morsi, S. E.; El Azmirly, A. *Egypt. J. Phys.* **1977**, *8*, 111 and Metzger, R. M.; Kuo, C. S.; Arafat, E. S. *J. Chem. Thermodyn.* **1983**, *15*, 841). Taking into account the different experimental uncertainty intervals, the weighted average is 78.8 ± 0.9 kJ mol^{-1}. Since 1972 there have been ten measurements of the enthalpy of sublimation with an unweighted average of 72.4 ± 0.4 kJ mol^{-1}.

53. Of the many solid enthalpy of formation values determined for anthracene, ref. 4 chose only one, at that time the newest. A newer solid enthalpy of formation (see ref. 52) later became available, which although somewhat less positive than the earlier value is compatible with it because of the relatively large uncertainty interval ($\Delta H_f(c) = 125.54 \pm 5.6$ kJ mol^{-1}). The weighted average is 128.9 ± 3.4. The unweighted average of six enthalpies of sublimation (since 1972) is 99.6 ± 2.8 kJ mol^{-1}.

54. After the archival ref. 4 was published, a new enthalpy of formation of solid phenanthrene was determined ($\Delta H_f(c) = 109.76 \pm 1.6$ from Steele, W. V.; Chirico, R. D.; Nguyen, A.; Hossenlopp, I. A.; Smith, N. K. *Am. Inst. Chem. Eng. Symp. Ser.* **1990**, 138. The two values differ by almost 7 kJ mol^{-1} and are not identical within the error bars. Three enthalpy of sublimation measurements differ by ca. 1.5 kJ mol^{-1}. The weighted average of the derived gaseous enthalpies of formation is used here.

55. The inconsistency of the cited enthalpies of sublimation of naphthacene is a problem. Of five citations since 1951 including the newest, three are ca. 125 kJ mol^{-1}. A 1952 determination is 8 kJ mol^{-1} less and a 1980 determination is 18 kJ mol^{-1} higher. There is only one solid enthalpy of formation. Ref. 4 recommends an average (291.4 ± 9.4 kJ mol^{-1}) of the two gaseous enthalpies of formation calculated using the highest enthalpy of sublimation as well as the median value.

56. There have been five independent measurements of the enthalpy of sublimation of chrysene during the last 45 years. Of these, two are identically 131.0 ± 4.0 and the other three average 118.0 ± 0.7. Combining each of these with the only solid enthalpy of formation according to Eq. 32 gives two gaseous enthalpies of formation: 262.9 ± 4.7 and 276.3 ± 4.6 kJ mol^{-1}. Ref. 4 averages these, as do we.

57. There is only one published value for benz[a]anthracene, but two values for the enthalpy of sublimation. The weighted average is recommended by ref. 4.

58. Cox, J. D.; Pilcher, G. *Thermochemistry of Organic and Organometallic Compounds*; Academic Press: New York, 1970.

59. The number of carbon atoms was chosen as the independent variable, but we could choose any structural feature which constitutes an homologous increment, such as the number of rings or double bonds, etc.

60. The regression analysis is not weighted by taking into account the enthalpy-of-formation uncertainty intervals for two reasons. The extreme variation of the uncertainty intervals causes too much influence to be given to one of the three data points and the somewhat arbitrary way in which enthalpy of formation and enthalpy of sublimation data were chosen affects the calculated uncertainty intervals. The resulting regression constants and errors are not too different, but they do affect any subsequent extrapolation.

61. Although the error in the y-intercept seems large, it is a consequence of the long extrapolation from the first data point ($n_c = 10$) to the intercept ($n_c = 0$).

62. Kistiakowsky, G. B.; Ruhoff, J. R.; Smith, H. A.; Vaughan, W. E. *J. Amer. Chem. Soc.* **1936**, *58*, 146.

63. Eliel, E. L.; Wilen, S. H. *Stereochemistry of Organic Compounds*; Wiley: New York, 1994.

64. The same calculated hydrogenation energy must be used for both *o*- and *p*-xylene. While *o*-xylene might be better modeled by [2 · ΔH_{H2} (cyclohexene) + 1 · ΔH_{H2} (1,2-dimethylcyclohexene)], there is no measured enthalpy of formation value for the disubstituted cyclohexene.

65. Slayden, S. W.; Liebman, J. F. In *Supplement A3: The Chemistry of the Double-Bonded Functional Groups*; Patai, S., Ed.; Wiley: Chichester, 1997.

66. Politzer, P.; Lane, P. In: Ref. 1, *Conformational Studies of Structures and Properties of Energetic Difluoramines*, pp. 269–285.

67. A quantum-chemically derived value is suggested in Sana, M.; Leroy, G.; Peeters, D.; Wilante, C. *J. Mol. Struct. (Theochem)* **1988**, *164*, 249.

68. Pankratov, A. V.; Zercheninov, Z. N.; Chesnokov, V. I.; Zhadanaova, N. N.; *Russ. J. Phys. Chem.* **1969**, *43*, 212.

69. Note, this is for the *cis/trans* mixture of H–O–N=O rather than for each form separately, see ref. 5.

70. Justice, B. H.; Carr, I. H. *The Heat of Formation of Propellant Ingredients*, Dow Report No. AR-T0009-1S-67; The Dow Chemical Company, Midland, MI, 1967. We thank Dr. Malcolm Chase of NIST, formerly of Dow Chemical Company, for acquainting us with this data compendium and note that this source has information seemingly found nowhere else.

71. Carpenter, G. A.; Zimmer, M. F.; Baroody, E. E.; Robb, R. A. *J. Chem. Eng. Data* **16**, *1971*, 46.

72. Liebman, J. F.; Campbell, M. S.; Slayden, S. W. In: *The Chemistry of Functional Groups Supplement F2: The Chemistry of Amino, Nitroso, Nitro and Related Groups*; Patai, S., Ed.; Wiley: Chichester, 1996, pp. 337–378.

73. Tornieporth-Oetting, I. C.; Klapötke, T. M. In: Ref. 1, *Chemical Properties and Structures of Binary and Ternary Se-N and Te-N Species: Application of X-Ray and Ab Initio Methods*, pp. 287–311.

74. For a thermochemical study of some comparable (generally valence isoelectronic sets of) organo-sulfur, selenium, and tellurium compounds, see Voronkov, M. G.; Klyuchnikov, V. A.; Kolabin, S. N.; Shvets, G. N.; Varushkin, P. I.; Deryagina, E. N.; Korchevin, N. A.; Tsvebtniskaya, S. I. *Dokl. Phys. Chem.* **1989**, *307*, 650. Exemplary of the problems in selenium and tellurium thermochemistry are the differences between the enthalpies of formation of gaseous dimethyl selenide and telluride, 33.9 ± 11.0 kJ mol^{-1}, and between divinyl selenide and telluride, 58.5 ± 13.1 kJ mol^{-1}. The error bars allow for these two differences either to be essentially identical or to differ by ca. 50 kJ mol^{-1}.

INDEX

Advances in Molecular Structure Research

Edited by **Magdolna Hargittai,** *Structural Chemistry Research Group, Hungarian Academy of Sciences, Budapest, Hungary* and **István Hargittai,** *Institute of General and Analytical Chemistry, Budapest Technical University, Budapest, Hungary*

Volume 1, 1995, 352 pp.　　　　$109.50/£70.00
ISBN 1-55938-799-8

Volume 2, 1996, 255 pp.　　　　$109.50/£70.00
ISBN 0-7623-0025-6

J
A
I

P
R
E
S
S

Volume 3, 1997, 344 pp. $109.50/£70.00
ISBN 0-7623-0208-9

CONTENTS: Preface, *Magdolna Hargittai and István Hargittai*. Determination of Reliable Structures from Rotational Constraints, *Jean Demaison, Georges Wlodarczak, and Heinz Dieter Rudolph.* Equilibrium Structure and Potential Function: A Goal to Structure Determination, *Victor P. Spiridonov.* Structures and Conformations of Some Compounds Containing C-C, C-N, C-O, N-O, and O-O Single Bonds: Critical Comparison of Experiment and Theory, *Hans-Georg Mack and Heinz Oberhammer.* Absorption Spectra of Matrix-Isolated Small Carbon Molecules, *Ivo Cermak, Gerold Monninger, and Wolfgang Krätschmer.* Specific Intermolecular Interactions in Organic Crystals: Conjugated Hydrogen Bonds and Contacts of Benzene Rings, *Peter M. Zorky and Olga N. Zorkaya.* Isostructurality of Organic Crystals: A Tool to Estimate the Complementarity of Homo- and Heteromolecular Associates, *Alajos Kálmán and László Párkányi.* Aromatic Character of Carbocyclic p-Electron Systems Deduced from Molecular Geometry, *Tadeusz Marek Krygowski and Michal Cyránski.* Computational Studies of Structures and Properties of Energetic Difluoramines, *Peter Politzer and Pat Lane.* Chemical Properties and Structures of Binary and Ternary SE-N and TE-N Species: Application of X-Ray and *Ab Initio* Methods, *Inis C. Tornieporth-Oetting and Thomas M. Klapötke.* Some Relationships between Molecular Structure and Thermochemistry, *Joel F. Liebman and Suzanne W. Slayden.* Index.

FACULTY/PROFESSIONAL discounts are available in the U.S. and Canada at a rate of 40% off the list price when prepaid by personal check or credit card and ordered directly from the publisher.

JAI PRESS INC.

100 Prospect Street, P.O. Box 811
Stamford, Connecticut 06904-0811
Tel: (203) 323-9606   Fax: (203) 357-8446